To my wife and best friend, Jen.

Books are to be returned on or before
the last date below.

7 – DAY
LOAN

LIBREX–

The Social Lens

An Invitation to Social and Sociological Theory

Kenneth Allan
University of North Carolina at Greensboro

SAGE Publications
Thousand Oaks ▪ London ▪ New Delhi

For information:

Pine Forge Press
An Imprint of Sage Publications, Inc.
2455 Teller Road
Thousand Oaks, California 91320
E-mail: order@sagepub.com

Sage Publications Ltd.
1 Oliver's Yard
55 City Road
London EC1Y 1SP
United Kingdom

Sage Publications India Pvt. Ltd.
B-42, Panchsheel Enclave
Post Box 4109
New Delhi 110 017 India

Printed in the United States of America

Library of Congress Cataloging-in-Publication Data

Allan, Kenneth
The social lens : an invitation to social and sociological theory /
Kenneth Allan.
        p. cm.
Includes bibliographical references and index.
ISBN 1-4129-1409-4 or 978-1-4129-1409-3 (cloth)
ISBN 1-4129-1410-8 or 978-1-4129-1410-9 (pbk.)
        1. Sociology. 2. Sociology—Philosophy. 3. Sociologists. I. Title.

HM586.A44 2007
301.01—dc22   2006025247

This book is printed on acid-free paper.

06   07   08   09   10   10   9   8   7   6   5   4   3   2   1

Acquisitions Editor:	Benjamin Penner
Editorial Assistant:	Camille Herrera
Production Editor:	Libby Larson
Copy Editor:	Teresa Herlinger
Typesetter:	C&M Digitals (P) Ltd.
Proofreader:	Theresa Kay
Cover Designer:	Candice Harman

Contents

Chapter 7: Conflict and Critical Theories 211

Prelude

Imagine Society

> *Imagination is more important than knowledge.*
>
> —Albert Einstein

My road to sociology was indirect and rather bumpy. After two careers, I decided to go back to school. It was important for me to get into a profession in which I could help people and make a difference; so I picked psychology as my major. But while taking a Sociology of Education class for liberal arts requirements, I heard a sociologist speak about his involvement in Nicaragua. Nicaragua had been undergoing economic, political, and social reforms since 1979. Previously, the country had known 40 years of dictatorship under the Somoza family. But during the 1980s, the United States government supported a rebel group (the Contras) that plunged Nicaragua into a civil war. During this politically and militarily risky time, this professor adopted a village. Two to three times a year he took a group of students down to build houses, dig ditches, bring medical supplies, and care for the community. Here was an individual who was making a difference. His academic knowledge wasn't sterile; it affected his life and those around him. And that was exciting to me.

Two things struck me about the presentation. First, here was a man who walked the talk. He didn't hold knowledge in an unfruitful way; what he knew made him responsible. He cared about the human condition, and that wasn't something that I had come across in the university before. The second thing that struck me about the presentation was Karl Marx. I have always hated oppression. But as a good American I had been told to hate Marx, too; he was the father of communism and communism was bad (I grew up during the Cold War). But this man whom I respected because of his fight against human suffering told us that he was motivated by the theories of Marx. That day in class I found out that Marx told a fascinating story about economic oppression. This story takes elements from the past, along with social factors in modernity, and predicts what will happen in society. It tells how (not just why) people are economically oppressed.

I was hooked. This teacher was a man who made a difference and knew why he had to make a difference. I wondered if it had something to do with sociology. The professor certainly claimed that it did. For two weeks, I met with my professors in psychology and went door-to-door in the sociology department talking to everybody I could. At the end of those two weeks, I changed my major to sociology. Yep, I was hooked on sociology. But what I didn't realize at the time was that I was hooked on theory as well.

I had an important "theory moment" on my first day of graduate school. I had just moved back to Southern California, having spent a good deal of my adult life away. I was on my way to class, driving on the freeway, and noticed a car ahead of me. The car was chopped, or lowered so that the body was only a few inches off the ground. It had wide tires and rims. The side windows were tinted and there was a conspicuous religious symbol painted on the back window. The car was painted a mesmerizing purplish color that changed as the sun hit it from different angles. As I pulled closer to the car, I could see small, soft, yarn-like balls—what we used to call "dingle balls"—hanging from the back window. I thought, "Things haven't changed a bit." I smiled to myself with that comfort that comes from nostalgia, until I pulled directly next to the car and looked at the driver. He was a white kid! In fact, there were two white kids in the car, which left out the possibility that he was driving his friend's car (at least in my mind it did). It was a white kid driving what I had thought was a Chicano's car.

There may be a number of lessons that can be drawn from this—such as don't stereotype—but for me the experience left me with questions and problems. I grew up in Southern California, and when I was in high school and early college, a car like the one I saw would only have been driven by a Chicano. The cultural boundaries surrounding identity were clear. White kids fixed their cars in certain ways and wore certain clothes, and the same with Chicanos and blacks. The kids I hung out with weren't really prejudiced. We would interact with one another, party with one another, and have friends in the different groups. But we all clearly claimed our identities through culture, and the cultures were all different, with distinct boundaries. This car blew all that cultural understanding out the window. Apparently, things *had* changed.

But why and how did they change? These questions are why I called that experience on the freeway a theory moment. Questions, problems, and oddities are the cauldron out of which theory is born. It wasn't until a couple of years later that I actually figured out the answers to those questions. It took awhile in school before I came across theories that spoke of cultural fragmentation and the effects that increases in markets, advertising, and communication and transportation technologies can have on culture.

I was thrilled when I found my answer. Using theory, I was able to see what was hidden. Theory is that which lifts the veil and connects the dots. It lifts the veil because it can show us what is going on beneath the surface. Like the way Marx's theory exposes the workings of a system (capitalism) that we grew up taking for granted, theory connects the dots because it can help us make sense of the social world around us. Seeing through the prisms of theory can be a wonderful experience. Even the mundane can become fascinating—like standing in a line.

I used to hate standing in line. Actually, I probably still do, but I have also been enthralled by it ever since I discovered a theory called ethnomethodology (see Chapter 9). Ethnomethodology let me see that lines, or queues, are ongoing social accomplishments. We take the ability to line up for granted, but how do we accomplish it? The other day I stood in line for an hour for show tickets. When I got through the line and got my tickets, I noticed that the line was just as long as it had been when I got in it, and it was in almost the exact same form. The line wound around this large room like a snake and then went out a door. Now here's the interesting thing: there was no one there telling us how to line up, and there were no ropes to guide us. The new queue contained all different people, yet the line looked exactly the same as when I had been in it. How did we do it, and then how did they do it again? Theory gives us the eyes to be able to connect those dots.

Theory and Perspectives

Have you ever thought that you and your parents live in separate worlds? I know I did when I was younger. And you know what? You're right. Your parents may listen to your favorite band and say, "Why are they screaming? That's not even music." They might then point to the "beautiful voice" of their favorite singer and tell you, "Now that's the way singing is supposed to sound; that's music!" It's kind of amazing, but people can look at the same thing and see, or hear, very different worlds.

Think about this story: There once was a group of blind men who examined an elephant, and each, by his individual experience of the animal, drew very different conclusions about the nature of an elephant. One determined that an elephant is like a very smelly rope (based on the tail); another said that an elephant is like a tree trunk (based on a leg); yet another concluded that an elephant is like a brick wall (based on the side); and the last one said that an elephant is like a hose (based on the trunk). Forrest Gump may be right, that life is like a box of chocolates, but the world is also like an elephant—what you see is what you get, and what you see depends on where you stand.

Because human reality is a cultural reality, perspectives are an essential and unavoidable part of our existence. Joel Charon (2001) explains it this way:

> Perspectives sensitize the individual to see parts of reality, they desensitize the individual to other parts, and they guide the individual to make sense of the reality to which he or she is sensitized. Seen in this light, a perspective is an absolutely basic part of everyone's existence, and it acts as a filter through which everything around us is perceived and interpreted. There is no possible way that the individual can encounter reality "in the raw," directly, as it really is, for whatever is seen can be only part of the real situation. (p. 3)

In other words, we never directly experience the world; we encounter it through our perspectives. For a trained sociologist, theory is our perspective—it is a way of seeing and not seeing the world.

Theories are useful precisely because they do call our attention to certain elements in the social environment. I never realized what an amazing social achievement lining up outside a theater is until I understood ethnomethodology— and I never understood how people in different religions could all be convinced they are right until I read Durkheim. In sociology, we have a large number of theoretical perspectives that tell us what to see so that we gain insight into the social world; among them are functionalism, conflict theory, interactionist theory, critical theory (including feminism, race, and queer theory, postmodernism, and so on), exchange theory, rational choice theory, dramaturgy, ethnomethodology, structuration, network theory, ecological theory, social phenomenology, and on and on.

All of these perspectives are based on assumptions and values, and they contain ideas, concepts, and language. They tell us how to see the social world and how to talk about it. Each of these becomes particularly important to the theorist. Sets of words help order the human world in general. Think about when and what you eat, for example. Biologically we are all driven to eat. But what and when we eat is socially and linguistically ordered. Thus, restaurants like Denny's serve "breakfast all day." What is meant by that phrase is that certain foods that are normally associated with eating in the morning are available for "lunch" or "dinner." The sociological perspective has specific sets of words associated with it as well, and these words order and make the world comprehensible in a specific way. Thus, part of what you will be learning in this book are sets of words—languages that help to organize the world in a specific manner.

Perspectives also contain assumptions: things that we suppose to be true without testing them. All theories and theorists make certain assumptions. They form the bedrock upon which theory can be built. Two things are extremely important when thinking about assumptions: assumptions are never proven or disproved, and without them it is impossible to theorize (or even live). They thus inform everything that a theorist says. Let's take human nature, for example. Almost every theorist makes certain assumptions about what constitutes human nature. If a theorist assumes that human nature is individualistic and self-seeking, what then is she going to think about social structures? Chances are she will think that social structures are good and necessary to create social order. On the other hand, if a theorist thinks that human nature is intrinsically good and altruistic, then social structures can be bad things that oppress people. The idea of human nature cannot be tested (can you think why?), yet it is impossible to theorize without it, and once we assume something about human nature, it influences every aspect of our thoughts. (By the way, functionalists generally assume that human nature is egocentric and in need of social control, and Marxists generally assume that human nature is social and thus structures can be oppressive.) Remember: every theory is a perspective and is founded upon and contains assumptions, values, concepts, and languages.

Introducing This Book

Seeing theory as informed by perspectives has several implications for this book. First, the book has many voices—it's *polyvocal*. I try as much as possible to let the voice of the theorist come through. I don't intentionally impose a scheme on the theories. If the theory is critical, I present it as critical, and if it is scientific, I present it as scientific. This is a fairly unique feature of this book; and I want you to keep it in mind as we journey together through these great minds. Society is a multifaceted gem that comes to life as we allow different perspectives to give us eyes to see and ears to hear.

Second, I don't believe in any of these theories more or less than the others. That might sound a little odd, but I think it's a good thing. If you are a sociology major, you undoubtedly have heard people define themselves as "functionalist" or "interactionist." That kind of approach is extremely limiting and it strikes me as a bit dogmatic. The approach that I am advocating in this book is a tool approach. Theory is like a toolbox, and the more diverse tools you have at your disposal, the greater will be your insights. Please don't limit yourself to one or two perspectives. As you read through this book, let each theorist expand your mind and challenge you to new insights about the social world in which we live.

The third implication is that I don't offer many critiques of these theories. Every theory in here is absolutely amazing! And the real thrill of theory is when that first lightbulb goes on. If I can get you to "get it"—well, let me just say that getting you to think a new thought is an ultimate high for me. In my view, critical, reflexive thinking is what makes us most human.

Obviously, the next step in theory is to be able to critique, but you should first have a lightbulb experience. They won't all hit you like this, but some of these theories will remake your world. In fact, it is my explicit intent to disturb and unsettle you, and the theories all by themselves will do that. However, the critiques of the theories are actually within the book. Each theory offers a different perspective from the one before it. For example, what is society? Is it a system, like Niklas Luhmann claims, or is it made up of exchange networks, like Peter Blau argues, or joint actions, as Herbert Blumer argues? Is the subject dead, as Jean Baudrillard claims, or is the self the central organizing feature of the encounter, as Erving Goffman claims? My point is that once you understand a theory, you will automatically think differently about other theories and form your own critiques.

There are a couple of organizational features I'd like to point out as well. The book is divided into three sections: The Modern Agenda; Theory Cumulation and Schools of Thought in the Mid-Twentieth Century; and Contemporary New Visions and Critiques. The first section contains what are usually thought of as the classical statements—the people who first gave us our sociological lens. The second section is organized around schools of thought. Generally speaking, the perspectives that we usually think of, such as functionalism, conflict theory, and interactionism, were systematized during this period. The last period contains more contemporary theories, and they tend to be more critical than the theories contained in either of the first two sections. These sections roughly correspond to three time periods as well: the nineteenth century, the early to mid-twentieth century, and the latter part of the twentieth century. As I say, these are rough chronological divisions and there is some overlap. For example, Jürgen Habermas appears in the middle section but he is a contemporary theorist. But thematically he fits in with the project of the Frankfurt School and critical theory, which is in the middle section of the book.

The book also has some structural features I'd like to bring to your attention. Every chapter has an "Essential" box at the beginning of the discussions of each theorist, and a "Building Your Theory Toolbox" feature at the end of the chapter. The "Essential" box gives you a brief biographical sketch, as well as two other sections that I hope will give you a quick handle on what's going on: Passionate Curiosity (the central questions the theorist is interested in) and Keys to Knowing (central concepts to keep your eyes open for).

The "Building Your Theory Toolbox" section has a number of resources to help you with the material in this book. The first part, "Knowing," gives you an exhaustive list of terms and concepts as well as study questions for reviewing the text (hint: this section can be helpful in preparing for exams). "Learning More" will guide you to further sources for studying a particular theorist. There are always primary sources listed (those written by the theorist) and sometimes secondary sources (those written *about* the theorist).

The next part of the toolbox, "Theory You Can Use," will ask you to consider some of the theoretical issues raised in the chapter in light of your own life. Sometimes in this section I ask you to compare and contrast certain ideas or theories. These questions are actually important as they are a beginning step to your own theorizing. Finally, when appropriate, the toolbox will give you some suggestions for Web-based research. One more feature of the book may be of interest to you. Important terms and concepts are either highlighted in italics or marked in bold. The italicized words are those that you should be aware of, and the terms in bold are defined in the Glossary of Terms found in back of the book.

One final note before I turn you loose: this book is written in a conversational tone. I try to speak in academic jargon as little as possible, and I use personal examples to convey my experience with theory. The reason for this approach is that I want to invite you to have a thought or two with me. If I could, I would sit down with each one of you individually, maybe over a cup of coffee or a glass of beer, and I would find out what interests you. Then, I would bring up an idea or two, hopefully something you'd never thought of, and we'd sit and talk some more. This is how I intend this book: it's an invitation to have a thought. We'll think again about having a thought at the end of our time together. But for now, let's move on to imagining the social lens.

Acknowledgments

I write this at the end of a three-year project with Pine Forge Press. Out of it came three books, the last of which you hold in your hands. It has been an intense yet productive period of time, one that I will always value. My team at Pine Forge has been amazing. They are the most creative, flexible, and patient people I've yet worked with. My thanks especially to Jerry Westby and Ben Penner who saw the vision and believed in me. After Jerry and Ben, the person who has had the greatest impact on this project is my copyeditor, Teresa Herlinger. Thank you, Teresa, for your questions, suggestions, and corrections. But thanks especially for making it fun. My thanks also to Camille Herrera, for patience and guidance through permissions and production—your name should be up in lights. Also, my gratitude to Annie, Katja, and Laureen for all your help through the various stages of this project. You are all and always will be the A-Team. And I am particularly grateful to the many reviewers that took time out of busy lives to give critical and creative feedback. Finally and continually, I thank you the reader for consenting to have a thought or two with me. I hope that we can someday meet and share some of these ideas that have shaped and unsettled our world.

Pine Forge Press gratefully acknowledges the contributions of the following individuals:

Stephen R. Couch
The Pennsylvania State University

Douglas Degher
Northern Arizona University

David R. Dickens
University of Nevada, Las Vegas

Stephan Groschwitz
University of Cincinnati

Laura Mamo
University of Maryland

Philip Manning
Cleveland State University

Anne Szopa
Indiana University East

Section I
Introduction

The Modern Agenda: Nineteenth-Century Theorizing

Michel Foucault, about whom we will talk in Chapter 14, tells a wonderful story about how he came to write one of his books. He says he was reading a short piece by another author, Jorge Luis Borges, and he burst out laughing. What Foucault found funny was a fabled Chinese categorization scheme that separated animals into such types as sirens; frenzied, sucking pigs; and those belonging to the emperor. How different from the way we think of animals! We type them as being mammals, reptiles, amphibians, and so on. After laughing about it, Foucault (1966/1994b) wrote about it, and here's what he had to say:

> In the wonderment of this taxonomy, the thing we apprehend in one great leap, the thing that, by means of the fable, is demonstrated as the exotic charm of another system of thought, is the limitation of our own, the stark impossibility of thinking *that*. (p. xv, emphasis original)

Look at what Foucault found funny. It wasn't so much the silliness of this fabled Chinese system that was amusing; it was *the limitations of his own thinking*. Foucault realized that he would never have thought to categorize animals in such a way. And then something profound hit him—what else would it never occur to him to think? Here's my point in telling this story: What we know and how we think are both made possible and circumscribed by history and culture.

We can see this historical contingency in social knowledge. At one time, American society as a whole "knew" (that is, held to be true) that people of color—such as blacks, Indians, Chinese, and Irish—were inferior. In the United States, we also used to "know" that homosexuality was a disease and that women were rationally inferior to men. But the historic specificity of knowledge isn't limited to social relations. Think about this: Up until the time of Copernicus, we "knew" that the earth was the center of the universe. And, as recently as the early 1800s, we see that doctors "knew" that all disease was created by an imbalance of bodily fluids or humors: blood, saliva, urine, and feces. In order to heal people, then, doctors would try to balance out the humors, which is why they would purge or bleed their patients. But we "know better" now, don't we? Both science and society have progressed, right? Maybe, but how do we know?

Modern Thinking

In terms of understanding social and sociological theory, *it is vitally important for us to grasp this relationship between history and thought.* I've highlighted that phrase because I want you to keep it in mind as we travel together through this book. The theorists in this section are theorists of modernity. What they thought, how they thought, and why they thought were related to the time period in which they lived—the dawn of modernity. The time period that we call modernity is generally defined by the movement of populations from small local communities to large urban settings, a high division of labor, high commodification and use of rational markets, and large-scale integration through nation-states. In general, the defining moments of modernity are the rise of nation-states, capitalism, mass democracy, science, urbanization, and mass media; the social movements that set the stage for modernity are the Renaissance, Enlightenment, Reformation, American and French Revolutions, and the Industrial Revolution.

The Enlightenment was a European intellectual movement that began around the time Sir Isaac Newton published *Principia Mathematica* in 1686, though the beginnings go back to Bacon, Hobbes, and Descartes. The Enlightenment is particularly important because the people creating this intellectual revolution felt that the use of reason and logic would enlighten the world in ways that fate and faith could not. The principle targets of this movement were the church and the monarchy, and the ideas central to the Enlightenment were progress, empiricism, freedom, and tolerance.

The ideas of progress and empiricism are especially significant. Prior to the Enlightenment, the idea of *progress* wasn't particularly important. The reason for this is that the dominant worldview was religious. In general, religion sees the world in terms of fate, faith, and revelation. Any change under a religious system comes not because of human effort and progress, but, rather, because God has revealed some new truth. Purely religious systems thus tend toward status quo and honoring tradition. On the other hand, the belief in progress puts human beings at the center (humanism) and holds that people can take control of their lives and the environment in order to make things better. This humanistic system privileges change, and change is defined as progress.

The idea of progress puts human beings in the driver's seat, and it implies that the universe and the social worlds are empirical. *Empiricism* is the assumption that there are no spiritual or unseen forces in back of the physical or social worlds; rather, real things exist as facts that are discernable through at least one of the five senses. Together, the ideas of progress and empiricism helped to form a uniquely modern philosophical orientation: positivism.

One of the easiest ways to understand *positivism* is to compare it to fatalism. Fatalism, or having a fatalistic attitude, has a negative connotation today, but it wasn't always so. Fatalism refers to a belief in the Fates. In antiquity, the Fates were believed to be goddesses who oversaw destiny and determined the course of human existence. This kind of idea is also found in Christianity as well, in the doctrine of predestination—God is in charge of the universe and human destiny. A less specific form of fatalism is practiced today by anyone who believes there is some spiritual force in back of the universe. You can see that fatalism isn't necessarily negative.

Positivism is the opposite of fatalism: It assumes that the universe is empirical (without spiritual force); operates according to law-like principles; and that humans can discover those laws and use them to understand, control, and predict the forces that influence their lives. Fatalism puts a spiritual force at the center of existence; positivism puts humanity at the center of existence. Thus, it is positive in a humanistic sense. Positivism, then, is a philosophy that confines itself to sense data; denies any spiritual forces or metaphysical considerations; and emphasizes the ability of human beings to affect their own fate, generally through science.

Harriet Martineau: Theorist of Modernity

The United States are the most remarkable examples now before the world of the reverse of the feudal system—its principles, its methods, its virtues and vices. In as far as the Americans revert, in ideas and tastes, to the past, this may be attributed to the transition being not yet perfected—to the generation which organised the republic having been educated amidst the remains of feudalism. (Martineau, 1838/2003, p. 46)

Before we get to what are usually thought of as the classical theorists, I want to introduce you to Harriet Martineau. Her work very clearly demonstrates the intent behind modernist theorizing. Martineau is actually one of the earliest sociologists. In fact, her work predates everyone else's in this book. She is best known for her translation of Auguste Comte's *Positive Philosophy*, one of the major foundation stones of sociology. Martineau, however, did more than translate. She revised Comte's work down from six volumes to two. So taken was Comte with Martineau's version, that he adopted it rather than his own and had the revision translated back into French. If you've read Comte, chances are you've read Martineau's version. Of Martineau, Comte said, "looking at it from the point of view of future generations, I feel sure that your name will be linked to mine, for you have executed the only one of those works that will survive amongst all those which my fundamental treatise has called forth" (as quoted in Harrison, 1913, p. xviii).

One of the guiding lights of modernity that directed Martineau's work is the idea of natural law. *Natural law* is the notion that, apart from human institutions, there are laws and rights to which every human being adheres. The United States Declaration of Independence contains this idea in the phrase, "We hold these truths to be self-evident, that all men are created equal, that they are endowed by their Creator with certain unalienable Rights, that among these are Life, Liberty and the pursuit of Happiness." This was Martineau's (1838/2003) belief as well: "every element of social life derives its importance from this great consideration—the relative amount of human happiness" (p. 25). Happiness—and its prerequisite, freedom—are thus a touchstone and concern for modern sociological analysis. We can see this especially in the work of Karl Marx, Max Weber, Émile Durkheim, and Georg Simmel.

Early positivists, such as Martineau, were impressed with the need to evaluate how well any modern society was doing with regard to its purpose. Again, let me remind you that fundamental shifts occurred in modernity. One of the major changes had to do with government: the shift from rule by monarchy to rule by democracy administered through the nation-state. This shift also implied that people went from being subjects to being citizens with the inalienable rights of life, liberty, and the pursuit of happiness. The Declaration of Independence continues, "to secure these rights, Governments are instituted among Men." Thus, the purpose of the state in modernity is to secure and safeguard civil rights for its citizens. Here is the critical part for our discussion: early social thinkers felt compelled to evaluate society's progress. Martineau had a number of ways she did this, chiefly through gender, religion, and education.

Gender in Modernity

The home is important for several reasons in Martineau's scheme. It is the central part of what she calls the domestic state. The primary socialization of children takes place within the family, and it is where the values and practices of a people, what Martineau calls "morals and manners," can best be observed. Further, the treatment of women in marriage is one of the most important keys for understanding how a society measures up to the universal law of happiness. Keep in mind that happiness is a basic human right and that marriage is a fundamental social institution. Thus, unhappy marriages are a strong indicator of the moral state of an entire society.

Martineau also argues that women's participation in the workforce, in terms of both level and type of work, influences the moral codes of a society in two ways. First, to the degree that women are left out of the workforce, they are seen as "helpless" and men "vicious." There is a relationship between personal power or efficacy and money. This relationship makes perfect sense, of course, in a society where one's survival is linked to work. In societies where a living wage is guaranteed to all, or all people are given equal opportunity, this association would be very weak. Women would not be dependent upon men for their survival, and the power between the genders would be equal. However, in a society where living is dependent upon working, and women are denied equal access to jobs, then men will have greater power not only in society but also in personal relationships.

The second result of unequal workforce participation is that the institution of marriage is "debased." The contemporary ideal of marriage is that it is a union freely entered into by two people in love. However, in a society where marriage is the chief way that women get ahead or survive, it can never be the case that marriage is secure "against the influence of some of these motives even in the simplest and purest cases of attachment" (Martineau, 1838/2003, p. 183). Martineau actually frames this issue more pointedly. Under nineteenth-century American patriarchy, the principal or only aim of women in society was to marry. When the chief goal of women is to marry and have children, then the values associated with that goal are enculturated in the very young. Young girls are socialized to value being attractive to men (yet chaste), marrying, and having children. The toys, games, and images provided for girls generally have these themes. Thus, when grown, women sense that they are unfulfilled and incomplete unless they are married and have children. One's femininity becomes synonymous with marriage and childbearing. In such a society, "the taint is in the mind before the [marital] attachment begins, before the objects meet; and the evil effects upon the marriage state are incalculable" (Martineau, 1838/2003, p. 183).

Religion in Modernity

Religion is another important social structure for measuring society's progress in modernity. Martineau isn't concerned with the origins of religion, the way Durkheim and Weber are, but she does say that religion evolves toward more inclusive, less judgmental forms. The *forms* of religion are of chief interest for Martineau. She gives us three types of religion: licentious, ascetic, and moderate. *Licentious religions* are those in which the gods are seen as part of nature and having human passions, as in the early Greek pantheons. During the early stages of this type of religion, licentious or immoral conduct is encouraged. This form changes as people come to be seen as less than the gods, and the power for doing good or harm is attributed to the gods. When this happens, the idea of propitiation (an act or gift given to appease the wrath of a god) enters and ritual worship begins. Even in its advanced form, this kind of religion is still preoccupied with licentious behavior, albeit in a negative way: the rituals to appease an almighty god satisfy the ill done by sin.

Ascetic religions are highly ritualized as well. Emphasis here is on rituals that provide proofs of holiness. Martineau argues that spiritual vices rather than immoral acts of the flesh are the center of this type of religion. The spiritual vices of "pride, vanity, and hypocrisy,—are as fatal to high morals under this state of religious sentiment as sensual indulgence under the other" (Martineau, 1838/2003, p. 79). Martineau sees religious ritual as misplaced emphasis. In this way of thinking, rituals are an empty form that takes our attention away from "self-perfection, sought through the free but disciplined exercise of their whole nature" (p. 80). *Moderate religions* are the least ritualized of the three and focus on self-actualization through broad education that involves the entire being.

The importance of the different religious forms is that religion plays a key role in shaping the morals of any society. If licentious religious forms dominate in society, then the political and domestic morals will be very low or contain mostly ritual. Where asceticism governs, the natural behaviors and affections of humanity are judged to be sin. Moderate religion encourages a culture of freedom and acceptance. Martineau argues that it is a "temper" rather than a pursuit, a settled state of mind focused on inner peace, authenticity, and social affirmation rather than personal behaviors, whether righteous or sinful.

Martineau places her thinking about religion under the research agenda of the "idea of liberty." The way a society is structured around the idea of liberty is extremely important to Martineau. It is clearly related to the idea of happiness. Thus, liberty can be seen as the positive counterpart to oppression in measuring the structured potential of happiness in any society. In many ways, though, the most important indicator of the idea of liberty is the condition of the education system in society.

Education in Modernity

According to Martineau, there are two great social powers—force and knowledge—and the story of human progress is the movement away from one toward the other. Social relations began through physical force and domination and the idea that "might makes right." In such societies the past is everything. The kind of knowledge that is of most value then is traditional knowledge. The tastes, ideas, sentiments, and ambitions are those that come from antiquity. This kind of knowledge is nonreflexive and maintains the status quo; it "falls back upon precedent, and reposes there" (Martineau, 1838/2003, p. 45).

Martineau (1938/2003) argues that society moves away from force toward education as the basis for moral power in response to increases in commerce, professionalization, communication, and urbanization (p. 46). Under these conditions, knowledge cannot stay fixed in the past. It is opened up to new horizons through increases in communication and commerce with other societies. Urbanization and commerce tend to expand the limits of knowledge as well as push for constant change. City living also brings people more in contact with diversity, which in turn opens up the boundaries of knowledge. Professionalization creates expert knowledge systems and, as more and more jobs and trades become professions, people in general begin to value expert knowledge over traditional information.

Concerning the education system and its relation to freedom, Martineau gives us both interesting (like the use of uniforms: it's "always suspicious") and more obvious measures (such as the distribution of social types—gender, race, religion, class—and the general level of educational attainment). However, two of the most important indicators of education and freedom are the educational system's financial base and the position of the university. The influence of education's financial base should be apparent, but it bears mentioning. The first issue is the extent of *free education*. The level to which a society supports public education is an unmistakable measure of its support of the ideas of freedom and democracy. In civilized countries, education is the way to socially advance; it's the legitimate way to get ahead. A high level of support for free, public education indicates a high level of support for equal opportunity. The level of supported equality is linked to the kinds of jobs or careers one can get with a given level of education. In other words, all we have to do to understand the level of equality that society supports is look at the kinds of job opportunities free and public education provides. For example, if a society supports public education only through high school, it indicates that the level of equality the state is interested in providing is equal to the jobs that require a high school education.

The second indicator of the place of education in society is the *position of the university*. Martineau (1838/2003) claims that "in countries where there is any popular Idea of Liberty, the universities are considered its stronghold" (p. 203). The reason for this link between liberty and universities might surprise you. Weber tells us that in societies where bureaucracy is the principal form of social organization, education will serve to credential a workforce. Martineau's function for the university is different from Weber's. Of it she says, "It would be an interesting inquiry how many revolutions warlike and bloodless, have issued from seats of learning" (p. 203).

According to Martineau, the function of the university in its relationship to liberty is to inspire critical thinking. A university education should make you question authority and society. Freedom must be won anew by every generation. Almost by definition, democracy is alive and expansive. A university that doesn't challenge its students with new ideas and doesn't provide an atmosphere wherein new principles can be founded is "sure to retain the antique notions in accordance with which they were instituted, and to fall into the rear of society in morals and manners" (Martineau, 1838/2003, p. 203).

Not only are the purpose, content, and environment of the university important, so are its students, in particular their motivation for study. To the degree that students are motivated to obtain a university education for a job, to that degree is education for freedom dead in a society. Martineau makes a comparison between students in Germany and those in the United States (Martineau is British and thus has no vested interest). German students are noted by a quest for knowledge: the German student may "remain within the walls of his college till time silvers his hairs." The young American student, on the other hand, "satisfied at the end of three years that he knows as much as his neighbors . . . plunges into what alone he considers the business of life" (Martineau, 1838/2003, p. 205). These words were written almost 170 years ago. Yet Martineau had great hope for the American system of education: "a literature will grow up within her, and study will assume its place among the chief objects of life" (p. 206). I will let the reader judge the fulfillment of her hope.

The most important thing I want you to glean from Martineau is her intent—the intent of her theorizing. Her purpose was in keeping with the project of modernity; the time in which she lived. She was interested in understanding how society measured up to the expectations of the Enlightenment. And she felt duty bound as a positivistic social thinker to work toward social progress.

I won't generally spend this much time introducing a section. But this section is important because it sets the tone for the entire book. If you look ahead, you'll see that the last theorist we will cover is Jean Baudrillard—a postmodernist. Modernity and postmodernity, then, are the bookends that hold the ideas in this volume together. The big story that we will be unfolding as we move through this book is the shift from modernist to postmodernist theorizing. In between these bookends is a vast middle section, where things change and multiple visions open up to us. Let's begin the journey with Chapter 1 and an introduction to Karl Marx.

Building Your Theory Toolbox

Knowing the Modern Agenda

After reading and understanding this chapter, you should be able to define the following terms theoretically and explain their importance to the modern agenda:

Modernity; progress; humanism; empiricism; positivism; natural law; licentious, ascetic, and moderate forms of religion

After reading and understanding this chapter, you should be able to

- Explain how modernity, progress, empiricism, and positivism are related to the social sciences and theory

- Explain how Martineau exemplifies modernist theorizing

- Discuss how gender, religion, and education are indicators of how well a society, especially its government, is fulfilling the social project of modernity

CHAPTER 1

Capitalism and Engines of Social Change

Karl Marx (1818–1883)

Photo: © Bettmann/Corbis.

Seeing Further: The Effects of Capitalism

Almost everything about our lives is shaped and governed by capitalism, from our inner desires concerning self-image, career, and personal possessions to our network of friends, the totality of material goods that surround us, and the global relations that give context and meaning to it all. Further, capitalism is without a doubt one of the most powerful social forces humankind has ever created. Beginning in Western Europe, it has swept across the face of the earth, bringing almost every nation under its sway. Obviously, capitalism is the financial core that helped establish the military and economic might of the United States. But even more indicative of the power of capitalism are the effects it is having on the economic, social, and political makeup of China, the last great socialist superpower.

Probably the most important feature of China's move toward capitalism has been the establishment of over 2,000 Special Economic Zones (SEZs). The economic development of these zones is market driven rather than directed by the state as in classic socialism. These SEZs have created massive urbanization, have dramatically increased the Chinese worker's standard of living, and are generally responsible for quadrupling China's Gross Domestic Product (GDP) since 1978, when markets were first introduced. In 2004, China had the world's second-largest economy with a GDP (purchasing power parity) of US$7.124 trillion.

But what else is going on with capitalism? One of the things I hope you take away from this book is the knowledge that all social factors have effects that extend beyond the obvious—effects that are generally unintended consequences. What all does capitalism bring with it? That in essence is the question that Karl Marx asks. And Marx's questions and his answers are so significant that they form the basis of quite a bit of contemporary theorizing.

The Essential Marx

Biography

Karl Marx was born on May 5, 1818, in Trier, one of the oldest cities in Germany, to Heinrich and Henrietta Marx. Both parents came from a long line of rabbis. His father was the first in his family to receive a secular education. At 17, Karl Marx enrolled in the University of Bonn to study law. It was there that he came in contact with and joined the Young Hegelians, who were critical of Prussian society (specifically, because it contained poverty, government censorship, and religious discrimination). Marx eventually moved to the Friedrich-Wilhelms-Universität in Berlin where he studied philosophy.

Because of his political affiliations, Marx was denied a university position by the government. Marx turned to writing and editing, but had to battle government censorship continually. Over the next several years, Marx moved from place to place, including stays in Brussels, Paris, and Germany. Much of his movement was associated with revolutions that broke out in Paris and Germany in 1848. Finally, in 1849, Marx moved to London where he remained.

The workers' movements were quiet after 1848, until the forming of the First International in 1864. Founded by French and British labor leaders at the opening of the London Exhibition of Modern Industry, the union soon had members from most industrialized countries. Its goal was to replace capitalism with collective ownership. Marx spent the next decade of his life working with the International. The movement continued to gain strength worldwide until the Paris Commune of 1871. The Commune was the first worker revolution and government, and it governed Paris for a brief time. Three months after its formation, Paris was attacked by the French government. Thirty thousand unarmed workers were massacred.

After the failure of the Commune, Marx continued to study but never produced another major work. Marx died in his home on March 14, 1883.

Passionate Curiosity

The obvious driving force in Marx's work is capitalism. But to simply say that and nothing else would strip Marx's ideas of their true depth. Marx was a man driven to understand why humanity is so often mired in misery. He has been seen as anti-spiritual, but this is far from the truth. Rather than being a cold materialist, Marx gives us a "spiritual existentialism in secular language" (Fromm, 1961, p. 5). He was driven to understand and contest humanity's inhumanity. "Marx's philosophy is one of protest; it is a protest imbued with faith in man, in his capacity to liberate himself, and to realize his potentialities" (Fromm, 1961, p. vi).

Keys to Knowing

Alienation, bourgeoisie, class consciousness, commodification, commodity fetish, material dialectic, division of labor, exchange value, exploitation, false consciousness, ideology, industrialization, markets, means and relations of production, private property, proletariat, species-being, surplus value

Marx's Perspective: Human Consciousness and History

Let's start off with a couple of questions. What is human nature? And, what is unique about human consciousness? These kinds of questions have been in the back of a great deal of philosophical and religious discussions for thousands of years. There are also a few sociologists who have wondered about the same things. Obviously, one of the things that a sociologist is going to assume about human nature is that it is social. But social in what way? Further, how can we understand the unique features of human consciousness from a sociological position? How does society affect, or perhaps even create, the mind?

We will visit these questions a few more times as we progress through this book—and when we do, keep in mind all the different things our theorists have to say. This cumulative work forms the basis of a robust understanding of how things work socially. But for now we want to turn our attention to Marx, who argues that human consciousness is based on something he calls species-being—the essence of human nature.

Species-Being

Most people accept that humans are aware in a way that other animals are not. First, we are not simply aware of the environment; we are also conscious of our own awareness of it. And second, we can be conscious of our own existence and give it meaning. Marx proposes a rather unique answer to the problem of human consciousness: species-being. "By means of it nature appears as *his* work and his reality. The object of labor is, therefore, *the objectification of man's species life* . . . and he sees his own reflection in a world which he has constructed" (Marx, 1932/1995, p. 102, emphasis original).

Marx argues that the unique thing about being human is that we create our world. All other animals live in a kind of symbiotic relationship with the physical environment that surrounds them. Zebras feed on the grass and lions feed on the zebras, and in the end, the grass feeds on both the zebras and the lions. The world of the lion, zebra, and grass is a naturally occurring one, but not so for the human world. Humans must create a world in which to live. They must in effect alter or destroy the natural setting and construct something new. The human survival mechanism is the ability to change the environment in a creative fashion in order to produce the necessities of life. Thus, when humans plow a field or build a skyscraper, they have created something new in the environment that in turn acts as a mirror through which humans can come to see their own nature. Self-consciousness as a species that is distinct from all others comes as the human being observes the created human world. There is, then, an intimate connection between producer and product: *the very existence of the product defines the nature of the producer.*

Here's an illustration to help us think about this: Have you ever made anything by hand, like clothing or a woodworking project, or perhaps built a car from the ground up? Remember how important that thing was to you? It was more meaningful than something you buy at the store simply because you had made it. You had invested a piece of yourself in it; it was a reflection of you in a way that a purchased commodity could never be.

But this is a poor illustration because it falls short of what Marx truly has in mind. Marx implies that human beings in their natural state live in a kind of immediate social consciousness. Initially, human beings had a direct relationship with everything in their world. There weren't supermarkets or malls. If they had a tool or a piece of clothing, they had made it or they knew the person who did. When they looked into the world they had produced, they saw themselves; they saw a clear picture of themselves as being human (creative producers), and they also saw intimate and immediate social relations with other people.

Notice a very important implication of Marx's species-being: human beings by their nature are social and altruistic. Marx's vision of humans is based on the importance of society in the survival of our species. We survive collectively and individually because of society. Through society, we create what is needed for survival; if it were not for society, the human animal would become extinct. We are not equipped to survive in any other manner. What this means, of course, is that we have a social nature—we are not individuals by nature. Species-being also implies that we are altruistic—by nature we have an uncalculated commitment to others' interests. If human survival is based on collective cooperation, then it would stand to reason that our most natural inclination would be to serve the group and not our self.

Marx's theory of species-being also has implications for knowledge and consciousness. In the primitive society that we've been talking about, humans' knowledge about the world was objective and real; *they held ideas that were in perfect harmony with their own nature.* According to Marx, human ideas and thought come about in the moment of solving the problem of survival. Humans survive because we creatively produce, and our clearest and truest ideas are grounded in this creative act. In species-being, people become truly conscious of themselves and their ideas. Material production, then, is supposed to be the conduit through which human nature is expressed, and the product ought to act as a mirror that reflects back our own nature.

If we understand this notion of species-being, then almost everything else Marx says falls into place. To understand species-being is to understand alienation (being cut off from our true nature), ideology (ideas not grounded in creative production), false consciousness (self-awareness that is grounded in anything other than creative production), and we can also understand why Marx placed such emphasis on the need for class consciousness in social change. This understanding of human nature is also why Marx places such importance on the economy and capitalism specifically. The economy is the substructure from which all other structures (superstructure) of human existence come into being and have relevance.

Before moving on to the next section, I want to call your attention to an important point about Marx. Collectively, Marx talks about these issues of species-being, consciousness, and the importance of economic structures in terms of what he calls *humanistic materialism.* In using the word "materialism," Marx is referencing an important philosophical debate about the nature of human reality. In Marx's time, there were two important ways of understanding the issue of reality: idealism and materialism. Idealism posits that reality only exists in our idea of it. While there may indeed be a material world that exists in and of itself, that world exists for humans only as it appears to us as ideas. On the other hand, materialism argues that all reality may be reduced to physical properties. In materialism, our ideas about the world are simple reflections; those ideas are structured by the innate physical characteristics of the universe.

Please notice that Marx's idea of materialism is vastly different from this philosophical one. Marx's is a *social materialism*. The material world that matters to people is the world of production—the material goods that we creatively produce. This materialism is humanistic in two senses. The most obvious one is that it puts humans at the center. The material world for humans is the one that comes through our species existence: creative production. The second way Marx's materialism is humanistic is that his theoretical work is devoted entirely to improving the condition of humanity.

Material Dialectic

Why do things change? "Well," you might say, "things always change. That's just what they do." True, but *why*? What is the dynamic that drives things to change? Think of it this way: If I were to ask you what makes your car move forward, what would you say? You might point to the wheels or the transmission or maybe even the gas pedal, but none of those things are what actually impels the car forward. The power that pushes the car forward is the motor. While the car may use the wheels, transmission, and gas pedal to move, it is the motor that is the actual power in back of movement. So, why do societies change? What is the dynamic (motor) that drives historic change? Marx's answer to these questions is material dialecticism.

Marx begins his theorizing about social change with an idea borrowed from Georg Wilhelm Friedrich Hegel. Hegel was a religious idealist who wanted to understand how the revelation of deity changes over time. For Hegel, this kind of change occurs through dialectical elements enclosed in ideas. A dialectic contains different elements that are naturally antagonistic to one another. Hegel called these the thesis and antithesis.

The dialectic is like an argument or a dialogue between elements that are locked together. (The word "dialectic" comes from the Greek word *dialektikos,* meaning discourse or discussion.) For example, to understand "good," you must at the same time understand "bad." To comprehend one, you must understand the other: Good and bad are locked in a continual dialogue. Hegel argued that these kinds of conflicts would resolve themselves into a new element or synthesis, which in turn sets up a new dialectic: every synthesis contains a thesis that by definition has conflicting elements. So, in our example, if you truly understand that good is defined by the presence of bad—if the dialectic becomes active for you—then your comprehension of the good/bad issue changes. After such an insight, it can never be the case that good triumphs over evil, because they are mutually dependent. A new ethic would be required as a result of perceiving the dialectic between good and bad.

Marx liked the historical process implied in Hegel's dialectic, but he disagreed with its ideational base. Marx, as we have seen, argues that human beings are unique because they creatively produce materials to fill their own material needs. Since the defining feature of humanity is production, not ideas and concepts, Marx found Hegel's notion of idealism to be false. The dialectic is oriented around material production and not ideas—the material dialectic. Thus, the dynamics of the historical dialectic are to be found in the economic system, with each economic system inherently containing antagonistic elements (see Figure 1.1). Please notice that the dynamic or "motor" within the dialectic comes from tensions that must be resolved. As the antagonistic elements work themselves out, they form a new economic system.

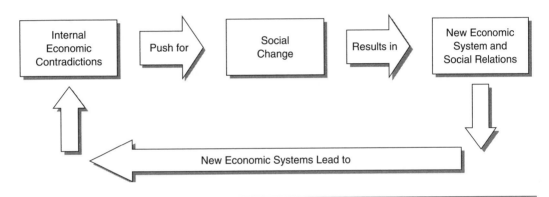

Figure 1.1   Marx's Material Dialectic

We can think of this way of seeing society as an upheaval model. According to this perspective, society is not like an organism that gradually and peacefully becomes more complex in order to increase its survival chances, as in functionalism. Rather, society is filled with human beings who exercise power to oppress and coerce others. Periods of apparent peace are simply times when the powerful are able to dominate the populace in an efficient manner. However, according to this model, the suppressed will become enabled and will eventually overthrow and change the system. Social change, then, occurs episodically and through social upheaval.

In mapping out the past of the historical dialectic, Marx categorizes five different economic systems (means of production along with their relations of production): preclass societies, Asiatic societies, ancient societies, feudal societies, and capitalist societies. Preclass societies are like hunter–gatherer groups. These were small groups of people with a minimal division of labor (one that Marx termed the natural division of labor) and communal ownership of property (termed primitive communism). Ancient societies such as Rome developed around large urban centers. Private property and slave labor came into existence, as well as significant class inequality. Ancient societies were replaced by feudal systems wherein the primary economic form was serf labor tied to the land of the aristocracy. Feudal systems were later replaced by capitalist systems.

Eventually, the capitalist system will be replaced by socialism and that in turn by communism. The specific dynamics that Marx says caused these shifts in economic systems are not important in our consideration right now. What is important to see is that for Marx, social change comes about because of inherent contradictions in the economic structure. In the final analysis then, Marx argues that social changes come through structural dynamics—it is the structural relations in society that bring change, not individual actions.

Concepts and Theory: The Basic Dynamics of Capitalism

Class and Class Structure

When Marx speaks about production, he is concerned with three intertwined issues: the means or actual process of production, the social relationships that form because of production, and the end result of production—the product. The **means of production** refers to the methods and materials that we use to bring into being those things we need to survive. On a small scale, we might think of the air-hammers, nails, wood, concrete mixers, and so forth that we use to produce a house. Inherent within any means of production are the *relations of production:* in this case, the contractor, subcontractor, carpenter, financier, buyer, and so forth. In the U.S. economy, we organize the work of building a house through a contracting system. The person who wants the house built has to contract with a licensed builder who in turn hires different kinds of workers (day laborers, carpenters, foremen, etc.). The actual social connections that are created through particular methods of production are what Marx wants us to see in the concept of the relations of production.

Of course, Marx has something much bigger in mind than our example. In classic feudalism, for instance, people formed communities around a designated piece of land and a central manor for provision and security. At the heart of this local arrangement was a noble who had been granted the land from the king in return for political support and military service. At the bottom of the community hierarchy was the serf. The serf lived on and from the land and was granted protection by the noble in return for service. Feudalism was a political and economic system that centered on land—land ownership was the primary means of production. People were related to the land through oaths of homage and *fealty* (the fidelity of a feudal tenant to his lord). These relations functioned somewhat like family roles and spelled out normative obligations and rights. The point here, of course, is that the way people related to each other under feudalism was heavily influenced by that economic system and was quite different from the way we relate to one another under capitalism: most of us don't think of our boss as family.

For Marx, human history is the history of class struggles. Marx identifies several different types of classes, such as the feudal nobility, the bourgeoisie, the petite bourgeoisie, the proletariat, the peasantry, the subproletariat, and so on. As long as these classes have existed, they have been antagonistic toward each other. Under capitalism, however, two factors create a unique class system.

The first thing capitalism does is lift economic work out of all other institutional forms. Under capitalism, the relationships we have with people in the economy are seen as distinctly different from religious, familial, or political relations. For most of human history, all these relationships overlapped. For example, in agriculturally based societies, family and work coincided. Fathers worked at home and all family members contributed to the work that was done. Capitalism lifted this work away from the farm, where work and workers were embedded in family, and placed it in urban-based factories. Capitalism thus separates the worker from his or her social networks. This separation is part of what helped define labor as pure commodity.

Additionally, contemporary gender theorists point out that this movement created dual spheres of home and work, each controlled by a specific gender. Marx, on the other hand, sees gender inequality reaching further back than capitalism, though capitalism has certainly accentuated gender problems. Marx, and more specifically Engels, was actually among the first to write on the issue of gender. In 1877, an American anthropologist, Lewis H. Morgan, published a book that argued for the matrilineal origins of society. Both Marx and Engels felt that this was a significant discovery. Marx planned on writing a thesis based on Morgan's work and made extensive notes along those lines. Marx never finished the work. Friedrich Engels, however, using Marx's notes, did publish *The Origin of the Family, Private Property, and the State* in 1884. The argument is fairly simple, yet it presents one of the basic ways in which we understand how gender inequality and the oppression of women came about.

As with all of Marx and Engels's work, Engels starts with a conception of primitive communist beginnings. In this setting, people lived communally, sharing everything, with monogamy being rarely, if ever, practiced. Under such conditions, family is a social concern rather than a private issue, with children being raised by the community at large, rather than by only two parents. As a matter of fact, because identifying the father with any certainty was impossible, in premodern society, paternity itself wasn't much of an issue. Thus, in Marx's way of thinking, "the communistic household implies the supremacy of women" (Engels, 1884/1978, p. 735).

The key in the transition from matrilineal households to patriarchy is wealth. Primitive societies generally lived "from hand to mouth," but as surplus began to be available, it became possible to accumulate goods. As men began to control this wealth, it was in their best interest to control inheritance, which meant controlling lineage. Engels (1884/1978) says that the way this happened is lost in prehistory, but the effect "was the world-historic defeat of the female sex" (p. 736). In order to control the inheritance of wealth, men had to control fertilization and birth, which meant that men had to have power over women. Control over women's sexuality and childbirth is why, according to Engels, we have developed a dual morality around gender—women are considered sluts if they sleep around, whereas men are studs if they do so. Marriage, monogamy, and the paired family (husband and wife) were never intended to control men's sexuality. They were created to control women's sexuality in order to assure paternity. This control, of course, implied the control of the woman's entire life. She became the property of the man so the man could control his property (wealth). Quoting Marx, Engels (1884/1978) concludes,

"The modern family contains in embryo not only slavery. . . . It contains within itself in miniature all the antagonisms which later develop on a wide scale within society and its state" (p. 737).

The second unique feature of class under capitalism is its bipolarization. That is, under capitalism, class tends to be structured around two positions—the *bourgeoisie* (owners) and the *proletariat* (workers). Marx does talk about other classes in capitalism, but they have less importance. The *petite bourgeoisie* is the class of small land and business owners and the *lumpenproletariat* is the underclass (such as the homeless). While the lumpenproletariat played a class-like position in early French history, Marx argues that because they have no relationship to economic production at all, they will become less and less important in the dynamics of capitalism. The petite bourgeoisie, on the other hand, does constitute a legitimate class in capitalism. However, Marx argues that this class shrinks in number and becomes less and less important, as they are bought out and pushed aside by powerful capitalists.

The increasing size of the working class along with the centralization of capital into fewer and fewer hands is brought about by dialectic elements within capitalism (see Figure 1.2). Capitalism is defined by the reinvestment of profit to make more profit. This intrinsic part of capitalism brings with it certain elements that are destructive to capitalism. As capitalists reinvest capital, the demand for labor goes up. The increased demand for labor causes the labor pool—the number of unemployed—to shrink. As with any commodity, when demand is greater than the supply, the price goes up—in this case, the amount of wages being paid out. The increase in the amount of wages causes profits to go down. As profits go down, capitalists cut back production, which precipitates a crisis in the economy. The crisis causes more workers to be laid off and small businesses (petite bourgeoisie) to fold. These small businesses are bought out by the larger capitalists and the once small-scale capitalists become part of the working class. The result of this process being repeated over time is that the class of dependent workers increases and capital is centralized into fewer and fewer hands. Thus, the existing capitalists accumulate additional capital and the entire cycle starts again.

Figures that measure wealth and ownership over time are difficult to produce. Wealth in the United States is usually well hidden and ferreting out ownership lines is thorny at best. Nevertheless, we can find numbers that seem to substantiate the effects that Marx is talking about. While these are generally indirect measures, it is interesting to note that according to *Forbes* magazine (see Kennickell, 2003), the average wealth of the 400 richest people in the United States rose from $921 million in 1989 to $2.1 billion in 2002. The distribution of income over time has likewise shifted upwards in the United States, according to census data. In 1967, the top fifth of the population received 43.8% of the distributed income and the bottom fifth received just 4%. In 2001, the top fifth was paid 50.1% of the available income and the bottom fifth received 3.5%. The big drop in income between 1967 and 2001 actually was in the middle fifth of the population: they dropped from 17.3% to 14.6%. Like I said, these figures don't prove Marx's theory, but they do provide some indication that the control of capital is centralizing into fewer and fewer hands.

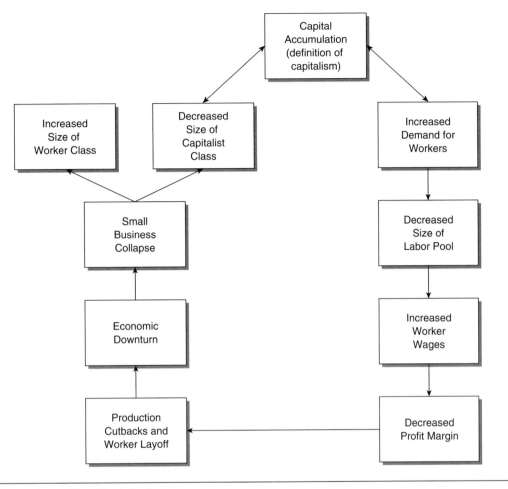

Figure 1.2   Class Bipolarization

> But not only has the bourgeoisie forged the weapons that bring death to itself;
> it has also called into existence the men who are to wield those weapons—the
> modern working class—the proletariat. (Marx & Engels, 1848/1978, p. 478)

As a result of this business cycle, three things occur. First, the size of the capitalist class shrinks. For Marx, that means that when the revolution does come about,
it will be easy to take over power from fewer capitalists. Of course, the second thing
that occurs is that the size of the working class army increases, which will give them
more power. And finally, the gap between the owners and the workers becomes
more and more apparent, leading to a bipolarization of conflict (the reduction of
conflict to that between two parties).

The bipolarization of conflict is a necessary step for Marx in the process of social
change, and an important one for conflict theory as a whole. Overt and intense
conflict are both dependent upon bipolarization; crosscutting interests, that is, having more than one issue over which groups are in conflict, tend to pull resources

(emotional and material) away from conflict. For example, during WWII it was necessary for the various nations to coalesce into only two factions, the Axis and the Allied powers. So the United States became strange bedfellows with the Soviet Union in order to create large-scale, intense conflict. Note that each time the United States has engaged in a police action or war, attempts are made to align resources in as few camps as possible. This move isn't simply a matter of world *opinion*—rather, it is a *structural necessity* for violent and overt conflict. The lack of bipolarized conflict creates an arena of crosscutting interests that can drain resources and prevent the conflict from escalating and potentially resolving.

Industrialization, Commodification, and Overproduction

Another dynamic in the dialectic of capitalism occurs as a result of the interconnected relations among industrialization, markets, and commodification. Industrialization is the process through which work moves from being performed directly by human hands to having the intermediate force of a machine. Industrialization increases the level of production, which in turn expands the use of markets, because the more product we have, the more points of purchase we need. The relationships among industrialization, production, and markets are reciprocal so that they are mutually reinforcing. If a capitalist comes up with a new "labor-saving" machine, it will increase production, and increased production pushes for new or expanded markets in which to sell the product. These expanding markets also tend to push for increased production and industrialization. Likewise, if a new market opens up through political negotiations (like with Mexico or China, for the United States) or the invention of a new product, there will be a corresponding push for increased production and the search for new machinery.

An important point to note here is that capitalism requires expanding markets, which implies that markets and their effects may be seen as part of the dialectic of capitalism. In capitalism, there is always a push to increase profit margins. To increase profits, capitalists can expand their markets horizontally and vertically, in addition to increasing the level of surplus labor. In fact, profit margins would slip if capitalists did not expand their markets. For example, one of the main reasons that you likely own an MP3 player is that the market for cassette tape players and compact discs bottomed out (and there was a push for newer technologies). Most people who were going to buy a CD player had already done so, and the only time another would be purchased is for replacement. So, capitalists invented something new for you to buy so that their profit margin would be maintained.

The speed at which goods and services move through markets is largely dependent upon a *generalized medium of exchange*, something that can act as a universal value. Barter is characterized by the exchange of products for one another. The problem with bartering is that it slows down the exchange process because there is no general value system. For example, how much is a keg of beer worth in a barter system? We can't really answer that question because the answer depends on what it is being exchanged for, who is doing the exchanging, what their needs are, where the exchange takes place, and so on. Because of the slowness of bartering, markets tend to push for more generalized media of exchange—such as money. Using money, we can give an answer to the keg question, and having such an answer speeds up the exchange process quite a bit. Note that in this context, "money" refers to any generalized medium of exchange, such as the use of credit cards. Capitalists are thus driven to find new and more inventive means for media of exchange. Marx argues that as markets expand and become more important in a society, and the use of money becomes more universal, money becomes more and more the common denominator of all human relations. As Marx (1932/1978a) says,

> By possessing the *property* of buying everything, by possessing the property of appropriating all objects, *money* is thus the *object* of eminent possession. The universality of its *property* is the omnipotence of its being. It therefore functions as the almighty being. Money is the *pimp* between man's need and the object, between his life and his means of life. But that which mediates *my* life for me, also *mediates* the existence of other people *for me*. For me it is the other person. (p.102, emphasis original)

The concept of **commodification** describes the process through which more and more of the human life-world is turned into something that can be bought or sold. So, instead of creatively producing the world as in species-being, people increasingly buy (and sell) the world in which they live. This process of commodification becomes more and more a feature of human life because it continually expands. Think of a farming family living in the United States around 1850 or so. That family bought some of what they needed, and they bartered for other things, but the family itself produced much of the necessities of life. Today, the average American family buys almost everything they want or need. The level of commodification is therefore much higher today.

As you can tell, once in place, these elements of capitalism are self-reinforcing. Money facilitates exchanges in markets; money and the drive for profit push the size and exchange rate of markets, which in turn pushes for increased production and commodification. Notice that, because of human nature, there is no natural limit to this process. One of the things that Marx (1932/1978b) points out is that human beings have the unusual ability to create their own needs: "life involves before everything else eating and drinking, a habitation, clothing and many other things . . . that the satisfaction of the first need . . . leads to new needs; and this production of new needs is the first historical act" (p. 156). In other words, human beings can create additional or secondary "needs." And, because people can create new needs, there is no limitation placed upon the proliferation of commodities. Thus, the potential for the production of commodities is endless.

Now think about what this means: Capitalists are driven to endlessly accumulate capital (it's the definition of modern capitalism). They use this capital to invest in human labor and industrialization, and to endlessly create new commodities. Added to these factors is the expansive nature of markets. Together, these normal

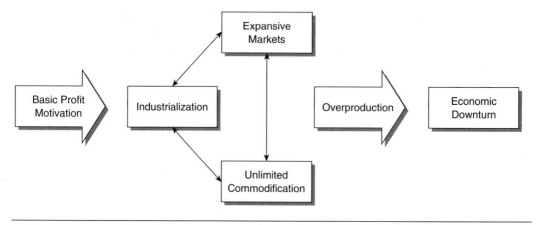

Figure 1.3 Overproduction

forces of capitalism inevitably lead to *overproduction*—too much production for the current demand. Like the business cycle that produces class bipolarization, overproduction is a fundamental property of capitalism that inexorably leads to production cutbacks, worker layoffs, and, in due course, economic downturns. I've pictured these relationships in Figure 1.3.

Interestingly, the issue of overproduction has entered the mindset of popular culture. Commenting on a slow holiday season, *Time* magazine carried an article saying that

> Karl Marx theorized that capitalism was condemned to repeated depressions because of "cycles of overproduction." . . . if Marx had hit the shopping malls last week and seen the heavy discounting—or looked on the Internet and seen the emergence of cut-rate sites like Amazon.com's new outlet store—he would no doubt have felt vindicated. (Cohen et al., 2001, p. 21)

Value and Exploitation

Marx forms much of his theory of capitalism in direct contrast to the political economists of his day, for whom commodities, value, profit, private property, and the division of labor were seen as natural effects of social evolution. Marx takes a critical perspective and sees these same processes as instruments of oppression that dramatically affect people's life chances.

One of the important issues confronting early political economists concerned the problem of value. We may have a product, such as a car, and that product has value, but from where does its value come? More importantly, why would anybody pay more for the car than it is worth? Well, you say, only a sucker would pay more for a car than it is worth. Yet, as you'll see, there is a way in which we all pay more for every commodity or economic good than it is worth. That was what struck the early economists as a strange problem to be solved.

In order to understand this issue, Adam Smith, an early economist and author of *The Wealth of Nations* (1776), came up with some useful concepts. He argues that every commodity has at least two different kinds of values: use-value and exchange-value. **Use-value** refers to the actual function that a product contains. This function gets used up as the product is used. Take a car, for example. The use-value of a car is transportation. As the car is used for its purpose, its use-value diminishes and you will eventually have to buy another car. The car also has exchange-value that is distinct from its use-value. *Exchange-value* refers to the rate of exchange one commodity bears when compared to other commodities. Let's say I make a pair of shoes. Those shoes could be exchanged for 1 leather-bound book or 5 pounds of fish or 10 pounds of potatoes or 1 cord of oak wood, and so on.

This notion of exchange-value poses a question for us: what do the shoes, books, fish, and potatoes have in common that allow them to be exchanged? I could exchange my pair of shoes for the leather-bound book and then exchange the book for a shirt. The shirt might have a use-value for me where the leather book does not; nonetheless, they both have exchange-value. This train of exchange could be extended indefinitely with me never extracting any use-value from the products at all, which implies that exchange- and use-values are distinct. So, what is the common denominator that allows these different items to be exchanged? What is the source of exchange-value?

Smith argues, and Marx agrees, that the substance of all value is human labor: "Labour, therefore, is the real measure of the exchangeable value of all commodities" (Smith, 1776/1937, p. 30). There is labor involved in the book, the fish, the shoes, the potatoes, and in fact everything that people deem worthy of being exchanged. It is labor, then, that creates exchange-value. If we stop and think for a moment about Marx's idea of species-being, we can see why Smith's notion appealed to him: the value of a product is the "humanness" it contains.

This explanation is termed the *labor theory of value*, and it is the reason why Marx thinks money is so insidious. The book, the shoes, and the potatoes can have exchange-value because of their common feature—human labor. If we then make all those commodities equal to money—the universal value system—then labor is equated with money: "This physical object, gold [or money] . . . becomes . . . the direct incarnation of all human labour" (Marx, 1867/1977, p. 187), which of course adds to the experience of alienation from species-being.

In employing the two terms, Smith tends to collapse them, focusing mainly on exchange-value. Nevertheless, Marx maintains the distinction and argues that the difference between use-value and exchange-value is where profit is found (we pay more for a product than its use-value would indicate). Smith eventually argued that profit is simply added by the capitalist. Profit in Smith's theory thus becomes arbitrary and controlled by the "invisible hand of the market." However, in analyzing value, Marx discovers a particular kind of labor—surplus labor—and argues that profit is better understood as a measurable entity that he calls *exploitation*.

Like Smith, Marx distinguishes between the use-value of a product and its exchange-value in the market. Capital, then, is created by using existing commodities to create a new commodity whose exchange-value is higher than the sum of the original resources used. This situation was odd for Marx. From where did the added value come? His answer, like Smith's original one, is human labor. But he takes Smith's argument further. Human labor is a commodity that is purchased for less than its total worth. According to the value theory of labor, the value of any commodity is determined by the labor time necessary for the production or replacement of that commodity. So, what does it cost to produce human labor? The cost is reckoned in terms of the necessities of life: food, shelter, clothing, and so on. Marx also recognizes that a comparative social value has to be added to that list as well. What constitutes a "living wage" will thus be different in different societies. Marx calls the labor needed to pay for the worker's cost of living *necessary labor*. The issue for Marx is that the cost of necessary labor is less than that of what the worker actually produces.

The capitalist pays less for a day's work than its value. I may receive $75.00 per day to work (determined by the cost of bare sustenance plus any social amenities deemed necessary), but I will produce $200.00 worth of goods or services. The necessary labor in this case is $75.00. The amount of labor left over is the **surplus labor** (in this case, $125.00). The difference between necessary labor and surplus labor is the rate of exploitation. Different societies can have different levels of exploitation. For example, if we compare the situation of automobile workers in the United States with those in Mexico, we will see that the level of exploitation is higher in Mexico (which is why U.S. companies are moving so many jobs out of the country—they can employ workers at lower cost). Surplus labor and exploitation are the places from which profit comes: "The rate of surplus-value is . . . an exact expression for the degree of exploitation . . . of the work by the capitalist" (Marx, 1867/1977, p. 326).

By definition, capitalists are pushed to increase their profit margin and thus the level of surplus labor and the rate of exploitation. There are two main ways in which this can be done: through absolute and relative surplus labor. The capitalist can directly increase the amount of time work is performed by lengthening the workday, say from 10 to 12 hours; or he or she can remove the barriers between "work" and "home," as is happening as a result of increases in communication (computers) and transportation technologies. The product of this lengthening is called *absolute surplus labor*. The other way a capitalist can increase the rate of exploitation is to reduce the amount of necessary labor time. The result of this move is called *relative surplus labor*. The most effective way in which this is done is through industrialization. With industrialization, the worker works the same number of hours, but his or her output is increased through the use of machinery. These different kinds of surplus labors can get a bit confusing, so I've compared them in Figure 1.4.

The figure starts off with the type of surplus labor employed. Since there is always exploitation (you can't have capitalism without it), I've included a "base rate" for the purpose of comparison. Under this scheme, the worker has a total output of $200; he or she gets paid $75 of the $200 produced, which leaves a rate of exploitation of about 63%, or $125. The simplest way to increase profit, or the rate of exploitation, is to make the worker work longer hours or take on added responsibilities without raising pay (as a result of downsizing, for example). In our

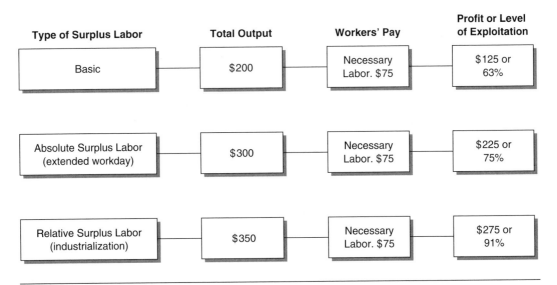

Type of Surplus Labor	Total Output	Workers' Pay	Profit or Level of Exploitation
Basic	$200	Necessary Labor. $75	$125 or 63%
Absolute Surplus Labor (extended workday)	$300	Necessary Labor. $75	$225 or 75%
Relative Surplus Labor (industrialization)	$350	Necessary Labor. $75	$275 or 91%

Figure 1.4   Surplus Labor

hypothetical case, the wage of $75 remains, but the profit margin (rate of exploitation) goes up to 75%. By automating production, the capitalist is able to extract more work from the worker, thus increasing the total output and the level of exploitation (91%). If this seems natural to us (capitalists have a right to make a profit), Marx would say that it is because we have bought into the capitalist ideology. We should also keep in mind that in their search for maintaining or increasing the rate of exploitation, capitalists in industrialized nations export their exploitation—they move jobs to less developed countries.

Concepts and Theory: The Effects of Capitalism

Alienation, Private Property, and Commodity Fetish

We often think of alienation as the subjective experience of a worker on an assembly line or at McDonald's. This perception is partly true. To get a little better handle on it, however, let's think about the differences between a gunsmith and a worker in a Remington plant. The gunsmith is a craftsperson and would make every part of the gun by hand—all the metal work, all the woodwork, everything. If you knew guns, you could tell the craftsperson just by looking at the gun. It would bear the person's mark because part of the craftsperson was in the work, and the gunsmith would take justifiable pride in the piece.

Compare that experience to the Remington plant worker. Perhaps she is the person who bolts the plastic end piece on the butt of the rifle, and she performs the task on rifle after rifle, day after day. It won't take long until the work becomes mind-numbing. It is repetitive and noncreative. Chances are, this worker won't feel the pride of the craftsperson but will instead experience disassociation and depression. We can see in this example some of the problems associated with a severe division of labor and over-control of the worker. There's no ownership of the product or pride as there is in craftsmanship. We can also see how the same issues would apply to the worker at McDonald's. All of it is mind-numbing, depressive work, and much of it is the result of the work of scientific management.

Frederick Taylor was the man who applied the scientific method to labor, at the beginning of the twentieth century. He was interested in finding the most efficient way to do a job. Efficiency here is defined in terms of the least amount of work for the greatest amount of output. Under this system, the worker becomes an object that is directly manipulated for efficiency and profit. Taylor would go out into the field and find the best worker. He and his team would then time the worker (this is the origin of time-management studies) and break the job down into its smallest parts. In the end, Taylorism created a high division of labor, assembly lines, and extremely large factories.

The problem with understanding our Remington and McDonald's examples as alienation is that it focuses on the subjective experience of the worker. It implies that if we change the way we control workers—say, from Taylorism to the Japanese management system, as many American companies have—we've solved the problem of alienation. For Marx, that wouldn't be the case. While alienation implies the subjective experience of the worker (depression and disassociation), it is more accurate to think of it as an objective state. Thus, workers under the Japanese system are not structurally less alienated because they are more broadly trained, are able to rotate jobs, work holistically, and have creative input. According to Marx, workers are alienated under all forms of capitalism, whether they feel it or not. Alienation is a structural condition, not a personal one. Workers are of course more likely to revolt if they experience alienation, but that is a different issue.

Alienation always exists when someone other than the worker owns the means of production and the product itself. Having said that, we can note that Marx actually talked about four different kinds of alienation. **Alienation** in its most basic sense is separation from one's own awareness of being human. We know we are human; Marx doesn't mean to imply that we don't know we are human. What he means is that our idea is wrong or inaccurate. Recall Marx's argument of species-being: that which makes us distinctly human is creative production, and we become aware of our humanity as our nature is clearly reflected back to us by the mirror of the produced world. At best, the image is distorted if we look to see our nature in the things we own. It's the right place to look, but according to Marx, products should be the natural expression of species-being and production should result in true consciousness.

Under capitalism, the production process and the resulting commodity are owned and controlled by someone other than the worker. The commodity itself, then, does not reflect humanity, does not result in true consciousness, is a fetish in the sense that humans are irresistibly drawn to the commodity but left empty, and it causes people to misrecognize the oppression inherent in the capitalist system.

Further, commodities aren't produced to meet human need. Most of the "needs" we experience are produced through the existence of the commodity: the idea for the commodity came from capitalists seeking profit and our awareness and desire for the commodity were created through advertising.

Thus, literally everything about the commodity is wrong. At worst, the image is simply false. Marx argues that if we look elsewhere for our definition and knowledge of human nature (such as using language; having emotions; possessing a soul, religion, rationality, free-will, and so on), it is not founded on the essential human characteristic—free and creative production. Thus, the human we see is rooted in false consciousness.

From this basic alienation, three other forms are born: alienation from the work process, alienation from the product itself, and alienation from other people. In species-being, there is not only the idea of the relationship between consciousness and creative production, there is also the notion that human beings are related to one another directly and intimately through species-being. We don't create a human world individually; it is created collectively. Therefore, under conditions of species-being, humans are intimately and immediately connected to one another. The reflected world around them is the social, human world that they created. However, when we are alienated from our own species-being, because someone else controls the means and ends of production, we are estranged from other humans as well.

> We arrive at the result that man (the worker) feels himself to be freely active only in his animal functions—eating, drinking, and procreating, or at most also in his dwelling and in personal adornment—while in his human functions he is reduced to an animal. The animal becomes human and the human becomes animal. (Marx, 1932/1995, p. 99)

Finally, of course, we are alienated from the process of work and from the product itself. There is something essential in the work process, according to Marx. The kind of work we perform and how we perform that work determines the kind of person we are. Obviously, doctors are different from garbage collectors, but that isn't what Marx has in mind. An individual's humanity is rooted in the work process. When humans are cut off from controlling the means of production (the way in which work is performed and for what reason), the labor process itself becomes alienated. There are three reasons for this alienated labor, which results in the alienation of the product.

First, when someone else owns the means of production, the work is external to the worker; that is, it is not a direct expression of his or her nature. So, rather than being an extension of the person's inner being, work becomes something external and foreign. Second, work is forced. People don't work because they want to; they work because they have to—under capitalism, if you don't work, you die. Concerning labor under capitalism Marx (1932/1995) says, "Its alien character is clearly shown by the fact that as soon as there is no physical or other compulsion it is avoided like the plague" (pp. 98–99). And third, when we do perform the work, the thing that we produce is not our own; it belongs to another person.

Further, alienation, according to Marx, is the origin of *private property*—it exists solely because we are cut off from our species-being; someone else owns the means and ends of production. Underlying markets and commodification—and capitalism itself—is the institution of private property. Marx felt that the political economists of his day assumed the fact of private property without offering any explanation for it. These economists believed that private property was simply a natural part of the economic process. But for Marx, the source of private property was the crux of the problem. Based on species-being, Marx argues that private property emerged out of the alienation of labor.

Marx also argues that there is a reciprocal influence of private property on the experience of alienation. As we have seen, Marx claims that private property is the result of alienated labor. Once private property exists, it can exert its own influence on the worker and it becomes "the realization of this alienation" (Marx, 1932/1995, p. 106). Workers then become controlled by private property. This notion is most clearly seen in what Marx describes as **commodity fetish:** workers become infatuated with their own product as if it were an alien thing. It confronts them not as the work of their hands, but as a commodity, something alien to them that they must buy and appropriate.

Commodity fetish is a difficult notion and a hard one to illustrate. Workers create and produce the product, yet we don't recognize our work or ourselves in the product. So we see it outside of us and we fall in love with it. We have to possess it, not realizing that it is ours already by its very nature. We think it, the object, will satisfy our needs, when what we need is to find ourselves in creative production and a socially connected world. We go from sterile object to sterile object, seeking satisfaction, because they all leave us empty. To use a science fiction example, it's like a male scientist who creates a female robot, but then forgets he created it, falls in love with it, and tries to buy its affection through money. As I said, it is a difficult concept to illustrate because in this late stage of capitalism, *our entire world and way of living are the examples*. In commodity fetish, our perception of self-worth is linked with money and objects in a vicious cycle.

In addition, in commodity fetish we fail to recognize—or in Marxian terms, we **misrecognize**—that there are sets of oppressive social relations in back of both the perceived need for and the simple exchange of money for a commodity. We think that the value of the commodity is simply its intrinsic worth—that's just how much a Calvin Klein jacket is worth—when, in fact, hidden labor relations and exploitation produce its value. We come to need these products in an alienated way: We think that owning the product will fulfill our needs. The need is produced through the commodification process, and in back of the exchange itself are relations of oppression. In this sense, the commodity becomes reified: it takes on a sense of reality that is not materially real at all.

Contemporary Marxists argue that the process of commodification affects every sphere of human existence and is the "central, structural problem of capitalist society in all its aspects" (Lukács, 1923/1971, p. 83). Commodification translates all human activity and relations into objects that can be bought or sold. In this process, value is determined not by any intrinsic feature of the activity or the relations, but by the impersonal forces of markets, over which individuals have no control. In this expanded view of commodification, the objects and relations that will truly gratify human needs are hidden, and the commodified object is internalized and accepted as reality. So, for example, a young female college student may internalize the commodified image of thinness and create an eating disorder such as anorexia nervosa that rules her life and becomes unquestionably real. Commodification, then, results in a consciousness based on reified, false objects. It is difficult to think outside this commodified box. There is, in fact, a tendency to justify and rationalize our commodified selves and behaviors.

False Consciousness and Religion

Alienation, false consciousness, and ideology go hand in hand. In place of a true awareness of species-being comes false consciousness—consciousness built on any foundation other than free and creative production. Humans in false consciousness thus come to think of themselves as defined through the ability to have ideas, concepts, and abstract thought, rather than through production. When these ideas are brought together in some kind of system, Marx considers them to be ideology. Ideas function as ideology when they are perceived as independent entities that transcend historical, economic relations: ideologies contain beliefs that we hold to be true and right, regardless of the time or place (like the value of hard work and just reward). In the main, ideologies serve to either justify current power arrangements (like patriarchy) or to legitimate social movements (like feminism) that seek to change the structure.

Though Marx sometimes appears to use the terms interchangeably, I think it is important to keep the distinction between false consciousness and ideology clear. Ideologies can change and vary. For example, the ideology of consumerism is quite different from the earlier ideologies of the work ethic and frugality, yet they are all capitalist ideologies. The ideologies behind feminism are different from the beliefs behind the racial equality movement, but from Marx's position, both are ideologies that blind us to the true structure of inequality: class. Yet false consciousness doesn't vary. It is a state of being, somewhat like alienation in this aspect. We are by definition in a state of false consciousness because we are living outside of species-being. The very way through which we are aware of ourselves and the world around us is false or dysfunctional. The very *method of our consciousness* is fictitious.

> [I]t is clear that the more the worker spends himself, the more powerful the alien objective world becomes which he creates over-against himself, the poorer he himself—his inner world—becomes, the less belongs to him as his own. The more man puts into God, the less he retains in himself. (Marx, 1932/1978a, p. 72)

Generally speaking, false consciousness and ideology are structurally connected to two social factors: religion and the division of labor. For Marx, *religion* is the archetypal form of ideology. Religion is based on an abstract idea, like God, and religion takes this abstract idea to be the way through which humans can come to know their true nature. So, for example, in the evangelical Christian faith, believers are exhorted to repent not only their sinful ways but also their sinful nature, and to be born again with a new nature—the true nature of humankind. Christians are thus encouraged to become Christ-like because they have been created in the image of God. Religion, then, reifies thought, according to Marx. It takes an abstract (God), treating it as if it is materially real, and it then replaces species-being with non-materially based ideas (becoming Christ-like). It is, for Marx, a never-ending reflexive loop of abstraction, with no basis in material reality whatsoever. Religion, like the commodity fetish, erroneously attributes reality and causation. We pour ourselves out, this time into a religious idea, and we misrecognize our own nature as that of a god or devil. Religion is ideological because it is based in and reifies ideas (**reification**).

There is also a second sense in which religion functions as ideology. Marx uses the term ideology as *apologia*, or a defense of one's own ideas, opinions, or action. In this kind of ideology, the orientation and beliefs of a single class, the elite, become generalized and seem to be applicable to all classes. Here the issue is not so much reification as class consciousness. The problem in reification is that we accept something as real that isn't. With ideology, the problem is that we are blinded to the oppression of the class system. This is in part what Marx means when he claims that religion is "the opium of the people." As we have seen, Marx argues that because most people are cut off from the material means of production, they misrecognize their true class position and the actual class-based relationships, as well as the effects of class position. In the place of class consciousness, people accept an ideology. For Marx, religion is the handmaiden of the elite; it is a principal vehicle for transmitting and reproducing the capitalist ideology.

Thus, in the United States we tend to find a stronger belief in American ideological ideas (such as meritocracy, equal opportunity, work ethic, poverty as the result of laziness, free enterprise, and so on) among the religious (particularly among Christian denominations in the United States). We would also expect to see religious people being less concerned with the social foundations of inequality and more concerned with patience in this life and rewards in the next. These kinds of beliefs, according to Marx, dull the workers' ability to institute social change and bring about real equality.

It is important to note that this is a function of religion in general, not just Christian religions. The Hindu caste system in India is another good example. There are five different castes in the system: Brahmin (priests and teachers), Kshatriya (rulers), Vaishya (merchants and farmers), Shudra (laborers and servants), and Harijans (polluted laborers, the outcastes). Position in these different castes is a result of birth; birth position is based on karma (action); and karma is based on dharma (duty). There is virtually no social mobility among the castes. People are taught to accept their position in life and perform the duty (dharma) that their caste dictates so that their actions (karma) will be morally good. This ideological structure generally prevents social change, as does the Christian ideology of seeking rewards in heaven.

For Marx, then, religion simultaneously represents the furthest reach of humanity's misguided reification and functions to blind people to the underlying class conditions that produce their suffering. Yet for Marx (1844/1978c), "*Religious* suffering is at the same time an *expression* of real suffering and a *protest* against real suffering. Religion is the sigh of the oppressed creature, the sentiment of a heartless world, and the soul of soulless conditions. It is the *opium* of the people" (p. 54, emphasis original). Marx thus recognizes that religion also gives an outlet to suffering. He feels that religion places a "halo" around the "veil of tears" that is present in the human world. The tears are there because of the suffering that humans experience when they don't live communally and cooperatively. Marx's antagonism toward religion, then, is not directed at religion and God per se, but at "the *illusory* happiness of men" that religion promises (p. 54).

Marx also sees ideology and alienation as structurally facilitated by the **division of labor** (how the range of duties is assigned in any society). Marx talks about several different kinds of divisions of labor. The most primitive form of separation of work is the "natural division of labor." The natural division was based upon the individual's natural abilities and desires. People did not work at something for which they were ill suited, nor did they have to work as individuals in order to survive. Within the natural division of labor, survival is a group matter, not an individual concern. Marx claims that the only time this ever existed was in preclass societies. When individuals within a society began to accumulate goods and exercise power, the "forced division of labor" replaced the natural division. With the forced division, individual people must work in order to survive (sell their labor), and they are forced to work at jobs they neither enjoy nor have the natural gifts to perform. The forced division of labor and the *commodification of labor* characterize capitalism.

This primary division of labor historically becomes extended when mental labor (such as that performed by professors, priests, philosophers) is divided from material labor (workers). When this happens, reification, ideology, and alienation reach new heights. As we've seen, Marx argues that people have true consciousness only under conditions of species-being. Any time people are removed from controlling the product or the production process, there will be some level of false consciousness and ideology. Even so, workers who actually produce a material good are in some way connected to the production process. However, with the separation of mental from material labor, even this tenuous relationship to species-being is cut off. Thus, the thought of those involved with mental labor is radically cut off from what makes us human (species-being). As a result, everything produced by professors, priests, philosophers, and so on has some reified ideological component and is generally controlled by the elite.

Class Consciousness

So far we have seen that capitalism increases the levels of industrialization, exploitation, market-driven forces like commodification, false consciousness, ideology, and reification, and it tends to bifurcate the class structure. Marx also argues that these factors have dialectical effects and will thus push capitalism inexorably toward social change. Conflict and social change begin with a change in the way we are aware of our world. They begin with class consciousness.

Marx notes that classes exist objectively, whether we are aware of them or not. He refers to this as a "class in itself," that is, an aggregate of people who have a common relationship to the means of production. But classes can also exist subjectively as a "class *for* itself." It is the latter that is produced through class consciousness. **Class consciousness** has two parts: the subjective awareness that experiences of deprivation are determined by structured class relations and not individual talent and effort, and the group identity that comes from such awareness.

Industrialization has two main types of effect when it comes to class consciousness. First, it tends to increase exploitation and alienation. We've talked about both of these already, but remember that these are primarily objective states for Marx. In other words, these aren't necessarily subjectively felt—alienation isn't chiefly a feeling of being psychologically disenfranchised; it is the state of being cut off from species-being. Humans can be further alienated and exploited, and machines do a good job of that. What happens at this point in Marx's scheme is that these objective states can produce a sense (or feeling) of belonging to a group that is disenfranchised—that is, class consciousness.

As you can see from Figure 1.5, industrialization has positive relationships with both exploitation and alienation. As capitalists employ machinery to aid in labor, the objective levels of alienation and exploitation increase. As the objective levels increase, so does the probability that workers will subjectively experience them, thus aiding in the production of class consciousness. Keep in mind that industrialization is a variable, which means that it can increase (as in when robots do the work that humans once did on the assembly line—there is a human controlling that robot, but that person is far, far removed from the labor of production) or decrease (as when "cottage industries" spring up in an economy).

The second area of effect is an increase in the level of worker communication. Worker communication is a positive function of education and ecological concentration. Using more and more complex machines requires increasing levels of technical knowledge. A crude but clear example is the different kinds of knowledge needed to use a horse and plow compared to a modern tractor. Increasing the use of technology in general requires an increase in the education level of the worker (this relationship is clearly seen in today's computer-driven U.S. labor market).

In addition, higher levels of industrialization generally increase the level of worker concentration. Moving workers from small guild shops to large-scale machine shops or assembly lines made interaction between these workers possible in a way never before achievable, particularly during break and lunch periods when hundreds of workers could gather in a single room. Economies of scale tend to increase this concentration of the workforce as well. So, for a long time in the United States, we saw ever-bigger factories being built and larger and larger office buildings (like the Sears Tower and World Trade Center—and it is significant that terrorists saw the World Trade Center as representative of American society). These two processes, education and ecological concentration, work together to increase the level of communication among workers. These processes are supplemented through greater levels of communication and transportation technologies. Marx argues that communication and transportation would help the worker movement spread from city to city.

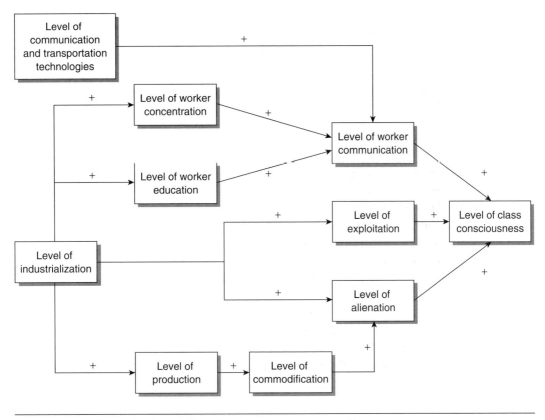

Figure 1.5   The Production of Class Consciousness

So, in general, class consciousness comes about as workers communicate with each other about the problems associated with being a member of the working class (like not being able to afford health care). The key to Marx's thinking here is that these things occur due to structural changes brought about simply because of the way capitalism works. Capitalists are driven to increase profits. As a result, they use industrialization, which sets in motion a whole series of processes that tend to increase the class consciousness of the workers. Class consciousness increases the probability of social change. As workers share their grievances with one another, they begin to doubt the legitimacy of the distribution of scarce resources, which in turn increases the level of overt conflict. As class inequality and the level of bipolarization increase, the violence of the conflict will tend to increase, which in turn brings about deeper levels of social change.

However, class consciousness has been difficult to achieve. There are a number of reasons given as to why this is true, but many Marxist approaches focus on the relationships among a triad of actors: the state, the elite, and workers. Marx felt that as a result of the rise of class consciousness, and other factors such as the business cycle, workers would unite and act through labor unions to bring about change. Some of the work of the unions would be violent, but it would eventually lead to a successful social movement. In the end, the labor movement would bring about socialism.

Marx sees two of the actors in the triad working in collusion. He argues that under capitalism, the state is basically an arm of the elite. It is controlled by capitalists and functions with capitalist interests in mind. Many of the top governing officials come from the same social background as the bourgeoisie. A good example of this in the United States is the Bush family. Of course, electing a member of the capitalist elite to high public office not only prejudices the state toward capitalist interests; it is also an indicator of how ideologically bound a populace is.

In addition, as C. W. Mills (1956) argues, the elite tend to cross over, with military officers serving on corporate boards and as high-placed political appointees, and CEOs functioning as political advisees and cabinet members, and so on. A notable example of these kinds of interconnections is Charles Erwin Wilson, president of General Motors from 1941 to 1953. He was appointed Secretary of Defense under President Dwight D. Eisenhower. At his senate confirmation hearing, Wilson spoke the words that epitomize the power elite: "For years, I thought what was good for our country was good for General Motors—and vice versa."

In response to the demands of labor unions, certain concessions were ultimately granted to the working and middle classes. Work hours were reduced and health care provided, and so forth. Capitalists working together with a capitalist-privileging state granted those concessions. Thus, even when allowances are granted, they may function in the long run to keep the system intact by maintaining the capitalists' position, silencing the workers, and preventing class consciousness from adequately forming.

The state is also active in the production of ideology. Marx sees the state as somewhat ill defined. In other words, where the state begins and ends under capitalist democracy is hard to say. The state functions through many other institutions, such as public schools, and its ideology is propagated through such institutions, not only through such direct means as the forced pledge of allegiance in the United States but also through indirect control measures such as specific funding initiatives. A good deal of the state's ideology is, of course, capitalist ideology. As a result, the worker is faced with a fairly cohesive ideology coming from various sources. This dominant ideology creates a backdrop of taken-for-grantedness about the way the world works, against which it is difficult to create class consciousness.

The movement of capitalist exploitation across national boundaries, which I mentioned before, accentuates the "trickle-down" effect of capitalism in such countries as the United States. Because workers in other countries are being exploited, the workers in the United States can be paid an inflated wage (inflated from the capitalists' point of view). This is functional for capitalism in that it provides a collection of buyers for the world's goods and services. Moving work out from the United States and making it the world's marketplace also changes the kind of ideology or culture that is needed. There is a movement from worker identities to consumer identities in such economies.

In addition, capitalists use this world labor market to pit workers against one another. While wages are certainly higher due to exported exploitation, workers are also aware that their jobs are in jeopardy as work is moved out of the country. Capitalism always requires a certain level of unemployment. The business cycle teaches us this—that zero unemployment means higher wages and lower profits.

The world labor market makes available an extremely large pool of unemployed workers. So, workers in an advanced industrialized economy see themselves in competition with much cheaper labor. This global competition also hinders class consciousness by shifting the workers' focus of attention away from the owners and onto the foreign labor market. In other words, a globalized division of labor pits worker against worker in competition for scarce jobs. This competition is particularly threatening for workers in advanced capitalist countries, like the United States, because the foreign workers' wage is so much lower. These threatened workers, then, will be inclined to see their economic problems in terms of global, political issues rather than class issues.

Further, the workers divide themselves over issues other than class. We tend to see ourselves not through class-based identities, which Marx would argue is the identity that determines our life chances; instead, we see ourselves through racial, ethnic, gender, and sexual preference identities. Marxists would argue that the culture of diversity and victimhood is part of the ideology that blinds our eyes to true social inequality, thus preventing class consciousness.

Marx Summary

- Marx's perspective is created through two central ideas: species-being and the material dialectic. Species-being refers to the unique way in which humans survive as a species—we creatively produce all that we need. The material dialectic is the primary mechanism through which history progresses. There are internal contradictions within every economic system that push society to form new economic systems. The idea of the dialectic is found in a number of theorists in this book. From our discussion of Marx, we can pull out three important and interrelated elements: (1) a dialectic contains different elements that are necessarily locked together in some kind of tension; (2) a dialectic implies a dynamic process; and (3) a dialectic inevitably brings change that comes out of and is related to the initial tensions. When we come across the idea of a dialectic again, keep these elements in mind.

- Every economic system is characterized by the means and relations of production. The means of production in capitalism is owned by the bourgeoisie and generally consists of commodification, industrial production, private property, markets, and money. One of the unique features of capitalism is that it will swallow up all other classes save two: the bourgeoisie and the proletariat. This bifurcation of class structure will, in turn, set the stage for class consciousness and economic revolution.

- Capitalism affects every area of human existence. Through it, individuals are alienated from each aspect of species-being and creative production. The work process, the product, other people, and even their own inner being confront the worker as alien objects. As a result, humankind misrecognizes the truth and falls victim to commodity fetish, ideology, and false consciousness. However, because capitalism contains dialectical elements, it will also produce the necessary ingredient for economic revolution: class consciousness. Class consciousness is the result of workers becoming aware that their fate in life is determined primarily by class position. This awareness comes as alienation and exploitation reach high levels and as workers communicate with one another through increasing levels of education, worker concentration in the factory and city, and communication and transportation technologies.

Looking Ahead

A knowledge of the writings of Marx and Engels is virtually indispensable to an educated person in our time. . . . For classical Marxism . . . has profoundly affected ideas about history, society, economics, ideology, culture, and politics; indeed, about the nature of social inquiry itself. . . . Not to be well grounded in the writings of Marx and Engels is to be insufficiently attuned to modern thought, and self-excluded to a degree from the continuing debate by which most contemporary societies live insofar as their members are free and able to discuss the vital issues. (Tucker, 1978, p. ix)

The above quote isn't exaggerated in the least, especially when it comes to contemporary social and sociological theories. A good way to talk about his influence is to see that Marx made two kinds of arguments, one philosophical and the other structural. Philosophically, Marx was concerned with consciousness. He argued that human beings are uniquely conscious because of their mode of existence. Humans creatively produce what they need. Economic production, then, is the key to understanding almost everything about people. In terms of consciousness, economic production acts like a mirror that reflects human nature. Marx argued that because people are cut off from both the means of production and the product itself, their consciousness is false and the ideas they do hold are ideological.

Marx's structural argument was based on his view of history. The engine of history—the dynamic that brings historical change—is the dialectic inherent in noncommunistic economic forms. Structurally, communism (not Stalinism or Maoism) is in keeping with human nature. It's human nature to produce collectively, and communism is based on collective production. According to Marx, all other forms are out of sync with human nature. But here the issue isn't philosophical, it's structural. It's like putting a car together incorrectly. It may look like a car, and it may even function as a car for awhile, but eventually the mechanical problems will make the car break down. It's the same with the economy, for Marx.

Capitalism in particular is problematic, because more than any other system, it structurally separates the worker from the labor process. The classic Marxian contradiction is overproduction. Capitalist economies tend to overproduce. When they do, an economic crisis occurs. Overproduction occurs because of elements that are basic to capitalism. In brief, the profit motivation and the centralization of capital push the continued expansion of markets, commodification, production, and secondary needs (perceived needs created through advertising). We don't want to get hung up on the mechanics of this; we simply want to note that it is mechanical. The dynamics of historical change reside in structural contradictions within the economy (itself a structure).

These two distinct approaches of Marx's form the basic distinctions between critical and conflict theories. Conflict theory is generally oriented toward a structural analysis of social inequality. This perspective assumes that structures are objective entities that operate according to their own laws. The goal of conflict theory, then, is to discover those laws and principles. The purpose of discovery may or may not be to improve society; the goal could simply be knowledge for knowledge's sake, as with pure science. We will see this influence of Marx in the conflict theories found in Chapter 7 (Lewis Coser, Ralf Dahrendorf, and Randall Collins), in Janet Chafetz's work on gender and William Julius Wilson's work on race (Chapter 8), in Immanuel Wallerstein's world systems theory (Chapter 12), as well as in Pierre Bourdieu's theory of class culture (Chapter 11). Bourdieu locates the structure of class in a rather unique place, as we'll see when we discuss his theory.

Critical theory, on the other hand, is indebted to the philosophical Marx. It focuses on the ways in which human consciousness is formed and influenced. The key assumption is that many of the factors are hidden, especially in capitalist societies. Rather than focusing on social structures as conflict theory does, critical theory looks to culture and the production of and indoctrination in knowledge. Of particular importance are such cultural systems as art, mass media, advertising, science, and the human disciplines (sociology included).

Critical theory also sees an inseparable relationship between knowledge and interests. In other words, all culture and knowledge are produced by people with specific ideological beliefs and values. It is thus impossible to separate human knowledge from ideology. The intent of critical knowledge, then, is to expose the distortions, misrepresentations, and political values found in our culture and knowledge. In addition, where conflict theory may say it is only interested in pure knowledge (a claim critical theorists would dispute), critical theory is aimed at praxis: the reflexive and intentional use of critical theory in daily life and practice. And because of praxis, critical theorists want to intentionally engage public dialogue. Almost every critical theorist, then, is a public sociologist. We'll find this influence of Marx in the work of Jürgen Habermas and the Frankfurt School (Chapter 7), Dorothy E. Smith's gender standpoint theory (Chapter 13), and Jean Baudrillard's vision of postmodernism (Chapter 14).

Building Your Theory Toolbox

Knowing Marx

After reading and understanding this chapter, you should be able to define the following terms theoretically and explain their importance to Marx's theory:

species-being, false consciousness, substructure and superstructure, material dialectic, natural division of labor, primitive communism, class, means and relations of production, bourgeoisie and proletariat, capital accumulation, bipolarization of conflict, labor theory of value, use-value and exchange-value, exploitation, necessary and surplus labor, absolute and relative surplus labor, industrialization, markets, commodification, money, alienation, private property, commodity fetish, misrecognition, ideology, religion, reification, division of labor, commodification of labor, class consciousness

After reading and understanding this chapter, you should be able to

- Explain human nature through the idea of species-being and apply the idea to understanding your social world

- Define dialectical processes and analyze the structural dialectics of capitalism

- Explain the unique features of class struggle under capitalism

- Explain Marx's theory of exchange-value

- Describe Marx and Engel's theory of gender inequality

- Apply the idea of exploitation to current economic relations, including the globalization of exploitation

- Explain how industrialization, markets, and commodification are intrinsically expansive in capitalism and apply this dynamic to understanding contemporary capitalism and your or your family's buying patterns

- Explain how alienation and commodity fetish are created and how they affect the lives of people living under capitalism

- Explain Marx's theory of religion and ideology

- Describe labor false consciousness and explain how it is produced through dialectical elements within capitalism

- Explain how class consciousness develops and analyze why it has been slow to develop

Learning More: Primary Sources

With Marx, it is best to start off with a reader. My favorite is this one:
Tucker, R. C. (Ed.). (1978). *The Marx–Engels reader.* New York: W.W. Norton.

Learning More: Secondary Sources

Appelbaum, R. P. (1988). *Karl Marx.* Newbury Park, CA: Sage. (Part of the *Masters of Social Theory* series; short, book-length introduction to Marx's life and work)

Bottomore, T., Harris, L., & Miliband, R. (Eds.). (1992). *The dictionary of Marxist thought* (2nd ed.). Cambridge, MA: Blackwell. (Excellent dictionary reference to Marxist thought)

Fromm, E. (1961). *Marx's concept of man.* New York: Continuum. (Insightful explanation of Marx's idea of human nature)

McClellan, D. (1973). *Karl Marx: His life and thought.* New York: Harper & Row. (Good intellectual biography)

Theory You Can Use (Seeing Your World Differently)

- Using Google or your favorite search engine, type in "job loss." Look for sites that give statistics for the number of jobs lost in the past five years or so (this can be a national or regional number). How would Marx's theory explain this?

- Using Google or your favorite search engine, type in "class structure in U.S." Read through a few of the sites. Based on your research, what is happening to the class structure in the United States? How would Marx explain this?

- Get a sense about the frequency and uses of plastic surgery in this country. You can do this by watching "makeover" programs on TV, or by doing Web searches, or by reading through popular magazines, or in any number of ways. How can we understand the popularity and uses of plastic surgery using Marx's theory? (Hint: think of commodification and commodity fetish.) Can you think of other areas of our life for which we can use the same analysis?

Further Explorations—Web Links

http://www.marxists.org/archive/marx/works/index.htm

http://www.uta.edu/huma/illuminations

http://en.wikipedia.org/wiki/Karl_Marx

http://cepa.newschool.edu/het/profiles/marx.htm

http://www.marxists.org/archive/index.htm

CHAPTER 2

Rationality and the Bureaucratic Society

Max Weber (1864–1920)

Photo: © Granger Collection.

Seeing Further: The Effects of Bureaucratic Organization

What do you think is the most powerful social influence in your life? I began Chapter 1 by asking you to consider the dominance of capitalism in our lives. There Marx showed us the underbelly of capitalism when viewed from a humanistic perspective. As we saw, there is more to capitalism than growing stock portfolios and profit margins. Using capitalism as our chief economic structure influences who and what we are. But is there something else that might be just as powerful?

Weber would say yes. Your life, in more ways than you can imagine, is influenced by bureaucracy. Most of us probably take bureaucracy for granted—unless, of course, someone has pointed out that the reason you have to fill out all that paperwork in triplicate is because the bureaucracy demands it. But the influences of bureaucracy are much more subtle and pervasive than that. Think about this: Ever since you were five years old and went off to school for the first time, the greater part of your life has been spent within the confines of and being socialized by bureaucracies. In fact, it's probably not too far off to say that you are who you are because of bureaucracies.

But the influence of bureaucracy extends much further than your immediate life. Let me ask you a question: Why are you in school? If you're like most students, you're in school, taking this class, and reading this book in order to get a good job.

Now, let me ask you another question: What do this book and this class have to do with the job you want to get? Why do you have to go to school to get a good job when most of the stuff that you learn in your major will not directly apply to your occupation? There is, of course, a lot you learn in school that prepares you in more general ways. But Weber gives us an incisive understanding into what another theorist calls the credential society (see Collins, 1979).

Weber also gives us a different understanding of class relations. Marx gave us a theory of the relations of production, which he saw as slowly yet irresistibly moving toward class bipolarization—the creation of two classes and class interests. What Weber sees that Marx didn't is a growing middle class that stands between the elite and the workers. Where did this middle class come from? What does Weber see that Marx missed? If we think at all about the middle class, most of us probably assume that it simply grew out of capitalism, just like the capitalist elite and the working class. However, that isn't exactly right. A different social force had a significant impact on the growth of the middle class. And that's right where Weber centers his work.

The Essential Weber

Biography

Max Weber was born on April 21, 1865, in Erfurt, Germany (Prussia). His father, Max, was a typical bourgeois politician of the time. His mother, Helene Fallenstein, was a devoutly religious (Calvinist) woman. In 1882, Weber entered the University of Heidelberg, where he studied law. After a year's military service in 1883, Weber returned to school at the University of Berlin, and in 1893, Weber married Marianne Schnitger (who became an early feminist thinker). Weber began teaching at Heidelberg in 1896 as a professor of economics. While at Heidelberg, Max and Marianne's home became a meeting place for the city's intellectual community. Marianne was active in these meetings, which at times became significant discussions of gender and women's rights. Georg Simmel frequently attended. During this time, Weber worked as a professor, lawyer, and public servant.

In 1897, Weber suffered a complete emotional and mental breakdown and was unable to write again until 1903; he left the university and didn't teach again for almost 20 years. During his convalescence, Weber read the works of Wilhelm Dilthey and Heinrich Rickert. After his breakdown, Weber wrote the majority of the works that he is best known for: his methodological writings date from this period (now found in *The Methodology of the Social Sciences*), as well as *The Protestant Ethic and the Spirit of Capitalism, The Sociology of Religion, The Religion of China, The Religion of India,* and *Ancient Judaism.* (Several of these weren't finished until later, but were begun during this time.)

In 1918, Weber accepted a position at the University of Vienna, where he once again began to teach. He started working on *Economy and Society,* which was to be the definitive outline of interpretive sociology. His *General Economic History* came from lectures given during this time.

On June 14, 1920, Max Weber died of pneumonia.

Passionate Curiosity

Weber is one of the most diverse writers of sociology, and he was a workaholic. He thus produced a vast body of literature that covers a wide range of subjects. Randall Collins (1986a), a contemporary Weberian scholar, characterizes Weber's writings as "somewhat schizophrenic. . . . [I]n his voluminous works, one can find almost anything one looks for" (p. 11). So, it is difficult to say that Weber was passionately curious about any one thing. Rather, Weber was curious about *everything* and adamant about explaining everything in its real-world complexity. Thus, in all his writings, Weber is interested in explaining the relationships among cultural values and beliefs (generally expressed in religion), social structure (overwhelmingly informed by the economy), and the psychological orientations of the actors.

Keys to Knowing

Rationalization; legitimacy; bureaucracy; class, status, and power; the spirit of capitalism; ideal types

---

Weber's Perspective: Cultural Sociology

Max Weber is one of sociology's most intricate thinkers. Part of this complexity is undoubtedly due to the breadth of his knowledge. Weber was a voracious reader with an encyclopedic knowledge. In addition, Weber was in contact with a vast array of prominent thinkers from diverse disciplines. As Lewis Coser (2003) comments, "In leafing through Weber's pages and notes, one is impressed with the range of men with whom he engaged in intellectual exchanges and realizes the widespread net of relationships Weber established within the academy and across its various disciplinary boundaries" (p. 257). This social network of intellectuals in diverse disciplines helped create a flexible mind with the ability and tendency to take assorted points of view.

Culture and Social Science

Another reason why I think Weber's writings are complex is due to the way he views the world. Weber sees that human beings are animals oriented toward meaning, and meaning has pronounced subjective qualities. Weber also understands that all humans are oriented toward the world and each other through values. Further, Weber sees the primary level of analysis to be the social action of individuals. For Weber, individual action is social action only insofar as it is meaningfully oriented toward other individuals; Weber sees these meaningful orientations as produced within a unique historical context. Weber's (1949) theoretical questions, then, are oriented toward understanding "on the one hand the relationships and the cultural significance of individual events in their contemporary manifestations and on the other hand the causes of their being historically *so* and not *otherwise*" (p. 72, emphasis original). What this means is that Weber contextualizes individual social action within the historically specific moment. He then asks the question, why does this cultural context exist and not another one? How is it that out of all the possible cultural worlds, this one exists right here, right now, and not a different one?

Weber's perspective, then, is a cultural one that privileges individual social action within a historically specific cultural milieu. This orientation clearly sets him apart from Durkheim and Marx who were much more structural in their approaches. It also means that Weber's (1949) explanations are far more complex and tentative: "There is no absolutely 'objective' scientific analysis of culture . . . [because] . . . all knowledge of cultural reality . . . is always knowledge from particular points of view" (pp. 72, 81).

It's important to note that Weber's view of culture is not determinative—he doesn't see culture as determining human action. According to Weber (1948), people are very much motivated by economic and cultural interests, but culture can act like a switch on railroad tracks and actually change the course of the train:

> Not ideas, but material and ideal interests directly govern men's conduct. Yet very frequently the "world-images" that have been created by "ideas" have, like switchmen, determined the tracks along which action has been pushed by the dynamic of interest. (p. 280)

One of the tasks, then, of Weberian sociology is to *historically explain* which factor is more critical at any given time and why, rather than predicting an outcome. So, for example, we could never have truly predicted what "African American" would mean in 2007 based on what Negro meant in 1950, but we can explain the historical, social, and cultural processes through which it came about. That being the case, the kind of knowledge that Weberians construct about the world is decidedly different from that proposed by the general scientific model.

Weber also sees another culture-based issue for researchers. He recognizes that to ask a question about society or humans is itself a cultural act: it requires us to place value on something. In other words, for us to even see a problem to study, we must have a value that helps us to see it. For example, it would have been almost impossible for us to study spousal abuse 300 years ago (it would have been difficult even 50 years ago). It isn't that the behaviors weren't present; it is simply that the culture would not have allowed us to define them as abusive, at least not very easily. And the same is true about everything social scientists study (and laboratory scientists, too, for that matter). Humans can ask questions only insofar as they have a culture for it, and culture by definition is a value orientation toward the world: "Empirical reality becomes 'culture' to us because and insofar as we relate it to value ideas" (Weber, 1949, p. 76).

So, if scientific knowledge is defined as being empirical and nonevaluative, then you can see why creating a social *science* might be a problem. Human reality is meaningful, not empirical; it is historical and thus concerned with unique configurations of values, and all the questions we ask are strongly informed by our culture and thus are value-laden. However, Weber is also convinced that a kind of social science is possible, but there are certain caveats. Because human existence is a subjective one, creating an objective science about people is difficult. Knowledge about people must be based upon an interpretation of their subjective experience. And because people are self-aware free agents, the law-like principles that science wants to discover are provisional and probabilistic at best (people can always decide to act otherwise). So the kind of knowledge that we produce about people will be different from that produced in the laboratory, though Weber feels that objective knowledge is still possible. One key in creating this objective-like knowledge is that social scientists have to be reflexively and critically aware of their values in forming and researching their questions.

Weber's Method

Weber also argues that knowledge about humans in society can be made object-like through the use of *ideal types*. Ideal types are analytical constructs that don't exist anywhere in the real world. They simply provide a logical touchstone to which we can compare empirical data. Ideal types act like a yardstick against which we can measure differences in the social world. These types provide objective measurement because they exist outside the historical contingency of the data we are looking at. According to Weber, without the use of some objective measure, all we can know about humans would be subjective.

There are two main kinds of ideal types: historical and classificatory. Historical ideal types are built up from past events into a rational form. In other words, the researcher examines past examples of whatever phenomenon he or she is interested in and then deduces some logical characteristics. Weber uses this form in *The Protestant Ethic and the Spirit of Capitalism*. In that work, Weber constructs an ideal type of capitalism in order to show that the capitalism in the West is historically unique. Classificatory ideal types, on the other hand, are built up from logical speculation. Here the researcher asks him- or herself, what are the *logically possible* kinds of _____ (fill in the research interest)?

In his methodology, Weber also emphasizes understanding of the subjective meanings of the actions to the actors by contextualizing it in some way. Weber advocates the use of *verstehen*, the German word meaning to understand. It is important to note that when Weber talks about meaning in this context, he has in mind the motivations of the actor. These motives may be intellectual in the sense that the actor has an observable and rational motive for his or her actions in terms of means and ends; or they may be emotional in the sense that the behavior may be understood in terms of being motivated by some underlying feeling like anger. So we can understand it when Sam hits John if we know that Sam is angry with John for cheating him in a business deal—the meaning of the action comes out of our knowledge of the motivation.

Weberian sociologists, then, will see the world in terms of ideal types (abstract categorical schemes), broad historical and cultural trends, or from the point of view of the situated subject (interpretive sociology). Weber's causal explanations have to do with understanding how and why a particular set of historical and cultural circumstances came together, and his general explanation is always subject to case-specific variations. There are a couple of things that this implies.

Rationalization

The historical and cultural trends that interest Weber the most, and continue to be a focus of Weberian sociology, concern the broad sweeping movement toward rationalization and rational–legal legitimation (note that these are cultural concerns). Weber argues that one of the prime forces bringing about modernity is the process of rationalization. He uses the word **rationalization** in at least three different ways: He uses it to talk about means–ends calculation, in which rationality is individual and specific. Rational action is action based on the most efficient means to achieve a given end. Second, Weber uses the term to talk about bureaucracies. The bureaucratic form is a method of organizing human behavior across time and space. Initially we used kinship to organize our behaviors, using the ideas of extended family, lineages, clans, moieties, and so forth. But as the contours of society changed, so did our method of organizing. Bureaucracy is a more rational form of organization than the traditional and emotive kinship system.

Finally, Weber uses the term rationality in a more general sense. One way to think about it is to see rationalization as the opposite of enchantment: "The fate of our times is characterized by rationalization and intellectualization and, above all, by the 'disenchantment of the world'" (Weber, 1948, p. 155). Specifically, an enchanted world is one filled with mystery and magic. *Disenchantment*, then, refers to the process of emptying the world of magical or spiritual forces. Part of this, of course, is in the religious sense of secularization.

Peter Berger (1967) provides us with a good definition of secularization: "By secularization we mean the process by which sectors of society and culture are removed from the domination of religious institutions and symbols" (p. 107). If we think about the world of magic or ancient religion, one filled with multiple layers of energies, spirits, demons, and gods, then in a very real way, the world has been subjected to secularization from the beginning of religion. The number of spiritual entities has steadily declined from many, many gods to one; and the presence of a god has been removed from immediately available within every force (think of the gods of thunder, harvest, and so on) to completely divorced from the physical world, existing apart from time (eternal) and space (infinite). In our more recent past, secularization, and disenchantment and rationalization, have of course been carried further by science and capitalism.

This general process of rationalization and disenchantment extends beyond the realm of religion. Because of the prominence of bureaucracy, means–ends calculation, science, secularization, and so forth, our world is emptier: the organizational, intellectual, and cultural movements toward rationality have emptied the world of emotion, mystery, tradition, and affective human ties. We increasingly relate to our world through economic calculation, impersonal relations, and expert knowledge. Weber (1948) tells us that as a result of rationalization, the "most sublime values have retreated from public life" and that the spirit "which in former times swept through the great communities like a firebrand, welding them together" is gone (p. 155). Even our food is subject to rationalization, whether it is the McDonaldized experience (Ritzer, 2004) or the steak dinner that is subjected to the "fact" that it contains in excess of 2,000 calories and 100 grams of fat. Thus, for Weber, the process of modernization brings with it a stark and barren world culture.

Legitimation

Another issue arises from seeing the social world culturally: legitimation. When you see things through a cultural lens, as does Weber, you realize that for society or a social structure to work, people have to *believe* in it. **Legitimation** refers to the process by which power is not only institutionalized but more importantly is given moral grounding. Legitimations contain discourses or stories that we tell ourselves that make a social structure appear valid and acceptable. The strongest legitimations will make social structure appear inevitable and beyond human control (for example, the essentialist arguments surrounding gender—women cannot think logically and men cannot nurture children because of the nature of their sex). Weber argues that all oppressive structures, and, in fact, all uses of power, must exist within a legitimated order.

A legitimated order creates a unified worldview and is based on a complex mixture of two kinds of legitimations: subjective (internalized ethical and religious norms) and objective (having the possibility of enforced sanctions from the social group [conventions] or an organizational staff [law]). Weber indicates that subjective legitimacy is assumed in the presence of the objective. Underlying both subjective and objective legitimacy are three different kinds of belief systems or authority (charismatic, traditional, and rational-legal). Legitimacy works only because people believe in the rightness of the system. So, for example, your professor tells you that you will be taking a test in two weeks. And in two weeks you show up to take the test. No one has to force you; you simply do it because you believe in the right of the professor to give tests. That's Weber's point: social structures can function because of belief in a cultural system.

Concepts and Theory: The Rise of Capitalism

For Weber, there are three main factors that influenced the rise of capitalism as an economic form: religion, nation-states, and transportation and communication technologies. But, as with most of Weber's work, it is difficult to disengage their effects. Not only is there quite a bit of overlap when he talks about these issues, he is also interested in explicating the preconditions for capitalism rather than determining a causal sequence. So, what we have are a number of social factors that overlap to create the bedrock out of which capitalism could spring, but that did not necessarily cause capitalism.

The Religious Culture of Capitalism

The Protestant Ethic and the Spirit of Capitalism is probably Weber's best-known work. It is a clear example of his methodology. In it, he describes an ideal type of spirit of capitalism, he performs a historical-comparative analysis to determine how and when that kind of capitalism came to exist, and he uses the concept of *verstehen* to understand the subjective orientation and motivation of the actors. Weber had three interrelated reasons for writing the book. First, he wanted to counter Marx's argument concerning the rise of capitalism—Weber characterizes Marx's historical materialism as "naïve." The second reason is very closely linked to the first: Weber wanted to argue *against* brute structural force and argue *for* the effect that cultural values could have on social action.

The third reason that Weber wrote *The Protestant Ethic* was to explain why rational capitalism had risen in the West and nowhere else. Capitalism had been practiced previously, but it was traditional, not rational capitalism. In *traditional capitalism,* traditional values and status positions still held; the elite would invest but would spend as little time and effort doing so as possible, in order to live as they were "accustomed to live." In other words, the elite invested in capitalistic ventures in order to maintain their lifestyle. It was, in fact, the existence of traditional values and status positions that prevented the rise of rational capitalism in some places. *Rational capitalism,* on the other hand, is practiced to increase wealth for its own sake and is based on utilitarian social relations.

Weber's argument is that there are certain features of Western culture that set it apart from any other system, thus allowing capitalism to emerge. As we talk about this culture, it is important to keep in mind that Weber is describing the culture of capitalism in its beginning stages (rational capitalism). In many ways, the United States is now experiencing a different form of capitalism, and some of the spirit that Weber is describing may be lost to one degree or another. Also keep in mind that what he describes is an ideal type.

Weber's first task was to define the **spirit of capitalism.** The first thing I want you to notice is the word "spirit." Weber is concerned with showing that a particular cultural milieu or mindset is required for rational capitalism to develop. This culture or mindset is morally infused: the spirit of capitalism exists as "an *ethically*-oriented maxim for the organization of life" (Weber, 1904–1905/2002, p. 16, emphasis original). This culture, then, has a sense of duty about it, and its individual components are seen as virtues. Weber's argument is that the culture of modern capitalism provides us with certain principles, values, maxims, and morals that act as guideposts telling us how to live. In my reading of Weber, I see three such prescriptions in the spirit of capitalism.

Weber begins his consideration of the spirit of capitalism with a lengthy quote from Benjamin Franklin:

> Remember, that time is money . . . that credit is money . . . that money is of the prolific, generating nature. . . . After industry and frugality, nothing contributes more to the raising of a young man in the world than punctuality and

justice in all his dealings. . . . The sound of your hammer at five in the morning, or eight at night, heard by a creditor, makes him easy six months longer. . . . [Keep] an exact account for some time, both of your expenses and your income. (Franklin, as quoted in Weber, 1904–1905/2002, pp. 14–15)

From these sayings, Weber gleans the first maxim of rational capitalism: Life is to be lived with a specific goal in mind. That is, it is good and moral to be honest, trustworthy, frugal, organized, and rational because it is useful for a specific end: making money, which has its own end—the acquisition of money, and more and more money. The culture of modern capitalism says that money is to be made but not to be enjoyed. Immediate gratification and spontaneous enjoyment are to be put off so that money can be "rationally used." That is, it is invested to earn more money. The making of money then becomes an end in itself and the purpose of life: "People do not wish 'by nature' to earn more and more money. Instead, they wish simply to live, and to live as they have been accustomed and to earn as much as is required to do so" (Weber, 1904–1905/2002, p. 23).

The second prescription is that each of us should have a vocational calling. Of course, another word for vocation is job, but Weber isn't simply saying that we should each have a job or career. The emphasis is on our *attitude toward* our vocation, or the way in which we carry out our work. There are two important demands to this attitude. The first is that we are obligated to pursue work: we have a *duty* to work. In the spirit of capitalism, work is valued in and of itself. People have always worked, but generally speaking, we work to achieve an end. The spirit of capitalism, however, exalts work as a moral attribute. We talk about this in terms of a work ethic, and we characterize people as having a strong or weak work ethic. We can further see the moral underpinning when we consider that the opposite of a strong work ethic is laziness. We still see laziness as a character flaw today. In the culture of capitalism, *work becomes an end and moral value in itself, rather than a means to an end.* More than that, it becomes the central feature of one's life, overshadowing other areas such as family, community, and leisure.

We not only have duty to work, we also have duty *within* work. Weber (1904–1905/2002) says that "*competence and proficiency* is the actual alpha and omega" of the spirit of capitalism (p. 18, emphasis original). Notice the religious reference: in the New Testament, Jesus is referred to as the alpha and omega. Weber is again emphasizing that this way of thinking about work is a moral issue. Our duty within work is to organize our lives "according to *scientific* vantage points" (p. 35, emphasis original). That is, under the culture of the spirit of capitalism, *we are morally obligated to live our lives rationally.*

The rationally organized life is one that is not lived spontaneously. Rather, all actions are seen as stepping stones that bring us closer to explicit and valued goals. Let me give you a contrasting example to bring this home. A number of years ago, a (non-Hawaiian) friend of mine managed a condominium complex on the island of Maui. In the interest of political and cultural sensitivity, he hired an all-Hawaiian crew to do some construction. He tells of the frustrations of having a crew come to work whenever they got up, rather than at the prescribed time of 8 AM. What's more, periodically during the day the crew would leave at the shout of "surf's up!"

The work got done and it was quality work, but the Hawaiians that he supervised did not organize their lives rationally. They lived and valued a more spontaneous and playful life. In contrast, most of us have Day Runners and Palm Pilots that guide our life and tell us when every task is to be performed in order to reach our life goals.

The third prescription or value of the spirit of capitalism, according to Weber (1904–1905/2002), is that life and actions are legitimized "on the basis of strictly *quantitative* calculations" (p. 35, emphasis original). Weber makes an interesting point with regard to legitimation or rationalization—humans can rationalize their behaviors from a variety of ultimate vantage points. And we always *do* legitimize our behaviors. We can all tell stories about why our behaviors or feelings or prejudices are right. And those stories can be told from various religious, political, or personal perspectives. Weber's point here is that the culture of capitalism values quantitative legitimations. That is, *capitalist behaviors are legitimized in terms of bottom-line or efficiency calculations.* So, for example, *Roger and Me*, a film by Michael Moore, depicts the closing of the General Motors (GM) plants in Flint, Michigan, resulting in the loss of over 30,000 jobs and the destruction of Flint's economy. The film asks about GM's social responsibility, but from GM's position, the closing was legitimated through bottom-line, financial portfolio management.

These cultural directives find their roots in Protestant doctrine and practice. But Weber doesn't mean to imply that these tenets of capitalist culture are themselves religious—far from it. What we can see here is how culture, once born, can have unintended effects, independent of its creating group. Protestantism did not directly produce capitalism, but it did create a culture that, when cut loose from its social group, influenced the rise of capitalism.

The most important religious doctrine behind the spirit of capitalism is Luther's notion of a *calling*. Prior to Luther and the advent of Protestantism, a calling was seen as something peculiar to the priesthood. Men could be called out of daily life to be priests and women to be nuns, but the laity was not called. Luther, however, taught that every individual can have a personal relationship with God; people didn't need to go through a priest. Luther also taught that individuals were "saved" based on personal faith. Prior to this, the church taught that salvation was a property of the church—people went to heaven because they were part of the Bride of Christ, the Church.

With this shift to the individual also came the notion of a calling. If each person stood before God individually, and if priests aren't called to intercede, then the ministry belongs to the laity. Each person will stand before God on Judgment Day to give an account of what he or she did in this life. That means that God cares about what each individual does and has a plan for each life. God's plan involves a calling. Luther argued that every person in the church is called to do God's will. So it is not simply the case that ministers are called to work for God—carpenters are called to work for God as well, using their skills as carpenters. One calling is not greater than another; each is a religious service. This doctrine, of course, leads to a *moral organization of life.*

This idea of a calling was elaborated upon and expanded by John Calvin. Calvin took the idea of God's omniscience seriously: If God knows something, then God has always known it. Calvin also took the "sin nature" of humanity seriously. According to the sin nature doctrine, every human being is born in sin, reckoned sinful under Adam. That being the case, there is nothing anyone can do to save him- or herself from hell. We are doomed because of our very nature. Salvation, then, is from start to finish a work of God. Taking these ideas together, we come up with the Calvinistic idea of predestination. People are born in sin, there is nothing they can do to save themselves (neither faith nor good works), salvation is utterly a work of grace, and God has always known who would be saved. We are thus predestined to go to heaven or hell.

This doctrine had some interesting effects. Let's pretend you're a believer living in the sixteenth century under Calvin's teaching. Heaven and hell are very real to you, and it is thus important for you to know where you are headed. But there isn't anything you can do to assuage your fears. Because salvation is utterly of God, joining the church isn't going to help; neither is being baptized or evangelizing. Confessing faith isn't going to help either, because you are either predestined for heaven or you are not. And you can't go by whether or not you *feel* saved—feelings were seen as promoting "sentimental illusions and idolatrous superstition."

"*Restless work in a vocational calling* was recommended as the best possible means to *acquire* the self-confidence that one belonged among the elect" (Weber, 1904–1905/2002, p. 66, emphasis original). Good works weren't seen as a path to salvation, but they were viewed as the natural fruit: If God has saved you, then your life will be lived in the relentless pursuit of His glory. In fact, only the saved would be able to dedicate their entire lives in such a way. The emphasis was thus not on singular works, but on an entire life organized for God's glory.

Thus, diligent labor became the way of life for the Calvinist. Every individual was called to a job and was to work hard at that job for the Lord. Even the rich worked hard, for time belonged to the Lord and glorifying Him was all that mattered. Everything was guarded and watched and recorded. Individuals kept journals of daily life in order to be certain that they were continuing to do good works. Their entire lives became rational and systematically ordered.

Further, if one was truly chosen for eternal salvation, then God would bless the individual and the fruits of one's labor would multiply. In other words, in response to your labor, you could expect God to bless you economically. Yet, asceticism became the rule, for the world was sinful and the lusts of the flesh could only lead to damnation. The pleasures of this world were to be avoided. So the blessings of God were reinvested in the work that God had called you to.

Taken together, then, the Protestant ethic commits each individual to a worldly calling, places upon him or her the responsibility of stewardship, and simultaneously promises worldly blessings and demands abstinence. This religious doctrine proved to be fertile ground for rational capitalism: a money-generating system that values work, rational management of life, and the delay of immediate gratification for future monetary gain.

Structural Influences on Capitalism

Thus, Weber argues that rational capitalism in the West found a seedbed in a culture strongly influenced by Protestantism. Yet there are other preconditions for the emergence of capitalism. I've illustrated these in Figure 2.1. They are divided into institutional, structural, and cultural influences, but these demarcations are not clear-cut. I've pictured them as rather shapeless, overlapping preconditions, as that's how Weber talks about them. All of these processes mutually reinforce one another. With capitalism in particular, there is movement back and forth between culture and structure in terms of causal influence.

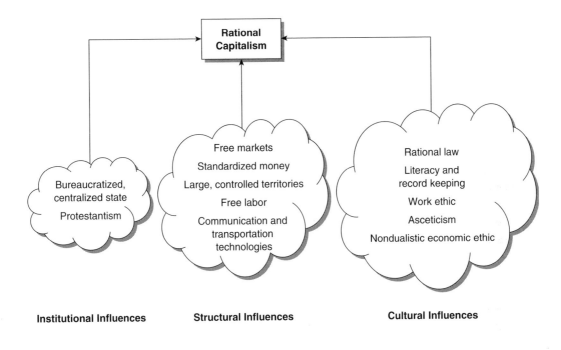

Figure 2.1   Preconditions of Capitalism

We've seen the influence of Protestantism, but in terms of institutions, there also had to be a centralized, bureaucratized state, as well as significant changes in modes of transportation and communication. I will be spending far less time explaining these structural influences than I did the effects of religious culture. The structural issues are more clear-cut than the cultural ones (which is generally the case). But don't let brevity fool you: these effects are equally important and you must understand how each of them works to comprehend the basis for rational capitalism.

Nation-states are relatively recent inventions. Up until the nineteenth century, the world was not organized in terms of nation-states. People were generally organized ethnically, with fairly fluid territorial boundaries. They didn't have nations as we think of them today. A nation is a collective that occupies a specific territory, has a common history and identity, and sees itself as sharing a common fate. The widespread use of the idea of a nation for organizing people was necessary for capitalism. Nations are responsible for controlling large territories, standardizing money, organizing social control, and facilitating free and open markets. All of these factors allowed for easier exchanges of goods and services involving large populations of people. These kinds of exchanges are, of course, necessary for rational capitalism to exist. Labor also had to be freed from social and structural constraints. Capitalism depends on a labor force that is free to sell its labor on the open market. Workers can't be tied down to apprenticeships or guild obligations, nor can they be attached to land obligations as in feudalism.

Along with increases in communication and transportation technologies, nation-states and Protestantism helped form an objectified, rationalized culture, wherein written records were kept, people practiced a strong work ethic coupled with asceticism, and the traditional dualistic approach to economic relations would break down. This latter issue is particularly important for Weber. In traditional societies, groups usually had two ethical frameworks when participating in exchanges. There were restrictions having to do with ritual and fairness when dealing with group members, but people outside the group could be exploited without measure. Both of these frameworks needed to be lifted in order for capitalism to flourish. All business transactions and loans needed to be rationalized so that there could be continuity.

Concepts and Theory: Class, Authority, and Social Change

As we can see with capitalism, social change for Weber is a complex issue, one involving a number of variables coming together in indeterminate ways. Social change always involves culture and structure working together, and it involves complex social relations. We turn our attention now to stratification and change. Again, we will see the interplay of culture and structure as well as complex social categories.

Weber's understanding of stratification is more complex than Marx's. Marx hypothesized that in capitalist countries, there will be only one social category of any consequence—class. And in that category, Marx saw only two types: owners and workers. For Weber, status and power are issues around which stratification can be based. Most importantly, class, status, and power do not necessarily covary. That is, a person may be high on one of those dimensions and low on another (like a Christian minister, typically high in terms of status but low in terms of class). These crosscutting life circumstances or affiliations can prevent people from forming into conflict groups and bringing about social change. For example, in the United States, a black man and a white man may both be in the poor class, but their race (status) may prevent them from seeing their life circumstances as being determined by similar factors.

Weber also argues that all systems are socially constructed and require people to believe in them. Marx did say that capitalism has a cultural component that holds it together (if not for ideology, the proletariat would immediately overthrow the system), but true communism requires no such cultural reinforcement, because it corresponds to our species-being nature. For Weber, legitimation is the glue that holds not only society together but also its systems of stratification. Thus, for Weber, issues of domination and authority go hand in hand (in fact, Weber used just one German word to denote them both—*Herrschaft*). So, people must have some level of belief in the authority (culture) of those who are in charge, and they must cooperate with the system to some degree in order for it to work.

Class

Weber's definition of class is different from Marx's. Marx defines class around the ownership of the means of production. Weber (1922/1968), on the other hand, says that a "class situation" exists where there is a "typical probability of 1. procuring goods 2. gaining a position in life and 3. finding inner satisfaction, a probability which derives from the relative control over goods and skills and from their income-producing uses within a given economic order" (p. 302). In other words, Weber defines class based on your ability to buy or sell goods or services that will bring you inner satisfaction and increase your life chances (how long and healthy you will live).

Weber also sees class as being divided along several dimensions as compared to Marx's two. Marx acknowledges that there are more than two class elements, but he also argues that the other classes (such as the lumpenproletariat or petite bourgeoisie) become less and less important due to the structural squeeze of capitalism. Weber also speaks of two main class distinctions, yet they are constructed around completely different issues, each with "positively privileged," "negatively privileged," and "middle class" positions. The property class is determined by property differences, either owning (positively privileged) or not owning (negatively privileged). Rentiers—people who live off property and investments—are clear examples of owners, whereas debtors—those who have more debt than assets—are good examples of the negatively privileged. The middle property classes are made up of those who do not acquire wealth or surplus from property, yet they are not deficient either.

The commercial class is determined by the ability to trade or manage a market position. Those positively related to commercial position are typically entrepreneurs who can monopolize and safeguard their market situation. Negatively privileged commercial classes are typically laborers. They are dependent on the whim of the labor market. The in-between or middle classes that are influenced by labor market positions are those such as self-employed farmers, craftspeople, or low-level professionals who have a viable market position yet are not able to monopolize or control it in any way.

There's an extremely important point here that we need to expand: Commercial classes are able to dominate market positions around skills and knowledge. This kind of class becomes increasingly significant in what are sometimes called "service economies." Capitalism has moved from large national corporations to larger international corporations to multinational conglomerates. Along with the movement of capital comes the movement of the workforce—primary production moves from core nations to peripheral or third-world nations. Economies in such countries as the United States are becoming less based on production and more based on consumption and service. For example, in terms of raw numbers, the United States lost approximately 3 million manufacturing jobs between 1995 and 2005, yet gained 4 million jobs in education and health services, 2 million in government, and 1.4 million in financial activities (Marcus, 2005).

Obviously, these numbers are only suggestive, but they do give us a picture of the movement from a (working) class-based economy to a service-based one. These new jobs and careers exist within bureaucracies and speak of the increasing bureaucratization of societies such as the United States; they are exactly what Weber is talking about with the commercial class. This is the middle class that has been created because of the increasing demand for "white collar" skills. Combining Weber's ideas of class with his theories of rationalization and bureaucratization gives us a clear understanding of the rise and importance of the middle classes.

In Figure 2.2, I've given us a picture of Weber's ideas about class. We can see that there are two axes to class: property and market position. People can have a positive, negative, or middle position with respect to each of these issues. I've conceptualized this as a typology, because Weber spoke of people holding a position on both. I have also provided some contemporary examples in this typology. Today, those in upper management, such as CEOs, are not only paid large salaries, they are also given stock options that translate into ownership, which places them high in both the property and market dimensions (they hold a monopoly with the skills they have). For example, *Forbes* magazine lists Jeffrey C. Barbakow of Tenet Healthcare as having been the highest paid American CEO in 2003. He received about $5.4 million in salary, plus he had stock gains of $111 million. Even the lowest-paid major CEO in 2003 received $1 million in salary and owned $28.2 million in stock. Weber's typology of class also allows us to conceptualize certain kinds of knowledge as a resource or market position and thus a class. Those who monopolize such skills and knowledge in our society include medical doctors and other such professionals. Stock owners as well as Weber's rentiers use the control of property to attain wealth, and we find them in the upper right corner of the picture. Those who have little or no control over property and market are the poor.

Weber sees that holding a specific class position does not necessarily translate into being a member of a group. Unless and until people develop a similar identity that focuses on their class differences, they remain a statistical aggregate. In other words, the census bureau may know you are in the middle class, but you may not have a group identity around your class position. So, in addition to the property and commercial classes, Weber also talks about social class. His emphasis here seems to be on the social aspect—the formation of unified group identity.

Weber identifies at least three variables for the formation of social class. He argues that class-conscious organization will most easily succeed if the group meets the following criteria:

> *It is organized against immediate economic groups.* There must be some immediate market position or control issue, and the other group must be perceived as relatively close. Thus, in today's economy, workers are more likely to organize against management rather than ownership.

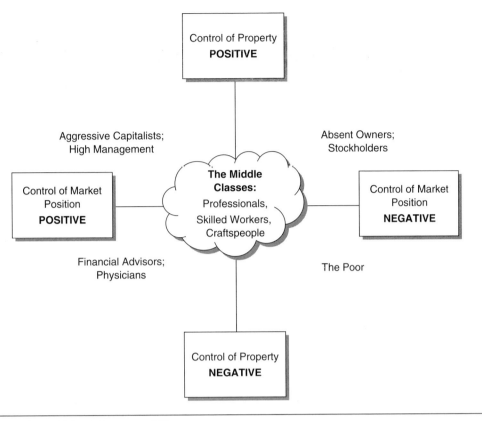

Figure 2.2   Weber's Concept of Class

Large numbers of people are in the same class position. The size of a particular group can give it the appearance, feel, and unavoidability of an object.

The technical conditions of organization are met. Weber recognizes that organization doesn't simply happen; there's a technology needed to organize people. People must be able to communicate and meet with one another; there must be recognized and charismatic leadership; and there must be a clearly articulated ideology in order to organize.

Keep in mind that these are all variables. A group may have more or less of any of them. As each of these increases, there will be a greater likelihood of the formation of a social class.

Status and Party

Weber sees stratification as more complex than class. Money isn't the only thing in which people are interested. People also care about social esteem or honor. Weber termed this issue status. *Status*, for Weber (1922/1968), entails "an effective claim to social esteem in terms of positive or negative privileges" (p. 305). Thus, status groups are hierarchically ranked by structural or cultural criteria and imply differences in honor and privilege.

Status may be based or founded on one or more of three things: a distinct *lifestyle,* formal *education,* or differences in *hereditary or occupational prestige.* Being a music fan may entail a distinct lifestyle—people who listen to jazz are culturally different from those who listen to rock—and there are positive and negative privileges involved as well: you can get a degree in jazz at my university, but not in rock. Those who are homosexual can also be seen in terms of lifestyle status groups. One's education level can provide the basis of status. Being a senior at the university is better, status wise, than being a freshman, and you have greater privileges as a senior as well. Professors are also examples of prestige based on education. They are good examples because they usually rate high on occupational prestige tables but low in class and power. Race and gender are also examples of status founded on perceived heredity.

Status groups maintain their boundaries through particular practices and symbols. Boundary maintenance is particularly important for status groups because the borders are more symbolic than actual. The differences between jazz and rock, for example, are in the ear of the listener. Generally speaking, Weber argues, we maintain the boundaries around status through marriage and eating restrictions, monopolizing specific modes of acquisition, and different traditions—all of which are described in greater detail below.

The norm of marriage restriction is fairly intuitive; we can think of religious and ethnic groups who practice "endogamy" (marriage within a specific group as required by custom or law). But eating restrictions may seem counterintuitive. It may help us to realize that most religious groups have special dietary restrictions and practice ritual feasting. Feasts are always restricted to group members and are usually seen as actively uniting the group as one (the traditional Jewish Yom Kippur is a good example). We can also think of special holidays that have important feast components, like Thanksgiving and Christmas here in the United States. And, if you think back just a few years, I'm sure you can recall times in junior high when someone your group didn't like tried to sit at your lunch table. For humans, eating is rarely the simple ingestion of elements necessary for biological survival. It is a form of social interaction that binds people together and creates boundaries.

We also maintain the symbolic boundaries around status groups through monopolizing or abhorring certain kinds and modes of acquisition, and by having certain cultural practices or traditions. The kinds of things we buy obviously set our status group apart. That's what we mean when we say that a BMW is a status symbol. Status groups also try and guard the modes of acquisition. Guilds, trade unions, and professional groups can function in this capacity. And, of course, different status groups have different practices and traditions. Step concerts, pride marches, Fourth of July, Kwanzaa, Cinco de Mayo, and so on are all examples of status-specific traditions and cultural practices.

When sociologists talk about Weber's three issues of stratification, they typically refer to them as class, status, and power. However, power is not the word that Weber uses; instead he talks about *party.* What he has in mind, of course, is within the sphere of power: "the chance of a man or a number of men to realize their own will in a social action even against the resistance of others who are participating in the action" (Weber, 1922/1968, p. 926). Yet, it is important to note that power always involves social organization, or, as Weber calls it, party. Weber uses the word party to capture the *social practice of power*. The social groups that Weber would consider parties are those whose practices are oriented toward controlling an organization and its administrative staff. As Weber puts it, a party organizes "in order to attain ideal or material advantages for its active members" (p. 284). The Democratic and Republican parties in the United States are obvious examples of what Weber intends. Other examples include student unions or special interest groups like the tobacco lobby, if they are oriented toward controlling and exercising power.

Authority and Social Change

One of Weber's enduring contributions to conflict theory is this tripartite distinction of stratification. Marx rightly argues that overt conflict is dependent upon bipolarization. The closer an issue of conflict gets to having only two defined sides, the more likely is the conflict to become overt and violent. But Weber gives us an understanding of stratification that shows the difficulty in achieving bipolarization.

Every individual sits at a unique confluence of class, status, and party. I have a class position, but I also have a variety of status positions and political issues that concern me. While these may influence one another, they may also be somewhat different. To the degree that these issues are different, it will be difficult to achieve a unified perspective. For example, let's say you're white, gay, male, the director of human resources at one of the nation's largest firms, and Catholic. Some of these social categories come together, like being white, male, and in upper management. But some don't. If this is true of you as an individual, then it is even more so in your association with other people. You'll find very few white, gay, male, Catholic upper managers to hang around with. And if you add in other important status identities, such as Southern Democrat, it becomes even more complex. Thus, Weber is arguing that the kind of conflict that produces social change is a very complex issue. Different factors have to come together in unique ways in order for us to begin to formulate groups and identities capable of bringing about social change.

At the heart of the issue of stratification and social change is legitimacy, and here we see Weber's emphasis on culture. In order for a system of domination to work, people must believe in it. Part of the reason behind this need is the cost involved in the use of power. If people don't believe in authority to some degree, they will have to be forced to comply through coercive power. The use of coercive power requires high levels of external social control mechanisms, such as monitoring (you have to be able to watch and see if people are conforming) and force (because they won't do it willingly). To maintain a system of domination not based on legitimacy costs a great deal in terms of technology and manpower. In addition, people often ultimately respond to the use of coercion by either rebelling or giving up—the end result is thus contrary to the desired goal.

Authority, on the other hand, implies the ability to require performance that is based upon the performer's belief in the rightness of the system. Because authority is based on socialization, the internalization of cultural norms and values, authority requires low levels of external social control. We can thus say that any structure of domination can exist in the long run if and only if there is a corresponding culture of authority. Weber identifies three ideal types of authority. (Keep in mind that these are ideal types and may be found in various configurations in any society.)

- **Charismatic authority**—belief in the supernatural or intrinsic gifts of the individual. People respond to this kind of authority because they believe that the individual has a special calling. (Good examples of this type of authority include Susan B. Anthony, Adolf Hitler, Martin Luther King Jr., John Kennedy, Golda Meir, and Jesus—notice that it is people's belief in the charisma that matters; thus, we can have Hitler and Jesus on the same list.)

- **Traditional authority**—belief in time and custom. People respond to this kind of authority because they honor the past and they believe that time-proven methods are the best. (Good examples of this type of authority are your parents and grandparents, the pope, and monarchies.)

- **Rational-legal authority**—belief in procedure. People respond to this kind of authority because they believe that the requirements or laws have been enacted in the proper manner. People see leaders as having the right to act when they obtain positions in the procedurally correct way. (A good example of this type is your professor—it does not matter who the professor is, as long as he or she fulfills the requirements of the job.)

These diverse types of authority interact differently in the process of social change. According to Weber, the only kind of authority that can instigate social change is a charismatic one. Traditional and rational-legal authorities bring social stability—they are each designed to maintain the system: "Charisma, on the other hand, may effect a subjective or internal reorientation born out of suffering, conflicts, or enthusiasm. It may then result in a radical alteration of the central attitudes and directions of action with a completely new orientation of all attitudes toward the different problems of the 'world'" (Weber, 1922/1968, p. 245). Charismatic individuals come to bring social change, yet charismatic authority is also inherently precarious. Because charisma is based on belief in the special abilities of the individual, every instance of charismatic authority will fail within that person's lifetime—the gifts die with the person. Thus, every charismatic authority will someday have to face the **problem of routinization** (making something routine and thus predictable). Every social movement based upon charismatic authority—and Weber argues that they all are—routinizes the changes by either using traditional authority or rational-legal authority.

The case of the Christian church might give us insight. Jesus was a charismatic leader. When he died, the church was faced with the problem of continuing his leadership. Though somewhat a gloss, it can be said that the Catholic Church is based upon the traditional authority of the pope and that Protestant churches are based upon rational-legal authority. You cannot study to become a pope, but you can study to become a Protestant minister. As I said, this is a gloss: people do study to become priests, and ministers are perceived as charismatically ordained by God.

So, these authority systems are mixed (as is always the case with ideal types). But their systems are *stabilized* through the use of tradition and bureaucracy.

Combining Weber's ideas of class, status, and party with his argument concerning authority and social change, we can put together a theory of conflict and social change (see Figure 2.3). Social change will occur only if the legitimacy of the system of stratification is questioned. Conflict and change are likely to occur when there are clear breaks between the systems or limited upward mobility in one of them—I've depicted this in the figure as "perceived group boundary." Groups will tend to perceive their boundaries when in close proximity to another competing group or when mobility is limited. It is possible for change to occur in only one of three areas, but that change will be limited due to the crosscutting influences of the other systems. In the United States, for example, many people questioned the legitimacy of the systems of race and gender during the 1960s and 1970s. These systems are primarily built around and understood through status. So, while there have been some real structural changes, it appears that *most* of the change in race and gender is cultural—the status of each has been improved. (Please note the "appears" in my statement; this issue needs close empirical research.)

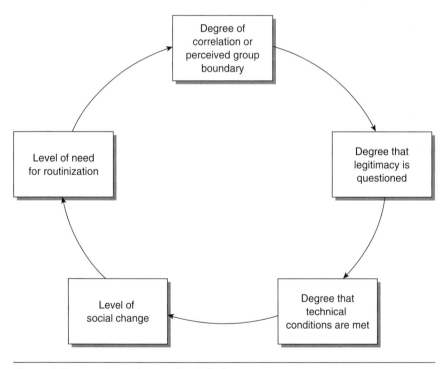

Figure 2.3   Weber's Theory of Social Change

In a Weberian sense, significant social change will occur if and when there is perceived correlation among class, status, and power ("degree of correlation" in Figure 2.3). In other words, if people perceive that certain groups are high on all three stratification systems and certain groups are low, then they are more likely to question the legitimacy of the whole system rather than just one part. This *delegitimation* is particularly likely when the correlation is perceived as arbitrary. As legitimacy is questioned, the technical conditions of group conflict tend to be met. These include charismatic leadership, clearly articulated goals and ideology, and the ability to meet and communicate. Note that there is a reciprocal effect between questioning legitimacy and technical conditions: each will tend to reinforce the other. As technical conditions are met, the group will become more effectual in bringing about social change. The degree of social change in turn will impact the need for routinization.

Weber argues that routinization will in the long run lead to a kind of stratification that again sets up the conditions for conflict and change. In this respect, Weber's model is more dialectical than Marx's, which stops with the advent of communism. Weber sees conflict and change as ubiquitous features of every social system. For Weber, social systems will move cyclically through routinization and charismatic change. At different times, for different reasons, social groups will question the legitimacy of domination.

Concepts and Theory:
The Process of Rationalization

To be human is to be social, and to be social is to be organized. Thus, organization is primary to what it means to be human. Human beings have been organized over time by different social forms, and these different organizational forms have clear consequences for us. In the not too distant past, we organized much of our lives around affective (emotionally-based) systems, such as kinship. Kinship linked people through blood and marriage. In such organizations, people saw each other in terms of familial obligations and rights, which they felt as emotional ties.

Bureaucracies tend to rationalize and *routinize* all tasks and interactions. Tasks and interactions are rationalized in terms of means–ends efficiency, and they are routinized in the sense that they may be carried out without thought or planning or dependency on individual talents. Thus, each person and task is met in the same efficient and equal manner; personal issues and emotions do not have a place in a bureaucracy.

Preconditions for Bureaucratic Organization

While bureaucracies have been around for quite awhile, for much of their history they were not the primary way in which humans organized. Certain social factors came about that simultaneously pushed out affective systems of organization (traditional) and set the stage for systems based more on reason than emotion. For Weber, these social factors are preconditions for bureaucracy, not causal forces. As these different features lined up, they created an environment ripe for rational organization.

According to Weber (1922/1968, pp. 217–226, 956–958), there are at least six preconditions for bureaucracy, and they don't necessarily occur in any specific order. These include increases in

- the size and space of the population being organized
- the complexity of the task being performed
- the use of markets and the money economy
- communication and transportation technologies
- the use of mass democracy
- the volume of complicated and rationalized culture

All these factors created needs for more objective and rational social relations and culture, thus driving a society to use rational-legal authority and bureaucracy almost exclusively. As society became larger and spread out over vast spans of geographic space, it became increasingly difficult to use personal relationships as a method of organization. However, increases in population size and geographic space aren't sufficient to make bureaucratic organization necessary. For example, China was able to primarily use an elaborate kinship system rather than bureaucracy to organize most of their behaviors for many years. Other factors are necessary to push a society toward bureaucratic organization, such as increasing complexity of the task being performed. In England and elsewhere, this occurred through the Industrial Revolution, urbanization, and high division of labor. The increasing use of money and open markets for exchange also increased the need for rationalized and speedy calculations, thus adding to rational culture.

Increasing levels of communication and transportation technologies created the demand for faster and more predictable reactions from the governing state. Insightfully, Weber (1922/1968) says that with "traditional authority it is impossible for law or administrative rule to be deliberately created by legislation" (p. 227). Remember that traditional authority is based on history and time, so the only way a new rule or law can be legitimized through tradition is to say that it is according to the wisdom of the ages. Rational-legal authority, on the other hand, can create new rules simply because it is expedient to do so. So, as states came into more frequent contact with one another and dealt with more complex problems, it became necessary to respond quickly to new situations with new rules, thus pushing forward a rational-legal authority and bureaucracy.

Two additional forces created demands for rational and objective standards. The first was democracy for the masses, rather than simply for the elite, which created demands for equal treatment before the law. Initially, the idea of democracy was limited to the educated and powerful. Including the masses created the demand for equal treatment regardless of power or prestige. This, of course, led to the idea that rules and laws should be blind to individual differences. The other additional social factor that pushed for bureaucratic organization was the increase in complex and rational culture. Due to the increases in knowledge that came with science and technology, a new social identity came into existence and rose to prominence—the expert, bureaucracy's manpower: "Bureaucratic administration means fundamentally domination through knowledge" (Weber, 1922/1968, p. 225).

Effects of Bureaucratic Organization

Weber gives us an ideal type for bureaucracy. Remember that this ideal type is not intended to tell us what the perfect bureaucracy should look like; rather, Weber uses the ideal type as an objective yardstick against which we may measure different subjective and cultural states. Weber talks about bureaucracy in a number of different places. But if we combine his lists, we come up with six important features to the ideal type of *bureaucracy:* an explicit division of labor with delineated lines of authority, the presence of an office hierarchy, written rules and communication, accredited training and technical competence, management by rules that is emotionally neutral, and ownership of both the career ladder and position by the organization rather than the individual. Each of these characteristics is a variable; organizations will thus be more or less bureaucratized.

Bureaucracies have unintended and largely unavoidable consequences as well. They influence both the people in the bureaucracies and the social system as a whole. One general effect of bureaucracy and rationalization is a shift from traditional and affective types of social ties and actions to more rational ones. In order to talk about this, Weber developed an ideal typology of social action that contains four categories of social action. *Instrumental-rational* action is behavior in which the means and ends of action are rationally related to each other. So, your action in coming to the university is instrumental-rational in that you see it as a logical means to achieve an "end," that is, a good job or career.

Value-rational behavior is action that is based upon one's values or morals. If there is no way you could get caught paying someone to write your term paper for you, then it would be instrumentally rational for you to do so. It would be the easiest way to achieve a desired end. However, if you don't do that because you think it is dishonest, then your behavior is being guided by values or morals and is value-rational. *Traditional action* is action that is determined or motivated by habit, and *affectual action* is determined by people's emotions in a given situation.

Because of the general processes of rationalization and bureaucratization, people tend to engage in instrumental-rational behavior rather than traditional, affective, or even value-rational action. Recall the point we made about how rational capitalism is legitimated. Under the spirit of rational capitalism, decisions and behaviors are legitimized through quantitative calculations. This is the same kind of point except applied to all areas of life, not just capitalist enterprises. Due to the ever-increasing dominance of bureaucratic organization, most of our connections to other people are rational and our behaviors motivated by instrumental-rationality and not based in emotion, ethical values, or tradition.

Contemporary theorists point out that living in a society that organizes through bureaucracy can produce the **bureaucratic personality**. There are at least four characteristics of this kind of temperament. First, individuals tend to live more rationally due to the presence of bureaucracy, and not just at work. People generally become less and less spontaneous and less emotionally connected to others in their lives. They understand goals, the use of time and space, and even relationships through rational criteria. Second, people who work in bureaucracies also tend to identify with the goals of the organization. Workers at levels that are less bureaucratized tend to complain about the organization; on the other hand, management who exist at more bureaucratized levels tend to support and believe in the organization. Again, this isn't something that we just put on at work—we *become* bureaucratic ourselves.

Third, because bureaucracies are based on technical knowledge, people in bureaucratic societies tend to depend on expert systems for knowledge and advice. In traditionally based societies, people would trust the advice of those they loved, or those who had extensive experience, or those who stood in a long lineage of oral discipleship. Conversely, in societies like the United States, we look to those who have credentials to help us. Honor in modern society is given to those with credentials; age and experience are of no consequence.

Lastly, bureaucracies lead to sequestration of experience. By that I mean that different life experiences are separated from one another, such as dying from living. In traditional society, life was experienced holistically. People would see birth, sickness (emotional, mental, and physical), and death as part of their normal lives. Children were born and grandparents died in the same home. Today, most of those experiences are removed from us and occur in bureaucratic settings where we don't see them as part of our normal and daily life; such settings include hospitals, rest homes, asylums, and so on. The world has thus become tidy, clean, and rational.

Society is affected as well by bureaucracy. There are two main effects. The first is the **iron cage of bureaucracy**. Once bureaucracies are in place, they are virtually inescapable and indestructible for several reasons: they are the most efficient form of organizing large-scale populations, they are value-free, and they are based upon expert knowledge. We've just seen that individuals within a rationalized society become more and more dependent upon expert knowledge; the same is true with leaders. Whether the leaders are in charge of political, religious, or economic organizations, they become increasingly dependent upon rationally trained personnel and expert knowledge in the bureaucratic information age. The experts themselves engage in secrecy and mystification in order to avoid inspection and secure their position. Further, one of the definitions of a professional is self-administration, which means that bureaucratic experts become a self-recruiting and self-governing class, existing apart from any other organizational control. Thus, neither the people ruled, nor the rulers, nor the experts themselves can escape the domination of the bureaucratic form.

There is one further factor to note: bureaucracies are value-free, which means they can be used for any purpose, from spreading the gospel to the eradication of ethnic minorities. This implies that bureaucracies are quite good at co-optation. To co-opt, in this context, means to take something in and make it part of the group, which on the surface might sound like a good thing. But because bureaucracies are value- and emotion-free, there is a tendency to downplay differences and render them impotent. For example, one of the things that our society has done with race and gender movements is to give them official status in the university. One can now get a degree in race or gender relations. Inequality is something we now study, rather than the focus of social movements. In this sense, these movements have been co-opted.

Bureaucracies also accelerate the process of *credentialing*. Remember that position within a bureaucracy is achieved through diplomas and certification. It isn't supposed to be *who* you know but *what* you know that determines rank in the organization. That being the case, society needs a legitimated process through which credentials can be conferred. The United States uses the education system. But one of the effects of that decision is that education is no longer simply about learning. The education system is used to credential technical expertise rather than cultivate an informed citizenry. Universities are thus becoming populated by professional schools—the school of business, nursing, social work, computer technology, criminal justice, and so on—and there is mounting tension between the traditional liberal arts and these professional schools. Many students express this tension (and preference for credentialing) when they ask, "How will this course help me get a job?"

The emphasis on credentials coupled with the American view of mass education and the use of education to give credentials has created *credential inflation*. Every year, more and more people are going to college to get their degree so that they can be competitive in the job market. Using data from the National Center for Education Statistics, we can get a sense of how this is working. In the United States, between the mid-eighties and mid-nineties, there was a 20% increase in the number of bachelor degrees, 37% increase in master's degrees, and 44% increase in doctorates. The result is that there are too many people with advanced degrees, which, in turn, decreases the value of those degrees.

Weber Summary

- To think like Weber is to take seriously the ramifications of culture. Weberians focus on the historical, cultural, and social contexts wherein the subjective orientation of the actor takes place. To think like Weber, then, means to use ideal types and *verstehen* to explain how these contexts came to exist rather than others. To think like Weber also means paying attention to the process of rationalization and the need for legitimation.

- The cultural foundations of rational capitalism were laid by Protestantism. This religious movement (through the doctrines of predestination and abstention, and the idea of a calling) indirectly created a rationalizing, individuating culture wherein money could be made for the purpose of making more money, rather than for immediate enjoyment. The establishment of nation-states structurally paved the way for rational capitalism by creating a free labor force, controlling large territories, standardizing money, and protecting free global markets.

- Social stratification is a complex of three scarce resources: class, status, and party (or power). These three systems produce crosscutting interests that make social change difficult and multifaceted. Large-scale social changes become increasingly likely only as class, status, and power are seen to correlate; the legitimacy of the system is questioned; and the technical conditions of organization are met. Since social change is led by charismatic authority, each change will need to be routinized through traditional or rational-legal authority, which, in the long run, will once again set up conditions for conflict and social change.

- Bureaucratic forms of organization became prominent as societies became larger and more democratic, as tasks and knowledge became more complex, as communication and transportation technologies increased, and as markets became more widespread through the use of money. The extent of bureaucratic organization can be measured through an ideal type consisting of six variables: explicit division of labor, office hierarchy, written rules and communication, accreditation for position, affectless (without emotion or emotional connection) management by rule, and the ownership of career ladders and position by the organization. The use of bureaucracy as the chief organizing technology of a society results in the bureaucratic personality, the iron cage of bureaucracy, and social emphasis on credentials.

Looking Ahead

Looking forward to Weber's influence on social thinking is a daunting task. As I mentioned previously, Weber's thinking is quite complex and, as a result, Weber probably inspired our thinking in more areas than we realize. Thus, any list I construct will be partial and will neglect some important ways in which Weber has influenced contemporary theory. I do, however, want to give you a taste of Weber's abiding inspiration.

One place where Weber's influence is felt is one that you probably already know. It should be clear from reading this chapter that Weber is the founding thinker in the sociology of organizations, specifically bureaucratic organizations. While the sociology of organizations is a massive field, the specific work that you may be aware of is George Ritzer's *The McDonaldization of Society*. As Ritzer (1998) explains, "The theoretical starting point for the McDonaldization thesis is, of course, the work of Max Weber on rationalization" (p. 2). Ritzer sees McDonald's as archetypical of what Weber has in mind with his ideas of efficiency, calculability, predictability, and bureaucratic control. And, like Weber, Ritzer (2004) sees that this form of organization "suffers from the *irrationality of rationality*" and that bureaucratic organization results in an iron cage "in the sense that people are trapped in them, their basic humanity denied . . . a society of people locked into a series of rational structures, who could move only from one rational system to another" (pp. 27, 28, emphasis original).

It should also be apparent to you that Weber is an important contributor to contemporary conflict theory. The most obvious contribution is his idea of crosscutting influences of different systems of stratification (class, status, power). Though she doesn't cite Weber as a source, this idea comes to fruition in the hands of Patricia Hill Collins (Chapter 13). Collins argues that inequality can only truly be understood in terms of intersectionality and matrices of domination. Collins's basic argument is that every system of inequality—such as gender, race, class, nationality, age, and sexual identity—operates differently. People occupy different positions where these forces come together and produce diverse matrices of domination. We'll also find Weber's influence on conflict theory in the work of Ralf Dahrendorf and Randall Collins (Chapter 7).

Interestingly enough, Weber is also a major influence on twentieth-century structural-functionalism (Chapter 6). Talcott Parsons, who next to Durkheim is undoubtedly the most important functionalist that has ever lived, took his basic theoretical problem from Weber's notion of voluntaristic action: what are the conditions under which people make free choices about behavior in such a way that there is little uncertainty experienced in social life? Parsons argues that human action involves cultural elements such as norms, values, and beliefs; situational factors such as peer pressure; known goals that are informed by both cultural and situational factors; and choices about means and ends that are likewise influenced.

This Weberian notion of **action theory** also becomes the focal point for Alfred Schutz and what is called social phenomenology—a core intellectual source for the work of Harold Garfinkel (Chapter 9) and Dorothy E. Smith (Chapter 13). Schutz criticizes Weber's understanding of meaning attribution. Of particular concern for Schutz (1967) is the way in which Weber defines meaningful behavior: "when Weber talks about meaningful behavior, he is thinking about rational behavior and, what is more, 'behavior oriented to a system of discrete individual ends' *(zweckrational)*. This kind of behavior he thinks of as the archetype of action" (pp. 18–19). The implication is that for Weber, the meaning of an action is determined beforehand—action is meaningful if it is motivated by explicit goals, and motivations and goals occur before the action takes place.

Schutz, however, argues that meaning is always produced *after the fact* through a backward glance—through the conscious act of an individual picking out from the stream of experience a particular object as the focus of attention. For Weber, meaning and action are rather linear, in the sense that it works something like this: goal motivation → action → meaning. Schutz, on the other hand, sees meaning as much more contextual. That is, when we think or talk about our actions, they are always past, and the way we give meaning to them is strongly influenced by the present social context. It's this ongoing conceptuality that influences both Garfinkel's and Smith's later work.

Building Your Theory Toolbox

Knowing Weber

After reading and understanding this chapter, you should be able to define the following terms theoretically and explain their importance to Weber's theory:

ideal types, instrumental-rationality, value-rationality, traditional action, affective action, *verstehen*, rationalization, disenchantment, legitimation, traditional capitalism, rational capitalism, spirit of capitalism, Protestant calling, doctrine of predestination, legitimation, property class, commercial class, status, party, crosscutting stratification, authority, charismatic authority, traditional authority, rational-legal authority, ideal type bureaucracy, bureaucratic personality, iron cage of bureaucracy, credentialing, credential inflation

After reading and understanding this chapter, you should be able to

- Discuss the relative importance of culture and social structure and examine the implications of a Weberian cultural sociology approach

- Describe the effects of religious culture and structural changes on the emergence of capitalism

- Differentiate class, status, and power and analyze their effects on the possibility of social change

- Explain Weber's theory of social change

- Discuss the historical shift to bureaucracy as the chief organizing principle and its effects

- Analyze an organization using the ideal type of bureaucracy

Learning More: Primary Sources

Weber wrote extensively and thus choosing a short list of his works is difficult. However, the following will provide a strong background in Weber's thought.

Weber, M. (1948). *From Max Weber: Essays in sociology* (H. H. Gerth & C. Wright Mills, Trans. & Eds.). London: Routledge and Kegan Paul.

Weber, M. (1949). *The methodology of the social sciences* (E. A. Shils & H. A. Finch, Trans. & Eds.). New York: The Free Press. (Originally published 1904, 1906, 1917–1919)

Weber, M. (1968). *Economy and society* (G. Roth, & C. Wittich, Eds.). Berkeley, CA: University of California Press. (Originally published 1922)

Weber, M. (2002). *The Protestant ethic and the spirit of capitalism* (S. Kalberg, Trans.). Los Angeles: Roxbury. (Original work published 1904–1905)

Learning More: Secondary Sources

Bendix, R. (1977). *Max Weber: An intellectual portrait.* Berkeley: University of California Press. (The standard Weber reference)

Collins, R. (1986). *Weberian sociological theory.* New York: Cambridge University Press. (Systemization of Weber's theory by a prominent contemporary theorist)

Turner, B. S. (1992). *Max Weber: From history to modernity.* New York: Routledge. (Explanation of Weber's theories of modernity)

Turner, S. P. (2000). *The Cambridge companion to Max Weber.* New York: Cambridge University Press. (Excellent resource concerning Weber's works and influence)

Weber, M. (1988). *Max Weber: A biography* (H. Zohn, Trans.). New Brunswick, NJ: Transaction Books. (Definitive biography, written by Weber's wife)

Theory You Can Use (Seeing Your World Differently)

- Weber presents us with our first critique of social science. He takes seriously the ideas of culture, value, and free will. However, Weber is still convinced that we can create object knowledge through the use of ideal types. His most famous one is bureaucracy. I'd like for you to create an ideal type with at least five points of comparison, and use it to think about the social world—use gender or another topic that interests you (perhaps one as important as "criminal behavior" or as mundane as "professor"). How did you form your ideal type (remember, Weber said there are two ways)? What were some of the difficulties you ran into? What do you think you can learn about the social world using your ideal type? For example, if you used gender, how do people vary from the ideal type? Under what external conditions do the variations take place? What do the ideal type and the empirical variations tell you about the cultural expectations concerning gender? What does this tell you about the way society is organized? (Notice I didn't ask what it told you about the person.) What are the drawbacks to using ideal types?

- One of the central themes in Weberian theory is rationalization (as a contemporary example, see George Ritzer's *The McDonaldization of Society*). Take a look at your life: In what ways has rationality influenced you? Do you think your life is more or less rationalized than your parents' was at your age? Let's take this one step further. Go to a place of business, like a fast-food restaurant or mall, and observe behaviors for at least two hours. How rationalized were the actions you observed? Overall, do you think that life is becoming increasingly rationalized? What are the benefits and drawbacks to rationalization? How do you think Weber felt about this process?

- Get a sense of the kinds of jobs you can get today with a college education and the jobs available with the same education 50 years ago. You can do this by using your Internet search engine, going to http://nces.ed.gov/ and searching the data, or by asking your parents and grandparents. Explain your findings using Weber's theory. What do you think society can or should do in response to these changes? Using Weber's theory, do you think this trend will continue or abate? Do you think the purpose of education has changed in this country? To what level is education completely funded by government (that is, to what level is education free)? Why to that level, do you think?

- Think about Weber's ideal type of the spirit of capitalism. Are those traits more or less present in the United States today? Does this imply anything about capitalism in this country? Are we perhaps practicing a different kind of capitalism? If so, what would you call it?

- In addition to being spiritual centers, churches are social organizations. As such, Weber would argue that the type of authority and the concurrent organizational type that a church uses will influence the church and its parishioners. Using either your own experiences or by calling various churches in your area, what kind of authority and organization do you find to be most prevalent? Using Weber's theory, how do you think the church is being affected?

Further Explorations—Web Links

http://www.faculty.rsu.edu/~felwell/Theorists/Weber/Whome.htm

http://www2.fmg.uva.nl/sociosite/topics/weber.html

CHAPTER 3

Cultural Diversity and Social Integration

Émile Durkheim (1858–1917)

Photo: © Bettmann/Corbis.

Seeing Further: Social Integration and Cultural Diversity

Cultural diversity is a byword in modern society. It generally refers to racial or ethnic diversity. However, if we think about cultural diversity theoretically, the phrase "racial or ethnic diversity" begs the question: How is it that racial or ethnic groups come to have different cultures? Most people simply assume that different races and ethnic groups have diverse cultures. Yet there is no necessary relationship between what we think of as race and cultural diversity. In fact, race itself is a cultural designation. For example, did you know that at one time in the United States, "Irish" was considered a "black" racial group? They were referred to derogatorily as the "black Irish."

Theoretically and sociologically, then, it is much better to ask how cultural diversity is created rather than simply assuming it exists. Besides, cultural diversity is much broader than merely race and ethnicity. For example, it is quite possible that the cultural differences between the elite and the poor are greater than the differences between racial groups within the same society. So, how is cultural diversity created? More specifically, what are the general processes through which cultural differences are created, whether among racial, ethnic, class, or gender groups? Émile Durkheim provides us with answers to these kinds of questions.

Yet Durkheim is actually concerned with a more important issue, one that few people think about when considering cultural diversity. His concern is based on the insight that every society needs a certain level of cultural integration and social solidarity to exist and function. Durkheim's main concern is this: How much cultural diversity can a society have and still function? Think about an extreme situation as an example: Picture two people who speak totally different languages. How easy would it be for them to carry on a conversation? If it was necessary, they undoubtedly could find a way, but what they could talk about would be limited and it would take a great deal of time to have even the simplest of conversations.

The same is true with cultural diversity. Cultural diversity includes language, but it also encompasses nonverbal cues, dialects, values, normative behaviors, beliefs and assumptions about the world, and so on. The more different people are from one another, the more difficult it will be for them to work together and communicate, which is the basis of any society. Durkheim, then, specifically asks, how can a diverse society create social solidarity and function?

One of the reasons that Durkheim is concerned with cultural diversity and moral integration is due to his assumptions about human nature. Where Marx assumes that humans are social and naturally altruistic, Durkheim assumes that people apart from society are self-centered and driven by insatiable desires. While Durkheim gives us an answer to the question of integration in the face of cultural diversity, he also addresses the deeper problem of human egoism. If we assume, as Durkheim does, that individuals tend to go off each in his or her own direction, then how can this thing called society work? Durkheim came up with an ingenious answer: the collective consciousness. Today, sociologists usually talk about norms, values, and beliefs, but in back of those terms lies Durkheim's idea of the collective consciousness.

The Essential Durkheim

Biography

Emile Durkheim was born in Epinal, France, on April 15, 1858. His mother, Melanie, was a merchant's daughter, and his father, Moïse, was a rabbi, descended from generations of rabbis. Durkheim did well in high school and attended the prestigious *Ecole Normale Supérieure* in Paris, the training ground for the new French intellectual elite. In 1887, Durkheim was appointed as *Chargéd'un Cours de Science Sociale et de Pédagogie* at the University of Bordeaux. Durkheim thus became the first teacher of sociology in the French system.

Between 1893 and 1897, Durkheim completed and published his French dissertation (*The Division of Labor*), *The Rules of Sociological Method,* and *Suicide,* in addition to a Latin thesis on Montesquieu. Having thus established the basis for French sociology, in 1898 he founded a sociological journal for his work,

L'Année Sociologique. The journal flourished for many years and became the leading journal of social thought in France.

In 1902, Durkheim took a post at the Sorbonne and by 1906 was appointed Professor of the Science of Education, a title later changed to Professor of Science of Education and Sociology. In this position, Durkheim was responsible for training the future teachers of France and served as chief advisor to the Ministry of Education.

In December of 1915, Durkheim received word that his son, André, had been declared missing in action (WWI). André had followed in his father's footsteps to the *Ecole Normale* and was seen as an exceptionally promising social linguist. Durkheim had hoped his son would complete the research he had begun in linguistic classifications. The following April, Durkheim received official notification that his son was dead. Durkheim withdrew into a "ferocious silence." After only a few months following his son's death, Durkheim suffered a stroke; he died at the age of 59 on November 15, 1917.

Passionate Curiosity

Durkheim is intensely concerned with understanding how social solidarity and integration could be preserved in modernity. He recognizes that society is built on a foundation of shared values and morals. Yet he also realizes that there are structural forces at work in modernity that relentlessly produce cultural diversity, something that could tear away this foundation of social solidarity. His project, then, is to discover and implement the necessary social processes that could create a new kind of unity in society, one that would allow the dynamics of modernity to function within a context of social integration.

Keys to Knowing

Social facts, society sui generis, collective consciousness, religion, sacred and profane, collective effervescence, the division of labor, mechanical and organic solidarity, social pathologies, social differentiation, anomie, suicide, the cult of the individual

Durkheim's Perspective: The Reality of Society and Collective Consciousness

Organismic Analogy

An analogy is a way of explaining something through comparison. Analogies help us understand new ideas. So, for example, we may not understand the difference between the brain and the mind, but someone might be able to help us by using the analogy of a computer—the brain is like the hardware and the mind like the software. Early functionalists, like Durkheim, employed the *organismic analogy* to understand how society works.

The **organismic analogy** is a way of looking at society that understands the form of society and the way society changes as if it were an organism. The fundamental idea taken from this analogy is that organisms have *requisite needs*. These needs push the organism to select and create internal structures in order to meet those needs. For example, in order for you to survive, you need oxygen, and you get your oxygen from air. Because of that need and the way it is met, your body has a specific organ or structure—your lungs. Other organisms such as fish don't have lungs because they don't get oxygen from air. By analogy, the same is true for society: different social structures meet specific needs.

The organismic analogy also implies evolution. Evolution basically occurs to enhance an organism's survivability. Generally speaking, the more complex an organism is, the greater will be its chances of surviving. Complexity is defined in terms of structural differentiation and specialization. Initial organisms were single-celled; that is, they only had one structure. As the evolutionary processes continued, organisms became increasingly more complex. They developed different structures to meet specific needs, which, in turn, enhanced the organism's ability to survive. The human body, for example, is made up of many different kinds of structures (such as heart, lungs, liver, and bones) and many different subsystems (digestive system, respiratory system, nervous system, and so on).

The same is true for society. As society evolves, it becomes more structurally differentiated and specialized. Generally speaking, social structures are made up of connections among sets of positions that form a network. The interrelated sets of positions in society are generally defined in terms of status positions, roles, and norms. These social and cultural elements create and manage the connections among people, and it is the connections that form the structure. **Structural differentiation** in society, then, is the process through which social networks break off from one another and become functionally specialized. For example, the goals, values, roles, status positions, and even language of education are significantly different from those of religion or government.

The idea of structural differentiation implies that society works like a system. Systems are defined in terms of relatively self-contained wholes that are made up of diverse interdependent parts. That's a mouthful, but it really isn't as daunting as it might seem. Your body is a system, and it's made up of various parts and subsystems that are mutually dependent upon one another, each part contributing to the good of the whole. For example, your heart could not do its job apart from the rest of the circulatory system; in turn, the circulatory system is dependent upon the digestive system, and so on. In society, religion, education, and government are structurally different but they also depend upon and complement one another.

However, structural differentiation and social evolution can create problems. For Durkheim, the most important problem is phrased in terms of social differentiation and social integration. Social integration concerns how well coordinated separate social structures and actors are. Thinking again of our organismic analogy, there are certain disabilities, such as cerebral palsy, that disturb muscular coordination and speech. In cases such as these, we can say that the different systems in the body aren't well integrated. Like the human body, society can't function unless it is integrated.

Now think about what we just learned regarding social evolution and structural differentiation. Social evolution implies that the structures in society become more complex and functionally separated and specialized. Structural differentiation thus sets up a situation where integration may be a problem: society has to come up with a way of coordinating the actions of discrete social structures and actors. If it can't, certain social pathologies will develop.

As we'll see later in this chapter, structural differentiation creates social differentiation. *Social differentiation* is the process through which people become socially different from one another—the term is generally used in the same way we use cultural diversity. Durkheim's argument is that people become socially distinct, or culturally diverse, as they interact in different structural environments. And, just like structural differentiation, cultural diversity may cause problems of integration within a society.

The organismic analogy implies one further issue for functionalism: equilibrium. **Equilibrium** is a state of balance between or among opposing forces or processes resulting in the absence of change. Most organic systems will tend toward equilibrium because it is the natural state of life. If we look at the physical or animal world, we see a great deal of overall stability. In fact, that is one of the problems with which evolutionists are faced: We don't see much change in our lifetime or even over many millennia.

One of the reasons for this slowness is undoubtedly due to the fact that all things appear to exist within interrelated systems and subsystems. Because systems are interrelated, sudden change would probably bring chaos instead of order. In the same way, sudden change in one societal subsystem will tend to bring chaos, unless the change is countered with equal changes in other subsystems. In other words, unless social changes are met with equilibrating pressures, they will lead to the demise of that society.

Durkheim's theory is clearly informed by the organismic approach. His central, burning question concerns the structural and social differentiation that comes with modernity. If the changes that modernity brings with it come too quickly, then chances are good that society will have problems with integration and this will cause social imbalances and pathologies. Durkheim argues that the most important requisite need society has is the collective consciousness.

Collective Consciousness

For Durkheim, the **collective consciousness** is the totality of ideas, representations, beliefs, and feelings that are common to the average members of society. There does exist, of course, the individual consciousness. However, whatever unique ideas, feelings, beliefs, impressions, and so forth an individual might have are by definition idiosyncratic. In other words, "individual consciousnesses are actually closed to one another" (Durkheim, 1912/1995, p. 231). The collective consciousness, on the other hand, does allow us a basis for sharing our awareness of the world. Yet the function of this body of culture is not simply to express our inner states to one another; the collective consciousness contributes to the making of our individual subjective states. It is through the collective consciousness that society becomes aware of itself and we become aware of ourselves as social beings.

Durkheim divides the collective consciousness into two basic features: cognitive and emotional. Durkheim argues that the collective consciousness contains primary symbolic categories (time, space, number, cause, substance, and personality). These categories are primary because we can't think without using them. They form our basic cognitions or consciousness of the world around us. These categories are, of course, of a social origin for Durkheim, originating with the physical features of society. (The way the population is dispersed in space, for example, influences the way we conceive of space.) Durkheim published some work (*Primitive Classification*) in this area of cognitive categories, especially with his pupil and nephew Marcel Mauss. His work in this area also influenced Ferdinand De Saussure, the founder of French structuralism (which led to poststructuralism and influenced postmodernism—see Chapter 14). However, I think for Durkheim the more important aspect of the collective consciousness is emotional. Social emotions or sentiments "dominate us, they possess, so to speak, something superhuman about them. At the same time they bind us to objects that lie outside our existence in time" (Durkheim, 1893/1984, p. 56).

If, as Durkheim supposes, humans are naturally self-serving, then why will self-centered human beings act collectively and selflessly? Durkheim argues that rational exchange principles are not enough. Because our entire being is involved in action, we need to be emotionally bound to our culture. We have to have an emotional sense of something greater than ourselves. This feeling of something greater is what underlies morality. We act socially because it is moral to do so. While we can always give reasons for our actions, many of our actions—especially social actions—generally come about because of *feelings* of responsibility: "Whence, then, the feeling of obligation? It is because in fact we are not purely rational beings; we are also emotional creatures" (Durkheim, 1903/1961, p. 112).

Social Facts

One of Durkheim's abiding influences on social thinking is the way he conceptualizes society. Durkheim (1895/1938) argues that society has an empirical or factual existence and that its *facticity* (ability to exist as a fact) "consists of ways of acting, thinking and feeling, external to the individual, and endowed with a power of coercion, by reason of which they control him" (p. 3). According to Durkheim, society exists as an objective fact because it does not depend upon individuals, it exists apart from individuals, and it exerts power over individuals. We may not be able to smell, see, taste, or touch society, but we can *feel* its objective influence and existence. We feel the facticity of society every time we either think about or actually break a social norm. We feel it as we go through our daily rounds with the expectations that institutions, organizations, and people place upon us. The social fact is "the very essence of the idea of social constraint; for it merely implies that collective ways of acting or thinking have a reality outside the individuals who, at every moment of time, conform to it" (p. lvi).

There's an interesting implication that comes from thinking that society exists as sets of social facts. Durkheim says that society exists *sui generis*—a Latin term meaning "of its own kind." Durkheim (1912/1995) uses the term to say that society exists in and of itself, not as a "mere epiphenomenon of its morphological base" p. 426). Society is more to Durkheim than simply the sum of all the individuals within it. Society exists as its own kind of entity, obeying its own rules and creating its own effects: "*The determining cause of a social fact should be sought among the social facts preceding it and not among the states of the individual consciousness*" (Durkheim, 1895/1938, p. 110, emphasis original). The most important of all social facts for Durkheim is the collective consciousness. I want you to read carefully the following quote and notice what the collective consciousness can do because it exists *sui generis*. Once you truly understand what Durkheim is saying, ask yourself, how can society have this kind of power and life? (We'll see Durkheim's answer later in the chapter.)

> If collective consciousness is to appear, a *sui generis* synthesis of individual consciousnesses must occur. The product of this synthesis is a whole world of feelings, ideas, and images that follow their own laws once they are born. They mutually attract one another, repel one another, fuse together, subdivide, and proliferate; and none of these combinations is directly commanded and necessitated by the state of the underlying reality. Indeed, the life thus unleashed enjoys such great independence that it sometimes plays about in forms that have no aim or utility of any kind, but only for the pleasure of affirming itself. (Durkheim, 1912/1995, p. 426)

Let's think about these ideas for a moment; they're pretty profound. Durkheim believes that society is held together by a moral, collective consciousness that exists as a social fact. Social facts transcend the individual. They exist outside of the individual and they profoundly impact the individual's behavior. And, as we've seen, social facts and society exist independently of the individuals that make them up. There is yet another, perhaps more important way that society transcends the individual. Society provides a sense of transcendence, an experience of something greater than our individual selves. How can society exist *sui generis*? How can it impose itself upon us and provide a sense of transcendence? Durkheim's answer to these kinds of questions is in the religious roots of society.

Concepts and Theory: The Religious Roots of Society

> Fundamentally, then, there are no religions that are false. (Durkheim, 1912/ 1995, p. 2)

Durkheim had two major purposes in writing *The Elementary Forms of the Religious Life*. His primary aim was to understand the empirical elements present in all religions. He wanted to go behind the symbolic and spiritual to grasp what he calls "the real." Durkheim intentionally puts aside the issue of God and spirituality. He argues that no matter what religion is involved, whether Christianity or Islam, and no matter what god is proclaimed, there are certain social elements that are common to all religions. To put this issue another way, anytime a god does anything here on earth, there appear to be certain social elements always present. Durkheim is interested in discovering those empirical, social elements.

Religion and Science

Durkheim's second reason for writing his book is a bit trickier to understand. For Durkheim, religion is the most fundamental social institution. He argues that religion is the source of everything social. That's not to say that everything social is religious, especially today. But Durkheim is convinced that social bonds were first created through religion. We'll see that in ancient clan societies, the symbol that bound the group together and created a sense of kinship (family) was principally a religious one.

Further, Durkheim argues that our basic categories of understanding are of religious origin. Humans divide the world up using categories. We understand things in terms of animal, mineral, vegetable, edible, inedible, private property, public property, male, female, and on and on. Durkheim says that many of the categories we use are of what one might call "fashionable" origin, that is, culture that is subject to change. Durkheim argues that fashion is a recent phenomenon and that its basic social function is to distinguish the upper classes from the lower. There is a tendency for fashion to circulate. The lower classes want to be like the upper classes and thus want to use their symbols (we want to drive their cars, wear the same kinds of clothes they do, and so on). This implies, in the end, that fashionable culture is rather meaningless: "Once a fashion has been adopted by everyone, it loses all its value; it is thus doomed by its own nature to renew itself endlessly" (Durkheim, 1887/1993, p. 87).

Durkheim has little if any concern for such culture (though quite a bit of contemporary cultural theory and analysis is taken up with it). He is interested in primary "categories of understanding." He argues that these categories—time, space, number, cause, substance, and personality—are of social origin, but not the same kind of social origin that fashion has. Fashion comes about as different groups demarcate themselves through decoration, and in that sense it isn't tied to anything real. It is purely the work of imagination. The primary categories of understanding, on the other hand, are tied to objective reality. Durkheim argues that the primary categories originate *empirically and objectively in society,* in what he calls social morphology.

Merriam-Webster (2002) defines morphology as "a branch of biology that deals with the form and structure of animals and plants." So, when Durkheim talks about *social morphology,* he is using the organismic analogy to refer to the form and structure of society, in particular the way in which populations are distributed in time and space. Let's take time, for example. In order to conceive of time, we must first conceive of differentiation. Time, apart from humans, appears like a cyclical stream.

There is daytime and nighttime and seasons that endlessly repeat themselves. Yet that isn't how we experience time. For us, time is chopped up. Today, for example, the day I'm writing these words, is March 21, 2006. However, that date and the divisions underlying it are not a function of the way time appears naturally. So, from where do the divisions come? "The division into days, weeks, months, years, etc., corresponds to the recurrence of rites, festivals, and public ceremonies at regular intervals. A calendar expresses the rhythm of collective activity while ensuring that regularity" (Durkheim, 1912/1995, p. 10). The same is true with space. There is no up, down, right, left, and so on apart from the orientation of human beings that is itself social.

The important thing to see here is that Durkheim makes the claim that the way in which we divide up time and space, and the way we conceive of causation and number, is not a function of the things themselves, nor is it a function of mental divisions. Rather, the way we conceive of these primary categories is a function of the objective form of society—the ways in which we distribute and organize populations and social structures. Our primary categories of understanding come into existence through the way we distribute ourselves in time and space; they reflect our gatherings and rituals. Thus, all of our thinking is founded upon social facts, and these social facts, according to Durkheim, originated in religion.

Durkheim wants to make a point beyond social epistemology. He argues that if our basic categories of understanding have their roots in religion, then all systems of thought, such as science and philosophy, have their basis in religion. Durkheim (1912/1995) extends this theme, stating that "there is no religion that is not both a cosmology and a speculation about the divine" (p. 8). Notice that there are two functions of religion in this quote. One has to do with speculations about the divine—in other words, religion provides faith and ideas about God. Also notice the other function: to provide a cosmology. A *cosmology* is a systematic understanding of the origin, structure, and space–time relationships of the universe. Durkheim is right: every religion tells us what the universe is about—how it was created, how it works, what its purpose is, and so on. But so does science, and that's Durkheim's point. Speculations about the universe began in religion; ideas about causation began in religion. Therefore, the social world, even in its most logical of pursuits, was set in motion by religion.

Defining Religion

But how did religion begin? Here we turn back again to Durkheim's principal purpose: to explain the origins of religion. In order to get at his argument, we will consider the data Durkheim uses, his definition of religion, and, most importantly, how the sacred is produced. The data that Durkheim employs are important because of his argument and intent. He wants to get at the most general social features underlying all religions—in other words, apart from doctrine, he wants to discern what is common to all religions. To discover those commonalities, Durkheim contends that one has to look at the most primitive forms of religion.

Using contemporary religion to understand the basic forms of religion has some problems, most notably the natural effects of history and storytelling. You've probably either played or heard of the game "telephone," where people sit in a circle and take turns whispering a story to one another. What happens, as you know, is that the story changes in the telling. The same is true with religion, at least in terms of its origins. Basically what Durkheim is saying is that the further we get away from the origins of religion, the greater will be the confusion around why and how religion began in the first place. Also, because ancient religion was simpler, using the historical approach allows us to break the social phenomenon down into its constituent parts and identify the circumstances under which it was born. For Durkheim, the most ancient form of religion is *totemism*. He uses data on totemic religions from Australian Aborigines and Native American tribes for his research.

Conceptually, prior to deciding what data to use, Durkheim had to create a definition of religion. In any research, it is always of utmost importance to clearly delineate what will count and what will not count as your subject. For example, if you were going to study the institution of education, one of the things you would have to contend with is whether home schooling or Internet courses would count, or do only accredited teachers in state-supported organizations constitute the institution of education? The same kinds of issues exist with religion. Durkheim had to decide on his definition of religion before choosing his data sources—he had to know ahead of time what counts as religion and what doesn't, especially since he wanted to look at its most primitive form.

> A religion is a unified system of beliefs and practices relative to sacred things, that is to say, things set apart and forbidden—beliefs and practices which unite into one single moral community called a Church, all those who adhere to them. (Durkheim, 1912/1995, p. 44, emphasis original)

Note carefully what is missing from Durkheim's definition: there is no mention of the supernatural or God. Durkheim argues that before humans could think about the supernatural, they first had to have a clear idea about what was natural. The term *supernatural* assumes the division of the universe into two categories: things that can be rationally explained and those that can't. Now, think about this— when was it that people began to think that things could be rationally explained? It took quite of bit of human history for us to stop believing that there are spirits in back of everything. We had gods of thunder, forests, harvest, water, fire, fertility, and so on. Early humans saw spiritual forces behind almost everything, which means that the idea of *nature* didn't occur until much later in our history. The concept of nature—those elements of life that occur apart from spiritual influence—didn't truly begin until the advent of science. So, the idea of supernatural is a recent invention of humanity and therefore can't be included in a definition of religion, since there has never been society without religion—early societies would have had religion but no concept of supernatural.

Durkheim also argues that we cannot include the notion of God in the definition of religion. His logic here concerns the fact that there are many belief systems that are generally considered religion that do not require a god. Though he includes other religions such as Jainism, his principal example is Buddhism. The focus of Buddhist faith is the Four Noble Truths, and "salvation" occurs apart from any divine intervention. There are deities acknowledged by Buddhism, such as Indra, Agni, and Varuna, but the entire Buddhist faith can be practiced apart from them. The practicing Buddhist needs no god to thank or worship, yet we would be hard pressed to not call Buddhism a religion.

Thus, three things constitute religion in its most basic form: the sacred, beliefs and practices, and a moral community. The important thing to notice about Durkheim's definition is the centrality of the notion of the sacred. Every element of the definition revolves around it. The beliefs and practices are relative to sacred things and the moral community exists because of the beliefs and practices, which of course brings us back to sacred things. So at the heart of religion is this idea of the sacred.

Creating the Sacred

But what are sacred things? By that I mean, what makes something sacred? For most of us in the United States, we think of the cross or the Bible as sacred objects. It's easy for us to think they are sacred because of some intrinsic quality they possess. The cross is sacred because Jesus died on it. However, the idea of the intrinsic worth of the object falls apart when we consider all that humanity has thought of as sacred. Humans have used crosses, stones, kangaroos, snakes, birds, water, swastikas, flags, and yellow ribbons—almost anything—to represent the sacred. Durkheim's point is that sacredness is not a function of the object; sacredness is something that is placed *upon* the object.

> Since, in themselves, neither man nor nature is inherently sacred, both acquire sacredness elsewhere. Beyond the human individual and the natural world, then, there must be some other reality. (Durkheim, 1912/1995, pp. 84–85)

We could argue that sacredness comes through association; that's true at least in part. We think the cross is sacred because of its association with Jesus. But what makes the image of an owl sacred? It can't just be its association with the owl, so there must be something else in back of it. What, then, can make both the owl and the cross of Jesus sacred? (Remember, with a positivist like Durkheim, we are looking for *general* explanations, ones that will fit all instances.) Durkheim wouldn't accept the answer that there is some general spiritual entity in back of all sacred things. To begin with, the sacred things and their beliefs are too varied. But more importantly, Durkheim is interested in the *objective* reality behind religion. So, how can we explain the power of the sacred using objective, general terms? Durkheim begins with his consideration of totemic religion.

Totems had some interesting functions. For instance, they created a bond of kinship among people unrelated by blood. Each clan was composed of various hunting and gathering groups. These groups lived most of their lives separately, but they periodically came together for celebrations. The groups weren't related by blood, nor were they connected geographically. What held them together was that all the members of the clan carried the same name—the name of their totem—and the members of these groups acted toward one another as if they were family. They had obligations to help each other, to seek vengeance on behalf of each other, to not marry one another, and so forth based on family relations.

In addition to creating a kinship name for the clans, the totem acted as an emblem that represented the clan. It acted as a symbol both to those within the clan and those outside it. The symbol was inscribed on banners and tents and was tattooed on bodies. When the clan eventually settled in one place, the symbol was carved into doors and walls. The totem thus formed bonds, and it represented the clan.

In addition, the totem was used during religious ceremonies. In fact, Durkheim (1912/1995) tells us, "Things are classified as sacred and profane by reference to the totem. It is the very archetype of sacred things" (p. 118). Different items became sacred because of the presence of the totem. For example, the clans both in daily and sacred life would use various musical instruments; the only difference between the sacred and the mundane instruments was the presence of the totemic symbol. The totem imparted the quality of being sacred to the object.

This is an immensely important point for Durkheim: The totem represents the clan and it creates bonds of kinship. It also represents and imparts the quality of being sacred. Durkheim then used a bit of algebraic logic: If A = B and B = C, then A and C are equal. So, "if the totem is the symbol of both the god and the society, is this not because the god and the society are the same?" (Durkheim, 1912/1995, p. 208). Here Durkheim begins to discover the reality behind the sacred and thus religion. The empirical reality behind the sacred has something to do with society, but what exactly?

One of the primary features of the sacred is that it stands diametrically opposed to the profane. In fact, one cannot exist in the presence of the other. Remember the story of Moses and the burning bush? Moses had to take off his shoes because he was standing on sacred ground. These kinds of stories are repeated over and over again in every religion. The sacred either destroys the profane or the sacred becomes contaminated by the presence of the profane. So, one of Durkheim's questions is, how did humans come to conceptualize these two distinct realms? The answer to this will help us discover the reality behind sacredness.

Durkheim found that the aborigines had two cycles to their lives, one in which they carried on their daily life in small groups and the other in which they gathered in large collectives. In the small groups, they would take care of daily needs through hunting and gathering. This was the place of home and hearth. Yet each of these small groups saw themselves as part of a larger group: the clan. Periodically, the small groups would gather together for large collective celebrations.

During these celebrations, the clan members were caught up in *collective effervescence*, or high levels of **emotional energy**. They found that their behaviors changed; they felt "possessed by a moral force greater than" the individual. "The effervescence often becomes so intense that it leads to outlandish behavior. . . . [Behaviors] in normal times judged loathsome and harshly condemned, are contracted in the open and with impunity" (Durkheim, 1912/1995, p. 218). These clan members began to conceive of two worlds: the mundane world of daily existence where they were in control, and the world of the clan where they were controlled by an external force greater than themselves. "The first is the profane world and the second, the world of sacred things. It is in these effervescent social milieux, and indeed from the very effervescence, that the religious idea seems to have been born" (Durkheim, 1912/1995, p. 220).

In some important ways, Durkheim is describing the genesis of society. Let's assume, as Durkheim and many others do, that human beings are by nature self-serving and individualistic. How, then, is society possible? One answer is found here: In Durkheimian thought, humans are linked emotionally. Undoubtedly these emotions, once established, mediate human connections unconsciously. That is, once humans are connected emotionally, it isn't necessary for them to rationally see or understand the connections, though we will always come up with legitimations. This emotional soup that Durkheim is describing is the stuff out of which human society is built. Initially, emotions run wild and so do behaviors in these kinds of primitive societies. But with repeated interactions, the emotions become focused and specified behaviors, symbols, and morals emerge.

In general, Durkheim is arguing that human beings are able to create high levels of emotional energy whenever they gather together. We've all felt something like what Durkheim is talking about at concerts or political rallies or sporting events. We get swept up in the excitement. At those times, we feel "the thrill of victory and the agony of defeat" more poignantly than at others. It is always more fun to watch a game with other people such as at a stadium. Part of the reason is the increase in emotional energy. In this case, the whole is greater than the sum of the parts, and something emerges that is felt outside the individual. This dynamic is what is in back of mob behavior and what we call "emergent norms." People get caught up in the overwhelming emotion of the moment and do things they normally wouldn't do.

Randall Collins (1988), a contemporary theorist, has captured Durkheim's theory in abstract terms (I also refer you to Chapter 10 where we go into more detail with Collins's use of Durkheim). Generally speaking, there are three principal elements to the kind of interactions that Durkheim is describing: *co-presence*, which describes the degree of physical closeness in space (we can be closer or further away from one another); *common emotional mood*, the degree to which we share the same feeling about the event; and *common focus of attention*, the degree to which participants are attending to the same object, symbol, or idea at the same time (a difficult task to achieve, as any teacher knows).

When humans gather together in intense interactions—with high levels of co-presence, common emotional mood, and common focus of attention—they produce high levels of emotional energy. People then have a tendency to symbolize the emotional energy, which produces a sacred symbol, and to create *rituals* (patterned behaviors designed to replicate the three interaction elements). The symbols not only allow people to focus their attention and recall the emotion, they also give the collective emotion stability. These kinds of rituals and sacred symbols lead a group to become morally bounded; that is, many of the behaviors, speech patterns, styles of dress, and so on associated with the group become issues of right and wrong.

Groups with high moral boundaries are difficult to get in and out of. Street gangs and the Nazis are good examples of groups with high moral boundaries. One of the first things to notice about our examples is the use of the word "moral." Most of us probably don't agree with the ethics of street gangs. In fact, we probably think their ethics are morally wrong and reprehensible. But when sociologists use the term *moral,* we are not referring to something that we think of as being good. A group is moral if its behaviors, beliefs, feelings, speech, styles, and so forth are controlled by strong group norms and are viewed in terms of right and wrong. In fact, both the Nazis and street gangs are probably more moral, in that sense, than you are, unless you are a member of a radical fringe group.

This theory of Durkheim's is extremely important. First of all, it gives us an empirical, sociological explanation for religion and sacredness. One of the problems that we are confronted with when we look across the face of humanity is the diversity of belief systems. How can people believe in diverse realities? The Azande of Africa seek spiritual guidance by giving a chicken a magic potion brewed from tree bark and seeing if the chicken lives or dies. Christians drink wine and eat bread believing they are drinking the blood and eating the body of Christ. How can we begin to explain how people come to see such diverse things as real? Durkheim gives us a part of the puzzle.

The issues of reality and sacredness and morality aren't necessarily based on ultimate truth for humans. Our experience of reality, sacredness, and morality is based on Durkheimian rituals and collective emotion (see Allan, 1998). Let me put this another way. Let's say that the Christians are right and the Azande are wrong. How is it, then, that both the Christians and the Azande can have the same experience of faith and reality? Part of the reason is that human beings create sacredness in the same way, regardless of the correctness of any ultimate truth. One of the common basic elements of all religions, particularly during their formative times, is the performance of Durkheimian rituals. These kinds of rituals create high levels of emotional energy that come to be invested in symbols; such symbols are then seen as sacred, regardless of the meaning or truth-value of the beliefs associated with the symbol.

Another reason that this theory is so important is that it provides us with a sociological explanation for the experience that people have of transcendence—something outside of and greater than themselves. All of us have had these kinds of experiences, some more than others. Some have experienced it at a Grateful Dead concert, others at the Million Man March, others watching a parade, and still others as we conform to the expectations of society. We feel these expectations not as a cognitive dialogue, but as something that impresses itself upon us physically and emotionally. Sometimes we may even cognitively disagree, but the pressure is there nonetheless.

Concepts and Theory: Social Diversity and Morality

One of the big questions that drove Durkheim is, what holds modern industrial societies together? Up until modern times, societies stayed together because most of the people in the society believed the same things, acted the same, felt the same, and saw the world in the same way. However, in modern societies people are different from each other and they are becoming more so. What makes people different? How can all these different people come together and form a single society? In order to begin to answer these questions, Durkheim created a typology of societies.

A theoretical typology is a scheme that classifies phenomena into different categories. We aren't able to explain things directly by using a typology, but it does make things more apparent and more easily explained. Durkheim's typology reflects his central concern with social solidarity.

Mechanical and Organic Solidarity

Social solidarity can be defined as the degree to which social units are integrated. According to Durkheim, the question of solidarity turns on three issues: the subjective sense of individuals that they are part of the whole, the actual constraint of individual desires for the good of the collective, and the coordination of individuals and social units. It is important for us to notice that Durkheim acknowledges three different levels of analysis here: psychological, behavioral, and structural. Each of these issues becomes itself a question for empirical analysis: How much do individuals feel part of the collective? To what degree are individual desires constrained? And, how are activities coordinated and adjusted to one another? As each of these varies, a society will experience varying levels of social solidarity.

Durkheim is not only interested in the *degree* of social solidarity; he is also interested in the way social solidarity comes about. He uses two analogies to talk about these issues. The first is a mechanistic analogy. Think about machines or motors. How are the different parts related to each other? The relationship is purely physical and involuntary. Machines are thus relatively simple. Most of the parts are very similar and are related to or communicate with each other mechanistically. If we think about the degree of solidarity in such a unit, it is extremely high. The sense of an absolute relationship to the whole is unquestionably there, as every piece is connected to every other piece. Each individual unit's actions are absolutely constrained by and coordinated with the whole.

The other analogy is the organismic one. Higher organisms are quite complex systems, when compared to machines. The parts are usually different from one another, fulfill distinct functions, and are related through a variety of diverse subsystems. Organismic structures provide information to one another using assorted nutrients, chemicals, electrical impulses, and so on. These structures make adjustments based on the information that is received. In addition, most organisms are open systems in that they respond to information from the environment (most machines are closed systems). The solidarity of an organism when compared to a machine is a bit more imprecise and problematic.

By their very nature, analogies can be pushed too far, so we need to be careful. Nonetheless, we get a clear picture of what Durkheim is talking about. Durkheim says that there are two "great currents" in society: similarity and difference. Society begins with the first being dominant. In these societies, which Durkheim terms *segmented,* there are very few personal differences, little competition, and high egalitarianism. These societies experience **mechanical solidarity**. Individuals are mechanically and automatically bound together. Gradually, the other current, difference, becomes stronger and similarity "becomes channeled and becomes less apparent." These social units are held together through mutual need and abstract ideas and sentiments. Durkheim refers to this as **organic solidarity**. While organic solidarity and difference tend to dominate modern society, similarity and mechanical solidarity never completely disappear.

In Table 3.1, I've listed several distinctions between mechanical and organic solidarity. In the first row, the principal defining feature is listed. In mechanical solidarity, individuals are directly related to a group and its collective consciousness. If the individual is related to more than one group, there are very few and the groups tend to overlap with one another: "Thus it is entirely mechanical causes which ensure that the individual personality is absorbed in the collective personality" (Durkheim, 1893/1984, p. 242). People are immediately related to the collective consciousness by being part of the group that creates the culture in highly ritualistic settings. In these groups, the members experience the collective self as immediately present. They feel its presence push against any individual thoughts or feelings. They are caught up in the collective effervescence and experience it as ultimately real.

Table 3.1   Mechanical and Organic Solidarity

Mechanical Solidarity	*Organic Solidarity*
Individuals directly related to the collective consciousness with no intermediary	Individuals related to collective consciousness through intermediaries
Joined by common beliefs and sentiments (moralistic)	Joined by relationships among special and different functions (utilitarian)
Collective ideas and behavioral tendencies are stronger than individual.	Individual ideas and tendencies are strong and each individual has own sphere of action.
Social horizon limited	Social horizon unlimited
Strong attachment to family and tradition	Weak attachment to family and tradition
Repressive law: crime and deviance disturbs moral sentiments; punishment meted out by group; purpose is to ritually uphold moral values through righteous indignation.	Restitutive law: crime and deviance disturbs social order; rehabilitative, restorative action by officials; purpose is to restore status quo.

When an individual is mechanically related to the collective, all the rest of the characteristics we see under mechanical solidarity fall into place. In Table 3.1, the common beliefs and sentiments and the collective ideas and behavioral tendencies represent the collective consciousness. The collective consciousness varies by at least four features:

The degree to which culture is shared—how many people in the group hold the same values, believe the same things, feel the same way about things, behave the same, and see the world in the same way

The amount of power the culture has to guide individual's thoughts, feelings, and actions—a culture can be shared but not very powerful (A group where the members feel they have options doesn't have a very powerful culture.)

The degree of clarity—how clear the prescriptions and prohibitions are in the culture (For example, when a man and a woman approach a door at the same time, is it clear what behavior is expected?)

The content of the culture. Durkheim is referring here to the ratio of religious to secular and individualistic symbolism. Religiously inspired culture tends to increase the power and clarity of the collective consciousness.

As we see from Table 3.1, in mechanical solidarity, the social horizon of individuals tends to be limited. Durkheim is referring to the level of possibilities an individual has in terms of social worlds and relationships. The close relationship the individual has to the collective consciousness in mechanical solidarity limits the number of possible worlds or realities the individual may consider. In modern societies, under organic solidarity, we have almost limitless possibilities from which to choose. Media and travel expose us to uncountable religions and their permutations. Today you can be a Buddhist, Baptist, or Bahai, and you can choose any of the varied universes they present. This proliferation of possibilities, including social relationships, is severely limited under mechanical solidarity. One of the results of this limiting is that tradition appears concrete and definite. People in segmented societies don't doubt their knowledge or reality. They hold strongly to the traditions of their ancestors. At the same time, people express their social relationships using family or territorial terms.

But even under mechanical solidarity, not everyone conforms. Durkheim acknowledges this and tells us that there are different kinds of laws for the different types of solidarity. The function of these laws is different as well, corresponding to the type of solidarity that is being created. Under mechanical solidarity, punitive law is more important. The function of *punitive law* is not to correct, as we usually think of law today; rather, the purpose is expiation (making atonement). Satisfaction must be given to a higher power, in this case the collective consciousness. Punitive law is exercised when the act "offends the strong, well-defined states of the collective consciousness" (Durkheim, 1893/1984, p. 39). Because this is linked to morality, the punishment given is generally greater than the danger represented to society, such as cutting off an individual's hand for an act of thievery.

Organic solidarity, on the other hand, has a greater proportion of *restitutive law,* which is designed to restore the offender and to repair broken social relations.

Because organic social solidarity is based on something other than strong morality, the function of law is different. Here there is no sense of moral outrage and no felt need to ritualize the sacred boundaries. Organic solidarity occurs under conditions of complex social structures and relations. Modern societies are defined by high structural differentiation, with large numbers of diverse structures necessitating complex interconnections of communication, movement, and obligations. Because of the diversity of these interconnections, they tend to be more rational than moral or familial—which is why we tend to speak of "paying one's debt to society." The idea of "debt" comes from rationalized accounting practices; there is no emotional component, as there would be with moral or family connections. In addition, the interests guarded by the laws tend to be more specialized, such as corporate or inheritance laws, rather than generally held to by all, such as "thou shalt not kill." As an important side note, we can see that restitutive and punitive laws are material social facts that help us see the nonmaterial organic and mechanical solidarity within a society.

Organic solidarity thus tends to be characterized by weak collective consciousness: There are fewer beliefs and sentiments, and ideas and behavioral expectations tend to be shared. There is greater individuality and people and other social units (like organizations) are connected to the whole through utilitarian necessity. In other words, we need each other to survive, just like in an organism (my heart would die without its connection to my lungs and the rest of my biological system).

The Division of Labor

Earlier, I mentioned that Durkheim says that similarity and mechanical solidarity gradually become channeled and less apparent. That statement gives the impression that the change from mechanical to organic solidarity occurred without any provocation, and that's not the case. The movement from mechanical to organic solidarity, from similarity to difference, from traditional to modern was principally due to increases in the division of labor. The concept of the **division of labor** refers to a stable organization of tasks and roles that coordinate the behavior of individuals or groups that carry out different but related tasks. Obviously, the division of labor may vary along a continuum from simple to complex. We can build a car in our garage all by ourselves from the ground up (as the first automobiles were built), or we can farm out different manufacturing and assembly tasks to hundreds of subcontractors worldwide and simply complete the construction in our plant (as is done today). These examples illustrate the poles of the continuum, but there are multiple steps in between.

What kinds of processes tend to increase the division of labor in a society? Bear with me for a moment—I'm going to put together a string of rather dry-sounding concepts and relationships. The answer to what increases the overall division of labor is competition. Durkheim sees competition not as the result of individual desires (remember, they are curtailed in mechanical solidarity) or free markets, but rather as the result of what Durkheim variously calls dynamic, moral, or physical density. These terms capture the number and intensity of interactions in a collective taken as a whole. The level of dynamic density is a result of increasing population density, which is a function of population growth (birth rate and migration) and ecological barriers (physical restraints on the ability of a population to spread out geographically).

Durkheim's theory is based on an ecological, evolutionary kind of perspective. The environment changes and thus the organism must change in order to survive. In this case, the environment is social interaction. The environment changes due to identifiable pressures: population growth and density. As populations concentrate, people tend to interact more frequently and with greater intensity. The rate of interaction is also affected by increases in communication and transportation technologies. As the level of interaction increases, so does the level of competition. More people require more goods and services, and dense populations can create surplus workers in any given job category. The most fit survive in their present occupation and assume a higher status; the less fit create new specialties and job categories, thus bringing about a higher degree of division of labor.

The Problem of Modernity

Now, let me ask you a question. What kind of problem do you think that increasing the division of labor might cause for the collective consciousness? The answer to this question is Durkheim's problem of modernity. If people are interacting in different situations, with different people, to achieve different goals (as would be the case with higher levels of the division of labor), then they will produce more particularized than collective cultures. Therefore, because they contain different ideas and sentiments, the presence of *particularized cultures* threatens the power of the collective consciousness.

There's an old saying, "Birds of a feather flock together." Well, what Durkheim is telling us is slightly different: "Birds *become* of a feather *because* they flock together." In other words, the most prominent characteristics of people come about because of the groups they interact with. As we internalize the culture of our groups, we learn how to think and feel and behave, and we become socially distinct from one another. We call this process **social differentiation**. As people become socially differentiated, they, by definition, share fewer and fewer elements of the collective consciousness. This process brings with it the problem of integration. It's a problem that we here in the United States are very familiar with: how can we combine diverse populations into a whole nation with a single identity? But there's more to this problem of modernity: as the division of labor increases, so does the level of structural differentiation, which increases the problems of coordination and control.

Thus, we are confronted with the problem of modernity. Because groups are more closely gathered together, the division of labor has increased. In response to population pressures and the division of labor, social structures have differentiated to better meet societal needs. As structures differentiate, they are confronted with the problem of integration. In addition, as the division of labor increases, people tend to socially differentiate according to distinct cultures. As people create particularized cultures around their jobs, they are less in tune with the collective consciousness and face the problem of social integration. I've illustrated these relationships in Figure 3.1.

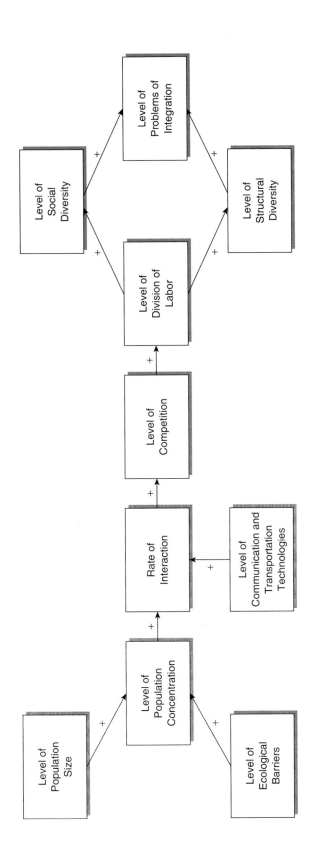

Figure 3.1　The Division of Labor and Problems of Integration

Organic Solidarity and Social Pathology

As social and structural differentiation create problems of integration, they simultaneously produce social factors that counterbalance these problems: intermediary group formation, culture generalization, restitutive law and centralization of power, and structural interdependency. Together, these four factors form *organic solidarity*. First, structures become more dependent upon one another as they differentiate (*structural interdependency*). For example, your heart can't digest food, so it needs the stomach to survive. It is the same for society as it differentiates. The different structures become dependent upon one another for survival. Further, in order to provide for the needs of other structures, they must be able to interact with one another. Thus, structural and social differentiations also create pressures for a more generalized culture and value system (*culture generalization*). Let's think about this in terms of communication among computers. I have a PC and my friend Jamie has an Apple. Each has a completely different platform and operating system. Yet almost every day my computer communicates with his. How can it do this? The two different systems can communicate with one another because there is a more general system that contains values broad enough to encompass both computers (i.e., the Internet).

Societies thus produce more general culture and values in response to the need for different subsystems and groups to communicate. In contemporary structural analysis, this is an extremely important issue. Talcott Parsons (see Chapter 6) termed this focus the "generalized media of exchange"—symbolic goods that are used to facilitate interactions across institutional domains. So, for example, the institutional structure of family values love, acceptance, encouragement, fidelity, and so on. The economic structure, on the other hand, values profit, greed, one-upmanship, and the like. How do these two institutions communicate? What do they exchange? How do they cooperate in order to fulfill the needs of society? These kinds of questions, and their answers, are what make for structural analysis and the most sociological of all research. They are the empirical side of Durkheim's theoretical concern.

Of course, in the United States we have successively created more generalized cultures and values. The idea of *citizen,* for example, has grown from white-male-Protestant-property-owner to include people of color and women. Yet generalizing culture is a continuing issue. The more diverse our society becomes, the more generalized the culture must become, according to Durkheim. For example, while we in the United States may say "in God we trust," it is now a valid question to ask "which god?" We have numerous gods and goddesses that are worshiped and respected in our society. To maintain this diversity, Durkheim would argue that there has to be in the culture a concept general enough to embrace them all. According to Durkheim, if the culture doesn't generalize, we run the risk of disintegration.

The fact is that generalized culture is often too broad to invest much moral emotion, so more and more of our relations, both structural and personal, are mitigated by law. This law has to be rational and focused on relationships, not morals. For example, I don't know my neighbors. The reasons for this have a lot to do with what

Durkheim is talking about: increases in transportation technologies, increasing divisions of labor, and so on. But if I don't know my neighbors, how can our relationship be managed? Obviously, if I have a problem with their dog barking or their tree limb falling on my house, there are laws and legal proceedings that manage the relationship. Increases in social and structural diversity thus create higher levels of *restitutive law* (in comparison to *restrictive/moral law*) and more centralized government to administer law and relations. Of course, one of the things we come to value is this kind of law, and we come to believe in the right of a centralized government to enforce the law.

Both social and structural diversity also push for the formation of intermediary groups. Durkheim always comes back to real groups in real interaction: cultures can have independent effects but they require interaction to be produced. Thus, the problem becomes, how do individual occupational groups create a more general value system if they don't interact with one another? Durkheim theorizes that societies will create *intermediary groups*—groups between the individual occupational groups and the collective consciousness. These groups are able to simultaneously carry the concerns of the smaller groups as well as the collective consciousness.

For example, I'm a sociology faculty member at the University of North Carolina. Because of the demands of work, I rarely interact with faculty from other disciplines (like psychology), and I only interact with medical doctors as a patient. Yet I am a member of the American Sociological Association (ASA), and the ASA interacts with the American Psychological Association and the American Medical Association, as well as many, many others. And all of them interact with the United States government as well as other institutional concerns. This kind of interaction amongst intermediary groups creates a higher level of value generalization, which, in turn, is passed down to the individual members. As Durkheim (1893/1984) says, "A nation cannot be maintained unless, between the state and individuals, a whole range of secondary groups are interposed. These must be close enough to the individual to attract him strongly to their activities and, in so doing, to absorb him into the mainstream of social life" (p. liv).

As we've seen, structural interdependency, culture generalization, intermediary group formation, and restitutive law and centralization of power together create organic solidarity (see Figure 3.2). Increasing division of labor systemically pushes for these changes: "Indeed, when its functions are sufficiently linked together they tend of their own accord to achieve an equilibrium, becoming self-regulatory" (Durkheim, 1893/1984, p. xxxiv). As a side note, part of what we mean by functional analysis is this notion of system pressures creating equilibrium. Notice the dynamic mechanism: it is the system's need that brings about the change. It's the social system itself that provides the impetus for change, not necessarily the individuals within it. In this way of thinking, society is a smart system, regulating its own requirements and bringing about changes to keep itself in equilibrium. However, if populations grow and/or become differentiated too quickly, the system can't keep up and these functions won't be "sufficiently linked together." Society can then become pathological or sick.

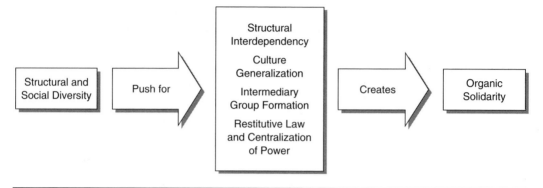

Figure 3.2   Organic Solidarity

Durkheim elaborates two possible pathologies. (Actually, Durkheim mentions three—anomie, forced division of labor, and "lack of coordination"—but he clearly elaborates only the first two.) The first is **anomie**—social instability and personal unrest resulting from insufficient normative regulation of individual activities. Durkheim argues that social life is impossible apart from normative regulation. People are naturally driven by individual appetites, and without norms to regulate interactions, cooperation is impossible. If it were necessary to "grope de novo for an appropriate response to every stimulus from the environing situation, threats to its integrity from many sources would promptly effect its disorganization . . . to this end, it is altogether necessary that the person be free from an incessant search for appropriate conduct" (Durkheim, 1903/1961, p. 37). The production of norms requires interaction. However, overly rapid population growth with excessive division of labor and social diversity hinders groups from interacting. Links among and between the groups cannot be formed when growth and differentiation happen too quickly. The result is anomie.

The other pathology that Durkheim considers at some length involves class inequality and the forced division of labor. Durkheim argues that the division of labor must occur "spontaneously," that is, apart from external constraint. Labor should divide because of organic reasons: population growth and density. The division of labor should not occur because powerful elite are driven by profit motivations. According to Durkheim, it is dysfunctional to force people to work in jobs for which they are ill suited. "We are certainly not predestined from birth to any particular form of employment, but we nevertheless possess tastes and aptitudes that limit our choice. If no account is taken of them, if they are constantly frustrated in our daily occupation, we suffer, and seek the means of bringing that suffering to an end" (Durkheim, 1893/1984, pp. 310–311), which of course would represent a threat to social solidarity.

This condition is similar to what Marx talks about, but it is different as well. For Marx, alienation is a state of existence that may or may not be subjectively experienced by the individual. Alienation for Marx is defined by separation from species-being, and we only become aware of it through critical or class consciousness. Durkheim, on the other hand, assumes that humans are egotistical actors without an essentially good nature from which to be alienated. Durkheim's alienation comes about only as a result of a pathological form of the division of labor—forced by capitalist greed rather than organic evolution. It exists as a subjective state—we are always aware of alienation when it occurs.

Durkheim's solution for this pathological state is "justice" enforced by the state—specifically, price controls based on the usefulness of labor (equal pay for equal work) and elimination of inheritance. Two of the most powerful tools in producing a structure of inequality are ascription and inheritance. *Ascription* assigns different status positions to us at birth, such as gender and race, and apart from legislation guaranteeing equal pay for equal work, they strongly influence inequality. For example, studies done in the United States consistently show that women earn about 70% of what a man makes for the same job with the same qualifications. Inheritance is an obvious way to maintain structural inequality; estimates are that by 2055 at least $41 trillion will be inherited in the United States (Havens & Schervish, 2003). Both ascription and inheritance make the perpetuation of class differences appear natural: our position is ascribed to us and we are wealthy or poor by birth. America may be the land of opportunity, but it is not the land of *equal* opportunity: it matters if you are male or female, black or white, or go to school in Harlem or Hollywood. On the other hand, if inheritance is done away with and laws are implemented that bring equal pay for equal work, structured inequality would have a difficult time surviving.

Concepts and Theory: Individualism in Modern Society

Suicide

In early societies, the human self was a more socially embedded self. People were caught up in and saw themselves in terms of the group. The self was an extension of the group just as certainly as your arm is an extension of your body. Yet as societies differentiated both structurally and socially, the self became more and more isolated and took on the characteristics of an individual. Thus, in modern societies, the individual takes on increasing importance. We can think of many benefits from this shift. We have increased freedom of choice and individual expression under conditions of organic solidarity. However, there are some dysfunctional consequences as well. Though Durkheim didn't phrase it in this way, in addition to the two pathologies of modern society listed above, we can include suicide.

There are two critical issues for the individual in modern society: the levels of group attachment and behavioral regulation. People need a certain level of group attachment. We are social creatures and much of our sense of meaning, reality, and purpose comes from having interpersonal ties (both in terms of number and density) and a sense of "we-ness" or collective identity. To illustrate, imagine having something be meaningful to you apart from language and feeling—what Durkheim would call collective representations and sentiments. You might object and say, "My feelings are my own." That's true, but what do you feel? Do you feel "anger"? Do you feel "love"? Or, do you simply, purely *feel*? We rarely, if ever, simply and purely feel. What are we doing when we say we feel anger? We are labeling certain physiological responses and giving them meaning. The label is linguistic and the meaning is social.

It should be clear by now that it is extremely difficult for us to untangle personal meanings and realities from social ones. Certainly, because we are human, we can create utterly individualistic realities and meanings, although we usually see those realities and those people as either strange or crazy. Most of us are aware, and even unconsciously convinced, that our meanings and realities have to be linked in some way to the social group around us—which is why, when group attachment is too low, Durkheim argues that *egoistic suicide* is likely: low group attachment leads to extreme individualism and the loss of a sense of reality and purpose.

However, extremely high group attachment isn't a good thing for the individual either. High attachment leads to complete fusion with the group and loss of individual identity, which can be a problem in modernity. Under conditions of high group attachment, people are more likely to commit *altruistic suicide*. The Kamikaze pilots during WWII are a good example. Some contemporary examples include religious cults such as The People's Temple and Heaven's Gate, and the group solidarity the United States government fosters in military boot camps. Under conditions of high group attachment, individual life becomes meaningless and the group is the only reality.

It's important to note here that modern societies are characterized by the presence of both mechanical and organic solidarity. It is certainly true that in general the society is held together by organic means—general values, restitutive law, and dependent opposites—yet it is also true that pockets of very intense, particularized culture and mechanical solidarity exist as well. These kinds of group interactions may in fact be necessary for us. This need may explain such intense interaction groups as dedicated fans of rock music or organized sports. Both of these groups engage in the kind of periodic ritual gatherings that Durkheim explains in *The Elementary Forms of the Religious Life*. During these gatherings (shows or events), rock and sports fans experience high levels of emotional energy and create clear group symbols. Yet extremes of either organic or mechanical solidarity can be dangerous, as Durkheim notes.

The other critical issue for individuals in modern society is the regulation of their behaviors. Because in the advanced, industrialized nations we believe in individualism, the idea of someone or something regulating our behavior may be objectionable. But keep in mind Durkheim's view of human nature. Apart from regulation, our appetites would be boundless and ultimately meaningless. The individual by him- or herself "suffers from the everlasting wranglings and endless friction that occur when relations between an individual and his fellows are not subject to any regulative influence" (Durkheim, 1887/1993, p. 24). Further, our behaviors must have meaning for us with regard to time. Time, as we think of it, is a function of symbols (the past and future only exist symbolically), and, of course, symbols are a function of group membership.

Thus, there are a variety of reasons why the regulation of behaviors is necessary. Behaviors need to be organized according to the needs and goals of the collective, but the degree of regulation is important. Under conditions of rapid population growth and diversity, anomie may result if the culture is unable to keep pace with the social changes. Under these conditions, it is likely there will be an increase in the level of *anomic suicide*. The lack of regulation of behaviors leads to a complete lack of regulation of the individual's desires and thus an increase in feelings of meaninglessness. On the other hand, overregulation of behaviors leads to the loss of individual effectiveness (and thus increases hopelessness), resulting in more *fatalistic suicide*.

I've listed the suicide types in Table 3.2. It's important to keep in mind that the motivation for suicide is different in each case, corresponding to group attachment and behavior regulation. Also note that the kinds of social pathologies we are talking about here are different from the ones in the previous section. Here we are seeing how modernity can be pathological for the *individual*, in the extremes of attachment and regulation. In the previous section, we looked at how modernity can be pathological for society as a whole and its solidarity.

Table 3.2   Suicide Types

	Group Attachment	*Behavior Regulation*
High	Altruistic Suicide	Fatalistic Suicide
Low	Egoistic Suicide	Anomic Suicide

The Cult of the Individual

Both egoistic and anomic suicide can be seen as a function of high levels of individuality, but individuality itself is not the problem. Individualism is in fact necessary in modernity. Before we go on, I need to make a distinction between this idea of individuality and what might be called egoistic hedonism or materialism (what most of us think of when we hear the term individualism: "I gotta be me."). From a Durkheimian view, individuals who are purely and exclusively out to fulfill their own desires can never form the basis of a group. Group life demands that there be some shared link that motivates people to work for the collective rather than individual welfare. As we've seen in our discussion of Durkheim, this kind of group life and awareness is dependent upon a certain degree of collective consciousness. The idea of the individual that we have in mind here is therefore not pure ego. Rather, what is at stake might better be understood as the idea of "individual rights" and how it has progressed historically.

To do the concept justice, we should really go back to early Greek and then Roman times. But in actuality, we need only go back to the founding of the United States. What does the following sentence mean? "We hold these truths to be self-evident, that all men are created equal, that they are endowed by their Creator with certain unalienable Rights, that among these are Life, Liberty, and the pursuit of Happiness." The history of the United States is the tale of working out the meaning of that line. Obviously, the central struggle concerns the term "all men." Initially, "all men" referred only to white, property-owning, heterosexual, Protestant males. Somewhere along the line, the United States government decided (often in response to fierce civil struggle, such as the fight for women's rights) that you didn't have to own property, and that you didn't have to be male, or white, or Protestant, though you still have to be heterosexual to have full access to civil rights. The reason there has been struggle over this sentence is the basis that is given for these rights. According to the quoted line, the basis for civil rights is simply being human. These rights can't be earned nor can they be taken away. They are yours not because of anything you've done or because of any group membership, but simply because of your birth into the human race. Thus, what matters isn't what group you belong to; it's you, as an *individual* human being. It is this moral idea of individualism that Durkheim has in mind and that has the potential for creating social solidarity.

Durkheim calls this new moral basis for society the "cult of the individual." The individual, as he or she is historically separated from the group, becomes the locus of social concern and solidarity. The *individual* becomes the recipient of social rights and responsibilities, rather than castes or lineages. Today we see the individual as perhaps the single most important social actor. Even our legal system here in the United States is occupied with preserving the civil rights of the perpetrator of a crime because it is the individual that is valued. The individual becomes the focus of our idea of "justice," which Durkheim sees as the "medicine" for some of the problems that come with pathological forms of the division of labor. For example, in a society such as the United States, the problems associated with labor issues, poverty, deviance, and depression (all of which can be linked to Durkheim's pathologies) are generally handled individually through the court system or counseling. Remember that Durkheim sees culture as the unifying force of society, so the importance of these kinds of cases for Durkheimian sociology has more to do with *the culture that the practices create and reproduce* than the actual legal or psychological effects. In this way, the individual becomes a ritual focus of attention, the symbol around which people can seek a kind of redress for the forced division of labor, inequality, and anomie that we noted above. Today the individual has taken on a moral life. But it is not the particular person *per se,* with all of his or her idiosyncrasies, that has value. Rather, it is the ethical and sacred *idea* of the individual that is important. As Durkheim (1957) says,

> This cult, moreover, has all that is required to take the place of the religious cultures of former times. It serves as well as they to bring about the communion of minds and wills which is a first condition of any social life. (p. 69)

Durkheim Summary

- Durkheim is extremely interested in what holds society together in modern times. In order to understand this problem, he constructs a perspective that focuses on three issues: social facts, collective consciousness, and the production of culture in interaction. Durkheim argues that society is a social fact, an entity that exists in and of itself, which can have independent effects. The facticity of society is produced through the collective consciousness, which contains collective ideas and sentiments. The collective consciousness is seen as the moral basis of society. Though it may have independent effects, the collective consciousness is produced through social interaction.

- Durkheim argues that the basis of society and the collective consciousness is religion. Religion first emerged in society as small bands of hunter–gatherer groups assembled periodically. During these gatherings, high levels of emotional energy were created through intense interactions. This emotional energy, or effervescence, acted as a contagion and influenced the participants to behave in ways they normally wouldn't. So strong was the effervescent effect that participants felt as if they were in the presence of something larger than themselves as individuals, and the collective consciousness was born. The emotional energy was symbolized and the interactions ritualized so that the experience could be duplicated. The symbols and behaviors became sacred to the group and provided strong moral boundaries and group identity.

- Because of high levels of division of labor, modern society tends to work against the effects of the collective consciousness. People in work-related groups and differentiated structures create particularized cultures. As a result, society has to find a different kind of solidarity than one based on religious or traditional collective consciousness. Organic solidarity integrates a structurally and socially diverse society through interdependency, generalized ideas and sentiments, restitutive law and centralized power, and through intermediary groups. These factors take time to develop, and if a society tries to move too quickly from mechanical to organic solidarity, it will be subject to pathological states, such as anomie and the forced division of labor.

- At the center of modern society is the cult of the individual. The ideal of individuality, not the idiosyncrasies of individual people, becomes one of the most generalized values a society can have. However, the individual can also be subject to pathological states, depending on the person's level of group attachment and behavioral regulation. If a society produces the extremes of either of these, then the suicide rate will tend to go up. Suicide due to extremes in group attachment is characterized as either egoistic or altruistic. Suicide due to extremes in behavioral regulation is characterized as either anomic or fatalistic.

Looking Ahead

In some ways, Durkheim has informed sociological theory in more profound yet diffuse ways than anyone in this book. The first way this affect can be seen is through his concept of social facts. As you know, sociology is fundamentally based on the idea that there are social factors that influence human life. In that these factors are perceived as institutions or structures, chances are good that the idea comes from Durkheim's idea of the social fact. We've also seen how his study of suicide informs the kind of methodology practiced in sociology. In fact, Durkheim's concern with social order and integration is one of the primary questions in sociology today.

Durkheim's specific influence can be seen in this book through the theories of Talcott Parsons, Erving Goffman, and Randall Collins. In addition to picking up Durkheim's questions of social order and integration, Parsons (Chapter 6) specifically follows Durkheim's argument for culture being the ultimate integrating force of society. We'll see this specially in Parsons's cybernetic hierarchy of control. Both Goffman and Collins take up Durkheim's notion of ritual. Goffman (Chapter 9) brings it out of the realm of religion and argues that daily life is filled with interaction rituals that preserve the sacred-self and interaction order. Collins is particularly powerful in his use of Durkheim's theory. Remember that I referred to him when we were talking about Durkheim's theory of rituals. Collins explicates Durkheim's variables more clearly than did Durkheim himself and he uses this theory of ritual to elaborate a theory of interaction ritual chains that provide a link between the micro and macro levels of society (Chapter 10). Moreover, Collins creatively uses Durkheim's theory to expand our understanding of conflict (Chapter 7): conflict is engendered and maintained through Durkheimian rituals.

I'd like to give you another example of a contemporary theorist who uses Durkheim's theory that is not covered in this book, Robert Bellah. One of the reasons I want to introduce you briefly to Bellah's work is that he is centrally concerned with social integration in what we might call late- or postmodernity—a time when society once again seems on the brink of disintegrating (see Chapter 14). Bellah (1970, 1980; Bellah, Madsen, Sullivan, Swidler, & Tipton, 1985, 1991) acknowledges the contemporary debate over the cohesiveness of national culture, but it's his position that moral debates over the dysfunctions of institutions are an intrinsic element of the culture of any society. Societies have always argued over the economy, education, religion, and so forth. He makes the case that these debates actually constitute the active component of any culture. He sees culture as a dynamic conversation or argument, something like a Durkheimian ritual. Thus, for Bellah, the debate over the condition of culture is itself a vehicle through which a kind of collective consciousness is produced.

However, Bellah also argues that religion has evolved in modernity to the point where the dualism between the sacred and the profane that was crucial to past religions and morals has collapsed. The collapse of the sacred is due to the loss of an institutional sphere where collective morals are self-consciously formed—institutional differentiation and size have grown to the point where individuals have lost the ability to collectively inform the general moral culture and have become passive recipients of institutional control.

Bellah decries the results of capitalistic exploitation and the colonization of everyday life and the public sphere by institutions that have failed to live up to their moral obligation. He makes the same point as many postmodernists: capitalist greed and exploding advertising have eroded the collective moral base. According to Bellah et al. (1991), what we need is to "move away from a concern for maximizing private interests" and to move toward "justice in the broadest sense—that is, giving what is due to both persons and the natural environment" (p. 143). This can be accomplished through the cultivation of face-to-face communities where people are "involved in the creation of regional cultures in some degree of harmony with the natural environment, where individuals, families, and local communities could grow in moral and cultural complexity" (p. 265). Bellah's argument clearly echoes Durkheim's concern with the creation of intermediate moral groups and the need to interact in face-to-face situations to produce moral culture.

These interactions will revolve around civil rather than religious moral discourses. Central to these civil moral discourses are *representational characters*—a kind of symbol. They are a way through which people define what is good and legitimate in a society. In the United States, Abraham Lincoln, Helen Keller, Eleanor Roosevelt, John Kennedy, Jacqueline Kennedy Onassis, Rosa Parks, and Martin Luther King Jr. are examples of such characters. They have assumed mythic significance and have the ability to generate a degree of social solidarity. Today we might also add such media and sports figures as Oprah Winfrey and Lance Armstrong. What's important about these characters is that they provide us with a focus for intense interactions around central themes in our society. As such, they are the touch point for Durkheimian rituals that produce social solidarity.

Central themes (such as, "Americans as the chosen people, the champions of freedom") along with representational characters can form a *civil religion*. In civil religion, sacred qualities that in previous times were the custody of religion are attached to certain civil institutional arrangements, historical events, and people. The theme of Bellah's (1975) work is a call to create such a civil religion because "only a new imaginative, religious, moral, and social context for science and technology will make it possible to weather the storms that seem to be closing in on us in the late 20th century" (p. xiv).

It is certainly the case that the United States has a great deal of cultural diversity and institutional differentiation. But are we left without a unifying story or grand narrative that can give us a collective identity? Or are we instead a nation with a civil religion that can bring social solidarity? Or are there layers of each, like Durkheim's continuum of mechanical and organic solidarity? If there are pockets of each, how do they relate to each other? How do they form a collective consciousness? If we are indeed as diverse and differentiated as we seem, what is holding us together? What values do we hold in common? How can large-scale collective projects, such as war, be carried out? Or is the diversity only skin deep? Are we modern or postmodern? I don't know the answers, but these are good Durkheimian (and important) questions.

Building Your Theory Toolbox

Knowing Durkheim

After reading and understanding this chapter, you should be able to define the following terms theoretically and explain their importance to Durkheim's theory:

Organismic analogy; social facts; *sui generis;* collective consciousness; religion; categories of understanding; fashion; social morphology; totemism; supernatural; deity; sacred and profane; emotional effervescence; co-presence, common emotional mood, common focus of attention; ritual; symbols; moral boundaries; social solidarity; mechanical and organic solidarity; punitive law; restitutive law; division of labor; dynamic density; particularized culture; social differentiation; structural interdependency; intermediate groups; culture generalization; social pathologies; anomie; forced division of labor; inequality; egoistic, altruistic, anomic, and fatalistic suicides; group attachment; behavioral regulation; cult of the individual

After reading and understanding this chapter, you should be able to

- Explain the organismic analogy and use it to analyze the relations among and between social structures
- Define social facts and explain how society exists *sui generis*
- Explain how society is based on religion
- Discuss how Durkheimian rituals create sacred symbols and group moral boundaries
- Define collective consciousness, social solidarity, and mechanical and organic solidarity
- Explain the problem of modernity and describe how organic solidarity creates social solidarity in modernity
- Describe how organic solidarity can produce certain social pathologies
- Define the cult of the individual and explain its place in producing organic solidarity

Learning More: Primary Sources

Almost every book of Durkheim's is worth reading. The following are indispensable:

Durkheim, É. (1938). *The rules of sociological method* (S. A. Solovay & J. H. Mueller, Trans.; G. E. G. Catlin, Ed.). Glencoe, IL: The Free Press. (Original work published 1895)

Durkheim, É. (1951). *Suicide: A study in sociology* (J. A. Spaulding & G. Simpson, Trans.). Glencoe, IL: The Free Press. (Original work published 1897)

Durkheim, É. (1984). *The division of labor in society* (W. D. Halls, Trans.). New York: The Free Press. (Original work published 1893)

Durkheim, É. (1995). *The elementary forms of the religious life* (K. E. Fields, Trans.). New York: The Free Press. (Original work published 1912)

Learning More: Secondary Sources

Alexander, J. C. (Ed.). (1988). *Durkheimian sociology: Cultural studies.* Cambridge, UK: Cambridge University Press. (Excellent collection of contemporary readings concerning Durkheim's contribution to cultural sociology)

Giddens, A. (1978). *Émile Durkheim.* New York: Viking Press. (Brief introduction to Durkheim's work by one of contemporary sociology's leading theorists)

Jones, R. A. (1986). *Émile Durkheim: An introduction to four major works.* Newbury Park, CA: Sage. (Part of the *Masters of Social Theory* series; short, book-length introduction to Durkheim's life and work)

Lukes, S. (1972). *Émile Durkheim, his life and work: A historical and critical study.* New York: Harper & Row. (The definitive book on Durkheim's life and work)

Meštrovic, S. G. (1988). *Émile Durkheim and the reformation of sociology.* Totawa, NJ: Rowman & Littlefield. (Unique treatment of Durkheim's work; emphasizes Durkheim's vision of sociology as a science of morality that could replace religious morals)

Theory You Can Use (Seeing Your World Differently)

- I'd like for you to go to a sporting event—football, basketball, or hockey would be best. Analyze that experience using Durkheim's theory of rituals. What kind of symbols did you notice? What kinds of rituals? Did the rituals work as Durkheim said they would? What do you think this says about religious rituals in contemporary society? Can you think of other events that have the same characteristics?

- There are a lot of differences between gangs and medical doctors. But there might also be some similarities. Using Durkheim's theory and perspective, how are gangs and medical doctors alike?

- Explain this event and the reactions from a Durkheimian perspective: On September 11, 2001, hijacked jetliners hit the World Trade Center in New York and the Pentagon outside Washington, D.C. News headlines around the world proclaimed "America Attacked" and people in places such as San Diego, CA, Detroit, MI, and Cornville, AZ, wept openly. What Durkheimian processes must have been in place for such an event to happen? And, how would Durkheim explain the fact that Americans had such a strong emotional reaction to the loss of people unknown to them? Also, explain the subsequent use of flags and slogans, and the "war on terrorism," using Durkheim's theory.

- Often in theory class, I will take the students on a walk. We walk through campus, through a retail business section, past a church, and through a residential area. Either think about such a walk or go on an actual walk yourself. Based on Durkheim's perspective (not necessarily his theory), what would he see? How would it be different from what Marx would see?

- In 1997, Tiger Woods not only won the Golf Masters, he broke several tournament records in the process—youngest to win (21 years old), lowest score for 72 holes, and widest margin of victory. He was also the first African American to win the Masters. Or was he? Why is this a Durkheimian question? What can Durkheim's theory tell us about the ways through which we construct race?

Further Explorations—Web Links

http://durkheim.itgo.com

http://www.relst.uiuc.edu/durkheim/

CHAPTER 4

The Individual in Modern Society

George Herbert Mead (1863–1931)

Georg Simmel (1858–1918)

Seeing Further: The Individual in Society

Thus far we have considered the impact of capitalism, rationalization, and cultural diversity on society. While Marx, Weber, and Durkheim have implicit theories concerning the person, we have yet to directly consider the individual in society. For example, how does society influence you? How do you affect society? Who are you? More importantly, what are you? These are deceptively simple questions and ones that we usually don't think much about, especially the last question. Yet these questions become increasingly important as we move into contemporary social theory. We begin our consideration of self in society with George Herbert Mead and Georg Simmel.

Mead gives us a very clear theory about the mechanism or process through which the individual becomes a social being. Sociologists usually call this *social-ization*: the way in which socially formed norms, beliefs, and values come to exist within the individual to the degree that these things appear natural. You've probably come across the idea of socialization in many of your classes. For example, if you've taken a sociology class on gender, one of the things that you've undoubtedly learned is that men and women are socialized differently. But have you ever wondered about precisely *how* socialization takes place? In other words, if society is outside and the person is inside, what exactly is the bridge that connects the two? What's the mechanism or conduit? Mead gives us answers to such questions, and they provide us with an interesting perspective about what you are as well.

However, while Mead gives us the foundation for the sociological understanding of the self, he does not give us a very good perspective concerning the contours of the social environment of the self. This is where Simmel comes in. Let me ask you a question: Where are you from? I have a friend from Germany and another from Russia, and I'm from Southern California. Some of our most notable differences and interesting conversations come from our diverse social backgrounds. But this isn't surprising; I think it's commonly held that where a person is from will influence who he or she is. But let me ask you another question: *When* are you from? You and I and the two friends I mentioned grew up during modern times, and that is a significant and unique force in making us who we are.

We were introduced to this idea in the chapter on Weber. As Weber argues, we and our relationships are more rational and calculative as a result of living in a bureaucratized society; we also tend to value symbolic goods such as credentials and status more in modern societies; and working in a bureaucracy, especially in middle management, can result in a bureaucratic personality. Simmel is going to add to this beginning outline of the person in modernity and show us that there is a great deal of personal freedom in modern society, but that there are some negative psychological effects as well. In this chapter, then, Mead will give us a theory about the self—what it is, how it is formed, and how it works—and Simmel will put the self in the context of modernity and show us that the context wherein the self is formed significantly impacts how the individual experiences him- or herself.

George Herbert Mead

Photo: Reprinted with permission of The Granger Collections.

The Essential Mead

Biography

George Herbert Mead was born on February 27, 1863, in South Hadley, Massachusetts. Mead began his college education at Oberlin College when he was 16 years old and graduated in 1883. After short stints as a school teacher and surveyor, Mead did his graduate studies in philosophy at Harvard. In 1893, John Dewey asked Mead to join him to form the Department of Philosophy at the University of Chicago, the site of the first department of sociology in the United States. Mead's major influence on sociologists came through his graduate course in social psychology, which he started teaching in 1900. Among his students was Herbert Blumer (Chapter 9). Those lectures formed the basis for Mead's most famous work, *Mind, Self, and Society,* published posthumously by his students. Mead died on April 26, 1931.

Passionate Curiosity

Mead is intensely interested in the social basis of meaning, self, and action. What is the self? Why are humans the only animal to have a self? Where is meaning and how is it created? How are people able to act rather than react?

Keys to Knowing

Pragmatism, action, meaning, social objects, interaction, mind, self, generalized other, institutions

Mead's Perspective: American Pragmatism and Action

Pragmatism

There were many influences on Mead's thinking. In fact, his work is an early example of theoretical synthesis, bringing together several different strands of thought to create something new (for Mead's influences, see Morris, 1962). But for our purposes, we will concentrate on Mead's debt to the philosophy of pragmatism.

Pragmatism is the only indigenous and distinctively American form of philosophy, and its birth is linked to the American Civil War (Menand, 2001). The Civil War was costly in the extreme: The number of dead and wounded exceeds that of any other war that the United States has fought, and the dead on both sides were family members and fellow Americans. This extreme cost left people disillusioned and doubtful about the ideas and beliefs that provoked the war. It wasn't so much the content of the ideas that was the problem, but, rather, the fact that ideas that appeared so right, moral, and legitimate could cause such devastation. It took the United States almost 50 years to culturally recover and find a way of thinking and seeing the world that it could embrace. That philosophy was pragmatism.

Pragmatism rejects the notion that there are any fundamental truths and instead proposes that truth is relative to time, place, and purpose. In other words, the "truth" of any idea or moral is not found in what people believe or in any ultimate reality. Truth can only be found in the actions of people; specifically, people find ideas to be true if they result in practical benefits. Pragmatism is thus "an idea about ideas" and a way of relativizing ideology (Menand, 2001, p. xi), but this relativizing doesn't result in relativism. Pragmatism is based on common sense and the belief that the search for "truth and knowledge shifts to the social and communal circumstances under which persons can communicate and cooperate in the process of acquiring knowledge" (West, 1999, p. 151).

Understanding pragmatism helps us see the basis of Mead's concern for meaning, self, and society. As we will see, Mead argues that the self is a social entity that is a practical necessity of every interaction. We need a self to act deliberately and to interact socially; it allows us to consider alternative lines of behavior and thus enables us to act rather than react. In pragmatism, human action and decisions aren't determined or forced by society, ideology, or preexisting truths. Rather, decisions and ethics emerge out of a consensus that develops through interaction—a consensus that is based on a free and knowing subject: the self.

Pragmatism also helps us understand another important idea of Mead's: **emergence**. In general, the word "emergence" refers to the process through which new entities are created from different particulars. For Mead, then, meaning emerges out of different elements of interaction coming together. Let's take a hammer as an example. People create social objects such as hammers in order to survive, and hammers only exist as such for humans (hammers don't exist for tigers, though they might sense the physical object). But the meaning of objects isn't set in stone, once and for all. While the hammer exists in its tool context, its meaning can vary by its use, and its use is determined in specific interactions. It can be an instrument of construction or destruction depending upon how it is used. It can also symbolize an individual's occupation or hobby. Or it could be used as a weapon to kill, and it could be an instrument of murder or mercy, depending upon the circumstances under which the killing takes place. Thus, the "true" meaning of an object cannot be unconditionally known; it is negotiated in interaction. The meaning pragmatically emerges.

Human Action

Humans act—they don't react. As Mead characterizes it, the distinctly human *act* contains four distinct elements: impulse, perception, manipulation, and consumption. For most animals, the route from impulse to behavior is rather direct—they react to a stimulus using instincts or behavioristically imprinted patterns. But for humans, it is a circuitous route.

After we feel the initial impulse to act, we perceive our environment. This perception entails the recognition of the pertinent symbolic elements—other people, absent reference groups (what Mead calls generalized others), and so on—as well as alternatives to satisfying the impulse. After we symbolically take in our environment, we manipulate the different elements in our imagination. This is the all-important pause before action; *this is where society becomes possible.* This manipulation takes place in the mind and considers the possible ramifications of using different behaviors to satisfy the impulse. We think about how others would judge our behaviors, and we consider the elements available to complete the task. After we manipulate the situation symbolically in our minds, we are in a position to consummate the act. I want you to notice something about human, social behavior: *action requires the presence of a mind capable of symbolic, abstract thought and a self able to be the object of thought and action.* Both the mind and self, then, are intrinsically linked to society. Before considering Mead's theory of mind and self, we have to place it within the more general context of symbolic meaning.

Concepts and Theory: Symbolic Meaning

No longer can man confront reality immediately; he cannot see it, as it were, face to face. Physical reality seems to recede in proportion as man's symbolic activity advances. Instead of dealing with the things themselves man is in a sense constantly conversing with himself. (Cassirer, 1944, p. 42)

Symbols and Social Objects

As the above quote indicates, Ernst Cassirer argues that human beings created a new way of existing in the environment through the use of signs and meaning. While this sounds a bit like what Marx might say, it is distinctly different. Marx argues that humans survive because of creative production, the work of our hands. This externalization of human nature then stands as a mirror through which ideas and consciousness come. Thus, ideas for Marx have a material base. Cassirer argues in the opposite direction: it is through ideas that we manipulate the environment.

Mead actually anticipated Cassirer's argument. According to Mead, language came about as the chief survival mechanism for humans. We use it to pragmatically control our environment, and this sign system comes to stand in the place of physical reality. As we've seen, animals relate directly to the environment; they receive sensory input and react. Humans, on the other hand, generally need to decide what the input *means* before acting. Distinctly human action, then, is based on the world existing symbolically rather than physically. In order to talk about this issue, Mead uses the ideas of natural signs and significant gestures.

A sign is something that stands for something else, such as your GPA that can represent your cumulative work at the university. It appears that many animals can use signs as well. My dog Gypsy, for example, gets very excited and begins to salivate at the sound of her treat box being opened or the tone of my voice when I ask, "Wanna trrrreeeeet?" But the ability of animals to use signs varies. For instance, a dog and a chicken will respond differently to the presence of a feed bowl on the other side of a fence. The chicken will simply pace back and forth in front of the fence in aggravation, but the dog will seek a break in the fence, go through the break, and run back to the bowl and eat. The chicken appears to only be able to respond directly to one stimulus, where the dog is able to hold her response to the food at bay while seeking an alternative. This ability to hold responses at bay is important for higher-level thinking animals.

These signs that we've been talking about may be called *natural signs.* They are private and learned through the individual experience of each animal. So, if your dog also gets excited at the sound of the treat box, it is because of its individual experience with it—Gypsy didn't tell your dog about the treat box. There also tends to be a natural relationship between the sign and its object (sound/treat), and these signs occur apart from the agency of the animal. In other words, Gypsy did not make the association between the sound of the box and her treats; I did. So, in the absolute sense, the relationship between the sound and the treat isn't a true natural sign. Natural signs come out of the natural experiences of the animal, and the meaning of these signs is determined by a structured relationship between the sign and its object, like smoke and fire.

Humans, on the other hand, have the ability to use what Mead calls *significant gestures* or symbols. According to Mead, other animals besides humans have gestures but none have *significant* gestures. A gesture becomes significant when the idea behind the gesture arouses the same response (same idea or emotional attitude) in the self as in others. For example, if I asked about your weekend, you would use a variety of significant gestures (language) to tell me about it. So, even though I wasn't with you over the weekend, I could experience and know about your weekend because the words call out the same response in me as in you. Thus, human language is intrinsically reflexive: the meaning of any significant gesture always calls back to the individual making the gesture.

In contrast to natural signs, symbols are abstract and arbitrary. With signs, the relationship between the sign and its referent is natural (as with smoke and fire). But the meaning of symbols can be quite abstract and completely arbitrary (in terms of naturally given relations). For example, "Sunday" is completely arbitrary and is an abstract human creation. What day of the week it is depends upon what calendar is used, and the different calendars are associated with political and religious power issues, not nature. Because symbolic meaning is not tied to any object, the meaning can change over time. For example, in the United States there have been several meanings associated with the category of "people with dark skin."

According to Mead (1934), the meaning of a significant gesture, or symbol, is its "set of organized sets of responses" (p. 71)—notice the influence of pragmatism. Symbolic meaning is not the image of a thing seen at a distance, nor does it exactly correspond to the dictionary definition; rather, the meaning of a word is the action that it calls out or elicits. For example, the meaning of a chair is the different kinds of things we can do with it. Picture a wooden object with four legs, a seat, and a slatted back. If I sit down on this object, then the meaning of it is "chair." On the other hand, if I take that same object and break it into small pieces and use it to start a fire, it's no longer a chair—it's firewood. So the meaning of an object is defined in terms of its uses, or legitimated lines of behavior.

Because the meaning—legitimated actions—and objective availability (they are objects because we can point them out as foci for interaction) of symbols are produced in social interactions, they are **social objects.** Any idea or thing can be a social object. A piece of string can be a social object, as can the self or the idea of equality. There is nothing about the thing itself that makes it a social object; an entity becomes an object to us through our interactions around it. Through interaction, we call attention to it, name it, and attach legitimate lines of behavior to it.

For example, because of certain kinds of interactions, a Coke bottle here in the United States is a specific kind of social object. But to Xi, a bushman from the Kalahari Desert (in the film *The Gods Must Be Crazy*), the Coke bottle becomes something utterly different as a result of his interactions around it. For Xi, the Coke bottle dropped from the sky—an obvious gift from the gods. But when he brought it to his village, this playful gift from the gods became a curse, because there was only one and everybody wanted it. It became a scarce resource that brought conflict. Eventually, Xi had to go on a religious quest because of this gift from the gods (to us, a Coke bottle).

Interaction and Meaning

Notice in this illustration that the meaning of the Coke bottle changed as different kinds of interactions took place. In this sense, meaning *emerges* out of interaction. As Mead (1934) says, "the logical structure of meaning . . . is to be found in the threefold relationship of gesture to adjustive response and to the resultant of the given social act" (p. 80). **Interaction** is defined as the ongoing negotiation and melding together of individual actions and meanings through three distinct steps. First, there is an initial cue given. Notice that the cue itself doesn't carry any specific meaning. Let's say you see a friend crying in the halls at school. What does it mean? It could mean lots of things. In order to determine (or more properly, create or achieve) the meaning, you have to respond to that cue: "Is everything alright?" But we still don't have meaning yet. There must be a response to your response. After the three phases (cue–response–response to response), a meaning emerges: "Nothing's wrong," your friend responds, "my boyfriend just asked me to marry him."

But we probably still aren't done, because her response will become yet another cue. Imagine walking away from your friend without saying a word after she tells you she's getting married. That would be impolite (which would actually be a response to her statement). So, what does her second cue mean? We can't tell until you respond to her cue and she responds to your response.

Notice that interactions are rarely terminal or closed off. Let's suppose you told your friend who was crying in the hall that marrying this guy was a bad idea. You saw him out with another woman last Friday night. At this point, the social object—marriage—which was a cue that caused her to cry in happiness, has become an object of anger. So the meaning that emerges is now betrayal and anger. She then takes that meaning and interacts with her fiancé. In that interaction, she presents a cue (maybe she's crying again, but it has a different meaning), and he responds, and she responds to his response, and so on. Maybe she finds out that you misread the cues that Friday night and the "other woman" was just a friend. So she comes back to you and presents a cue, ad infinitum. Keep this idea of emergent meaning in mind as we see what Mead says about the self (it, too, emerges).

Concepts and Theory: Living Outside the Moment

Have you ever watched someone doing something? Of course you have. Maybe you watched a worker planting a tree on campus, or maybe you watched a band play last Friday night. And while you watched, you understood people and their behaviors in terms of the identity they claimed and the roles they played. In short, when you watch someone, you understand the person as a social object. After watching someone, have you ever called someone else's attention to that actor? Of course you have, and it's easy to do. All you have to say is something like, "Whoa, check him/her/it out." And the other person will look and usually understand immediately what it is you are pointing out, because we understand one another in terms of being social objects.

People-watching is a pretty common experience and we all do it. We can do it because we understand the other in terms of being a social object. But let me ask you something. Have you ever watched yourself? Have you ever felt embarrassed or laughed at yourself? How is that different from watching other people? Actually, it isn't. But there is something decidedly odd about this idea of watching our self. It's easy to watch someone else, and it is easy to understand *how* we watch someone else. If I am watching a band play, I can watch the band because they are on stage and I am in the audience. We can observe the other because we are standing outside of them. We can point to them because they are *there* in the world around us. But how can we point to our self, call our own attention to our self, and understand our self as a social object? Do you see the problem? We must somehow *divorce our self from our self* so that we can call attention to our self, so that we can understand our self as meaningfully relevant as a social object. So, how is that done? "How can an individual get outside himself (experientially) in such a way as to become an object to himself? This is the essential psychological problem of selfhood or of self-consciousness" (Mead, 1934, p. 138).

Creating a Self

Role-taking is the key mechanism through which people develop a self and the capacity to be social, and it has a very specific definition: *Role-taking* is the process through which we place our self in the position (or role) of another in order to see our own self. Students often confuse role-taking with what might be called role-*making*. In every social situation, we make a role for ourselves. Erving Goffman wrote at length about this process and called it impression management (see Chapter 9). Role-taking is a precursor to effective role-making—we put ourselves in the position of the other in order to see how they want us to act. For example, when going to a job interview, you put yourself in the position or role of the interviewer in order to see how she or he will view you—you then dress or act in the "appropriate" manner. But role-taking is distinct from impression management, and it is the major mechanism through which we are able to form a perspective outside of ourselves.

A perspective is always a meaning-creating position. We stand in a particular point of view and attribute meaning to something. Let's take the flag of the United States, for example. To some, it means freedom and pride; to others, oppression and shame; and to still others, it signifies the devil incarnate (as for some fundamentalist sects). I want you to notice something very important here: The flag itself has no meaning. Its meaning comes from the perspective an individual takes when he or she views it. That's why something like the flag (or gender or skin color or ethnic heritage) can mean so many different things. Meaning isn't in the object; *meaning arises from the perspective we take and from our interactions.*

Here's the important point: The self is just such a perspective. It is a viewpoint from which to consider our behaviors and give them meaning, and by definition, a perspective is something other than the object. In this case, the self is the perspective and the object is our actions, feelings, or thoughts. Taking this perspective, the self is how all these personal qualities and behaviors become meaningful social objects. Precisely how we can get outside ourselves in this way is Mead's driving question.

The Mind

Before the self begins to form, there is a preparatory stage. While Mead does not explicitly name this stage, he implies it in several writings and talks about it more generally in this theory of the mind. For Mead, the *mind* is not something that resides in the physical brain or in the nervous system, nor is it something that is unavailable for sociological investigation. The mind is a kind of behavior, according to Mead, that involves at least five different abilities. It has the ability

- to use symbols to denote objects
- to use symbols as its own stimulus (it can talk to itself)
- to read and interpret another's gestures and use them as further stimuli
- to suspend response (not act out of impulse)
- to imaginatively rehearse one's own behaviors before actually behaving

Let me give you an example that encompasses all these behaviors. A few years ago, our school paper ran a cartoon. In it was a picture of three people: a man and a woman arm-in-arm, and another man. The woman was introducing the men to one another. Both men were reaching out to shake one another's hands. But above the single man was a balloon of his thoughts. In it he was picturing himself violently punching the other man. He wanted to hit the man, but he shook his hand instead and said, "Glad to meet you."

There are a lot of things we can pull out of this cartoon, but the issue we want to focus on is the disparity between what the man felt and what the man did. He had an impulse to hit the other man, perhaps because he was jealous. But he didn't. Why didn't he? Actually, that isn't as good a question as, *how* didn't he? He was able to not hit the other man because of his mind. His mind was able to block his initial impulse, to understand the situation symbolically, to point out to his self the symbols and possible meanings, to entertain alternative lines of behavior, and choose the behavior that best fit the situation. The man used symbols to stimulate his own behavior rather than going with his impulse or the actual world.

Mead (1934) argues that the "mind arises in the social process only when that process as a whole enters into, or is present in, the experience of any one of the given individuals involved in that process" (p. 134). Notice that Mead is arguing that the mind evolves as the social process—or, more precisely, the social interaction—comes to live inside the individual. The mind, then, is a social entity that begins to form because of infant dependency and forced interaction.

When babies are hungry or tired or wet, they can't take care of themselves. Instead, they send out what Mead would call "unconventional gestures," gestures that do not mean the same to the sender and hearer. In other words, they cry. The caregivers must figure out what the baby needs. When they do, parents tend to vocalize their behaviors ("Oh, did Susie need a ba-ba?"). Babies eventually discover that if they mimic the parents and send out a significant gesture ("ba-ba"), they will get their needs met sooner. This is the beginning of language acquisition; babies begin to understand that their environment is symbolic—the object that satisfies hunger is "ba-ba" and the object that brings it is "da-da." Eventually, a baby will understand that she has a symbol as well: "Susie." Thus, language acquisition allows the child to symbolize and eventually to symbolically manipulate her environment, including self and others. The use of reflexive language also allows the child to begin to role-take, which is the primary mechanism through which the mind and self are formed.

The Play Stage

The second stage in the process of self-formation is the *play stage*. During this stage, the child can take the role, or assume the perspective, of certain significant others. Significant others are those upon whom we depend for emotional and often material support. These are the people with whom we have long-term relations and intimate (self-revealing) ties. Mead calls this stage the play stage because children must literally play at being some significant other in order to see themselves. At this point, they haven't progressed much in terms of being able to think abstractly, so they must act out the role to get the perspective. This is important: *a child literally gets outside of him- or herself in order to see the self.*

Children play at being Mommy or being Teacher. The child will hold a doll or stuffed bear and talk to it as if she were the parent. Ask any parent; it's a frightening experience because what you are faced with is an almost exact imitation of your own behaviors, words, and even tone of voice. But remember the purpose of role-taking (notice this isn't role-*playing*): it is to see one's own self. So, as the child is playing Mommy or Daddy with a teddy bear, who is the bear? The child herself. She is seeing herself from the point of view of the parent, literally. This is the genesis of the self perspective: being able to get outside of the self so that we can watch the self as if on stage. As the child acts toward herself as others act, the child begins to understand self as a set of organized responses and becomes a social object to herself.

The Game Stage

The next stage in the development of self is the *game stage*. During this stage, the child can take the perspective of several others and can take into account the rules (sets of responses that different attitudes bring out) of society. But the role-taking at this stage is still not very abstract. In the play stage, the child could only take the perspective of a single significant other; in the game stage, the child can take on the role of several others, but they all remain individuals. Mead's example is that of a baseball game. The batter can role-take with each individual player in the field and determine how to bat based on their behaviors. The batter is also aware of all the rules of the game. Children at this stage can role-take with several people and are very concerned with social rules. But they still don't have a fully formed self. That doesn't happen until they can take the perspective of the generalized other: "it is this generalized other in his experience which provides him with a self" (Mead, 1925, p. 269).

The Generalized Other Stage

The **generalized other** refers to sets of attitudes that an individual may take toward him- or herself—it is the general attitude or perspective of a community. The generalized other allows the individual to have a less segmented self as the perspectives of many others are generalized into a single view. It is through the generalized other that the community exercises control over the conduct of its individual members.

Up until this point, the child has only been able to role-take with specific others. As the individual progresses in the ability to use abstract language and concepts, she or he is also able to think about general or abstract others. So, for example, a woman may look in the mirror and judge the reflection by the general image that has been given to her by the media about how a woman should look.

Another insightful example of how the generalized other works is given to us by George Orwell in his account of "Shooting an Elephant." Orwell was at the time a police officer in Burma. He was at odds with the job and felt that imperialism was an evil thing. At the same time, the local populace despised him precisely because he represented imperialistic control; he tells tales of being tripped and ridiculed by people in the town. One day an elephant was reported stomping through a village. Orwell was called to attend to it. On the way there, he obtained a rifle, only for scaring the animal or defending himself if need be. When he found the animal, he knew immediately that there was no longer any danger. The elephant was calmly eating grass in a field, and Orwell knew that the right thing to do was to wait until the elephant simply wandered off—but at the same time, he knew he had to shoot the animal.

As Orwell (1946) relates, "I realized that I should have to shoot the elephant after all. The people expected it of me and I had got to do it; I could feel their two thousand wills press me forward, irresistibly" (p. 152). He felt the expectations of a generalized other. Though contrary to his own will, and after much personal anguish, he shot the elephant. As Orwell puts it, at that time and in that place, the white man "wears a mask, and his face grows to fit it. . . . A sahib has got to act like a sahib" (pp. 152–153). Not shooting was impossible, for "the crowd would laugh at me. And my whole life, every white man's life in the East, was one long struggle not to be laughed at" (p. 153). We may criticize Orwell for his decision (it's always easy from a distance); still, each one of us has felt the pressure of a generalized other.

Society and the Self

For Mead (1934), there could be no society without individual selves: "Human society as we know it could not exist without minds and selves, since all its most characteristic features presuppose the possession of minds and selves by its individual members" (p. 227). We don't have a self because there is a psychological drive or need for one. We have a self because society demands it. To emphasize this point, Mead (1934) says that "the self is not something that exists first and then enters into relationship with others, but it is, so to speak, an eddy in the social current and so still a part of the current" (p. 26). Eddies are currents of air or water that run contrary to the stream. It isn't so much the contrariness that Mead wants us to see, but the fact that an eddy only exists in and because of its surrounding current. The same is true for selves: they only exist in and because of social interaction. The self doesn't have a continuous existence; it isn't something that we carry around inside of us. It's a mechanism that allows conversations to happen, whether that conversation occurs in the interaction or within the individual.

So, the self isn't something that has an essential existence or meaning. Like all social objects, it must be symbolically denoted and then given meaning within interactions. And like all social-symbolic objects, the meaning of the self is flexible and emergent. As an example, let me tell you about a conversation I had not too long ago. My brother-in-law (Jim) and my sister (Susan) both told me about an incident in back-to-back telephone conversations. Jim is a distance runner and was training for the Pike's Peak run. During one of his training sessions at the Peak, he experienced heart fibrillation on the way down. He didn't tell Susan at that time, and he ran the race a week later without a problem.

After the race, he told Susan about the training incident over dinner and margaritas. Susan got upset. She then told me about it. She said that Jim is too much into machismo posturing and doesn't deal with reality (her definition of Jim's self). The "reality" that Susan referred to is that he is 65 years old and had previously experienced a five-hour fibrillation problem at the doctor's office. In this interaction with me, Susan referred to herself as a "caregiver and an organizer." That's her self as she sees it vis-à-vis Jim. Jim told the same story, as far as the actual events are concerned. But the meaning of all the events changed, as did the definitions of the selves involved. Jim defined the "condition" as not life threatening, as one that is normal for athletes, and he said that lots of doctors say it's okay for an athlete in his condition to continue training. Jim defined himself as a "competitor and an optimist" and Susan as a "worrier that mothers too much."

The important thing to see in this example is not that every individual has an interpretation of an event—that would be more psychological than sociological. What's important to note is what is happening socially: these two people are negotiating with a third party over the meaning of an event as well as the kinds of selves that that event indicates they have. Out of this interaction and negotiation emerge a definition and a sense of self for each of the participants. The process through which that occurs is the focus of a Meadian way of perceiving the social world.

This emphasis of Mead's (1934) leads him to see society and social institutions as "nothing but an organization of attitudes which we all carry in us"; they are "organized forms of group or social activity—forms so organized that the individual members of society can act adequately and socially by taking the attitudes of others toward these activities" (pp. 211, 261–262). Society, then, doesn't exist objectively outside the concrete interactions of people, as Durkheim or Marx would have it. Rather, society exists only as sets of attitudes, symbols, and imaginations that people may or may not use and modify in an interaction. In other words, society exists only as sets of potential generalized others with which we can role-take.

The I and the Me

Thus far it would appear that the self is to be conceived of as a simple reflection of the society around it. But for Mead, the self isn't merely this social robot; the self is an *active process*. Part of what we mean by the self is an internalized conversation, and by necessity interactions require more than one person. Mead thus postulates the existence of two interactive facets of the self: the "I" and the "Me." The Me is the self that results from the progressive stages of role-taking and is the perspective that we assume to view and analyze our own behaviors. The "I" is that part of the self that is unsocialized and spontaneous: "The self is essentially a social process going on with these two distinguishable phases. If it did not have these two phases, there could not be conscious responsibility and there would be nothing novel in experience" (Mead, 1934, p. 178).

We have all experienced the internal conversation between the I and the Me. We may want to jump for joy or shout in anger or punch someone we're angry at or kiss a stranger or run naked. But the Me opposes such behavior and points out the social ramifications of these actions. The I presents our impulses and drives; the Me presents to us the perspectives of society, the meanings and repercussions of our actions. These two elements of our self converse until we decide on a course of action. But here is the important part: The I can always act before the conversation begins or even in the middle of it. The I can thus take action that the Me would never think of; it can act differently from the community.

Mead Summary

- There are basic elements, or tools, that go into making us human. Among the most important of these are symbolic meaning and the mind. We use symbols and social objects to denote and manipulate the environment. Each symbol or social object is understood in terms of legitimated behaviors and pragmatic motives. The mind uses symbolic-social objects in order to block initial responses and consider alternative lines of behavior. It is thus necessary in order for society to exist. The mind is formed in childhood through necessary social interaction.

- The self is a perspective from which to view our own behaviors. This perspective is formed through successive stages of role-taking and becomes a social object for our own thoughts. The self has a dynamic quality as well—it is the internalized conversation between the I and the Me. The Me is the social object, and the I is the seat of the impulses. When the self is able to role-take with generalized others, society can exist as well as an integrated self. Role-taking with generalized others also allows us to think in abstract terms.

- Society emerges through social interaction; it is not a determinative structure. In general, humans act more than react. Action is predicated on the ability of the mind to delay response and consider alternative lines of behavior with respect to the social environment and a pertinent self. Thus, mind, self, and society mutually constitute one another. Interaction is the process of knitting together different lines of action. Meaning is produced in interaction through the triadic relation of cue, response, and response to response. What we mean by society emerges from this negotiated meaning as interactants role-take within specific definitions of the situation and organized attitudes (institutions).

Georg Simmel

Photo: © Granger Collection.

The Essential Simmel

Biography

Georg Simmel was born in the heart of Berlin on March 1, 1858, the youngest of seven children. His father was a Jewish businessman who had converted to Catholicism before Georg was born. In 1876, Simmel began his studies (history, philosophy, psychology) at the University of Berlin, taking some of the same courses and professors as Max Weber would a few years later. In 1885, Simmel became an unpaid lecturer at the University of Berlin, where he was dependent upon student fees. At Berlin, he taught philosophy and ethics, as well as some of the first courses ever offered in sociology. In all probability, George Herbert Mead was one of the foreign students in attendance. Though Simmel wrote many sociological essays and articles, his most important work of sociology was published in 1900, *The Philosophy of Money.* All together, Simmel published 31 books and several hundred essays and articles. In 1910, he, along with Max Weber and Ferdinand Tönnies, founded the German Society for Sociology. In 1914, Simmel was offered a full-time academic position at the University of Strasbourg. However, as WWI broke out, the school buildings were given over to military uses and Simmel had little lecturing to do. On September 28, 1918, Simmel died of liver cancer.

Passionate Curiosity

Many of Simmel's writings and public lectures addressed what were then considered nonacademic issues, such as love, flirtation, scent, and women. His diverse subject matter was due not only to his position as an outsider but also to his intellectual bent—he was extremely interested in cultural trends and new intellectual movements. He was little involved in politics, and quite involved in the avant-garde philosophies of the day. Overall, Simmel was intensely interested in the ways in which modern, objective culture impacts the individual's subjective experiences.

Keys to Knowing

Social forms, subjective and objective cultures, urbanization, division of labor, web of group affiliations, money, organic motivation, rational motivation, role conflict, blasé attitude

Simmel is particularly significant for what he adds to Mead's theory of the self. As we've seen, Mead's theory sees the self as a perspective that comes out of interactions, and he sees the meanings of symbols, social objects, and the self as emerging from negotiated interactions. In general, Simmel would not take issue with Mead's analysis; but he does add a caveat. For Simmel, cultural entities—such as social forms, symbols, and selves—can exist subjectively and under the influence of people in interaction, just as Mead says. However, Simmel also entertains the possibility that culture can exist objectively and independent of the person and interaction. It's the influence of this objective culture on the person that interests Simmel.

Simmel's Perspective:
The Individual and Social Forms

At the core of Simmel's thought is the individual. In contrast to Mead, Simmel assumes there is something called human nature with which we are born. For example, Simmel feels that we naturally have a religious impulse and that gender differences are intrinsic. He also assumes that in back of most of our social interactions are individual motivations. This emphasis sets up an interesting problem and perspective for Simmel. If the individual and his or her motivations and actions are paramount, then how is society possible?

In formulating his answer, Simmel follows one of his favorite philosophers, Immanuel Kant. Kant did not ask about the possibility of society. Instead, he wondered how nature could exist as the *object* "nature" to science. Basically, Kant argued that the universe could exist as "nature" to scientists only because of the *category* of nature. Objects in the universe can only *exist* as objects because the human mind orders sense perception in a particular way. But Kant didn't argue that it is all in our heads; rather, he argued for a kind of synthesis: the human mind organizes our perceptions of the world to form objects of experience. So, nature can only exist as the object "nature" because scientists are observing the world through the a priori (existing before) category of nature. In other words, a scientist can see weather as a natural phenomenon, produced through processes that we can discover, only because she assumes beforehand (a priori) that weather does *not* exist as a result of the whim of a god.

Simmel wants to discover the a priori conditions for society. This was a new way of trying to understand society. During Simmel's time, there were two main theories about society: mechanical-atomistic and organic. The mechanical-atomistic approach argues that individuals are the only reality: individuals are self-sustaining and independent and society is simply the summation of their activities. This approach is like Mead and very similar to many exchange or rational choice theories today (see Chapter 10). The organismic approach sees society as an independent entity, distinct from and sometimes subjugating of the individual. Marx and Durkheim are two examples of theorists with this perspective.

Simmel, however, wants to maintain the integrity of the individual but at the same time recognize society as a true force. What Simmel argues is that society exists as **social forms** that come about through human interaction, and society continues to exist and to exert influence over the individual through these forms of interaction: "All the various ways in which man lives by his actions, knowledge, feelings, and creativity might be regarded as types or categories that are *imposed on existence*" (Simmel, 1997, p. 139, emphasis added). These forms or categories of behavior, Simmel argues, are the a priori conditions of society.

Because of the focus of this chapter, we aren't going to look at any specific social forms. We will, however, review two of Simmel's social forms later in the book. In Chapter 7 we will explore Simmel's influence on contemporary conflict theory, and in Chapter 10 we'll talk about his impact on exchange theory; both conflict and exchange are two of Simmel's social forms. For now, what we want to see is that the existence of social forms implies a problem: social forms may come to exist apart from the individual; they may take on an objective existence.

Now, think about this: If Simmel is primarily concerned about the individual, what are the implications for the person if social forms take on objective existence? There is a sense in which Mead doesn't see symbols as objective. If the meaning of symbols emerges through social interaction, then they are always subjective, at least to some degree. What Simmel wants us to consider is the possibility that signs, symbols, ideas, social forms, and so forth can exist independently of the person and exert independent effects. The question then becomes, how does objective culture impact the subjectivity of the person? Guy Oakes (1984) wrote concerning Simmel, "the discovery of objectivity—the independence of things from the conditions of their subjective or psychological genesis—was the greatest achievement in the cultural history of the West" (p. 3).

Concepts and Theory: Subjective and Objective Cultures

Simmel was the first social thinker to make the distinction between subjective and objective culture the focus of his research. Individual or *subjective culture* refers to the ability to embrace, use, and feel culture. Collectives can form group-specific cultures, such as the spiked Mohawk haircut of early punk culture. To wear such an item of culture immediately links the individual to certain social forms and types, and a group member would subjectively feel those links. Individuals and dyads are able to produce such culture as well. An individual could have special incense that she or he blends just for extraordinary, ritual occasions; or a couple could create a picture that would symbolize their relationship. This culture is very close to the individual and her or his psychological experience of the world.

Objective culture is made up of elements that become separated from the individual or group's control and reified as separate objects. Think about tie-dye T-shirts, for example. You can now go to any department store and buy such a shirt. You do not have to be a hippie to wear it, nor are you necessarily identified as a hippie, nor do you necessarily feel the connection to the values and norms of the hippie culture. It exists as an object separate from the individuals who produced it in the first place. Once formed, objective culture can take on a life of its own and it can exert a coercive force over individuals. For example, many of us growing up in the United States believe in the ideology and morality of democracy, though in truth we are far removed from its crucial issues, ideals, and practices.

The following diagram (Figure 4.1) pictures the relationship between subjective and objective culture. What we see on the far left is probably what Simmel has in mind under ideal conditions. People need culture to interact with others, and in small, traditional communities the culture can be kept graspable and thus subjective. The double-headed arrows indicate reciprocal relations—subjective culture influences and is affected by people and interactions. Objective culture, on the other hand, stands apart from the individual psychology. Notice that culture becomes more objective as interactions are extended to distant others and as certain features of modernity become more prominent. When this happens, a lag is produced between the individual and objective cultures. As the size and complexity of the objective culture increases, it becomes more and more difficult for individuals to embrace it as a whole. Individuals come to experience culture sporadically and in fragments. How individuals respond to this tension between subjective experience and culture is of utmost concern to Simmel.

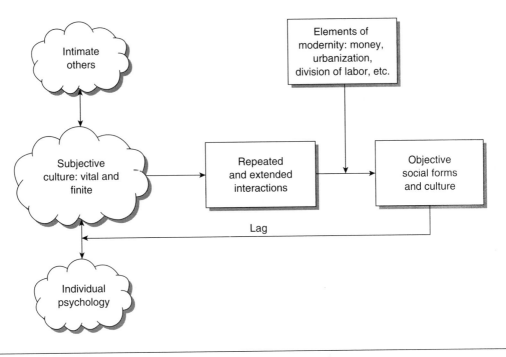

Figure 4.1   Subjective and Objective Culture

Simmel identifies three general variables of objective culture. As any of these variables increases, culture becomes more objective and less subjectively available to the individual. First, culture can vary in its *absolute size*. The pure bulk of cultural material can increase or decrease. In modernity, the amount of objective culture increases continuously. For example, in the year 2000, the world produced approximately 1,200 terabytes of scanned printed material. A terabyte contains over 1 trillion bytes. If we were to make a single book that contained just 1 billion characters, it would be almost 32 miles thick. A trillion is 1,000 billion. It has been estimated that to count to 1 trillion would take over 190,000 years, if we counted 24/7/365. The human race created over 1,200 *trillion* bytes of printed information in 2000. That's not counting the Internet. And that figure increases by 2 to 10% each year.

Culture can also vary by its *diversity of components*. Let's take fashion, for example. Not only are there simply more fashion items available (absolute size), there are also more fashion types or styles available—there are fashions for hip-hop, grunge, skater, hardcore, preppy, glam, raver, piercer, and so on, ad infinitum. Finally, culture can vary by its *complexity*. Different cultural elements can either be linked or unlinked. If different elements become linked, then the overall complexity of the objective culture increases. For example, when this nation first started, there were only a few different kinds of religions (a couple of different Protestant denominations and Catholicism). Due to various social factors, the objective culture of religion has increased in its size and diversity, resulting in any number of different kinds of religion in America today. The culture of religion has also become more complex, especially in the last few years. Today we find people who are joining together in what was previously thought to be antithetical forms of religion. So, for example, we can find Christian-Pagans in North Carolina. As these different forms become linked together, the religious culture becomes increasingly more complex.

For Simmel, cultural forms are necessary to achieve goals in a social setting. However, if these forms become detached from the lived life of the individual, they present a potential problem for the subjective experience of that individual. In an ideal world, there is an intimate connection between the personal experience of the individual and the culture that he or she uses. However, as the gap between the individual and culture increases, and as culture becomes more objective, culture begins to attain an autonomy that is set against the creative forces of the individual.

Concepts and Theory: The Self in Modernity

There are three interrelated forces in modernity that tend to increase objective culture in all three of its areas—urbanization, money, and the configuration of one's social network. Urbanization appears to be the principal dynamic. It increases the level of the division of labor and the extent that money and markets are used, and it changes one's web of affiliations from a dense, primary network to a loose, secondary one. As is typical with Simmel, we will find that these social processes bring some conflicting effects.

Urbanization

Simmel's (1950) concern with objective culture is nowhere clearer than in his short paper, "The Metropolis and Mental Life": "The most profound reason . . . why the metropolis conduces to the urge of the most individual personal existence . . . appears to me to be the following: the development of modern culture is characterized by the preponderance of what one may call the 'objective spirit' over the 'subjective spirit'" (p. 421). This increase in objective spirit happens principally because of two interrelated dynamics: the division of labor and the use of money, both of which are spurred on through **urbanization**—the process that moves people from country to city living.

Historically, people generally moved from the country to the city because of industrialization. As a result of the Industrial Revolution, the economic base of society changed and with it the means through which people made a living. As populations became increasingly concentrated in one place, more efficient means of providing for the necessities of life and for organizing labor were needed. This increase in the division of labor happened so that products could be made more quickly and the workforce could be more readily controlled. Simmel (1950) argues that the division of labor also increases because of worker-entrepreneur innovation: "The concentration of individuals and their struggle for customers compel the individual to specialize in a function from which he cannot be readily displaced by another" (p. 420).

The division of labor demands an "ever more one-sided accomplishment," and we thus become specialized and concerned with smaller and smaller elements of the production process. This one-sidedness creates objective culture: we are unable to grasp the whole of the product and the production process because we are only working on a small part. The worker in a highly specialized division of labor becomes "a mere cog in an enormous organization of things and powers which tear from his hands all progress, spirituality, and value in order to transform them from their subjective form into the form of purely objective life" (Simmel, 1950, p. 422).

Simmel claims that the consumption of products thus produced has an individuating and trivializing effect. It isn't difficult for us to see that we perform and create self out of things. I just recently bought a house and have spent a good bit of time decorating it. When I have friends over to see the new place, I glow when they say, "this place just fits you." In the same way, we are all particular about the clothes we wear, the car we drive, the CDs we possess, the sound system we own, the perfume we use, and so on. Notice a difference here between Simmel and Marx: for Simmel, there isn't anything intrinsically wrong with using commodities to create our self-concept and self-image.

On the other hand, Simmel also sees that the products we are using in modernity to produce and express the self are changing. They are becoming more and more divorced from subjectivity due to the division of labor and market economy: there are too many of them and they are too easily replaced (it is thus difficult to get attached to them); they have become trivialized in order to meet the demands of mass markets (they thus have very little meaning in the first place); and products are subject to the dynamics of fashion and diversification of markets, which leads to inappropriate sign use (for example, dressing like a rapper or a member of the upper class when you aren't a member of those groups).

Urbanization also increases the level of exchange in a society and thus the use of money-facilitated markets. The use of money increases the level of objective culture due to increasing demands for rational calculations. It also facilitates the processes we've just been talking about: the division of labor and trivialization of culture. In addition, we will see that money also has positive effects for the individual and society.

Money

As we saw in Chapter 1, money creates a universal value system wherein every commodity can be understood. Of necessity, this value system is abstract; that is, it has no intrinsic worth. In order for it to stand for everything, it must have no value in itself. The universal and abstract nature of money frees it from constraint and facilitates exchanges. But it also has other effects; we will talk about four. These effects increase the more money (or even more abstract systems such as credit) is used.

One effect of money is that it increases individual freedom by allowing people to pursue diverse activities (paying to join a dance club or a pyramid sales organization) and by increasing the options for self-expression (we can buy the clothes and makeup to pass as a raver this week and a business professional the next; we can even buy hormones and surgery to become a different sex). Second, even though we are able to buy more things with which to express and experience our self, we are less attached to those things because of money. We tend to understand and experience our possessions less in terms of their intrinsic qualities and more in terms of their objective and abstract worth. So, I understand the value of my guitar amplifier in terms of the money it cost me and how difficult it would be to replace (in terms of money). The more money I have, the less valuable my Sunn amplifier will be, because I could afford a hand-wired, boutique amp. Thus, our connection to things becomes more tenuous and objective (rather than emotional) due to the use of money.

Third, money also discourages intimate ties with people. Part of this is due to the universal nature of money. Because of its all-inclusive character, money comes to stand in the place of almost everything, and this effect spreads. When money was first introduced, only certain goods and services were seen as equivalent to it. Today in the United States, we would be hard pressed to think of many things that cannot be purchased with or made equivalent to money, and that includes relationships. Much of this outcome is due to indirect consequences of money: the relationships we have are in large part determined by the school or neighborhood we can afford. Some of money's consequences are more direct: we buy our way into country clubs and exclusive organizations.

Money further discourages intimate ties by encouraging a culture of calculation. The increasing presence of calculative and objectifying culture, even though spawned in economic exchange, tends to make us calculating and objectifying in our relationships. All exchanges require a degree of calculation, even barter, but the use of money increases the number and speed of exchanges. As we participate in an increasing number of exchanges, we calculate more and we begin to understand the world more in terms of numbers and rational calculations. Money is the universal value system in modernity, and as it is used more and more to assess the world, the world becomes increasingly quantified ("time is money" and we shouldn't "waste time just 'hanging out'").

Fourth, money also decreases moral constraints and increases anomie. Money is an amoral value system. What that means is that there are no morals implied in money. Money is simply a means of exchange, a way of making exchanges go easier. Money knows no good or evil: it can be equally used to buy a gun to kill school children as to buy food to feed the poor. So, as more and more of our lives are understood in terms of money, less and less of our lives have a moral basis. In addition, because moral constraints are produced only through group interactions, when money is used to facilitate group membership, it decreases the true social nature of the group and thus its ability to produce morality.

Thus, money has both positive and negative consequences for the individual. Money increases our options for self-expression and allows us to pursue diverse activities, but it also distances us from objects and people and it increases the possibility of anomie. In the same way, there are both positive and negative consequences for society. However, while it may seem that the negative consequences outweigh the positive for the individual, the consequences for society are mostly positive. We will look at a total of four social effects of the use of money.

The use of money actually creates exchange relationships that cover greater distances and last longer periods of time than would otherwise be possible. Let's think of the employment relationship as an illustration. If a person holds a regular job, she or he has entered a kind of contract. The worker agrees to work for the employer a given number of hours per week at a certain pay rate. This agreement covers an extended period of time, which should only be terminated by a two-week notice or severance pay. This relationship may cover a great deal of geographic space, as when the workplace is located on the West Coast of the United States and the corporate headquarters is on the East Coast.

This kind of relationship was extremely difficult before the use of money as a generalized medium of exchange, which is why many long-term work relationships were conceptualized in terms of familial obligations, such as the serf or apprentice. And, with more generalized forms of the money principle, as with credit and credit cards, social relations can span even larger geographic expanses and longer periods of time (for example, I am obligated to my mortgage company for the next 25 years, and I just completed an eBay transaction with a man living in Japan). What this extension of relations through space and time means is that the number of social ties increases. While we may not be connected as deeply or emotionally today, we are connected to more diverse people more often. Think of society as a fabric: the greater the number and diversity of ties, the stronger is the weave.

Second, money also increases continuity among groups. Money flattens, or generalizes, the value system by making everything equivalent to itself. It also creates more objective culture, which overshadows or colonizes subjective culture. Together these forces tend to make group-specific culture more alike than different. What differences exist are trivial and based on shifting styles. Thus, while the weave of society is more dense due to the effects of money, it is also less colorful, which tends to mitigate group conflicts.

Third, money strengthens the level of trust in a society. What is money, really? In the United States, it is nothing but green ink and nice paper. Yet we would do and give almost anything in exchange for enough of these green pieces of paper. This exchange for relatively worthless paper occurs every day without anyone so much as blinking an eye. How can this be? The answer is that in back of money is the U.S. government, and we have a certain level of trust in its stability. Without that trust, the money would be worthless. A barter system requires some level of trust, but generally speaking, you know the person you are trading with and you can inspect the goods. Money, on the other hand, demands a trust in a very abstract social form—the state—and that trust helps bind us together as a collective.

Lastly, in back of this trust is the existence of a centralized state. In order for us to trust in money, we must trust in the authority of a single nation. When money was first introduced in Greek culture, its use was rather precarious. Different wealthy landowners or city-states would imprint their image on lumps or rods of metal. Because there was no central governing authority, deceit and counterfeiting were rampant. The images were easily mimicked and weights easily manipulated. Even though money helped facilitate exchanges, the lack of oversight dampened the effect. It wasn't until there was a centralized government that people could completely trust money. Thus, when markets started using money for exchanges, they were inadvertently pushing for the existence of a strong nation-state. Centralized authority is the structural component to a society's trust and it binds us together.

Social Networks

As a further result of urbanization, Simmel argues that social networks (what he calls the **web of group affiliations**) have changed. When we talk about social networks today, what we have in mind are the number and type of people with whom we associate, and the connections among and between those people. Network theory is an established part of contemporary sociology, and Simmel was one of the first to think in such terms. For his part, Simmel is mostly concerned with why people join groups. While this issue may seem minor, there are in fact important ramifications.

In small rural settings, there are relatively few groups for people to join, and most of those memberships are strongly influenced by family. We tend to join the same groups as members of our family do. In these social settings, the family is a primary structure for social organization, and families tend not to move around much. So there are likely to be multiple generations present. As a result, the associations of the family become the associations of the child. A child reared in such surroundings will generally attend the same church, school, and work as his or her parents, grandparents, cousins, and so on. Further, most of these groups will overlap. For example, it would be very likely that a worker and her or his boss attend the same church and that they will have gone to the same school.

Simmel notes that people in these settings tend to join groups because of *organic motivations*—because they are naturally or organically connected to the group. Many of the groups with which a person affiliates in this setting are primary groups. **Primary groups** are noteworthy because they are based on ties of affection and personal loyalty, endure over long periods of time, and involve multiple aspects of a person's life. Under organic conditions, a person will usually be involved with mostly primary groups, and these groups have some association with one another. This kind of community will thus contain people who are very much alike. They will draw from the same basic group influences and culture, and the groups will possess a compelling ability to sanction behavior and bring about conformity.

On the other hand, people join groups in modern, urban settings out of *rational motivations*—group membership due to freedom of choice. The interesting thing to note about this freedom is that it is forced on the individual—in other words, there are few organic connections. In large cities, people usually do not have much family around and the personal connections tend to be rather tenuous. In Southern

California, for example, people move on average every five to seven years. Most only know their neighbors by sight, and the majority of interactions are work related (and people change jobs about as often as they change houses). What this means is that people join social groups out of choice (rational reasons) rather than out of some emotional and organic connectedness. These kinds of groups tend to have the characteristics of **secondary groups** (goal and utilitarian oriented, with a narrow range of activities, over limited time spans).

As a result of rational group affiliations, it is far more likely that individuals will develop unique personalities. A person in a more complex or rational web of group affiliations has multiple and diverse influences and groups' capacity to sanction is diminished. From Simmel's point of view, the group's ability to sanction is based on the individual's dependency upon the group. If there are few groups from which to choose, then individuals in a collective are more dependent upon those groups and the groups will be able to demand conformity. This power is crystal clear in traditional societies where being ostracized meant death. Of course, the inverse is also true: the greater the number of groups from which to choose and the more diverse the groups, the less the moral boundaries and **normative specificity** (the level of behaviors that are guided by norms). In turn, this decrease in sanctioning power leads to greater individual freedom of expression.

Many students, for example, are able to express themselves more freely after moving away from home to the university. This is especially true of students who move from rural to urban settings. Not only is the influence of the student's childhood groups diminished (family, peers, church), but there are also many, many more groups from which to choose. These groups often have little to do with one another. Thus, if one group becomes too demanding of time or emotion or behavior, you can simply switch groups. So, it may be the case that you experience yourself as a unique individual having choices, but it has little to do with you per se: it is a function of the structure of your network.

While decreased moral boundaries and normative specificity lead to greater freedom of expression, they can also produce anomie—also a concern of Durkheim's. For the individual, it speaks of a condition of confusion and meaninglessness. Unlike animals, humans are not instinctually driven or regulated. We can choose our behaviors. That also means that our emotions, thoughts, and behaviors must be ordered by group culture and social structure or they will be in chaos and will have little meaning. When group regulation is diminished or gone, it is easy for people to become confused and chaotic in their thoughts and emotions. Things in our life and life itself can become meaningless. So while we may think that personal freedom is a great idea, too much freedom can be disastrous.

Complex webs of group affiliations can have two more consequences: they can increase the level of role conflict a person experiences, and they contribute to the blasé attitude. **Role conflict** describes a situation in which the demands of two or more of the roles a person occupies clash with one another (such as when your friends want to go out on Thursday night but you have a test the next morning). The greater the number of groups with which one affiliates, the greater is the number of divergent roles and the possibility of role conflict. However, the tendency to keep groups spatially and temporally separate mitigates this potential. In other words, modern groups tend not to have the same members and they tend to gather at different times and locations. So we see the roles as separate and thus not in conflict.

Complex group structures also contribute to the **blasé attitude,** an attitude of absolute boredom and lack of concern. Every social group we belong to demands emotional work or commitment, but we only have limited emotional resources, and we can only give so much and care so much. There is, then, a kind of inverse relationship between our capacity to emotionally invest in our groups and the number of different groups of which we are members. As the number and diversity of social groups in our lives goes up, our ability to emotionally invest goes down. This contributes to a blasé attitude, but it also makes conflict among groups less likely because the members care less about the groups' goals and standards.

This blasé attitude is also produced by all that we have talked about so far, as well as overstimulation and rapid change. The city itself provides for multiple stimuli. As we walk down the street, we are faced with diverse people and circumstances that we must take in and evaluate and react to. In our pursuit of individuality, we also increase the level of stimulation in our lives. As we go from one group to another, from one concert or movie to another, from one mall to another, from one style of dress to another, or as we simply watch TV or listen to music, we are bombarding ourselves with emotional and intellectual stimulation. In the final analysis, all this stimulation proves to be too much for us and we emotionally withdraw.

Further, this stimulation is in constant flux. Knowledge and culture are constantly changing. We could point to a variety of changes in medicine or style or "common knowledge" over the past few years, but I think that one of the most poignant examples of this constant change is the MTV show, *So 5 Minutes Ago.* It is a show that looks at the changing culture of youth. The very existence of the show acknowledges and chronicles what Simmel is talking about for a specific segment of the culture. (I would invite you to check it out, but by the time you read this, the show itself will probably be "5 minutes ago.")

All that we have talked about in this section so far creates an interesting social-psychological need. Drowning in a sea of blasé and objectified culture, people feel they must exaggerate any differences that do exist in order to stand out and experience our personal selves. Thus, modernity increases our freedom of expression, but it also forces us to express it more dramatically with trivialized culture:

> On the one hand, life is made infinitely easy for the personality in that stimulations, interests, uses of time and consciousness are offered to it from all sides. . . . On the other hand, however, life is composed more and more of these impersonal contents and offerings that tend to displace the genuine personal colorations and incomparabilities. This results in the individual's summoning the utmost in uniqueness and particularization, in order to preserve his most personal cores. He has to exaggerate this personal element in order to remain audible even to himself. (Simmel, 1950, p. 422)

We have covered a great deal of conceptual ground in this section. It's been made all the more complicated because each of the things we have talked about brings both functional and dysfunctional effects, and the effects overlap and mutually reinforce one another. Normally, I would be tempted to draw all this out in a theoretical model; however, it would be way too complex for our purposes. Instead, I will finish this section with a list of the different effects (see Table 4.1). This list doesn't explain the theoretical dynamics, but it will provide you with a general and clear idea about how urbanization is influencing our lives. If in going over this list you find that you can't recall the theoretical explanation about why the effect is occurring, please go back and reread the text.

In general, then, modernity is characterized by a high level of urbanization. Urbanization brings with it three primary outcomes: increases in the division of labor, the use of money and markets, and rational group membership. Each of these has the following effects (note that ↑ denotes an increasing and ↓ a decreasing tendency):

Table 4.1   Effects of Urbanization

The Division of Labor	↑ specialized culture
	↑ production of trivialized products
Money Effects—Individual	↑ individual freedom
	↓ attachment to products
	↓ intimate ties with people
	↑ anomie
	↑ goal displacement
Money Effects—Societal	↑ number of social relationships
	↑ continuity between groups
	↑ social trust
	↑ centralization of power
Rational Social Networks	↑ unique personality
	↑ anomie
	↑ role conflict
	↑ blasé attitude

Simmel Summary

- There are two central ideas that form Simmel's perspective: social forms and the relationship between the subjective experience of the individual and objective culture. Simmel always begins and ends with the individual. He assumes that the individual is born with certain ways of thinking and feeling, and most social interactions are motivated by individual needs and desires. Encounters with others are molded to social forms in order to facilitate reciprocal exchanges. These forms constitute society for Simmel. Objective culture is one that is universal yet not entirely available to the individual's subjective experience. Thus, the person is unable to fully grasp, comprehend, or intimately know objective culture. The tension between the individual on the one hand and social forms and objective culture on the other is Simmel's focus of study.

- Urbanization increases the division of labor and the use of money, and it changes the configuration of social networks. All of these have both direct and indirect influence on the level of objective culture and its effects on the individual. The use of money increases personal freedom for the individual, yet at the same time it intensifies the possibility of anomie, diminishes the individual's attachment to objects, and increases goal displacement. People join groups based on either rational or organic motivations. Rational motivations are prevalent in urban settings and imply greater personal freedom coupled with less emotional investment and possible anomie and role conflict; organic motivations imply less personal freedom and greater social conformity coupled with increased personal and social certainty.

Looking Ahead

In this section, I want us to look both back and ahead. We will first look ahead and briefly examine Mead's and Simmel's influence on contemporary theory. After that, we will look back and bring elements of their theories together. There are two reasons for doing this.

First, I want to give you a hint of what theoretical synthesis looks like. There are a number of ways that we can create new theories. One of the most powerful and well-worn paths is through *synthesis*: taking elements from different theories and bringing them together to form a new, hopefully more powerful theory. In synthesis, we open up a theoretical space by simply contrasting and comparing ideas. Once this space has been opened, we can fill it through synthesizing theories. In it, we form a new theory by arranging elements from different theories in a creative way.

The second reason for doing this is that, together, Mead and Simmel give us a substantive question that in some ways far outstrips their direct influence on sociological theory. In other words, Mead's work forms the basis of symbolic interactionism (Chapter 9) and Simmel's work informs contemporary conflict and exchange theories (Chapters 7 and 10), but together they introduce us to an important question that has become a significant focus of contemporary social theory—especially for postmodernism (Chapter 14) and Anthony Giddens's theory of late-modernity (Chapter 11). It's a question that I think you will be most interested in: How does living in modern times influence the way you subjectively experience yourself?

Mead's and Simmel's Theoretical Impact

Mead's ideas form the base of the Chicago School of Symbolic Interaction. The Chicago School approaches human life as something that emerges through interaction. As a result, the interactionist approach, and thus Mead, has been the chief way through which we have become aware of the inner workings and experiences of diverse social worlds, such as Norm Denzin's (1993) work on alcoholics and Gary Alan Fine's (1987) analysis of Little League baseball. Symbolic interaction is also used as the principal perspective any time culture and the individual meet. For example, Arlie Hochschild (1983) used an interactionist approach to understand how emotions are socially scripted through "feeling rules." John Hewitt (1998) has recently used Mead's approach to understand how self-esteem is culturally and socially created and then internalized by the individual. In addition, Edwin Lemert (1951, 1967) and Howard Becker (1963) used Mead's ideas to understand deviance as an outcome of interaction and labeling.

Simmel has directly influenced contemporary theory in many ways. His ideas concerning culture are becoming increasingly important in the work of some postmodernists (see Weinstein & Weinstein, 1993). Simmel is also considered the father of formal sociology (see Ray, 1991). Formal sociology is concerned with social forms rather than content. The content of social action would be concerned with specific instances of social life, as in the actual actions and battles in a war or the specific causes and demonstrations of a social movement. As we've seen, forms are general categories of social action that must fit into a specific mold in order to occur. The general form of both the war and the social movement is conflict.

In this book, we will see Simmel's influence on exchange and conflict theory. Simmel was one of the first to explicate the implications of exchange on social encounters. Rather than theorizing about the structure of the economy per se, like Marx and Weber, Simmel is instead fascinated by the influence of the social form of exchange on human experience. As we'll see when we get to Chapter 10, Simmel is specifically concerned with how value is established and how it affects the use of power in social encounters.

Simmel has also had direct influence on contemporary conflict theory through Lewis Coser (Chapter 7). Before Simmel, conflict had been understood as a source of social change and disintegration. Simmel was the first to acknowledge that conflict is a natural and necessary part of society. Coser brought Simmel's idea to mainstream sociology, at least in America. From that point on, sociologists have had to acknowledge that "groups require disharmony as well as harmony" and that "a certain degree of conflict is an essential element in group formation and the persistence of group life" (Coser, 1956, p. 31).

The Self in Modernity: Bringing Mead and Simmel Together

Collectively, Mead and Simmel give us an insightful theory of how the self is created within the context of modernity. The specific point where they come together is Mead's idea of the Me and Simmel's notions of increasing objective culture and rationally based group affiliations. Recall that Mead argues that the self (Me) becomes integrated through role-taking with a generalized other. When role-taking with only significant others, the self will seem segmented, divided as it is among the different points of view. The generalized other is able to link all those individual perspectives into one abstract whole, thus giving the self a sense of integration. But what would happen if the generalized other was itself fragmented or constructed from vacuous images? This is where Simmel's theory comes in.

Simmel's theory implies three issues for the process of role-taking and self formation. First, Simmel says that an individual's relationship to groups is changing due to modern urbanization. As a result, people can pick and choose groups pretty much at will, and group "membership" is more and more mediated by money rather than by existing social relationships. Second, the groups themselves have changed in the sense that there is very little overlap or connection among groups, and group membership tends to lack consistency over time. Third, the general culture surrounding people and groups is becoming increasingly objective. That is, culture is becoming progressively more difficult for individuals to grasp, understand, and emotionally invest in.

Think for a moment about what these two theories imply about the self. Taken together, Simmel's three issues suggest that Mead's generalized others will tend to be fragmented, emotionally flat, and may lack the ability to guide behavior through role-taking in any significant way. The person's "Me" and his or her internal conversation is thus impacted: "Normally, within the sort of community as a whole to which we belong, there is a unified self, but that may be broken up. . . . Two separate 'me's' and 'I's,' two different selves, result, and that is the condition under which there is a tendency to break up the personality" (Mead, 1934, pp. 142–143). This possibility of a fragmented self is a central feature of postmodern theory (Chapter 14).

Building Your Theory Toolbox

Knowing Mead and Simmel

After reading and understanding this chapter, you should be able to define the following terms theoretically and explain their importance:

Pragmatism, emergence, mind, self, generalized other, symbols, natural signs, significant gestures, I and Me, role-taking, preparatory stage, play stage, game stage, generalized other stage, the act, interaction, institutions, a priori, social form, subjective and objective cultures, urbanization, division of labor, trivialization, money, web of group affiliations, organic motivation, rational motivation, primary groups, secondary groups, normative specificity, anomie, role conflict, blasé attitude

After reading and understanding this chapter, you should be able to

- Define pragmatism and apply the idea to meaning, truth, and self
- Explain how the mind and self are effects of social interaction
- Demonstrate how the mind and self are necessary for the existence of society
- Give the definition and functions of social forms and define and explicate a new social form
- Define objective culture and be able to explain how urbanization and the use of money increase the level of objective culture
- Identity and describe the effects of urbanization and rational group formation on the individual

Learning More: Primary Sources

George Herbert Mead: The chief source of Mead's theory is found in a compilation of student notes:

Mead, G. H. (1934). *Mind, self, and society: From the standpoint of a social behaviorist* (C. W. Morris, Ed.). Chicago: University of Chicago Press.

Georg Simmel: Unlike Mead, Simmel published quite a bit. I suggest that you start off with the first two readers, and then move to his substantial work on money:

Simmel, G. (1959). *Essays on sociology, philosophy, and aesthetics [by] Georg Simmel [and others]: Georg Simmel, 1858–1918* (K. H. Wolfe, Ed.). New York: Harper & Row.
Simmel, G. (1971). *Georg Simmel: On individuality and social forms* (D. N. Levine, Ed.). Chicago: University of Chicago Press.
Simmel, G. (1978). *The philosophy of money* (T. Bottomore & D. Frisby, Trans.). London: Routledge and Kegan Paul.

Learning More: Secondary Sources

To read more about Mead, I would recommend the following:

Baldwin, J. D. (1986). *George Herbert Mead: A unifying theory for sociology.* Beverly Hills, CA: Sage.
Blumer, H. (1969). *Symbolic interactionism: Perspective and method.* Englewood Cliffs, NJ: Prentice Hall.
Blumer, H. (2004). *George Herbert Mead and human conduct* (T. J. Morrione, Ed.). Walnut Creek, CA: AltaMira Press.
Cook, G. A. (1993). *George Herbert Mead: The making of a social pragmatist.* Urbana, IL: University of Chicago Press.

For Simmel, the following are excellent resources:

Featherstone, M. (Ed.). (1991). A special issue on George Simmel. *Theory, Culture & Society, 8*(3).
Frisby, D. (1984). *Georg Simmel.* New York: Tavistock.

Theory You Can Use (Seeing Your World Differently)

- More and more people are going to counselors or psychotherapists. Most counseling is done from a psychological point of view. Knowing what you know now about how the self is constructed, how do you think sociological counseling would be different? What things might a clinical sociologist emphasize?

- Using Google or your favorite search engine, enter "clinical sociology." What is clinical sociology? What is the current state of clinical sociology?

- Mead very clearly claims that our self is dependent upon the social groups with which we affiliate. Using Mead's theory, explain how the self of a person in a disenfranchised group might be different from one associated with a majority position. Think about the different kinds of generalized others and the relationship between interactions with generalized others and internalized Me's. (Remember, Mead himself doesn't talk about how we feel about the self.)

- How would Mead talk about and understand race and gender? According to Mead's theory, where does racial or gender inequality exist? From a Meadian point of view, where does responsibility lie for inequality? How could we understand class using Mead's theory? From Mead's perspective, how and why are things like race, class, gender, and heterosexism perpetuated (contrast Mead's point of view with that of a structuralist)?

- Remembering Simmel's definition and variables of objective culture, do you think we have been experiencing more or less objective culture in the last 25 years? In what ways? In other words, which of Simmel's concepts have higher or lower rates of variation? If there has been change, how do you think it is affecting you? Theoretically explain what the proportion of subjective to objective culture will be like for your children. Theoretically explain the effects you would expect.

- Perform a kind of network analysis on your web of group affiliations. How many of the groups of which you are a member are based on organic and rational motivations? In what kinds of groups do you spend most of your time? Over the next five years, how do you see your web of affiliations changing? Based on Simmel's theory, what effects can you expect from these changes?

Further Explorations—Web Links

Mead:

http://www.lib.uchicago.edu/projects/centcat/centcats/fac/facch12_01.html

http://www.iep.utm.edu/m/mead.htm

http://spartan.ac.brocku.ca/%7Elward

http://www.cla.sc.edu/phil/faculty/burket/g-h-mead.html

Simmel:

http://www.malaspina.com/site/person_1056.asp

http://socserv2.socsci.mcmaster.ca/~econ/ugcm/3113/simmel/society

http://www2.fmg.uva.nl/sociosite/topics/sociologists.html#simmel

CHAPTER 5

The Challenges
of Gender and Race

Charlotte Perkins Gilman (1860–1935)

W. E. B. Du Bois (1868–1963)

Seeing Further: Social Exclusion

In some ways, this chapter was the most difficult to put together. Including Marx, Weber, Durkheim, Simmel, and Mead in a section on classical theory is a no-brainer. These thinkers are part of our cultural capital and you'll need to know them to carry on a conversation in sociology. Not knowing about them would be like going to a sports bar and not knowing the difference between baseball and hockey. In addition, these men and their theories still hold sway. You can't really understand contemporary theory nor can you do your own theoretical work without knowing about their ideas and theories.

However, the human disciplines (sociology included) have systematically ignored the work of women and people of color. Hence, people like Harriet Martineau (refer to the introduction to Section I), Charlotte Perkins Gilman, and W. E. B. Du Bois don't have as clear a theoretical lineage nor do they form a major part of our cultural capital. Sad to say, but you could get through your entire graduate career in sociology and never need to know the name Harriet Martineau, but you would fall flat on your face in your first graduate class if you didn't know about Max Weber.

The reason for this discrepancy isn't because these people didn't say anything profound or inspiring. Quite the opposite is true: they have a lot to say to us. But social theory isn't simply built around the power of explanation. As Weber points out, values are always at the core of what we do. Thus, *what* is being explained is sometimes just as important as the efficacy of the theory, if not more. The questions we ask, and thus the answers we hold onto, are very much guided by our cultural values.

The point I want to make here is that social theory is contested terrain, and it should be. Critique and reevaluation can only make us and our theories better. I also want you to know that I view my representation of theory (this book) not as an end, or even as a beginning, but simply as a tentative statement in an ongoing dialogue. As Ritzer and Goodman (2000) say in *The Blackwell Companion to Major Social Theorists,* "although any list of theorists covered in a collection such as this one can be read as an official canon, this book is intended to be used as 'cannon fodder' in an open, contestable process of theory construction and reconstruction" (p. 2).

One of the things that studying the theories of Gilman and Du Bois ought to help us see better is the process of exclusion that is present in every modern society. Not only do the theories of Gilman and Du Bois explain how exclusionary practices work, their position with respect to the "canon" of theoretical scripture also speaks loudly of how even the most apparently noble pursuit of "knowledge" is biased and political, a theme we will explore further in the contemporary work of Michel Foucault (Chapter 14). As we move through this book, we will see these issues more and more clearly. After reading Chapter 8, we will be able to better see the structures of inequality surrounding race and gender; Chapter 13 will give us enhanced eyes to see how exclusionary tactics are practiced through matrices of domination that include not only gender and race, but also nation, sex, religion, and so forth; we'll also see how society creates gendered consciousnesses and the processes through which the social norm of heterosexism is materialized in our bodies; and in Chapter 14 we'll see that the way we use language creates discourses that exercise power over the subjective experiences of the powerless in society.

Charlotte Perkins Gilman

Photo: © Bettmann/Corbis.

The Essential Gilman

Biography

Charlotte Perkins Stetson Gilman was born Charlotte Perkins on July 3, 1860, in Hartford, Connecticut. Her father was related to the Beecher family, one of the most important American families of the nineteenth century. Gilman's great-uncle was Henry Ward Beecher, a powerful abolitionist and clergyman, and her great-aunt was Harriet Beecher Stowe, who wrote *Uncle Tom's Cabin,* which focused the nation's attention on slavery. Gilman's parents divorced early, leaving Charlotte and her mother to live as poor relations to the Beecher family, moving from house to house. Gilman grew up poor and was poorly schooled.

Gilman's adult life was marked by tumultuous personal relationships. Her first marriage in 1884, to Charles Stetson, ended in divorce. Gilman wrote a novella about the relationship, called *The Yellow Wallpaper,* which is still being read in women's studies courses today. The book tells the story of a depressed new mother (Gilman herself) who is told by both her doctor and husband to abandon her intellectual life and avoid any writing or stimulating conversation. The woman sinks deeper into depression and madness as she is left alone in the yellow-wallpapered nursery. Gilman's feeling of hopelessness against the tyranny of male-dominated institutions is heard in the repeated refrain of "but what is one to do?" Between her first and second marriages, a period of about 12 years, Gilman had a number of passionate affairs with women. However, her second marriage—to George Gilman, a cousin—proved to be at least a somewhat successful and important relationship for Charlotte: some of her best work was done during the courtship and after the marriage.

In her lifetime, Gilman wrote over 2,000 works, including short stories, poems, novels, political pieces, and major sociological writings. She also founded *The Forerunner,* a monthly journal on women's rights and related issues. Among her best-known works are *Concerning Children, The Home, The Man-Made World,* and *Herland* (a feminist utopian novel). However, Gilman's most important theoretical work is *Women and Economy.* In Gilman's lifetime, the book went through nine editions and was translated into seven different languages.

Gilman was also politically active. She often spoke at political rallies and was a firm supporter of the Women's Club Movement, which was an important part of the progress toward women's rights in the United States. It initially began as a place for women to share culture and friendship. Before the development of these clubs, most women's associations were either auxiliaries of men's groups or allied with a church. In contrast, many of the clubs of the Women's Club Movement became politically involved, but others were simply public and educational service organizations. One such club became the Parent–Teacher Association (PTA).

Sick with cancer, Charlotte Gilman died by her own hand on August 17, 1935, in Pasadena, California.

Passionate Curiosity

As should be apparent, Gilman is impassioned by the abuses that women suffer under patriarchy. Her basic questions, then, are how did patriarchy develop and how will it end? Interestingly enough, she approaches these issues from a scientific, social-evolutionary perspective.

Keys to Knowing

Social evolution, self-preservation, race-preservation, gynaecocentric theory, sexuo-economic relations, morbid excess in sex distinction

Gilman's Perspective: Evolution With a Twist

Critical-Functional Evolution

Gilman is an evolutionist. She generally understands the history of human society as moving from simple to complex systems, with various elements being chosen through the mechanism of survival of the fittest. Gilman argues that humans began as brute animals pursuing individual gain. Survival of the fittest at that point was based on individual strength and competition. The males would hunt for food and fight one another for sexual rights to the females. It was a day-by-day existence with no surplus or communal cooperation. Wealth was, of course, unknown, as it requires surplus and social organization. Competition among men was individual and often resulted in death. Women, as is the case with many other life forms, would choose the best fit of the men for mating. There was no family unit beyond the provision of basic necessities for the young, mostly provided by women. Life was short and brutal.

As competition between species continued, humans gradually developed social organization beyond its natural base. Very much like an organism, society began to form out of small cells of social organization that joined with other small cells. Gilman (1899/1975) says that "society is the fourth power of the cell" (p. 101). By that she means that once societies started to grow and structurally differentiate, the process continued in a multiplicative manner. The reason for this is simple: Just as in organic evolution, more complex social systems, because of the level of social cooperation and division of labor, have a greater chance of survival than do individuals or simple societies. As societies became more complex, and thus more social, structures and individuals had to rely on each other for survival. Just like the different organs in your body must depend upon other organs and processes, so every social unit (individuals and structures) must depend upon the others. (Recall the organismic analogy from Chapter 3.)

> The proposition is that Society is the whole and we are the parts: that the degree of organic development known as human life is never found in isolated individuals, and that it progresses to higher development in proportion to the evolution of the social relation. (Gilman, as cited in Lengermann & Niebrugge-Brantley, 1998, p. 142)

An interesting thing happened for humans because of the development of complex, social organization. All animals are influenced by their environment. If the environment turns cold, then the organism will either develop a way of regulating body temperature (polar bears are well-suited for the Arctic), move to a different environment, or die out. Humans at one time were equally affected by the environment. But after we organized socially, our organization and culture allowed us to live in a controlled manner with regard to the environment: humans can now exist in almost any environment, on or off the planet. We thus distanced ourselves from the natural environment, but at the same time we created a new environment in which to live. Gilman argues that human beings have become more influenced by the social environment of culture, structure, and relations than by the natural setting—we thus have a new evolutionary environment. Keep this idea of a new environment in mind; it becomes extremely important later on in Gilman's theory of gender.

Like Marx, Gilman argues that this new environment, this social structure that improved humanity's chances of survival, is the economy. All species are defined by their relationship to the natural environment. Stated in functionalist terms, structures in an organism are formed as the organism finds specific ways to survive (structure follows function); the same is true in society. The economy is the structure that is formed in response to our particular way of surviving, and it is thus our most defining feature. We are therefore most human in our economic relations. Marx argued that we are alienated from our nature as we are removed from direct participation in the economy, as with capitalism. Gilman also argues that alienation occurs, but the source of that alienation is different. Where Marx saw class, Gilman sees gender. Gilman, in fact, is quite in favor of capitalism—in no other system has our ability to survive been so clearly manifested. Even so, there are alienating influences that originate in our gendered relationship to the economy, most specifically the limitation of women's workforce participation.

As we will see in the following pages, Gilman understands the exclusion of women from the economy in evolutionary terms. This perspective yields amazing results. We will see in the section below on *self- and race-preservation* that removing women from economic participation threw off the balance between the individual and society. This disequilibrium from a functionalist's view means that the evolutionary path of humanity is not functional. Gilman explores the main point of this dysfunction in her use of *gynaecocentric theory,* which posits that the female is the basic social model rather than the male (which had been assumed for most of our scientific and philosophic history). Gilman argues that men and women have natural, intrinsic energies that are different from one another. Based on this postulate of different energies, Gilman contends that moving women out of the workforce actually ended up producing evolutionary advantages, which we'll cover in a moment. Thus, in the long run, it turned out to be functional to remove women from the workforce. However, it is certainly not the case that Gilman thinks everything is right and wonderful for women. In the section on *sexuo-economic effects,* we'll see that denying women access to the natural economic environment has had profound consequences.

Concepts and Theory: Dynamics of Social Evolution

Self- and Race-Preservation

I've diagrammed Gilman's basic model of social evolution in Figure 5.1. As you can see, she, like Marx, argues that the basic driving force in humanity is economic production: creative production is the way that we as a species survive. In all species, survival needs push the natural laws of selection, which results in a functional balance between self-preservation and race-preservation. Natural selection in **self-preservation** develops those characteristics in the individual that are needed to succeed in the struggle for self-survival. In the evolutionary model, individuals within a species fight for food, sex, and so on. Natural selection equips the individual for that fight.

Race-preservation, on the other hand, develops those characteristics that enable the species as a whole to succeed in the struggle for existence. Gilman is using the term "race" in the same way I'm using the term "species"; that is, she's using it in the more general sense, rather than to make racial distinctions. The most important point here is that the relationship between self- and race-preservation is balanced: individuals are selfish enough to fight for their own survival and selfless enough to fight for the good of the whole species.

Gilman's idea is interesting because most evolutionists assume that these two factors are one and the same; in other words, self-preservation is structured in such a way that it functions as species-preservation. This assumption is what allows social philosophers like Adam Smith to take for granted that severe individualism in systems like capitalism is natural and beneficial. Gilman insightfully perceives that these two functions can be at odds with one another. She thus creates an empirical research question where many people don't see any problem or make a value judgment, such as in the inherent goodness of capitalistic free markets.

The third box of Figure 5.1 is where human evolution becomes specific. Because of our overwhelming dependency on social organization for survival, we develop laws and customs that in turn support the equilibrium between self- and race-preservation, as noted by the feedback arrow. With the idea of "accumulation of precedent," Gilman has in mind art, religion, habit, and institutions other than the

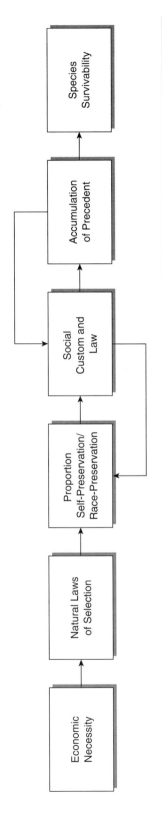

Figure 5.1   Gilman's Basic Model of Social Evolution

economy that tend to reinforce and legitimate our laws and customs. As you can see, there is a reinforcing cycle among the race-self-preservation proportion; social laws and customs; and the institutions, culture, and habits of a society. This social loop becomes humankind's unique environment. Of course, we still relate to and are affected by the natural world, but the social environment becomes much more important for the human evolutionary progress (or lack thereof). This entire process works together to increase the likelihood of the survival of the human race.

That's the model of how things are supposed to work. Figure 5.2 depicts how things actually work under patriarchy. As you can see, the model is basically the same. The same central dynamics are present: economic necessity pushes the laws of selection, which in turn produce the proportion of self- to race-preservation characteristics, which then create laws and customs that reinforce those character-istics and produce social behaviors, culture, and institutions that strengthen and justify the laws and customs. This time, however, the social loop doesn't result in greater species survivability, but, rather, in gender, sex, and economic dysfunctions.

I want to talk about those dysfunctions, but first we need to look at how Gilman sees the overall movement of evolution. Her conclusions are extremely interesting and perhaps startling, but in the overall way in which she makes her argument, she is years ahead of the functional arguments of her time. In fact, we can safely say that the kind of argument she presents wasn't generally found in sociology until after Robert K. Merton's work around 1950 (Chapter 6). In general, Gilman argues that dysfunctions in some subsystems can have functional consequences for the whole.

Gynaecocentric Theory

The dynamic difference between Figure 5.1 and Figure 5.2 is found in what Gilman calls a "morbid excess in sex distinction." Gilman argues that human sexu-ality and the kinship structure were evolutionarily selected. In the pre-dawn history of humankind, it became advantageous for us to have two sexes that joined together in monogamy. From an evolutionary perspective, the methods of reproducing the species are endless. They run from the extremes of hermaphroditism (both sexes contained in a single organism) to multiple-partnered egg hatcheries. Because human beings need high levels of culture and social organization to survive, the two-sex model was naturally selected along with monogamy. Thus, having two sexes that are distinct from one another is natural for us. In this model, the physi-cal distinctions between the sexes produce attraction and competition for mates, just as in other species. Eventually, however, because of the need for social organi-zation among humans, the brutal competition for mates was mitigated and the social bond extended past mother–child through monogamy.

This process, however, became tainted by patriarchy. Gilman presents a unique and fascinating argument as to why patriarchy came about. According to Gilman, there are distinct, natural male and female energies. The basic masculine character-istics are "desire, combat, self-expression; all legitimate and right in proper use" (1911/2001, p. 41). Female energy, on the other hand, is more conservative and is characterized by maternal instincts—the love and care of little ones. In this, Gilman

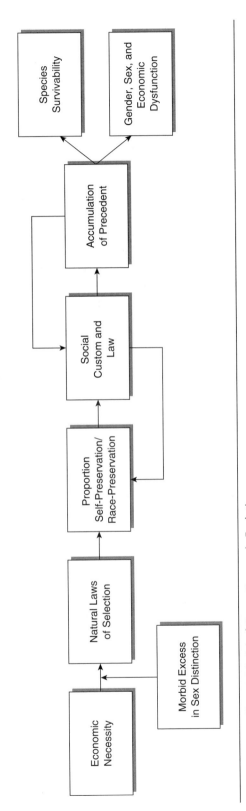

Figure 5.2 Gilman's Model of Sexuo-Economic Evolution

sees us as no different from most other species—these energies are true of many male and female creatures.

Yet the natural level of these energies posed a problem for human evolution. To survive as humans, we required a higher degree of social organization than simple male and female energy could provide. The male energy in particular was at odds with this need. As long as "the male savage was still a mere hunter and fighter, expressing masculine energy . . . along its essential line, expanding, scattering" (Gilman, 1899/1975, p. 126), social organization of any complexity was impossible. Men were aggressive and individualistic and thus had to be "maternalized" for social organization to grow.

In this argument, Gilman is using and expanding a theory from Lester F. Ward. Gilman acknowledged her debt by dedicated her book *The Man-Made World* to "Lester F. Ward, Sociologist and Humanitarian, one of the world's great men . . . and to whom all women are especially bound in honor and gratitude for his Gynaecocentric Theory of Life." In **gynaecocentric theory**—the word "gynaecocentric" means woman-centered—the female is the general race type while male is a sex type. The argument begins with the assumption that kinship is the basis of social organization. All species, including humans, are related to one another biologically through descent. In our case, we used those biological relations to build social connections. The basic kin relationship is found between mother and child. This is the one relationship that has always been clear: it's obvious from whom the child comes. In order to build social relations on that basic tie, fathers had to be included through some mechanism. Of all the possible forms of marriage, monogamy provides the clearest connection between father and child, and thus the strongest, most basic social relationship.

What Ward and Gilman add is the idea of intrinsic male and female energy. Male energy is at odds with family and social ties. The "giant force of masculine energy" had to be modified in order to add and extend the social tie between father and child. This was accomplished through the subjugation of women (I told you it was interesting). When men created patriarchy and dominated women, they also made women and children dependent upon men. In natural arrangements, both men and women produce economically, but as men dominated women, they also removed them from the world of economic production. Women could no longer provide for themselves, let alone their children. Men had to take on this responsibility, and as they did, they took on "the instincts and habits of the female, to his immense improvement" (Gilman, 1899/1975, p. 128). Therefore, men became more social as a result of patriarchy.

This functional need was a two-way affair. Not only did men need to take on some of the women's traits, the natural environment could be more thoroughly tamed by men rather than women. By taking women out of the productive sphere, and therefore forcing men to be more productive, "male energy . . . brought our industries to their present development" (Gilman, 1899/1975, pp. 132–133). Both men and human survival in general benefited from patriarchal arrangements.

Gilman draws two conclusions from gynaecocentric theory. First, women should not resent the past domination by men. Great harm to individuals and society has come through this system, but necessary and great good has come as well,

[I]n the extension of female function through the male; in the blending of facilities which have resulted in the possibility of our civilization; in the superior fighting power developed in the male, and its effects in race-conquest, military and commercial; in the increased productivity developed by his assumption of maternal function; and by the sex-relation becoming mainly proportioned to his power to pay for it. (Gilman, 1899/1975, pp. 136–137)

The second thing Gilman draws from this theory is that the women's movements, and all the changes that they bring, are part of the evolutionary path for humanity. While women's economic dependence upon men was functional for a time, its usefulness is over. In fact, Gilman tells us, the women's and labor movements are misnamed—they ought more accurately to be called human movements. The reason that we've become aware of the atrocities associated with male dominance at this time in our history is that the dysfunctions are now greater than the functions. The behaviors and problems of patriarchy have always existed, but the benefits outweighed the costs and it therefore remained functional. Now, because of the level of social and economic progress, it is time to cast off this archaic form.

Sexuo-Economic Effects

However, it should be clear that Gilman's purpose is not to justify the subordination of women. Gynaecocentric theory gives her a way of understanding the effects, including the women's movements, in functional, evolutionary terms. It is, I think, to her credit that she maintains theoretical integrity in the face of the obvious suffering of women, and her own personal anguish as a woman. That said, a great deal of Gilman's work is occupied with exposing and explaining the current state of women.

Gilman characterizes the overall system and its effects as sexuo-economic. In **sexuo-economic relations,** two structures overlap, one personal and the other public. The social structure is the economy and the private sphere is our sexual relationships. The basic issue here is that women are dependent upon men economically; as a result, two structures that ought to be somewhat separate intertwine. Historically, as women became more and more dependent upon men for their sustenance, they were removed further and further from economic participation.

As we've noted, the economy is the most basic of all our institutions: it defines our humanness. A species is defined by its mode of survival; survival modes, then, determine the most basic features of a species. In our case, that basic feature is the economy. The economy is where we as a species compete for evolutionary survival. It is where we as a species come in contact with the environment and where the natural laws of selection are at work. Thus, when workforce participation is denied to women, they are denied interaction with the natural environment of the species. That natural environment, with all of its laws and effects, is replaced with another one. Man (meaning males) becomes woman's economic environment, and, just as any organism responds to its environment, woman changed, modified, and adapted to her new environment (form and function).

Gilman talks about several effects from living in this new environment, but the most important are those that contribute to **morbid excess in sex distinction.** All animals that have two sexes also have sex distinctions. These distinctions are called secondary sex characteristics: those traits in animals that are used to attract mates. In humans, according to Gilman, those distinctions have become accentuated so much that they are morbid or gruesome.

In societies where they are denied equal workforce participation, the primary drive in women is to attract men: "From the odalisque with the most bracelets to the débutante with the most bouquets, the relation still holds good,—women's economic profit comes through the power of sex-attraction" (Gilman, 1899/1975, p. 63). Rather than improving her skills in economic pursuits, she must improve her skills at attracting a man. In fact, expertise in attraction skills *constitutes her basic economic skills,* which means that the secondary sex characteristics of women become exaggerated. She devotes herself to cosmetics, clothes, primping, subtle body-language techniques, and so on. Most of these are artificial, yet there are also natural, physical effects. Because women are pulled out from interacting with the natural environment, their bodies change: they become smaller, softer, more feeble, and clumsy.

The increase in sex distinctions leads to greater emphasis on sex for both men and women. The importance given to sex far exceeds the natural function of procreation and comes to represent a threat to both self- and race-preservation. The natural ordering of the sexes becomes perverted as well. In most animals, it is the male that is flamboyant and attractive; this order is reversed in humans. To orient women toward this inversion, the sexual socialization of girls begins early: "It is what she is born for, what she is trained for, what she is exhibited for. It is, moreover, her means of honorable livelihood and advancement. But—she must not even look as if she wanted it!" (Gilman, 1899/1975, p. 87).

In our time, young girls are taught that women should be beautiful and sexy, and they are given toys (such as Barbie) that demonstrate the accentuated sex distinctions, toy versions of makeup kits used to create the illusion of flamboyancy in women, and games that emphasize the role of women in dating and marriage. Another result of this kind of gender structure and socialization is that women develop an overwhelming passion for attachment to men at any cost; Gilman points to women who stay in abusive relationships as evidence. In the end, because of the excess in sex distinction brought on by the economic dependence of women, marriage becomes mercenary.

Obviously, the economic dependency of women will have economic results as well. Women are affected most profoundly as they become "nonproductive consumers." According to Gilman, in the natural order of things, productivity and consumption go hand in hand, and production comes first. Like Marx, Gilman sees economic production as the natural expression of human energy, "not sex-energy at all but race-energy." It's part of what we do as a species and there is a natural balance to it when everybody contributes and everybody consumes.

The balance shifts, however, as women are denied workforce participation. The consumption process is severed from the production process and the whole system is therefore subject to unnatural kinds of pressures. Non-economic women are focused on non-economic needs. In the duties of her roles of wife, homemaker, and mother, she creates a market that focuses on "devotion to individuals and their personal needs" and "sensuous decoration and personal ornament" (Gilman, 1899/1975, pp. 119–120).

The economy, of course, responds to this market demand. Contemporary examples of this kind of "feminized" market might include such things as fashion items (such as jewelry, chic clothing, alluring scents, sophisticated makeup, and so forth); body image products (such as exercise tapes, health clubs, day spas, dietary products, and the like); fashionable baby care products (such as heirloom baby carriages, designer nursery décor, and special infant fashions by the Gap, Gymboree, Baby B'Gosh, Tommy, Nike, Old Navy, ad infinitum); and decorative products for the home (such as specialized paints, wallpaper, trendy furnishings, and so on).

> As the priestess of the temple of consumption, as the limitless demander of things to use up, her economic influence is reactionary and injurious. . . . Woman, in her false economic position, reacts injuriously upon industry, upon art, upon science, discovery, and progress. (Gilman, 1899/1975, pp. 120–121)

Gilman argues that the creative efforts of men, which should be directed to durable commodities for the common good, are subject to the creation and maintenance of a "false market." Women are thus alienated from the commodities they purchase because they don't participate in the production process nor are their needs naturally produced, and men are alienated because they are producing goods and services driven by a false market.

As Figure 5.2 shows, the morbid excess in sex distinction disables the laws of natural selection and puts self- and race-preservation out of balance. As a result, the customs and laws of society respond, as do the social institutions and practices (accumulation of precedent). Institutions and practices that properly belong to humankind become gendered. For example, the governing of society is a "race-function," but in societies that limit women's workforce participation, it is seen as the duty and prerogative of men. Decorating is also distinctly human; it is a function of our species, but it is perceived as the domain of women. Religion, an obvious human function, is dominated by men and has been used to justify the unequal treatment of women, as have law, government, science, and so forth.

Yet Gilman does tell us that things are changing. She sees the women's movements as part of our evolutionary path of progress. Implied in this conclusion is the idea that equality is a luxury. Under primitive circumstances, only the strong survive. The weak, the feeble, and the old are left behind or killed. As social relationships come into existence, cooperation is possible, and as people cooperate, a surplus develops. For the most part, the powerful control the surplus and turn it into wealth, yet at the same time, it is then feasible to support some who had once been cut off. These people are integrated into society and society continues to grow. It becomes more complex and technically better able to control the environment and produce surplus. In turn, other groups are brought into the fold, and increasing numbers of disenfranchised groups are able to live and prosper. It is Gilman's position that just as technical progress is our heritage, so is ever-expanding equality. Evolution of the species not only involves economic advancement but increasing compassion as well; ethical and technical evolutions are inexorably linked.

At the present evolutionary moment, we are only hindered by our own blindness. We can afford the luxury of complete equality, but we fail to see the problem. Gilman says that this is due to a desire for consistency: we don't notice what we are used to. Even evil can become comfortable for us and we will miss it when it's gone. This is one of the reasons why slavery wasn't seen as evil for so many years of our history. We also have a tendency to think in individual terms rather than general terms. It's easier for us to attribute reasons for what we see to individuals rather than to broad social factors. This, of course, in C. Wright Mills's terms, is the problem of the sociological imagination. It is difficult to achieve.

> Being used to them, we do not notice them, or, forced to notice them, we attribute the pain we feel to the evil behavior of some individual, and never think of it as being the result of a condition common to us all. (Gilman, 1899/1975, p. 84)

Gilman Summary

- Gilman is a critical evolutionist. All species fight for survival and create organic structures to help them survive in their environment. In this case, the function produces the structure. For humans, the primary evolutionary structure is the economy. It is through the economy that we as a species survive.

- When women were removed from the economy through male domination, several things happened. First, women were taken out of the natural environment of economic production and given a false environment—men, and indirectly the home and family, became women's environment. Gilman is a social evolutionist and argues that, as every species will, woman changed in response to changes in her environment. This change in environment made the distinctions between the sexes more pronounced. Rather than being equipped to economically produce, women became equipped to pursue a husband (their survival depended upon it). Their bodies became smaller and softer from disconnection with the natural world, and women augmented these changes through artificial means (such as makeup, clothing, etc.). Human beings now have a morbid excess in sex distinction, and as a result, the natural order has been reversed. In most animals, it is the male that is marked for attraction; in humans, it is the female. In order to produce this artificial order, girls have to be socialized from birth to want to be attractive and to value having a husband over all else. Further, in societies where women's workforce participation is limited, women become unnatural consumers. The natural relationship is production followed by consumption. Women are simply consumers without production. As such, their perceived needs are radically changed and they produce a market focused on beauty, decoration, and private relationships. This market makes men produce dysfunctional goods, and they are thus alienated from the work of their hands; women, on the other hand, purchase goods that are not a natural part of their essential nature, thus alienating them from the commodity. Further, to support and legitimate this economy, the other institutions, laws, and customs in society have become gendered, and those things that are properly human (such as government, religion, and education) are now seen as the domain of one sex.

• This evolutionary path has also had functional consequences. The natural energy of men is aggressive, individualistic, and dominating. The male is, then, ill suited for social life. However, in making women and children dependent upon him, he also obligated himself to provide for them (to be socially involved and committed). Thus, through gender oppression, the natural energy of men has been modified: they now have maternal feelings and higher social abilities. Further, large societies require highly developed systems that can control the environment and extract resources. Gilman argues that men and women together could not have achieved the needed level of economic and technological development. In Gilman's evolutionary scheme, men are naturally more aggressive and objectifying. Thus, having men solely responsible for the economic sphere meant that society produced more and achieved higher levels of technological control. Thus, pulling women out of the workforce functioned to allow the economy to develop fully.

• Humans have, however, reached the point where the suppression of women is more dysfunctional than functional. The women's movements are sure signs that evolution is pushing us toward higher levels of equality.

W. E. B. Du Bois

Photo: © Bettmann/Corbis.

The Essential Du Bois

Extended Biography

As I mentioned earlier, W. E. B. Du Bois' life is at least as important as his theoretical writings. In keeping with that idea, I'm giving Du Bois an extended biographical treatment. William Edward Burghardt Du Bois was born in Great Barrington, Massachusetts, on February 23, 1868. His mother was Dutch African and his father French Huguenot-African. Du Bois' father, who committed bigamy in marrying Du Bois' mother, left the family when the boy was just two years old, but Du Bois grew up without experiencing the depth of suffering blacks felt at that time. Du Bois' education in the abject realities of blackness in America came during his college years at Fisk, in Tennessee; it was his first visit to the southern United States during "Jim Crow." In addition, he taught children for two summers at a rural Tennessee school and saw firsthand the legacy of slavery in the South. As a result, Du Bois became an ardent advocate of social change through protest. His belief in confrontation and dissent put Du Bois at odds with Booker T. Washington, a prominent black leader who favored accommodation. This disagreement with Washington influenced the early part of Du Bois' political life; he founded the Niagara Movement in large part to counter Washington's arguments.

Through most of his life, Du Bois generally favored integration, but toward the end he became discouraged at the lack of progress and increasingly turned toward Black Nationalism: He encouraged blacks to work together to create their own culture, art, and literature, and to create their own group economy of black producers and consumers. The cultural stand was directed at creating black pride and identity; the formation of a black economic community was the weapon to fight discrimination and black poverty. Du Bois was also a principal force in the Pan-African movement, which was founded on the belief that all black people share a common descent and should therefore work collectively around the globe for equality. In the latter part of his life, Du Bois became disheartened at the lack of change regarding the color line in the United States. In the end, he renounced his citizenship, joined the Communist Party, and moved to Ghana, Africa.

More than any other single person, Du Bois was responsible for black consciousness in America and probably the world during the twentieth century. His book, *The Souls of Black Folk*, defined the problem of the color line. He was a founding member of the National Association for the Advancement of Colored People (NAACP) and its chief spokesperson during its most formative years. Du Bois also produced the first scientific studies of the black condition in America.

I have listed the major points in Du Bois' life in the timeline below. I encourage you to not skip over this information; read it and let it sink in. The timeline is but an outline of a life that spanned almost a century and helped to change the face of society.

February 23,

1868:	Born in Great Barrington, Massachusetts
1888:	Received his B.A. from Fisk University, Nashville (turned down by Harvard)—taught two summers in rural Tennessee, which introduced him to the deep poverty of the Southern black person
1888–1895:	Completed second B.A. and Ph.D. at Harvard, during which time he spent two years in Berlin, attending lectures by Max Weber and becoming friends with both Marianne and Max
1894–1910:	Taught at Wilberforce, University of Pennsylvania, and Atlanta University (where he was a professor of economics and history for 14 years); published 16 research monographs
1899:	Published *The Philadelphia Negro: A Social Study* (the first ethnography of a black community in the United States)
1900:	Led the first Pan-African conference, in London
1903:	Published *The Souls of Black Folk*, which first set out the problem of the color line and Du Bois' opposition to Booker T. Washington's strategy
1905:	Founded the Niagara Movement, precursor to NAACP
1909:	Help found the NAACP; left academia and became editor of the association's journal, *The Crisis*
1910–1934:	Worked at the NAACP; edited *The Crisis;* was principal leader of blacks in America; advocated Black Nationalism
1919–1927:	Engineered four Pan-African congresses
1934–1944:	Left the NAACP due to disagreement over assimilation versus separation; returned to Atlanta University
1935:	Published *Black Reconstruction: An Essay Toward a History of the Part Which Black Folk Played in the Attempt to Reconstruct Democracy in America, 1860–1880* (Marxist interpretation of the Reconstruction)
1940–1944:	Chaired the Department of Sociology at Atlanta University; editor of social science journal, *Phylon*

1940:	Published *Dusk of Dawn: An Essay Toward an Autobiography of a Race Concept* (viewed his career and life as representative of black–white relations)
1944–1948:	Final association with the NAACP; in 1948, Du Bois left due to a disagreement with the director; continued discouragement and movement toward political left
1951:	Indicted as unregistered agent for a foreign power; acquitted
1961:	Joined the Communist Party; moved to Ghana, continued work on his long-term project, *Encyclopedia Africana*
1962:	Renounced American citizenship
August 27, 1963:	Died in Ghana at age 95

As Charles Lemert (2000) says,

> He died, as icons often do, at precisely the right moment—on the eve of the 1963 March on Washington. From the death of Frederick Douglass in 1895 to Martin Luther King Jr.'s coronation the day after Du Bois' death, no other person of African descent was as conspicuous in the worldwide work of contesting the color line. (p. 347)

Passionate Curiosity

In his work, Du Bois was passionate about one thing: the obliteration of the color line.

Keys to Knowing

Entered subject, seventh son, veil, grand narrative, history as ideology, representation, denotation, connotation, stereotypes, default assumptions, looking glass self, double consciousness, exploitation, dark nations, personal whiteness

Du Bois' Perspective: The Experience of Oppression and Critical Knowledge

The Centered Subject

There are a couple of things that strike the thinking reader as she or he spends time with Du Bois' texts. The first is that Du Bois centers the subject in his writing. One of the things most of us are taught in college English is that we should write from a de-centered point of view. We are supposed to avoid using "I" and "me" and should always write from an objective perspective. Yet Du Bois begins one of his most famous works, *The Souls of Black Folk*, with the phrase "Between me and the other world . . ." In another place he wonders, "Who and what is this I . . . ?" Du Bois isn't being self-centered nor is he unaware of the rules of composition. Du Bois is being quite deliberate in his use of personal pronouns and the centered subject. I believe one of the things he is telling us is that race is not something that can be understood through the cold, disassociated stance of the researcher. Race and all marginal positions must be experienced to be understood. Du Bois uses his life as the canvas upon which he paints the struggles of the black race in America and in the world.

The other thing that impresses the reader is that much of Du Bois' writing is a multimedia presentation. Du Bois moves back and forth among intellectual argumentation, song, prayer, poetry, irony, parable, data, riddles, analogy, and declaration. He weaves a tapestry for the reader, one that touches every part of the reader's being. He wants us to be able to understand the objective state of blackness as well as experience its soul. In this he reminds me of a colleague of mine. My colleague is a black man who has done some amazing work with Los Angeles gang members. One time I asked him to present a guest lecture to one of my classes. You know what a lecture looks like, right? Imagine my surprise when he asked us all to close our eyes and lay our heads down on the desks. His lecture consisted of a dramatic reading of a poem—complete with gunshots and cries—that left the class stunned and some in tears.

It might well be coincidence that both Du Bois, who is one of the few writers I've seen use such varied venues in writing, and the only professor I've seen break the mold of lecture so soundly, are African American. Yet it does give us pause and it provides me with a transition into one of Du Bois' important points. Though Du Bois undoubtedly sees the color line as something that is socially created, he also acknowledges and honors the unique characteristics of and contributions from the "souls of black folk." He maintains that American music and folklore, faith and reverence, light-hearted humility, literature and poetry, speech and styles of interaction, practices of sport, and our political fabric are all strongly influenced by African American culture. Yet it isn't a simple listing of accomplishments or influences alone that Du Bois has in mind; it is the soul of a particular kind of lived experience: "that men may listen to the striving *in the souls of* black folk" (Du Bois, 1903/1996a, p. 107, emphasis added).

Standpoint of the Oppressed

There are two things I think this subjective stance implies: the first is my own comment, the other is something I think Du Bois has in mind. First, any secondary reading of Du Bois, such as the book you have in your hands, falls short of the mark. This is generally true of any of the thinkers in this book—you would be much richer reading Durkheim than reading what somebody *says* about Durkheim—but it is particularly true of Du Bois. Part of what you can acquire from reading Du Bois is an experience, and that experience is a piece of what Du Bois wants to communicate.

The second implication of Du Bois' multidimensional, subjective approach is theoretical. Du Bois (1903/1996a) says that "the Negro is a sort of seventh son, born with a veil, and gifted with second-sight in this American World,—a world which yields him no true self-consciousness, but only lets him see himself through the revelation of the other world" (p. 102). Du Bois employs spiritual language here. The veil of which he speaks is the birth caul. In some births, the inner fetal membrane tissue doesn't rupture and it covers the head at delivery. This "caul" appears in about 1 in 1,000 births. Due to its rarity, some traditional cultures consider such a birth spiritually significant and the caul is kept for good luck. The same is true of the seventh son reference. The seventh son is considered to have special powers, and references to such are to be found in many folk and blues songs as well as in the Bible. The "second-sight" is a reference to clairvoyant or prophetic vision.

Thus, Du Bois is saying that because of their experiential position, African Americans are gifted with special insight—a prophetic vision—into the "American World." They see themselves not simply as they are; they also see their position from the perspective of the "other world"—the white social world around them. In other words, blacks and other oppressed groups have a particular point of view of society that allows them to see certain truths about the social system that escape others. This idea of critical consciousness goes back to Marx. Marxian philosophy argues that only those on the outside of a system can understand its true workings; it is difficult to critically and reflexively understand a system if you accept its legitimation. In other words, capitalists and those who benefit from capitalism by definition believe in capitalism. It is difficult for a capitalist to understand the oppressive workings of capitalism because in doing so the person would be condemning her- or himself.

Contemporary feminists argue that it is the same with patriarchy: men have a vested interest in the patriarchal system and will thus have a tendency to believe the ideology and have difficulty critiquing their own position. Dorothy Smith (Chapter 13), a contemporary feminist, refers to this as *standpoint theory*. In general, this refers to a theory that is produced from the point of view of an oppressed subject: "an inquiry into a totality of social relations beginning from a site outside and prior to textual discourses. Women's standpoint [is seen] here as specifically subversive of the standpoint of a knowledge of ourselves and our society vested in relations of ruling" (Smith, 1987, p. 212).

Du Bois is thus arguing that African Americans have, by virtue of their position, a critical awareness of the American social system. This awareness can be a cultural resource that facilitates structural change. As Du Bois (1903/1996a) puts it, "This, then, is the end of his striving: to be a co-worker in the kingdom of culture, to escape both death and isolation, to husband and use his best powers and his latent genius" (p. 102).

Generally, Du Bois' perspective is more cultural and critical than either Martineau's or Gilman's, and in many ways his is more in keeping with contemporary theories of difference. One of the things that oppression in modernity has done is deny the voice of the other. In the Durkheim chapter, we saw that one of the necessities for social solidarity is a collective consciousness—or what some contemporary theorists refer to as a *grand narrative*. Modern nation-states provide all-encompassing stories about history and national identity through a grand narrative. The purpose of these narratives is to offer a kind of Durkheimian rallying point for social solidarity. This sort of solidarity is necessary for nations to carry out large-scale programs, especially such things as colonization and war. The problem with such a narrative is that it hides inequities. For example, the grand narrative of equality in the United States was actually a story about white Anglo-Saxon Protestant males. Hidden in the national narrative and history was (and still is) the subjugation of Native Americans, African Americans, women, Mexican Americans, homosexuals, and so forth.

Contemporary theories of difference, then, focus on the subjective experience of the disenfranchised in contrast to this grand narrative. One of the things that is important in the fight for equality is allowing multiple voices to be heard; thus, we have recently moved in the United States from the cultural picture of the "melting pot" to that of the "salad bowl" (with each ingredient maintaining its own unique character). Du Bois' perspective is quite in keeping with this emphasis. In his own work, he uses the subjective mode to express the experience of the oppressed. He becomes a representative figure through which we might understand the plight of black people in America. Part of what this multiple-voice approach entails is valuing the outsider's point of view. Interestingly, Du Bois is much more in tune with the feminist idea of standpoint theory than either Martineau or Gilman, neither of whom privileges outsider knowledge. In that sense, Du Bois' work contains a more critical edge, again in keeping with much of contemporary analysis.

Du Bois was a prolific writer, as were Martineau and Gilman, so I have chosen to focus on particular points of his discourse about race. In addition, Du Bois was at least as much a political figure as he was an intellectual one, so I've had to be even more selective with his writings. In my opinion, I think that Du Bois' lasting contribution to social theory is his understanding of *cultural oppression*. In the section of this chapter on cultural oppression, we will see that it is just as necessary as structural oppression in the suppression of a social group. Du Bois' understanding of this process is quite good. He argues that cultural oppression involves exclusion from history, specific kinds of symbolic representations, and the use of stereotypes and their cultural logic of default assumptions. This cultural work results in a kind of *double consciousness* wherein the disenfranchised see themselves from two contradictory points of view. However, Du Bois isn't only interested in cultural oppression; he also gives us a race-based theory of world capitalism in the section called "Concepts and Theory: The Dark Nations and World Capitalism." There we will see that it isn't only the elite capitalists that benefit from the exploitation of blacks and other people of color; the middle class benefits as well.

Concepts and Theory: Cultural Oppression

Exclusion From History

If we can get a sense of the subjectivity that Du Bois is trying to convey, we might also get a sense of what horrid weight comes with cultural oppression. Undergirding every oppressive structure is cultural exclusion. While the relative importance of structure and culture in social change can be argued, it is generally the case that structural oppression is legitimated and facilitated by specific cultural moves—historical cultural exclusion in particular. In this case, African Americans have been systematically excluded from American history, and they have been deprived of their own African history.

History plays an important part in legitimating our social structures; this is known as *history as ideology*. No one living has a personal memory of why we created the institutions that we have. So, for example, why does the government function the way it does in the United States? No one personally knows; instead, we have an historical account or story of how and why it came about. Because we weren't there, this history takes on objective qualities and feels like a fact, and this facticity legitimates our institutions and social arrangements unquestionably. But, Du Bois tells us, the current history is written from a politicized point of view: because women and people of color were not seen as having the same status and rights as white men, our history did not see them. We have been blind to their contributions and place in society. The fact that we now have Black and Women's History Months underscores this historical blindness. Du Bois calls this kind of ideological history "lies agreed upon."

Du Bois, however, holds out the possibility of a *scientific* history. This kind of history would be guided by ethical standards in research and interpretation, and the record of human action would be written with accuracy and faithfulness of detail. Du Bois envisions this history acting as a guidepost and measuring rod for national conduct. Du Bois (1935/1996c) presents this formulation of history as a choice. We can either use history "for our pleasure and amusement, for inflating our national ego," or we can use it as a moral guide and handbook for future generations (p. 440).

It is important for us to note here that Du Bois is foreshadowing the contemporary emphasis on culture in studies of inequality. Marx argued that it is class and class alone that matters. Weber noted that cultural groups—i.e., status positions—add a complexity to issues of stratification and inequality. But it was not until the work of the Frankfurt School (see Chapter 1 on Marx) during the 1930s that a critical view of culture itself became important, and it was not until the work of postmodernists and the Birmingham School in the 1970s and 1980s that representation became a focus of attention. Yet Du Bois is explicating the role of culture and representation in oppression in his 1903 book, *The Souls of Black Folk*.

Representation

Representation is a term that has become extremely important in contemporary cultural analysis. Stuart Hall, for example, argues that images and objects by themselves don't mean anything. We see this idea in Mead's theory as well. The meaning has to be constructed, and we use representational systems of concepts and ideas to do so. *Representation,* then, is the symbolic practice through which meaning is given to the world around us. It involves the production and consumption of cultural items and is a major site of conflict, negotiation, and potential oppression.

Let me give you an illustration from Du Bois. Cultural domination through representation implies that the predominantly white media do not truly represent people of color. As Du Bois (1920/1996d) says, "The whites obviously seldom picture brown and yellow folk, but for five hundred centuries they have exhausted every ingenuity of trick, of ridicule and caricature on black folk" (pp. 59–60). The effect of such representation is cultural and psychological: the disenfranchised read the representations and may become ashamed of their own image. Du Bois gives an example from his own work at *The Crisis* (the official publication of the NAACP). *The Crisis* put a picture of a black person on the cover of the magazine. When the readers saw the representation, they perceived it (or consumed it) as "the caricature that white folks intend when they make a black face." Du Bois queried some of his office staff about the reaction. They said the problem wasn't that the person was black; the problem was that the person was *too black*. To this Du Bois replied, "Nonsense! Do white people complain because their pictures are too white?" (Du Bois 1920/1996d, p. 60).

While Du Bois never phrased it quite this way, Roland Barthes (1964/1967), a contemporary semiologist (someone who studies signs), explains that cultural signs, symbols, and images can have both denotative and connotative functions. Denotative functions are the direct meanings that can be looked up in an ordinary dictionary. Cultural signs and images can also have secondary, or connotative, meanings. These meanings get attached to the original word and create other, wider fields of meaning.

At times these wider fields of meaning can act like myths, creating hidden meanings behind the apparent. Thus, systems of connotation can link ideological messages to more primary, denotative meanings. In cultural oppression, then, the dominant group represents those who are subjugated in such a way that negative connotative meanings and myths are produced. This complex layering of ideological meanings is why members of a disenfranchised group can simultaneously be proud and ashamed of their heritage. Case in point: the black office colleagues to whom Du Bois refers can be proud of being black but at the same time feel that an image is *too black*.

Stereotypes and Slippery Slopes

In addition to history and misrepresentation, the cultural representation of oppression consists of being defined as a problem: "Between me and the other world this is ever an unasked question. . . . How does it feel to be a problem?" (Du Bois, 1903/1996a, p. 101). Representations of the group thus focus on its shortcomings, and these images come to dominate the general culture as stereotypes:

> While sociologists gleefully count his bastards and his prostitutes, the very soul of the toiling, sweating black man is darkened by the shadow of a vast despair. Men call the shadow prejudice, and learnedly explain it as the natural defense of culture against barbarism, learning against ignorance, purity against crime, the "higher" against the "lower" races. To which the Negro cries Amen! (Du Bois, 1903/1996a, p. 105)

I want to point out that last bit of the quote from Du Bois. He is saying that the black person *agrees* with this cultural justification of oppression. Here we can see one of the insidious ways in which cultural justifications can work. It presents us with an apparent truth that once we agree to can reflexively destroy us. Here's how this bit of cultural logic works: The learned person says that discrimination and prejudice are necessary. Why? They are needed to demarcate the boundaries between civilized and uncivilized, knowledge and ignorance, morality and sin, right and wrong. We agree that we should be prejudiced against sin and evil, and against uncivilized and barbarous behavior, and we do so in a very concrete manner. For example, we are prejudiced against allowing a criminal into our home. We thus agree that prejudice is a good thing. Once we agree with the general thesis, it can then be more easily turned specifically against us.

In this movement from general to specific, Du Bois hints at another piece of cultural logic that is used in oppression. Douglas R. Hofstadter (1985), in reference to gender issues, calls this the "slippery slope of sexism." Hofstadter argues that there can be a relationship between the general and specific use of a term, and therefore some of the connotations of each will rub off on the other. You are aware of examples of this process in gender, such as in the statements "all men are created equal" and "there was a four-man crew on board." Is the use of the masculine pronoun meant in its specific or its general meaning? Are we referring to men specifically or to mankind? When such slippery slopes of language occur, it is easy for society to obliterate or oppress a cultural identity, which is one reason why feminist scholars talk about the invisible woman in history.

Du Bois has a similar slope in mind, but obviously one that entails race. In the section from *The Souls of Black Folk* from which I have been quoting, Du Bois (1903/1996a) says that the Negro stands "helpless, dismayed, and well-nigh speechless" before the "nameless prejudice" that becomes expressed in "the all-pervading desire to inculcate disdain for everything black" (p. 103). In *Darkwater,* Du Bois (1920/1996b) refers to this slippery slope as a "theory of human culture" (p. 505) that has "worked itself through [the] warp and woof of our daily thought" (p. 505). We use the term "white" to analogously refer to everything that is good, pure, and decent. The term "black" is likewise reserved for things or people that are despicable, ignorant, and that instill fear. There is thus a moral, default assumption in back of these terms that automatically includes the cultural identities of white and black.

In our cultural language, we also perceive these two categories as mutually exclusive. For example, we will use the phrase "this issue isn't black or white" to refer to something that is undecided, that can't fit in simple, clear, and mutually exclusive categories. The area in between is a gray, no-person's land. It is culturally logical, then, to perceive unchangeable differences between the black and white races, which is the cultural logic behind the "one drop rule" (an historical slang term used to capture the idea that a person is considered black if he or she has any black ancestor). Again, keep in mind that this movement between the specific and the general is unconsciously applied. People don't have to intentionally use these terms as ways to racially discriminate. The cultural default is simply there, waiting to swallow up the identities and individuals that lie in its path.

Double Consciousness

Du Bois attunes us to yet another insidious cultural mechanism of oppression: the internalization of the **double consciousness.** With this idea, Du Bois is drawing on his knowledge of early pragmatic theories of self. He knew William James and undoubtedly came in contact with the work of both Mead and Charles Horton Cooley (1998). We can think of Mead's theory of role-taking and Cooley's looking-glass self (see below) and see how the double consciousness is formed.

According to Du Bois, African Americans have another subjective awareness that comes from their particular group status (being black and all that that entails), and they have an awareness based on their general status group (being an American). What this means is that people in disenfranchised groups see themselves from two perspectives.

Let's think about this issue using Cooley's notion of the *looking-glass self*. In Cooley's theory, the sense of one's self is derived from the perceptions of others. There are three phases in Cooley's scheme. First, we imagine what we look like to other people; we then imagine their judgment of that appearance; and finally, we react emotionally with either pride or shame to that judgment. Note that Cooley didn't say we actually *perceive* how others see and judge us; rather, we *imagine* their perceptions and judgments. However, this imagination is not based on pure speculation. It is based on social concepts of ways to look (cultural images); ways to behave (scripts); and ways we anticipate others will behave, based on their social category (expectations). In this way, Du Bois argues that African Americans internalized the cultural images produced through the ruling, white culture. They also internalize their own group-specific culture. Blacks thus see themselves from at least two different and at times contradictory perspectives.

In several places, Du Bois relates experiences through which his own double consciousness was formed. In this particular one, we can see Cooley's looking glass at work:

> In a wee wooden schoolhouse, something put it into the boys' and girls' heads to buy gorgeous visiting-cards—ten cents a package—and exchange. The exchange was merry, till one girl, a tall newcomer, refused my card,—refused it peremptorily, with a glance. Then it dawned upon me with a certain suddenness that I was different from the others; or like, mayhap, in heart and life and longing, but shut out from their world by a vast veil. (Du Bois, 1903/1996a, p. 101)

We can thus see in Du Bois' work a general theory of cultural oppression. That is, every group that is oppressed structurally (economically and politically) will be oppressed culturally. The basic method is as he outlines it: Define and label the group in general as a problem, emphasize and stereotype the group's shortcomings and define them as intrinsic to the group, employ cultural mechanisms (such as misrepresentation and default assumptions) so the negative attributes are taken for granted, and systematically exclude the group from the grand narrative histories of the larger collective.

Concepts and Theory: The Dark Nations and World Capitalism

Exporting Exploitation

Du Bois also sees the social system working in a very Marxian way. Like Marx, Du Bois argues that the economic system is vitally important to society and for understanding history. Du Bois (1920/1996b) says that "history is economic history; living is earning a living" (p. 500). He also understands the expansion to global capitalism in Marxian, world systems terms. Capitalism is inherently expansive. Once limited to a few national borders, it has now spread to a worldwide economic system. One of the things this implies is that exploitation can be exported:

> Thus the world market most wildly and desperately sought today is the market where labor is cheapest and most helpless and profit is most abundant. This labor is kept cheap and helpless because the white world despises "darkies." (Du Bois, 1920/1996b, p. 507)

Capitalism requires a ready workforce that can be exploited. Goods must be made for less than their market value, and one of the prime ways to decrease costs and increase profits is by cutting labor costs. In national capitalism, this level of exploitation is limited to the confines of the state itself. Consequently, as the national economy expands and the living wage increases, the level of profit goes down. In other words, the average real wage of workers within a country increases over time—the average worker in the United States makes more today than she or he did 100 years ago. This proportional increase in income would lower profit margins. However, with global capitalism, exploitation can be exported. It costs less for a textile company to make clothing in China than in the United States, yet the market value of the garment remains the same. Profits thus go up.

Du Bois (1920/1996b) argues that this expansion of exploitation is due to increasing levels of "education, political power, and increased knowledge of the technique and meaning of the industrial process" (p. 504). There are, according to Marx and Du Bois, certain inevitable "trickledown" effects of capitalism. As industrial technologies become more and more sophisticated, better-educated workers are needed. Increasing levels of education attune workers not only to the methods of production, but also to the ideologies behind the system (for example, your own higher education includes reading the critical theory in this book). Industrialization also brings workers closer together in factories and neighborhoods where they can communicate and become politically active.

The Need for Color

As a result, the economic system inexorably pushes against its national boundaries and seeks a labor force ready for exploitation. This notion of expanding systems and exporting exploitation is part of Immanuel Wallerstein's world systems theory (Chapter 12). He argues that core nations exploit periphery nations so places like the United States can have high wages and relatively cheap goods. Wallerstein sees nations, but Du Bois (1920/1996b) tells us that the reality of those nations is one of color. "There is a chance for exploitation on an immense scale for inordinate profit, not simply to the very rich, but to the middle class and to the laborers. This chance lies in the exploitation of darker peoples" (pp. 504–505). These are "dark lands," ripe for exploitation, "with only one test of success,—dividends!" (p. 505). In other words, middle class wages in advanced industrialized nations are based on there being racial groups for exploitation.

Du Bois thus sees race as a tool of capitalism. While Du Bois perceives distinct differences between blacks and whites that can be characterized in spiritual terms (the souls of black and white folk), he also argues that the color line is socially constructed and politically meaningful. The really interesting feature of Du Bois' perspective here is that he understands that both black and white social identities are constructed. Again, Du Bois beat contemporary social theory to the punch: we didn't begin to seriously think of "white" as a construct until the 1970s, and it didn't become an important piece in our theorizing until the late 1980s.

Du Bois argues that the idea of "personal whiteness" is a very modern thing, coming into being only in the nineteenth and twentieth centuries. Humans have apparently always made distinctions, but not along racial lines. Prior to modernity, people created group boundaries of exclusion by marking civilized and uncivilized cultures, religions, and territorial identity. As William Roy (2001) notes, these boundaries lack the essential features of race—they were not seen as biologically rooted or immutable and people could thus change. Race, on the other hand, is perceived as immutable and is thus a much more powerful way of oppressing people.

As capitalism grew in power and its need for cheap labor increased, indentured and captive slavery moved to chattel slavery. Slavery has existed for much of human history, but it was used primarily as a tool for controlling and punishing a conquered people or criminal behavior, or as a method of paying off debt or getting ahead. The latter is referred to as indentured servitude. People would contract themselves into slavery, typically for seven years, in return for a specific service, like passage to America, or to pay off debt. In most of these forms of slavery, there were obligations that the master had to the slave, but not so with chattel slavery. Under capitalism, people could be defined as property—the word "chattel" itself means property—and there are no obligations of owner to property. In this move to chattel slavery, black became not simply *a* race but *the* race of distinction. The existence of race, then, immutably determined who could be owned and who was free, who had rights and who did not.

Du Bois Summary

- Du Bois' perspective is that of a black man, and his subjective experience of race is central in his work. He often uses himself as a representational character, and he is vastly interested in drawing the reader into the experience of race. Du Bois also believes that being a member of a disenfranchised group, specifically African American, gives one a privileged point of view. People who benefit from the system cannot truly see the system. Most of the effects of active social oppression are simply taken for granted by the majority. They cannot see them.

- All structural oppression must be accompanied by cultural oppression. There are several mechanisms of cultural suppression: denial of history, controlling representation, and the use of stereotypes and default assumptions. As a result of cultural suppression, minorities have a double consciousness. They are aware of themselves from their group's point of view, which is positive, and they are conscious of their identity from the oppressor's position, which is negative. They are aware of themselves as black (in the case of African Americans), and they are aware of themselves as American—two potentially opposing viewpoints.

- Capitalism is based on exploitation: owners pay workers less than the value of their work. Therefore, capitalism must always have a group to exploit. In global capitalism, where the capitalist economy overreaches the boundaries of the state, it is the same: there must be a group to exploit. Global capitalism finds such a group in the "dark nations." Capitalists thus export their exploitation, and both capitalists and white workers in the core nations benefit, primarily because goods produced on the backs of sweatshop labor are cheaper.

Looking Ahead

For most of the people in this book, we can draw clear lines between them and contemporary theory; that isn't necessarily true for Gilman and Du Bois. As we've seen, both Gilman and Du Bois were enormously popular in their day, and the same holds true for Harriet Martineau. Martineau's first serialized piece, for example, outsold Charles Dickens. By the time of her death, Martineau had published over 70 books and around 1,500 newspaper articles. Yet, despite this popularity, the works of Martineau, Gilman, and Du Bois were never adopted into the canon of social theory. Even contemporary race and gender theory isn't explicitly built upon them—it is rare to find any contemporary gender or race theorist referencing them in the same way we see Marx and Weber referenced. Thus, we can't talk about the theoretical legacy of Gilman and Du Bois in quite the same way, though their political influence is a different story. Although their theories aren't referenced in the same way as Marx and Weber, they can certainly be credited as pioneers and forerunners of current race and gender theory.

Certainly the sociological theories of William Julius Wilson and Janet Chafetz (Chapter 8) could not have been written without the forerunners we have in this chapter. Both Wilson and Chafetz give us powerful, positivistic theories of how structural inequality works around the issues of race and gender, though neither draws directly from Gilman or Du Bois.

But rather than these structural theories of inequality, I would venture to say that the greatest impact of people such as Gilman and Du Bois is the development of theories of identity politics. *Identity politics* encompasses "a way of knowing that sees lived experiences as important to creating knowledge and crafting group-based political strategies. Also, [it is] a form of political resistance where an oppressed group rejects its devalued status" (P. H. Collins, 2000, p. 299). Notice that identity politics focuses on lived experiences rather than structural inequality. We will see this emphasis played out in this book through the works of Dorothy Smith and Patricia Hill Collins (Chapter 13).

Notice also that identity politics creates its own forms of political resistance. One specific form of resistance is "self-definition": expressing and insisting upon authentic representations of a group's lived experiences, an emphasis we see with Du Bois. Identity politics thus focuses on the shared cultural identity of members as defined by the members themselves, rather than oppressing groups. This sense of identity is intentionally cultivated, as an important part of identity politics is members' focused efforts to understand, explore, express, and claim the distinctive qualities of their group: What is it like to be black in this society? What is it like to be gay? What is it like to be a black *woman*? What are the lived experiences of the indigenous peoples of this society? Political resistance can take other forms as well, as with the idea of "safe places" that we will cover later with Patricia Hill Collins.

Building Your Theory Toolbox

Knowing Gilman and Du Bois

After reading and understanding this chapter, you should be able to define the following terms theoretically and explain their theoretical importance:

New evolutionary environment, women's workforce participation, self-preservation, race-preservation, natural selection, monogamy, sexuo-economic evolution, gynaecocentric theory, morbid excess in sex distinction, centered subject, standpoint of the oppressed, grand narrative, history as ideology, representation, stereotypes, default assumptions, double consciousness, exploitation, personal whiteness

After reading and understanding this chapter, you should be able to

- Explain the differences between self- and race-preservation and discuss their implications for gender

- Explain the sexuo-economic theory of evolution and its effects on the economy and gender

- Discuss the reasons why the perspectives of oppressed groups are able to give the kinds of critical insights necessary for social change

- Describe how history and representation can be used as tools in oppression and in silencing the voices of disenfranchised groups

- Explain how race is used in capitalism

Learning More: Primary Sources

More than with any of our other theorists, you should read Martineau, Gilman, and Du Bois, rather than read *about* them.

- For Du Bois, I recommend that you pick up *The Oxford W. E. B. Du Bois Reader*, edited by Eric Sundquist. It contains two of Du Bois' books in their entirety (*The Souls of Black Folk* and *Darkwater*) as well as a myriad of other important writings.

- For Gilman, start with *The Yellow Wallpaper*, and then move on to *Women and Economics, The Home, Human Work*, and *The Man-Made World, or Our Androcentric Culture.*

Each of these thinkers also published autobiographies (in fact, Du Bois wrote two, at different points of his life): For Du Bois, *The Autobiography of W. E. B. Du Bois: A Soliloquy on Viewing My Life From the Last Decade of Its First Century;* and *Dusk of Dawn: An Essay Toward an Autobiography of a Race Concept;* and for Gilman, *The Living of Charlotte Perkins Gilman: An Autobiography.*

In terms of introductions to other neglected theorists, I recommend Howard Brotz's reader, *Negro Social and Political Thought, 1850–1920;* and *The Women Founders: Sociology and Social Theory, 1830–1930,* by Patricia Madoo Lengermann and Jill Niebrugge-Brantley.

Theory You Can Use (Seeing Your World Differently)

- Go home and watch TV. Intentionally watch programs that focus on African Americans and pay particular attention to commercials that feature people of color. Using Du Bois' understanding of representation and Barthes' ideas of denotation and connotation, analyze the images that you've seen. What are some of the underlying connotations of the representations of blacks, Chicanos, Asians, and other minorities? How do you think this influences the consciousness of members of these groups?
- Evaluate the idea of standpoint theory or critical knowledge. If we accept the idea that knowledge is a function of a group's social, historical, and cultural position, then is this idea of critical knowledge correct? If so, what are the implications for the way in which we carry on the study of society? If you disagree with the idea of standpoint theory, from what position is true or correct knowledge formed?
- According to Du Bois, one of the ways cultural oppression works is by excluding the voices and contributions of a specific group in the history of a society. The Anti-Defamation League has a group exercise called "name five." The challenge is to name five prominent individuals in each category: Americans; male Americans; female Americans; African Americans; Hispanic Americans; Asian or Pacific Islander Americans; Native Americans; Jewish Americans; Catholic Americans; pagan Americans; self-identified gay, lesbian, or bisexual Americans; Americans with disabilities; and Americans over the age of 65. For which categories can you name five prominent people? For which can't you name the five? What does this imply about the way we have constructed history in this country? In addition to trying this activity on yourself or a friend, go to the Biography Channel's Web page (www.biography.com). There you will find a searchable database of over 25,000 people whose lives are deemed important. Try each of the categories that we mentioned. What did you find?

Further Explorations—Web Links

Du Bois:

http://members.tripod.com/~DuBois/

http://www.duboislc.org/index.html

Gilman:

http://www.cortland.edu/gilman/

http://www.womenwriters.net/domesticgoddess/gilman1.html

Section II
Introduction

Theory Cumulation and Schools of Thought in the Mid-Twentieth Century

The essence of science is precisely theory . . . as a generalized and coherent body of ideas, which explain the range of variations in the empirical world in terms of general principles. . . . [I]t is explicitly cumulative and integrating.

(R. Collins, 1986a, p. 1345)

A true science incorporates the ideas of its early founders in introductory texts and moves on, giving over the analysis of its founders to history and philosophy.

(Turner, 1993, p. ix)

T here are two ideas in the above quotes that are important for understanding what is going on with theory in our next section: the notion of generalized theory and that of theory cumulation. In order to get a good handle on these issues, let's back up a bit and remind ourselves about the assumptions of science. Recall that science is built upon positivism and empiricism. As such, science assumes that the universe is empirical, it operates according to law-like principles, and humans can discover those laws through rigorous investigation. Science also has very specific goals, as do most knowledge systems. Through discovery, scientists want to explain, predict, and control phenomena.

Please pay careful attention to the assumptions and goals of science. Every knowledge system is built upon assumptions and usually has implicit or explicit goals. For example, we can say that, in general, religion assumes that the universe is spiritual, it operates according to principles laid down by the Creator, and that humans have a very specific place in this universe: to receive the revelations of God and to live accordingly. I hope it's clear that I'm not arguing for or against either science or religion. That's not my intent here. My purpose is to drive home the point that the assumptions and goals of any knowledge system determine the kind of knowledge that is created. Thus, if we want to understand the business of sociology—a specific way of knowing—then we need to grasp the foundations upon which it is built.

Generalized Theory

By the time sociology in the United States reached the mid-twentieth century, it was busy organizing itself into a social science—as our two quotes point out, theory is the core of sociology and any scientific enterprise. In order to fulfill the goals of science, *scientific theory* is made up of at least three elements: concepts, definitions, and relationships. *Concepts* are the ideas we use to understand what we are looking at, and in science, concepts must be abstract. Think again about the goals of science. Prediction and control always require generalized knowledge, because it must be usable in many settings. In order for knowledge to be generalized out of a single empirical setting, it must be abstract.

For example, here is a quote from Marx that is tied too strongly to an empirical state: "Modern industry has established the world market, for which the discovery of America paved the way." As is, the statement can't be used to predict anything because America can only be discovered once. One way of making this statement more abstract is to change "discovery of America" to "geographic expansion." We then can make a usable, testable statement (this is called a *proposition*): The greater is the level of geographic expansion, the greater will be the level of market development. This is a very important point: *in order to fulfill the goals of science, theory must be abstract and cannot be tied to the context.* Therefore, the goals and assumptions of science necessitate a certain kind of theory.

Further, because a concept is an idea and not an object, we must be very careful to stake out the parameters of the concept through explicit and uniform *definitions.*

A rock, for example, is an object and we can usually tell where it begins and ends. But the case isn't as clear when we are talking about gender. Where does masculinity begin and end? What will count as masculine and what will not? Also, science strives to construct its knowledge in such a way that it is a public activity; that is, the very methods employed to construct the knowledge are explicit and known. Accordingly, scientific theory should contain explicit definitions of all the concepts used so that the knowledge constructed from the theory can be tested and replicated by others.

One further element we must consider is that scientific theory contains *relational statements*. These are statements that explain the relationships among and between the concepts. The relationship statements contain the variability of a theory. In other words, scientists know how a phenomenon will change because they understand the relationships among their concepts. Think again about the goals of scientific theory and you'll see that these relational statements are necessary. Explanation, prediction, and control imply something dynamic, not static (otherwise there wouldn't be anything to predict). For example, let's take two concepts, education and income. By themselves, they simply allow us to identify qualities that appear to be associated with two different entities. However, if we could put them together or relate them to each other in some way, then we might be able to make some predictions. Something like, the greater is the level of education, the greater will be the level of income. The phrases "the greater is" and "the greater will be" are relationship statements. In this case, the relationship is positive, that is, they both vary in the same direction in relation to each other (up or down); negative relations vary inversely.

Theory Cumulation

Scientific knowledge involves both theory synthesis and cumulation. Synthesis involves bringing together two or more elements in order to form a new whole. For example, water is the synthesis of hydrogen and oxygen. *Theoretical synthesis*, then, involves bringing together elements from diverse theorists so as to form a theory that robustly explains a broader range of phenomena. Cumulation refers to the gradual building up of something, such as the cumulative effects of drinking alcohol. *Theory cumulation* specifically involves the building up of explanations over time. This incremental building is captured by Isaac Newton's famous dictum, "If I have seen further it is by standing on the shoulders of giants." Yet, what isn't clear in Newton's quote is that the ultimate goal of theory cumulation is to forget its predecessors.

Robert K. Merton, whom we will meet in Chapter 6, makes the distinction between a scientific approach to theory and the humanities' method of understanding knowledge. There is a way in which sociology exists between the life and physical sciences on one hand and the humanities on the other. As such, it feels pressure from both sides to conform. In science, *obliteration by incorporation* is the rule (Merton, 1967, p. 27). The very nature of scientific work implies that each accomplishment will raise new questions and new problems to be solved, and each work will be overshadowed and outdated by the next. But in the humanities, each classic is viewed as distinct and viable on its own; and each work demands the direct experience of every scholar.

To make this clear, let's compare the writings of two authors, Edgar Allan Poe and Albert Einstein. Here's one of Poe's famous stanzas:

> Once upon a midnight dreary, while I pondered, weak and weary,
> over many a quaint and curious volume of forgotten lore,
> While I nodded, nearly napping, suddenly there came a tapping,
> As of someone gently rapping, rapping at my chamber door.
> "'Tis some visitor," I muttered, "tapping at my chamber door;
> Only this, and nothing more."

Here's one of Einstein's famous quotes:

$$E = mc^2$$

There are some obvious differences between these two quotes: one is poetry and the other a mathematical equation. But I want you to see a bit more. Does it matter who wrote "Once upon a midnight dreary"? Yes, it does. A large part of understanding poetry is knowing who wrote it—who they were, how they lived, what their other works are like, what style they wrote in, and so on. These issues are part of what makes reading Poe different from reading Emily Dickinson. Now, does it matter who wrote $E=mc^2$? Not really. You can understand everything you need to know about $E=mc^2$ simply by understanding the equation. The author in this sense is immaterial.

Recall the quotation with which we began this introduction, from Jonathan H. Turner. It comes from a book entitled *Classical Sociological Theory: A Positivist's Perspective*. Turner's (1993) goal in that book is "to codify the wisdom of the masters so that we can move on and *make books on classical theory unnecessary*" (p. ix, emphasis added). That last highlighted section is the heart of theory cumulation: cumulating theory implies that we do away with the individual authors and historic contexts and keep only the theoretical ideas that explain, predict, and control the social world. In that spirit, here's a theoretical statement from Turner's (1993) book:

> The degree of differentiation among a population of actors is a gradual s-function of the level of competition among these actors, with the latter variable being an additive function of
>
> A. the size of this population of actors,
>
> B. the rate of growth in this population,
>
> C. the extent of economical concentration of this population, and
>
> D. the rate of mobility of actors in this population. (p. 80)

Now, you and I might know from whom this proposition comes (Durkheim), but does it matter? No. Like Einstein's formula, it's immaterial. If we are doing social science, what matters is whether or not we can show this statement to be false through scientific testing. If we can't, then we can have a certain level of confidence that the proposition accurately reflects a general process in the social world. In science, authorship is superfluous; *it's the explanatory power of the theory that matters*. The cumulation of these general statements is one of the main goals of scientific theory.

Parsons's Project

We begin this section of our book with the work of Talcott Parsons. For much of the twentieth century, Parsons was "the major theoretical figure in English-speaking sociology, if not in world sociology" (Marshall, 1998, p. 480). As Victor Lidz (2000) notes, "Talcott Parsons . . . was, and remains, the pre-eminent American sociologist" (p. 388). Yet, it wasn't simply Parsons's theory that gave him this stature; it was the method he used to theorize and the vision he had for social science. Imagine the possibilities if we knew how society works just as well as we know how gravity or an automobile motor works.

Of his groundbreaking work, Parsons (1949) says, "*The Structure of Social Action* was intended to be primarily a contribution to systematic social science and not to history" (pp. A–B). His work is actually a synthesis of three theorists. Parsons (1961) notes how he used each one:

> for the conception of the social system and the bases of its integration, the work of Durkheim; for the comparative analysis of social structure and for the analysis of the borderline between social systems and culture, that of Max Weber; and for the articulation between social systems and personality, that of Freud. (p. 31)

Yet Parsons clearly wants us to forget the historical and personal origins of the theories—it's the power of the synthesized theory to illuminate and delineate social factors and processes that matters.

Parsons not only set the tone for sociology through his theoretical work, he also pointed the way administratively by establishing the Department of Social Relations at Harvard University. The department was made up by combining elements from sociology, psychology, and anthropology. Parsons's goal was to produce a general theory of social action that transcended disciplinary boundaries. The important thing to note is that Parsons exemplified what a scientific approach to understanding society ought to be about. His goal was to take elements from previous theories and combine them into a single abstract theory that explains the general processes in the social world. He also saw that the social world wasn't simply limited to sociology, and he worked to combine the various social disciplines into a general science of social relations.

We can see the continuation of Parsons's project throughout this section of the book, where many of our theorists are synthesizing various theoretical elements into more general and better-organized theories. An especially clear example of theory cumulation and synthesis is found in Chapter 8. There, Janet Chafetz brings together no less than six theorists to explain the general properties of gender stratification.

This kind of theorizing is also used in the three sociological perspectives or paradigms you were taught in your introduction to sociology courses: structural-functionalism, conflict theory, and interactionism. It was primarily during the mid-twentieth century that these various schools of sociological theory were defined and ordered. For example, the initial ideas of functionalism were generated by Auguste Comte, Herbert Spencer, and Émile Durkheim, but it was the work of Parsons, more than any other, that systematized functionalism into an identifiable school of thought. Conflict theory was also not clearly defined until the middle of the last century. While beginning in the work of Marx, Weber, and Simmel, conflict theory as a clear theoretical approach concerned with the general processes of conflict in society came about through the work of such people as Lewis Coser and Ralf Dahrendorf. Symbolic interactionism as well was based on previous theories but systematized in the twentieth century in the work of Herbert Blumer.

Enter Critical Theory

Not all theorizing during this time was scientific in nature, but then, having diverse perspectives isn't new for sociology. We specifically saw alternative approaches to theory in the previous section in the work of Max Weber and George Herbert Mead. Weber felt that, while we can certainly explain what is going on socially, we will have a difficult time predicting (let alone controlling) history and human behavior. We saw that his doubts concerning scientific theory revolved around the issues of culture, agency, and human values. Mead argued that because human beings are pragmatically oriented toward meaning, things like social institutions are best thought of as generalized others with which individuals may role-take in social encounters. Social encounters, then, are symbolic interactions through which meanings and society emerge.

These issues of agency and meaning continue to be concerns for mid-twentieth-century sociology; we can clearly see them coming up again in Chapter 9: Interactionist Theories. Yet there is another strain of theorizing that comes to fruition during this time, one that originated with Karl Marx. Marx clearly gave us a mechanistic or structural way of understanding society, and in this sense his theory lends itself readily to the scientific approach. There is, however, another vein in Marx's theory to be mined: his critical approach.

I will introduce you to critical theory in the second part of Chapter 7. But for now, let me give you two points that you can hang your hat on. First, *critical theory* "aims to dig beneath the surface of social life and uncover the assumptions and masks that keep us from a full and true understanding of how the world works"

(Johnson, 2000, p. 67). Critical theory doesn't simply explain how society operates. Rather, it uncovers the unseen or misrecognized ways in which "how society operates" actually works to oppress certain groups while maintaining the interests of others. The second thing I want to point out about critical theory is that its central principles "can perhaps be defined most clearly in contrast to some of the principles of twentieth-century positivism" (Marshall, 1998, p. 130). Thus, critical theory has a very clear agenda that stands in opposition to scientific sociology.

I am including critical theory in this section of the book for a number of reasons. The most obvious one is that critical theory began during the 1920s through the work of the Frankfurt School, so it falls within the general time frame of this section. Our critical theorist in the second part of Chapter 7 is Jürgen Habermas. Habermas is interesting because, while he is not a positivist, he is nonetheless a modernist—he clearly believes in the ideals of the modern project but doesn't think positivism can be used to achieve those goals.

Another, more important reason for including critical theory in this section is that it is a prelude to the "polyphony" that is contemporary theory. Most of us are familiar with the term symphony; it speaks of a harmony of sounds. Like Beethoven's Fifth Symphony, there are many parts, but all come together to express one musical piece. *Polyphony* is a musical composition that has melodically independent and individual parts or voices that are played simultaneously. Sometimes a polyphony can weave together the various strands into a pleasing whole, but other times the various parts of a polyphony are in competition with one another and the composition may sound harsh. Polyphonic music is a good analogy for contemporary theory in sociology; there are many voices that may be contesting with each other or providing us with a multifaceted view of social things.

One of the distinctions that people are making today in contemporary theory is between sociological and social theories. While definitions of debated topics such as this tend to be contested themselves, I think a recent issue of the American Sociological Association's *Theory Section* newsletter does a good job of outlining the issues:

> *Sociological theorists* are less concerned with criticizing and rebuilding society than with understanding it. They tend to be committed to a scientific sociology, as least in the broadest sense of the term. They may do general theory, or concentrate on formulating specific theories of particular substantive phenomena, and in some cases combine the two.
>
> *Social theorists* see themselves as social commentators and critics and as formulating theoretical critiques of modern society as much as, or more than, explaining social life. They are usually not committed to scientific sociology and are often strongly opposed to it. Their goals are primarily or even exclusively political. (Sanderson, 2005, pp. 2–3, emphasis added)

I'm sure you can see the similarity of ideas between critical theory and social theory. One way to think about it is to see critical theory, as it is specifically defined by such people as Habermas, as a type of social theory, with social theory being the broader category. In general, this book moves from more sociological theories to more social theories, from theories more concerned with empirically describing and explaining social behavior to those concerned with critiquing and rebuilding. I'm introducing critical, social theorizing in this section, but we will see it become much more pronounced in the last section of the book, Contemporary New Visions and Critiques.

Building Your Theory Toolbox

Knowing Sociological Schools of Thought

After reading and understanding this section introduction, you should be able to define the following terms theoretically and explain their importance to understanding sociological theory:

> scientific theory, concepts, definitions, relational statements, propositions, theory cumulation, theoretical synthesis, critical and social theories

After reading and understanding this introduction, you should be able to

- Explain the assumptions and goals of science
- Explain why the assumptions and goals of science demand a specific form of theorizing
- Discuss the basic differences between sociological and critical (social) theories

CHAPTER 6

Structural Functionalism

Talcott Parsons (1902–1979)

Robert K. Merton (1910–2003)

Seeing Further: Inter-Structural Relations and Society as a Whole

As you know, the United States isn't the only capitalist nation on earth. In fact, we could say that there are a number of what Jeffrey Hart (1993) calls "rival capitalists" in the world today. These nations are in the grips of a global competition. Currently, the United States is the top, or core, capitalist nation. However, because capitalism involves competition and there are rival capitalists, the United States may not always hold the top position. Is there a way that we could predict an outcome in this competition? How would you even begin to think about such a question?

We would have to be able to think of nations as social actors, and we would have to be able to conceptualize ways in which nations are different. In what ways can nations be different, especially nations that have similar economic practices? One of the best methods to compare and contrast nations is to look at the institutional arrangements within each nation. Jeffrey Hart did just that. He compared the different structural configurations among education, the economy, and government for five rival capitalists—the United States, France, Germany, Britain, and Japan. Based on this structural comparison, Hart predicts that one of these rival capitalists will emerge as the winner in the long run.

In order to find out who wins—and the answer may surprise you—you'll have to read his book. For now, what I want you to see is that thinking about large-scale social actors, such as nations or institutions, takes a very specific approach, one that is found in structural-functionalism. We were introduced to the functional perspective in Chapter 3 with the work of Émile Durkheim. There we saw that

Durkheim's concern was with structural differentiation and its influence on social differentiation and social solidarity.

As with Durkheim and Hart, true institutional or structural analysis will always be oriented toward the interconnections among and between the institutions. For example, if you want to study family, looking at the internal workings of kinship in your society is only a small part of understanding that institution. If you were to look at family in U.S. society, you might find that divorce rates and the number of "latchkey kids" have increased. So what? "Well," you say, "that's bad." How do you know? Unless you are simply going to make a moral argument, chances are that you will have to make a functional argument about how divorce and latchkey kids are dysfunctional for a specific type of social system.

You can't simply say that divorce is breaking apart the family and therefore it is bad. For example, let's say we lived several hundred years ago. At that point in time, we might look around us and see that our forms of government and economy are falling apart. We could look at any number of indicators (such as the use of standardized money, free markets, centralized taxation, bureaucratization, and so on) and say, "See? All these things are bad because they are leading to the downfall of feudalism." But we would have been misguided.

The downfall of feudalism was necessary for the advent of democracy and capitalism, and most of us seem pretty enamored with those. Thus, in order to understand anything like changes in government or changes in family, it is imperative that we place them in their context. In order to correctly understand how and why the kinship phenomena are occurring, you must place family firmly in its context, at least the interinstitutional relations that exist among family, economy, and government (you may want to add religion and education in there, too). This is the kind of perspective that functionalism gives us.

Defining Functionalism

Functionalism gives us the tools to look at society as a system of interrelated parts. It can be used in macro-level analysis of such things as society or global capitalism; and it is quite often used at the micro level to understand such small systems as the family unit (this is where the term "dysfunctional family" comes from). Using the term "structural functionalism" means that we are concerned with macro-level phenomena and the interrelationships among different social structures; it sets this approach off from the micro-level perspective of dysfunctional families.

Structural functionalism began with the work of Herbert Spencer in the nineteenth century and was systematized by Talcott Parsons in the twentieth. In general, functionalism is built around the **organismic analogy** and the evolutionary model. It has at least five defining features:

1. Every system has requisite needs that must be met in order for that system to survive.

2. Specialized structures **function** to satisfy the needs of the system. Social structures, functions, and the systemic whole are thus intrinsically related and affect one another.

3. Specialization of structures occurs through the evolutionary process of differentiation. That is, over time systems tend to become more complex through structural differentiation. Structural differentiation, in turn, makes the system more adaptive.

4. Differentiation creates problems of coordination and control, which, in turn, create evolutionary pressures for the selection of integrating processes, such as Durkheim's generalized values or Parsons's generalized media of exchange.

5. Integrating processes tend to keep the system in a state of equilibrium.

In this chapter, I will be introducing the work of two functionalists, Talcott Parsons and Robert K. Merton. Parsons sets the overall parameters of functionalist thinking in the twentieth century and gives us a very abstract, analytical model of functionalism. Merton was a student of Parsons, yet he also was critical of certain elements of his mentor's approach. As we'll see, Merton wanted to more firmly ground functionalist analysis in the empirical world, and he wanted to open the perspective up to alternative possibilities.

Talcott Parsons: Analytical Functionalism

Photo: © Granger Collection.

The Essential Parsons

Biography

Talcott Parsons was born December 13, 1902, in Colorado Springs, Colorado. As a young man, Parsons began his university studies at Amherst. He planned on becoming a physician but later changed his major to economics. Parsons received his B.A. in 1924. Beginning in that year, Parsons studied political economy abroad, first at the London School of Economics. There he came in contact with the anthropologist Bronislaw Malinowski, who was teaching a modified Spencerian functionalism. Parsons also met his wife-to-be, Helen Walker, while in London. Parsons then studied at the University of Heidelberg, where Max Weber had attended and taught. At Heidelberg, Parsons studied with Karl Jaspers, who had been a personal friend of Weber's.

After teaching a short while at Amherst, Parsons obtained a lecturing position at Harvard. He was one of the first instructors (along with Carle Zimmerman and Pitirim Sorokin) in Harvard's new sociology department in 1931. Parsons became department chair in 1942 and began work on the Department of Social Relations—formed by combining sociology, anthropology, and psychology. It was Parsons's vision to create a general science of human behavior. The department was in existence from 1945 to 1972 and formed the basis of other interdisciplinary programs across the United States.

After 10 years of work, Parsons's first book was published in 1937: *The Structure of Social Action*. More than any other single book, it introduced European thinkers to American sociologists and created the first list of "classical" theorists. Parsons was particularly responsible for bringing Max Weber to the attention of U.S. sociologists; Parsons translated several of Weber's works, including *The Protestant Ethic and the Spirit of Capitalism*.

Parsons died on May 8, 1979, while touring Germany on the 50th anniversary of his graduation from Heidelberg.

Passionate Curiosity

Parsons was a man with a grand vision. He wanted to unite the social and behavioral disciplines into a single social science and to create a single theoretical perspective. His desire, then, wasn't simply to understand a portion of human behavior; he wanted, rather, to comprehend the totality of the human context and to offer a full and complete explanation of social action. Parsons was possibly more passionate about this holistic view of social science than any before or after him.

Keys to Knowing

Voluntaristic action, action theory, the unit act, modes or orientation, adaptation, goal attainment, integration, latent pattern maintenance, socialization, cybernetic hierarchy of control, equilibrium, cultural strain, alienative motivational elements

Concepts and Theory: Making the Social System

If you were going to study society as a whole, where would you begin? You would probably start with what you know about society, or with your definition of society. Maybe you'd start with the idea of institutions or social structure. Like most of us, you'd probably start with a specific institutional problem, like institutionalized racism or class-based inequality, and study how it affects human behavior. You would, however, have to push beyond that level of analysis if you were going to study society as a whole—you would have to think more abstractly.

Let me put it to you this way: If you knew absolutely nothing about television and you were confronted with this box-shaped object that pictured moving people in it, how would you begin to understand it? You might begin with the picture itself, but the problem with that approach is that you would be looking at the effect rather than the object or process itself. That's kind of what we would be doing if we started thinking about the whole of society from the point of view of institutional effects. Another way to begin that makes sense is to start with one of the cords leading into the television, either the power cord or the cable. You might unplug one and then the other to figure out which one is more basic and begin with it, building up from the smallest component to the largest. That's exactly where Parsons begins his understanding of society—at its most basic level.

Voluntaristic Action

Parsons credits Max Weber with his beginning point for theory. This area of theorizing is referred to as action theory. **Action theory** references a group of theories that focus on human action rather than structure. Within this group are theories that center on meaning and interpretation (like Mead), and theories that are concerned with the nature of human action (like exchange theory, Chapter 10). Weber was concerned with both, but he explicitly developed a typology of social action.

According to Weber, simple behavior is distinguished from social action by the subjective orientation of the actor. If, in the action, the person takes into consideration the meaning of the act for others, then it is social action. Weber argued that there are four distinct types of social action: traditional, affective, value-rational, and instrumental-rational. The focus of Weberian action theory is on the latter two types, and it is concerned with the degree of rationality in human behavior. Exchange and rational choice theories take up these issues in particular, and argue that action is best understood in terms of people making rational choices in which they maximize their utilities (hence the name "utilitarianism"). In other words, people have clear preferences and try to get the most out of every encounter with other people by weighing costs and benefits. The question here becomes, how rational can people be in exchanges?

Parsons specifically names his approach "voluntaristic action theory." He isn't so much concerned about how meanings are negotiated in interaction, like Mead, but he wants to understand the context of human action. Like rational choice theories, voluntaristic action draws from utilitarianism in that it sees humans as making choices between means and ends. But it modifies utilitarianism by seeing these choices as circumscribed by the physical and cultural environments.

An example of voluntaristic action is your behavior right now. In order to read this book, you had to enroll in class, pick up the syllabus, buy the book, schedule time to read, and actually sit down and read it. All of this may be seen as voluntaristic action: you voluntarily acted, choosing among various ends and means—you could be drinking a beer and watching TV right now, but you selected this behavior. "But," you say, "I didn't volunteer to do this class work!" Yes, you did; nobody physically forced you. However, you volunteered under certain influences from the environment. The same with shaving or not shaving this morning, or the clothes you are wearing right now. They are all aspects of voluntaristic action; the questions have to do with how much freedom you have in making choices in action, and what goes into the decisions that you make in order to act.

Parsons's first theoretical work, *The Structure of Social Action*, explains this by giving us an analytical model of action. It's a framework or scheme through which we can view and understand human action. Parsons's scheme doesn't predict the kinds of actions in which people will engage; his thinking is more fundamental than that. He gives us an analytical model that we can take into any situation and begin to understand the myriad elements that go into human action.

The Unit Act

This analytical model in its completion is termed the *unit act* and is pictured in Figure 6.1. As you can see from the figure, Parsons argues that every act entails two essential ingredients: an agent or actor and a set of goals toward which the action is directed. We can see here Weber's notion of social action (in particular, instrumental rationality) and its influence on Parsons's theory. The initial state within which the actor chooses goals and directs his or her process of action has two important elements: the conditions of action and the means of action. The actor has little immediate agency or choice over the conditions under which action takes place. Parsons has in mind such things as the presence of social institutions or organizations, as well as elements that might be specific to the situation, such as the social influence of particular people or physical constraints of the environment. For example, being at a fraternity party (physical setting) will influence your action, but so will the presence of your parents (social influence) at the same party.

When it comes to the means, on the other hand, the actor does potentially have choice. Some situations allow quite a bit of freedom, but others, such as being in a jail, do not. However, notice that even in those situations where there is freedom of choice about the means and goals, the normative orientation of the action limits or defines the choices made. For Parsons, this is an extremely important point. Parsons

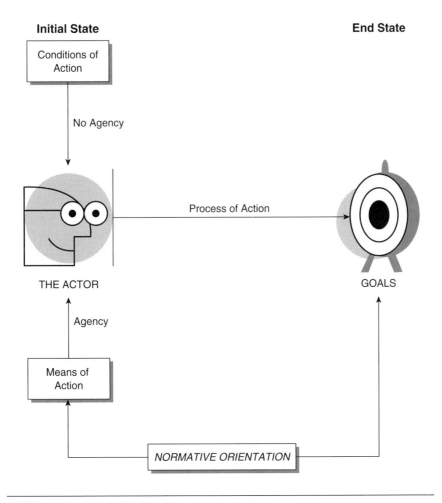

Figure 6.1   The Unit Act

says that the concept of human action demands that there be normative influence. In other words, Parsons is arguing that human action is distinctly *cultural* action. Remember that *norms* are behaviors that have sanctions attached to them, be they positive or negative. Such behaviors necessarily have social meaning attached to them, have a position on a value hierarchy, and are directly or indirectly related to some social group. Most of the rest of Parsons's work may be seen as an explication of the environment, particularly the cultural one, wherein social action takes place.

Notice also that Figure 6.1 does not contain any predicted relationships—none of the arrows have positive or negative signs. This kind of diagram is an analytical model. That is, its function is to call our attention to particular items in a given phenomenon. Thus, when a human being acts, we can understand that action by breaking it down into these different parts. For example, think about the reasons you decided (voluntaristic action) to come to college. What was your goal in such a decision? How were the methods you used influenced by cultural norms, values, and beliefs? Were there any conditions of action that were set for you by others, such as your parents or peers?

Constraining and Patterning Social Action

Having understood that human action is circumscribed by various conditions, Parsons sets out to understand those circumstances. In explaining the conditions, Parsons in the end creates an abstract theory of the social system. Let's begin by thinking about the situation. Remember, Parsons is thinking abstractly, so we don't want to be too specific. So, what do you bring with you to a social interaction, whether you are meeting to practice for a play or to study for a test? Parson argues that you bring two things: motivations and values. He refers to these as **modes of orientation** because they orient or position us within the situation, whatever it might be. Motives, of course, refer to something within a person (such as a need, idea, or emotion) that stimulates her or him to action. Motives are the energy for action: a person without motivation is like a car without gas—nothing happens. Value refers to a thing's worth and it is always a position on a hierarchy—some things are valued more highly than others. Thus, in every situation we are motivated to do something and we have certain ideas about what will be valued in the encounter with regard to the action.

According to Parsons (1990), all social action is understood in terms of some form of relation between means and ends: "This appears to be one of the ultimate facts of human life we cannot get behind or think away" (p. 320). The relationship between means and ends is formed through shared value systems. In this sense, cultural value systems function basically as a scale of priorities that contains the fundamental alternatives of selective orientations. This shared value system prioritizes means and ends, and, because it is shared, value hierarchies stabilize interactions across time and situations. Action and interaction would be disorganized without the presence of a value system that organizes and prioritizes goals and means. The evaluative aspect of culture is particularly important in this respect because it defines the patterns of role expectations and sanctions, and the standards of cognitive as well as appreciative judgments for any interaction. In other words, the values that we hold tell us the kinds of behaviors we can expect from others and how to judge those behaviors and other social objects.

Let me give you an example. I had coffee with a colleague the other day. We are in the planning stages of a book on sociological social psychology. There are a number of ways to look at social psychology from a sociological position, and in my department there are several people who hold these various views. I am not planning on writing a book with them, nor did I invite them to go to coffee. It's not that I don't like them; I do, but I have a particular perspective about social psychology based on valuing certain aspects of the literature more than others. My colleague shares those views. Because of that shared value system, we know what to expect from each other and how to value and appraise what is said and done, both intellectually and aesthetically.

Parsons sees value systems as having multiple levels that correspond to various degrees of commitment. Action may be stabilized through a shared system of meanings and priorities, but for a society to be integrated, people need to be committed to paying the costs necessary to preserve the system. In any functional interaction, we can name and prioritize the things that are important, but this discursive or cognitive accounting isn't enough. We must also be committed to some things more than others in terms of willingness to sacrifice. People are compelled to sacrifice when the collective *means* something to them, that is, when they have a significant level of emotional investment in the group; the more meaningful is the collective, the more willing people are to make sacrifices. Thus, the value system of any group varies by degree of commitment, with orientations and preferences at the most basic level and ultimate meanings and values at the highest level.

Like Durkheim, Parsons recognizes the basic human need for "ultimate" meanings, yet his argument concerning the need for ultimate significance is more tied to group identification and Weber's concern with legitimacy. Parsons argues that interaction requires individual actions to have meanings that are definable with reference to a common set of normative conceptions. In other words, our behaviors become meaningful because they are related to a group and its expectations.

Group identity is, of course, symbolic. In general, humans are not tied together by blood or instinct but by *meaningful* issues, such as the ideas of freedom and democracy in a national identity. By their very nature, symbols require legitimacy, grounds for believing in the meanings and system. No symbolic or normative systems are ever self-legitimating, nor are they legitimated by appeal to simple utilitarian issues. For example, we rarely hear someone justify the institution of American education by saying that it is necessary to the survival of the American system (unless you're in a sociology class). Legitimating stories always appeal to a higher source. Thus, Parsons (1966) argues that legitimation is always "meaningfully dependent" (p. 11) upon issues of ultimate meaning and therefore is always in some sense religious.

According to Parsons, there are three general kinds of values we hold: cognitive, appreciative, and moral. In other words, in any situation, we will place importance on empirical, factual knowledge (cognitive); standards of beauty and art (appreciative); or ultimate standards of right and wrong (moral). There are also three kinds of motives: cognitive, cathectic, and evaluative. *Cognitive motivation* refers to a need for information. You might be motivated to meet with your advisor because you need information about which courses to take. *Cathectic motivation* is the need for emotional attachment. You might feel the need to call home some weekend in order to experience emotional attachment to your family. With *evaluative motivation,* we are prompted to act because we feel the need for assessment, such as talking with your boss halfway to your year-end evaluation to find out where you stand.

There are also three types of cultural patterns. Culture acts as a resource for both our motivations and our values in action, so it shouldn't surprise us to find that Parsons's types of culture correspond to his types of motivations and values. Culture, then, contains a *belief system.* While we might think of beliefs in a religious sense, Parsons has in mind belief as cognitive significance. It's interesting that he would phrase cognitions in terms of belief. He's acknowledging that the ideas we hold in our head, through which we see and know the world around us, are in fact beliefs about the way things are. Culture also contains *expressive symbols.* Thus, culture not only provides the things we know, it also patterns the way we feel. These feelings are captured, understood, and expressed through symbols such as wedding rings to express love, or gang colors to symbolize aggression. Culture also contains systems of *value-orientation standards.* It is culture that tells us what to value and how to value it.

These different kinds of motives, values, and cultural patterns combine to produce three distinct types of social action. These function much like Weber's ideal types in that they are ways of understanding action, and none of them usually appears in its pure form. I've pictured how these ideal types are formed in Figure 6.2. Each type of action—strategic, expressive, and moral—is formed by combining a motivation with a value. Each of the specific culture systems provides information and meaning for each of the action types as well as the corresponding needs and values. As you can see, the ideal type of instrumental action is composed of the need for information and evaluation by objective criteria. Expressive action is motivated by the need for emotional attachment and the desire to be evaluated by artistic standards. Moral action is motivated by the need for assessment by ultimate notions of right and wrong.

In any social encounter, then, we will be oriented toward it with varying degrees of motivation and values. There are three different kinds of motivations and three types of value systems. The motivations and values will combine to create three different types of action. Any social action will have varying degrees of each type, depending on the specific combination of motives and values, and can be understood in terms of being closer to or further from any of the ideal types of action. However, the importance of this scheme of social action is not simply its potential for measurement. Parsons goes beyond Weber in proposing a typology of social action, and he uses it to form a broader theory of institutionalization.

What Parsons is doing is building from the ground floor to the top of the system, from the actor in the unit act to society. He starts with one small action and

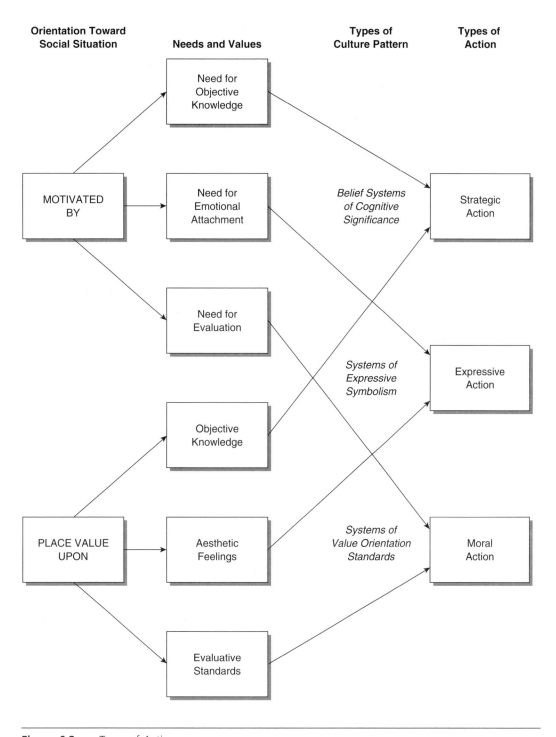

Figure 6.2   Types of Action

then argues that actors have discernable orientations toward their behavior. As a result, actions tend to fall into specific types. In brief, we have come this far: voluntaristic action → unit act → modes of orientation → types of action. From this point, Parsons argues that people tend to interact with others who share similar orientations and actors. So, if I want to engage in strategic, instrumental action, with whom will I be most likely to interact? For instance, if I'm interested in buying a guitar that has been advertised in the paper (strategic), then I'm not interested in interacting with someone who wants to talk about the evils of rock music (moral) or the beauty of a Vivaldi concerto (expressive).

Obviously, I will seek out others who want the same kind of thing out of the interaction. As we interact over time with people who are likewise oriented, we produce patterns of interaction and a corresponding system of status positions, roles, and norms. *Status positions* tell us where we fit in the social hierarchy of esteem or honor; *roles* are sets of expected behaviors that generally correspond to a given status position (for example, a professor is expected to teach); and *norms* are expected behaviors that have positive and/or negative sanctions attached to them. Together, these form a **social system**—an organization of interrelated parts that function together for the good of the whole. Society is composed of various social systems like these.

For this, Parsons gives us a theory of institutionalization. The notion of institutionalization is very important in sociology. Generally speaking, institutionalization is the way through which we create institutions. For functionalists such as Parsons, institutions are enduring sets of roles, norms, status positions, and value patterns that are recognized as collectively meeting some societal need. In this context, then, *institutionalization* refers to the process through which behaviors, cognitions, and emotions become part of the taken-for-granted way of doing things in a society ("the way things are").

I've diagrammed Parsons's notion of institutionalization in Figure 6.3. Notice that we move from voluntaristic action within a unit act and modes of orientation to social systems. In this way, Parsons gives us an aggregation theory of macro-level social structures. One of the classic problems in sociological theory is the link between the micro and macro levels of society. In other words, how are the levels of face-to-face interaction and large-scale institutions related? How do we get from one to the other? Most sociologists simply ignore the question and focus on one level or another for analysis. Here Parsons gives us the link through the process of institutionalization. Large-scale institutions are built up over time as individuals with particular motivations and values interact with like-minded people, thus creating patterns of interaction with corresponding roles, norms, and status positions.

There's a follow-up question to the micro–macro problem: once created, how do institutions relate back to the interaction? For Durkheim, the collective consciousness becomes an entity that can act independently upon the individual and the interaction. It does this through moral force: People feel the presence and pressure of something greater than themselves and conform. For Parsons, it's a bit different, even though he does acknowledge moral force. According to Parsons, all social arrangements, whether micro or macro, are subject to system pressures. Thus, institutions influence interactions not so much because of their independent moral force, but rather because interactions function better when they are

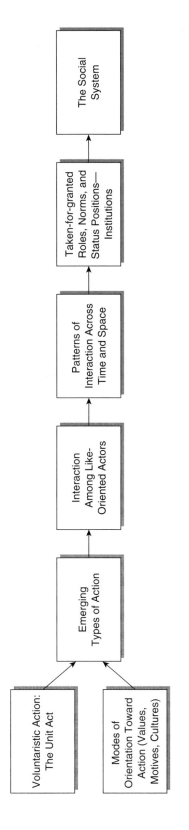

Figure 6.3 The Process of Institutionalization

systematically embedded in known and accepted ways of doing things. In addition, rather than being dependent upon individual people or interactions, or having its own whimsical nature as Durkheim would have it, society is subject to self-regulating pressures because it is a system.

Getting back to the actual process of institutionalization, Parsons argues that it has two levels: the structuring of patterned behaviors over time (this is the level we've been looking at) and individual internalization or socialization. Parsons understands internalization in Freudian terms. Freud's theory works like this: People are motivated by internal energies surrounding different need dispositions. As these different psychic motives encounter the social world, they have to conform in order to be satisfied. Conformity may be successful (well-adjusted) or unsuccessful (repressed), but the point to notice here is that the structure of the individual's personality changes as a result of this encounter between psychic energy and the social world. The superego is formed through these encounters.

For Parsons, the important point is that cultural traditions become meaningful to and part of the need disposition of individuals. The way we sense and fulfill our needs is structured internally by culture. For Parsons, then, the motivation to conform comes principally from within the individual through Freudian internalization patterns of value orientation and meaning. As the same set of value patterns and role expectations is internalized by others, that cultural standard is said to be, from the point of view of the individual, institutionalized.

It is worth pointing out that Parsons argues that the content of the institutional solutions to societal needs doesn't matter. So, for example, it doesn't matter if a collective perpetuates itself biologically through the institution of family (however it is defined) or through an institutionalized hatchery such as a chicken farm. What is important is that the perceived solutions are a set of highly ritualized behaviors that are seen as typical, belonging to particular settings (such as church rather than school), and are believed to solve collective problems. I've diagrammed Parsons's bottom-up theory of institutionalization in Figure 6.3.

Concepts and Theory: System Functions and Control

System Needs

We came across the idea of requisite needs in Chapter 3. You'll remember that functionalism argues that all societies have certain needs that must be met in order to function, just like the human body. With Durkheim we saw that society needs a certain level of solidarity, which is provided through the collective consciousness. Parsons not only gives us additional needs, he talks about them more abstractly than did Durkheim. Parsons argues that society is a system and that it functions like any other system. In other words, he contends that all systems have the same needs, whether social, biological, physical, cultural, or any other system. There are four such needs: adaptation, goal attainment, integration, and latent pattern maintenance.

To get us thinking, let me ask you a question: Are you hungry? Maybe you aren't now, but sooner or later you will be, because every body needs food to live. Yet it isn't really food per se that you need. You need the nutrients that are *in* the food to survive. When you eat something, your body has a system that extracts the necessary resources from the food and converts it into usable things (like protein). So, your body doesn't really need a steak; it needs what is in it. Parsons calls this function **adaptation** because it adapts resources and converts them into usable elements.

Let's use a larger illustration. Every organism, society, or system exists within and because of an environment. For example, ducks are not found at the South Pole but penguins are. Each of these organic systems has adapted to a given environment and extracts from the surroundings what it needs to exist. It is the same with society. In order to exist, each and every society must *adapt* to its environment by inventing ways of taking what is needed for survival (such as soil, water, seeds, trees, animals) and converting them into usable products (food, shelter, and clothing). Society must also move those products around so that they are available to every member (or at least most members). In society, we call this subsystem the economy. The economy extracts raw resources from the environment, converts them into usable commodities, and moves the commodities from place to place.

Be aware that the economy and adaptation are not the same thing. The economy is *the subsystem in society that fulfills the adaptation need*. In the body, it's the digestive subsystem. Yet the digestive system and the economy are obviously not the same things. They fulfill the same function but in different systems. The reason I'm taking such pains here is that it is important to see that Parsons's scheme is very abstract and can be used to analyze any system, so I want you to be clear on how to apply it. (In addition, if your professor asks you to explain the adaptation function and you say that it is the economy, you'll be wrong.)

Every system also needs a way of making certain that every part is energized and moving in the same direction or toward the same goal. Parsons refers to this subsystem as **goal attainment.** In the human body, the part of us that activates and guides all the parts toward a specific goal is the mind. The mind puts before us certain goals, things that we need or want to do. We feel motivated to action because our mind invests emotion into these goals.

Let's say that you have the goal of becoming the next Jimi Hendrix. So you set about listening to all of the legendary rock guitarist's CDs, you read all the books about Hendrix's style, and you practice six hours a day. You also work a job and save your money in order to buy the same kind of guitar and equipment that Hendrix used so you can sound just like him. You are motivated. Your mind has caught the image of yourself playing guitar and has controlled and coordinated your fingers, arms, and legs—in short, all your actions—to move you toward that goal. On the other hand, perhaps you aren't as motivated about school. After all, you're going to be a big rock star, so who needs school? So the different parts of your body are not energized and coordinated to meet the goal of doing well in school. In the body, it's the mind that coordinates all the different actions and subsystems to achieve a goal. In the social system, the institution that meets this need for goal attainment is government, or polity (same meaning, different word).

Systems also need to be integrated. By definition, systems do not contain a single part, but many different parts, and these parts have to be brought together to form a whole. Have you ever watched a flock of geese in flight? Rather than flying singularly in a haphazard manner, their actions are coordinated and integrated. The dictionary defines "integration" as meaning to form, coordinate, or blend into a functioning or unified whole; to unite with something else; and to incorporate into a larger unit. The geese are able to form into a larger unit mostly because of instincts. For human beings, it is a bit more complex.

Humans generally use norms, folkways, and mores to integrate their behavior. Norms can be informal (such as the norms surrounding our behavior in an elevator) or they can be formal and written down. Formal and written norms are called laws. Laws help to integrate our behaviors so that, rather than millions of individual units, we can function as larger units. Parsons refers to this function as **integration,** and in society that function is performed by the legal system. The legal system links the various components together and unites them as a whole. When, for example, Apple Computer crosses the boundaries of IBM, it is the legal system that makes them work together, even though they probably don't want to.

As should be apparent, polity and the legal system are intimately connected, because these two functions are closely related. In our bodies, for instance, the mind functions as the goal-attainment system and it uses the central nervous system to actually move the different parts of the body. But the mind and the central nervous system are two different things. A person can be completely paralyzed and still have full access to her or his mind, or the body can be in perfect working order with the mind completely gone. In the same way, polity and law are related but separate.

The final requisite function that Parsons proposes is **latent pattern maintenance.** Every system requires not only direct management, such as that performed by a government, but also indirect management, which Parsons terms latent pattern maintenance. Not everything that goes on in our body is directed through cognitive functions. Rather, some of these functions, like breathing, are managed and maintained through the autonomic nervous system, a subsystem that maintains patterns with little effort. Society is the same way. It is too costly to make people conform to social expectations through government and law; there has to be a method of making them *willing* to conform. For this task, society uses the processes of *socialization* (the internalization of society's norms, values, beliefs, cognitions, sentiments, etc.). The principal socializing agents in society are the structures that meet the requirement of latent pattern maintenance—structures such as religion, education, and family. (By the way, the word "latent," from which Parsons gets his term, means not visible, dormant, or concealed.)

Parsons argues that these four requirements can be used as a kind of scheme to understand any system. When beginning a study of a system, one of the first things that must be done is to identify the various parts and how they function. Parsons's scheme allows us to categorize any part of a system in terms of its function for the whole. In Figure 6.4, I have diagrammed the way this analytical scheme looks. The four functions are noted by the initials AGIL. The larger box represents that system as a whole, which, of course, needs the four functions. Because they function as systems themselves, each of the four subsystems can be analyzed in terms of the same scheme. I've used adaptation in this case, but the same can be done with each of them.

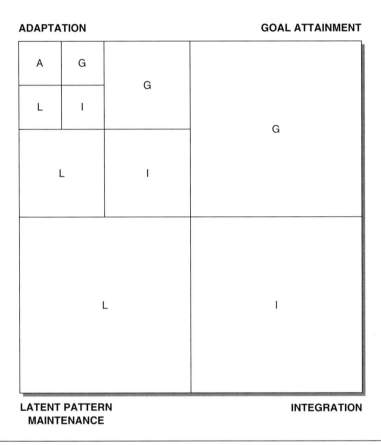

Figure 6.4   AGIL—Functional Requisites

Systemic Relations

Further, Parsons gives us a way of understanding interinstitutional relations. This is important but little explored ground. What usually passes as institutional analysis is *within* an institutional sphere rather than between. For example, we might look at the institution of family in the United States and see that changes are occurring. Some of these changes are societywide (such as the increase in single-parent families), some are the subject of much moral debate (such as whether or not to define gay couples as family), and some are present but not part of the public discourse (like an acquaintance of mine who introduced me to his wife, the mother of his children, and his girlfriend—three different women—who were all four living happily together in this arrangement). We can also look at marriage and divorce rates, birth and death rates, proportions of families under the poverty line, and so on.

These kinds of research agendas can be enlightening, but they are also quite limited. Studying the phenomenon of "latchkey kids" is important, but it only explains one small part of the institution of family and by itself says nothing about the instructional relations between the family and government. As we've noted, each subsystem is part of a whole and, as such, each is related to the other. If, for illustration, I put dirt in the fuel system of an automobile, it will affect the rest of the car, not just the fuel system itself. The same is true for society. If there are changes in one subsystem, those changes will ripple their way through all of society.

Family, in our society, is usually thought of as a married couple with 2.5 kids. Yet this model, called the nuclear family, has not always been the norm. In fact, it is a pretty recent model, historically speaking. Up through feudalism, marriage and family were far more important politically and economically. People got married to prevent wars or to seal economic commitments. As a result, the kinship structure was considerably more extensive and marriages were generally arranged because they were *socially* important. Marriages in the United States are not generally arranged; we conceptualize marriage as existing principally for the individual and as being motivated by love.

Thus, when we bemoan the loss of "family values," it is an historically specific set of values. These values came about because of changes in the rest of society. As institutions differentiated, the goal-attainment and adaptation functions were no longer dependent upon or related to family in the same ways. Bureaucratic nation-states emerged that were able to negotiate their interstate relations through treaty and war (using a standing army); the economy shifted to industrialized production and forced families to move from their traditional home to the city where most of the relationships that people have are not with or associated with family, as they were in traditional settings. Many other changes, such as the proliferation of capitalistic markets and the de-centering of religion, also influenced the definition, functions, and value of family.

The point I'm trying to make is that for us to truly understand an institution, we must see it in its institutional context, in its relationships to other institutions. From a systems or functionalist point of view, the environment for any institution is created by other institutions (subsystems); they mutually affect and sustain one another. Parsons conceptualizes subsystem relations using his AGIL scheme and the actual paths of influence as boundary exchanges. Just as the digestive subsystem in our bodies provides nutrients for the circulatory subsystem, and the circulatory system in exchange provides blood to the digestive system, so every social institution is locked in a mutual exchange. I've listed these boundary exchanges in Table 6.1. What you will see is that each relationship is defined in terms of what one subsystem gives to another. Both Spencer and Durkheim argued that as institutions differentiate, they become mutually dependent, but neither of them explicated the dependency. Here Parsons does that for us.

Table 6.1   Interinstitutional Relations

Originating Subsystem	Output →	Receiving Subsystem
Economy (A)	Productivity	Polity (G)
	New output combinations	Law (I)
	Consumer goods and services	Family (L)
Polity (G)	Imperative coordination	Law (I)
	Allocation of power	Family (L)
	Capital	Economy (A)
Law (I)	Motivation to pattern conformity	Family (L)
	Organization	Economy (A)
	Contingent support	Polity (G)
Family (L)	Labor	Economy (A)
	Political loyalty	Polity (G)
	Pattern content	Law (I)

As you can see, the table outlines what each subsystem or institution gives to the other three. Let's look at family as the originating subsystem for a moment in the table. You can read the list of its outputs in the center column and the receiving institutions on the right. The list shows the functions that family provides for the other subsystems. Through proper socialization, family provides political loyalty to the government; it provides a compliant pool of labor for the economy; and it influences the moral content of socialized patterns of norms that become law. I don't want to take us through each of these relations; you can do that on your own. The most important thing to glean is the idea of interinstitutional relations. In addition, while we may at some point become more sophisticated in our analysis of the associations, the place to begin is right where Parsons does: the functional dependencies.

Cybernetic Hierarchy of Control

Parsons develops an overall model of how the systems surrounding human life integrate. The model is called the general system of action or the **cybernetic hierarchy of control.** Cybernetics is the study of the automatic control system in the human body. The system is formed by the brain and nervous system and control is created through mechanical-electrical communication systems and devices.

In using the term "cybernetic," Parsons tells us that control and thus integration are achieved primarily through information. Also note that in cybernetics, control is achieved automatically, through what Parsons calls latent patterned maintenance.

I've outlined the control system in Figure 6.5. As you can see, the cybernetic hierarchy of control is understood through Parsons's AGIL system. (In the model, I have also expanded the social system to indicate what we have already seen: the social system is understood in terms of AGIL as well.) There are thus four systems that influence our lives: the culture, social, personality, and organic systems.

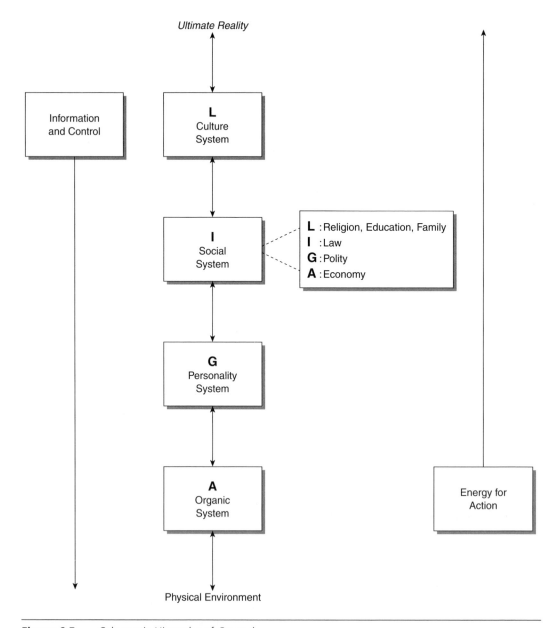

Figure 6.5   Cybernetic Hierarchy of Control

The culture system is at the top, indicating that control of human behavior and life is achieved through cultural information. This emphasis on culture would obviously not be true for most animals. Regardless of the recent news and debates about apes being able to use and possibly share sign language, culture is not the primary information system for any animal other than humans. For most animals, information comes generally through sensory data, instinctual predispositions, and habitual patterns of action.

The position of culture at the top also indicates that it requires the most energy to sustain. As information flows from the top down, energy moves from the bottom up. Culture has no intrinsic energy. It is ultimately dependent upon the systems that are lower in the hierarchy for its existence. Without such energy, culture will cease to exist. For example, anthropologists and archaeologists know that a Babylonian culture existed at one time. That knowledge of past existence is itself part of our culture, but the Babylonian culture has long since died because its support mechanisms have passed away.

Culture is most immediately dependent on the social system for its existence. It is also dependent upon the personality system, because it is individual humans who internalize and enact culture. Since the personality system is dependent upon the organic system (the human mind needs the human body), culture is indirectly dependent upon it as well. I'd like to pause here and mention one thing: Recent theorizing argues that culture is also directly reliant on the organic body. Pierre Bourdieu (1979/1984), for instance, argues that culture becomes embodied. There are not simply cognitive and emotional elements in culture. Culture also contains practices that form part of the way our bodies exist. Through our culture we develop tastes, dispositions, and automatic behaviors. For example, our taste in food is dependent upon our culture. Moreover, not only is our language cultural, but so is the way we speak it, as in regional accents.

Notice also that human life is contextualized by conditions of ultimate reality and the physical environment. Parsons never makes any comment about what ultimate reality is, but its understood existence is extremely important for the culture system. Remember that our most important values are framed in terms of ultimate truths, and these truths are religious in nature. Parsons, then, sees religion as an important influence on the culture system in general.

Overall, information moves down and energy moves up. Each system is embedded in and dependent upon the other—systems are reciprocally related to one another. One of the things that Parsons wants to point out with this kind of model is that differentiated, complex systems are dependent upon **generalized media of exchange** for facilitating communication and cooperation among and between the diversified parts. For example, in a complex society, each of the major structures has distinct goals, values, norms, and so forth. The capitalist economy has the goal of producing profit, while the education system has the goal of producing critical thinking. These value-oriented goals may at times clash, but a generalized medium of exchange will tend to keep the system in equilibrium. Parsons offers language as the prototype of such generalized media of communication and explicitly identifies money (from the adaptive subsystem), power (from goal attainment), and influence (integration) as other such media.

Concepts and Theory: Social Change

One of the critiques often leveled against Parsons is that he only sees systems in equilibrium and his theorizing thus maintains the status quo. The criticism is not entirely correct. Parsons does assume that systems are in a state of **equilibrium;** that is, the forces of integration and disintegration are balanced. He feels that any social system worth studying would have a fair degree of permanence, and thus, "there must be a tendency to maintenance of order except under exceptional circumstances" (Parsons & Shils, 1951, p. 107). Parsons (1951) calls this tendency toward equilibrium the "first law of social process" (p. 205) and the "law of inertia" (p. 482). However, it is not the case that Parsons ignores social change. He actually has a notion of revolutionary change in addition to slow evolutionary change.

Like Durkheim, Parsons argues that the principal dynamic of evolutionary change is differentiation, and he sees that structural differentiation brings about problems of integration and coordination. Parsons argues that these problems would create pressures for the production of an integrative, generalized value-culture and a generalized medium of exchange. Thus, like Durkheim, Parsons argues that culture is the most important facet of a complex social system. It is culture that provides the norms, values, and beliefs that allow us to interact, and it is culture that provides the general information that the social system needs in order to operate.

Cultural Strain

However, the process of culture generalization, Parsons (1966) notes, may also bring about severe conflicts: "To the fundamentalist, the demand for greater generality in evaluative standards appears to be a demand to abandon the 'real' commitments" (p. 23). For example, in U.S. society, the call to return to "family values" is just such an issue. Societies that are able to resolve these conflicts move ahead to new levels of adaptive capacity through innovation. Others may "be so beset with internal conflicts or other handicaps that they can barely maintain themselves, or will even deteriorate" (Parsons, 1966, p. 23).

It is at this point that revolutionary change becomes more likely. Culture generally allows people and other social units (like organizations) to interact. It provides us with a language and value system. When people or organizations begin to value different kinds of things or to speak different languages, the situation is ripe for conflict. Parsons sees this kind of problem as a type of strain; *strain* is defined as a disturbance of the cultural expectation system. When we have different values, we do not know what to expect in an encounter. Strain always sets up re-equilibrating processes, but these processes may take a long time to reach balance and the system may be substantially different as a result. Change due to revolution occurs in two phases: (1) the ascendancy of the movement; and (2) the adoption of the movement as "setting the tone" for the society—the re-equilibrating process (this latter is the part that Marx and critical theorists leave out of their theories of revolutionary change).

Revolution

There are four conditions that must be met for a revolutionary movement to be successful. First, the potential for change must exist; Parsons refers to this potential as the alienative motivational elements. People become motivated to change the system as the result of value inconsistencies. These inconsistencies are inevitable and continually present in an empirical system of action, particularly one that has been generalized to incorporate a number of diverse groups, such as in the United States. For example, the term "equality" has been stretched to include groups not intended by the founding documents (blacks weren't originally included in "all men are created equal"). The term has become more general, yet at the same time, the generality of the term sets up conflicts as more and more groups see themselves as disenfranchised, and others—the fundamentalists—see the generalization as movement away from traditional or received truth.

Second, dissatisfaction with the system is not enough to begin a revolutionary movement; the subgroup must also become organized. The organization of a group around a subculture enables members to evade sanctions of the main group, create solidarity, create an alternative set of normative expectations and sanctions, and it enables expressive leadership to arise.

Third, the organized group must develop an ideology that incorporates symbols of wide appeal and can successfully put forward a claim to legitimacy. The ability to develop an alternative claim to legitimacy is facilitated by two factors. One is that the central value system of large societies is often very general and is therefore susceptible to appropriation by deviant movements. The other factor is that serious strains and inconsistencies in the implementation of societal values create legitimacy gaps that can be exploited by the revolutionary group.

The fourth condition that must be met is that a revolutionary subgroup must eventually be connected to the social system. It is this connection that institutionalizes the movement and brings back a state of equilibrium. There are three issues involved: (1) The utopian ideology that was necessary to create group solidarity must bend in order to make concessions to the adaptive structures of society (e.g., kinship, education)—in other words, the revolutionary group must meet the reality of governing a social system; (2) the unstructured motivational component of the movement must be structured toward its central values—the movement must institutionalize its values both in terms of organizations and individuals; and (3) out groups must be disciplined vis-à-vis the revolutionary values that are now the new values of society.

Parsons Summary

- Parsons is usually the one credited with having clearly articulated a systems approach in sociology. This kind of theoretical method encourages us to see society in terms of system pressures and needs. Two issues in particular are important: the boundary between the system and its environment and the internal processes of integration. Parsons divides each of these into two distinct functions. External boundaries are maintained through adaptation and goal attainment; internal-process functions are fulfilled by integration and latent pattern maintenance. Systems theory also encourages us to pay attention to the boundaries between subsystems, in terms of their exchanges and communication. Because relatively smart or open systems have goal states, take in information, and contain control mechanisms, they tend toward equilibrium. Parsons conceptualizes society as just such a system. In addition, because Parsons sees everything as operating systemically, his theory is cast at a very abstract level and is intended to be applied to any and all systems.

- Parsons builds his theory of the social system from the ground up. He begins with voluntaristic action occurring within the unit act. Humans exercise a great deal of agency in their decisions; however, their decisions are also circumscribed by the situation and normative expectations. The normative expectations in particular are where human agency is most expressed and where culturally informed motives and values hold sway. These different motives and values orient the actor to the situation and combine to create three general types of action: strategic, expressive, and moral. People tend to interact socially with those who share their general types of action. As a result, interactions become patterned in specific ways, which in turn tends to create sets of status positions, roles, and norms. We may say that status positions, roles, and norms are institutionalized to the degree that people pattern their behaviors according to such sets and internalize the motives, values, and cultures associated with them.

- Different sets of institutionalized status positions, roles, and norms are clustered around different societal needs. Because society functions as a system, there are four general needs that must be met: adaptation, goal attainment, integration, and latent pattern maintenance. In complex, differentiated societies, these functions are met by separate institutional spheres. The different institutions are integrated through the system pressures of mutual dependency and generalized media of exchange. The social system itself is only one of four systems that surround human behavior. There are the cultural, social, personality, and physical systems, each corresponding to AGIL functional requisites. Because systems are dependent upon information, the culture system is at the top. Information flows from the top down, and the energy upon which culture is dependent flows from the bottom up. Parsons refers to this scheme as the cybernetic hierarchy of control.

- Systems tend toward equilibrium. They can, however, run amiss if the subsystems are not properly integrated. In the social system, this happens through cultural strain. As societies become more differentiated, the media of exchange must become more general. In this process, it is possible that some groups will seek to hold onto the dysfunctional culture. This case sets up a strain within the system, with some subsystems or groups refusing to change and other subsystems moving ahead. Motivation for social revolution is possible under these conditions. After people are motivated to change society, they must then create a subculture that can function to unite their group and create an alternative set of norms and values. This culture must eventually have wide enough appeal to successfully make a claim to legitimacy. In a revolution, either side could win (the reformers or the fundamentalists), but in either case, certain steps are systemically required to reintegrate the system. After the revolution, the subgroup must produce a culture that can unite the system. Institutionalization occurs at this point as it does at any other time: through behaviors patterned and people socialized around a set of status positions, roles, and norms.

Robert K. Merton: Empirical Functionalism

As you undoubtedly noticed, Parsons's work is very abstract. His desire was to create a scheme that could be used to analyze almost any social phenomenon in its broader context. We can call this kind of approach a grand theory. A *grand theory* is one that explains all phenomena through a single set of concepts. Grand theorists like Parsons generally assume that everything in the universe operates in system-like ways, and that all systems are built on the same principles and are subject to the same dynamics. For example, biology, psychology, sociology, and physics are all similar systems and can be ultimately explained using the same theory. Thinking as a grand theorist implies that you tend to see things working in very abstract and mechanistic terms. The theory has to be abstract in order to embrace all the phenomena in the universe. For instance, in a grand theory you couldn't use simple psychological terms to explain psychology because they wouldn't apply in sociology or biology— the terms of a grand theory must be more abstract than any one discipline.

Photo: Columbia University.

The Essential Merton

Biography

Robert K. Merton was born on July 4, 1910, in Philadelphia to Jewish immigrant parents. Merton's given name was Meyer R. Schkolnick, which he initially changed to better suit his amateur magician show. As a young man, Merton spent many hours at the Andrew Carnegie Library reading and studying history, science, and biographies. Merton studied and worked with Pitirim Sorokin, Talcott Parsons, and Paul Lazarsfeld, all significant names in sociology. He finished his Ph.D. from Harvard in 1936 and soon became a part of the faculty there. Merton moved to Tulane University (New Orleans) in 1939, and in 1941 took a position at Columbia University in New York, where he remained for the rest of his career.

Merton served as president of the American Sociological Association, the Eastern Sociological Society, and the Society for Social Studies of Science. His honors include more than 20 honorary doctoral degrees and membership in the National Academy of Sciences. In 1994, the president of the United States awarded Merton the National Medal of Science. In addition, Robert K. Merton was the father of Robert Cox Merton who won the Nobel Prize in economics (1997). Many of the concepts that Merton coined have made their way into popular culture, such as "self-fulfilling prophecy," "role model," "manifest and latent functions," and "unintended consequences."

Merton passed away on February 23, 2003.

Passionate Curiosity

Merton had an active, never-ending curiosity about the workings and intricacies of social life. He was above all a scientist, motivated by an intense desire to systematically build up our knowledge about how society works, piece by piece. This motivation involved working through middle-range theories and theory cumulation. Theory cumulation in particular meant obliteration by incorporation—forgetting the names, personalities, and historical backgrounds of theorists such as Marx and Weber (and Merton!) and focusing only on the concepts and theories.

Keys to Knowing

Middle-range theory, functional alternatives, manifest and latent functions, dysfunctions, unanticipated consequences, structural theory of deviance, sociological ambivalence

Concepts and Theory: Critiquing Parsons's Functionalism

Middle-Range Theories

Although his focus was ultimately on theory, Merton worked relentlessly to ground functionalism in the empirical world first, before making grand, abstract statements. His alternative approach is to work with middle-range theories. **Middle-range theories** "lie between the minor but necessary working hypotheses . . . and the all-inclusive systematic efforts to develop a unified theory that will explain all the observed uniformities of social behavior, social organization and social change" (Merton, 1967, p. 39). Merton argues that sociology is too young a discipline to be concerned with grand theories, and his intent with the concept is to ground sociological theory in "theoretically oriented empirical research" (p. 56).

There are five attributes to middle-range theories: (1) They consist of a limited set of assumptions that lead to specific hypotheses, which in turn are empirically confirmed; (2) they are capable of being brought together with other middle-range theories to form wider networks of theory; (3) middle-range theories can be generalized and applied to different situations (so the theory is not merely organized descriptive data); (4) because they don't address the more abstract assumptions of grand theory, middle-range theories can fit easily into different systems of theory, like Parsons's social systems or Marx's historical materialism; and (5) middle-range theories will typically be in harmony with the method of classical theorists.

Parsons's and Merton's approaches represent two methods of theory building: Parsons privileges reason and argues that theories should be logically deduced, and Merton's tactic privileges empirical data and induction. Interestingly, the roots of Parsons's approach can be found in ancient philosophy. Plato argued that sense perception is a problematic method for recognizing reality or truth. Knowledge gained through the five senses is constantly changing and particular to each individual. According to Plato, our senses only pick up shadows of the true reality, the reality of forms or ideas. This reality can only be accurately discerned through reason, not the physical senses. Because reason is subject to abstract laws of logic and not fluctuating physical impulses, it alone can perceive truth.

These kinds of logically deduced theories cannot be rejected or accepted on the basis of a single empirical test. The results of empirical tests can be due to simple happenstance and not theoretically significant relationships. Rather, the first and best test of theory is its *logical implications*. In other words, if we have done our theoretical homework properly, the logic of the theory is to be trusted more than its empirical tests, at least in the initial phases of testing and theory construction. We will see this kind of approach to theory building again in Peter Blau's work (Chapter 10).

Merton, on the other hand, argues that theory must be grounded in the empirical world. Of course, middle-range theories contain abstractions, "but they are close enough to observed data to be incorporated in propositions that permit empirical testing" (Merton, 1967, p. 39). Like Parsons's approach, this method goes back to early philosophers as well, this time to Aristotle's empiricism. As an empirical philosopher, he argued that all true knowledge begins with sense perception—those things that we can see, feel, taste, hear, or smell. Theory in this case moves upward from observations to middle-range theories to more general theoretical statements. According to Merton, this stepwise approach assures the validity of the theories as they are being built.

Middle-range theories usually come in the "theories of" form, such as theories of gender inequality, role conflict, deviance, and so forth. These middle-range theories don't try to explain society at large; rather, they explicate some small portion of it without necessarily connecting it to any other aspect of society. Merton's idea is that, as these mid-level theories are proposed and tested, they will be brought together to form a more comprehensive and empirically grounded general theory. For example, after sufficient empirical testing, we could take Lemert's labeling theory of deviance and Merton's structural strain theory and blend them together to form a more general theory of deviance.

My intent in explaining these approaches is not to argue for or against either one. I think both approaches are necessary and can work well together. As Jonathan H. Turner (1991) points out, "a body of useful abstract propositions . . . can serve as an inducement to middle-range theories to raise the level of abstraction," and "middle-range theories could potentially be one vehicle by which more abstract propositions . . . are attached to specific empirical problems" (p. 27).

My desire here is to stimulate your thinking about the different ways that theory can develop. Theorizing, especially in the beginning stages, is intuitive. Certain theories and theoretical approaches will have greater intuitive appeal to you than others. Go first with your intuition—let your mind be excited by ideas, possibilities, and approaches. Then, if necessary, worry about critique.

Critiquing Functionalism's Assumptions

Merton also raises questions about three of functionalism's major claims. Up until Merton, these assumptions seemed to be the defining features of functionalism. First, functionalism posits a *functional unity of society*. This postulate states that the social activities and cultural items that are patterned and standardized across society are functionally related. In other words, all working societies have a functional unity in which all parts of the system work together with a fairly high level of accord and consistency. For example, the functional unity of society principle would assume that the education, government, and religious sectors all work together harmoniously and without fail as parts of the same system. This assumption is what underlies the idea of system equilibrium.

The second assumption that Merton questions is the idea of *universal functionalism*. This supposition postulates that all social activities or cultural items have positive sociological functions. In other words, if there is a patterned feature in society, then it must be functional by the very fact that it exists. For example, finding a society such as the United States where male aggression has high value, the functionalist assumes that it must have positive benefits for the whole.

The last postulate of functionalism that Merton challenges is that of *indispensability*: every patterned part of society and culture fulfills a vital function within the society and is therefore indispensable. This postulate not only assumes there are certain functions that a society cannot do without, it also assumes that those functions must be fulfilled by certain cultural or social forms and those forms are therefore indispensable. For example, a functionalist may decide that every society needs a system of ultimate meanings that gives relevance to all other meanings. The functionalist may then assume that it is only religion that can provide those meanings. By such reasoning, religion itself becomes necessary for society, when in fact the function may be met through other institutions.

Merton sees these assumptions as potentially obscuring true research. Generally speaking, if we assume that something is true, we can usually find evidence for it. The first thing Merton wants to do, then, is change each of these assumptions into empirical questions. Rather than assuming that government and religion are positively integrated, make it a research question: In what ways are government and religion related to one another? Or, might not male aggression have effects that are other than positive? As you can probably tell, to ask such questions implies some ideas that Parsonian functionalism does not contain.

Concepts and Theory: Dynamic Functionalism

Alternate Outcomes

In order to help us think more clearly about functional relations in society, Merton gives us several new concepts: functional alternatives, manifest and latent functions, dysfunctions, and unanticipated consequences. The idea of *functional alternatives* conceptualizes the possibility that other kinds of structures may meet societal needs. For example, the function of biological reproduction doesn't necessarily have to be met through family. Or, for another example, could the meanings that religion provides be supplied through other institutions, such as a "civil religion"?

The concepts of **manifest and latent functions** specifically refer to the positive contributions that a social structure has for society, but the concepts allow us to see functions in a more complex light. Manifest functions are the known contributions and latent functions are the hidden or unacknowledged contributions of social structures. A good example is education: a manifest function of education is to pass on the cultural knowledge of a society; a latent function of education is to provide a marriage market that pairs people on several important dimensions (such as class).

Merton also wants us to be able to see that social structures or institutions may have negative effects, when seen from a functionalist position. A *dysfunction*, then, is a consequence of a social structure that leads to less adaptation and integration. The idea of a dysfunctional family is a micro-level phenomenon that comes from Merton's concept (here the dysfunctional consequences come from the performance of a social role, such as father or mother). In addition, dysfunctions may be manifest or latent.

Latent functions and dysfunctions are concepts that describe the outcomes of social structures. The idea of **unanticipated consequences,** on the other hand, generally refers to the effects of social acts of individuals. Unanticipated consequences, then, are those outcomes of social action that are not intended by the actor. Merton (1976) gives us several sources of unanticipated consequences. Among them are *ignorance* (at times, there is such a wide range of possible consequences that they can't all be known); *chance consequences* ("those occasioned by the interplay of forces and circumstance that are so numerous and complex that prediction of them is quite beyond our reach" [p. 151]); *error* (the most common source of which is habitual action); *imperious immediacy of interest* (concern for immediate consequences blocks out consideration of long-term effects); and *basic values* (the actor is concerned with subjective rightness rather than objective consequences: "Here is the essential paradox of social action—the 'realization' of values may lead to their renunciation" [p. 154]).

Emergent Social Change

With the idea of unanticipated consequences, Merton is opening up the social system for unpredictable change. Functionalism has usually seen social change in evolutionary terms—slow change over long periods of time that in the long run leads to increased complexity and thus survivability. As we've seen, Parsons opens this idea up a bit by giving us a functional theory of social revolution. The reasons it's a functional theory are that revolutions are only possible because of system strain, such as value inconsistency, and that all revolutions must reintegrate or equilibrate the system. Merton takes us a step further.

Not only are large systems susceptible to strain, as Parsons has it, but the behaviors of individual actors within the system are also susceptible to unanticipated consequences, because of ignorance, mistakes, values, failure to take the long run into account, and chance consequences. Unanticipated consequences, then, function as a wild card in the social system. Both the idea of dysfunctions and that of unintended consequences are particularly important for Merton's (1976) theory because they both lead to structural change. Structures change through "cumulatively patterned choices in behavior and the amplification of dysfunctional consequences resulting from certain kinds of strains, conflicts, and contradictions in the differentiated social structure" (p. 125), and through "unanticipated consequences of purposive social action" (p. 146).

There are two more issues that are important in structural change, according to Merton: deviance and ambivalence. Merton's structural theory of deviance argues that society values certain *goals* and the *means* to achieve those goals. In the United States, for example, we value economic success and believe that education and hard work are the proper means to achieve that goal. However, in any social system there are disenfranchised people who do not have equal access to the legitimate means to achieve success, yet are nonetheless socialized to value the same goals as the majority. This structural location puts these people in a position of tension or strain. Generally speaking, most people accept the goals and means of society; Merton calls this type of response *conformity* (acceptance of both goals and means). But for those in structural strain, there are four other possibilities: *innovation* (accept the goals but use innovative means, such as robbery), *ritualism* (abandon the goals and perform the means without any hope of success), *retreatism* (deny both the goals and the means and retreat through such things as alcoholism or drug abuse), or *rebellion* (the individual creates his or her own goals and means). As with unintended consequences, deviant behaviors may accumulate to the point where they influence the social structure.

We can think of this idea of accumulation within a system in terms of the heat thermostat in your home. If it's cold outside and you have your thermostat set at 65 degrees, the internal temperature of your house will trip the thermostat, which will turn your heat on. If you come into your house while the heat is on, chances are good that it will stay on—your body will have very little influence on the internal temperature. However, if you bring several of your friends and keep adding more people, the total mass of bodies within the house will eventually generate enough heat to trip the thermostat in the opposite direction and turn the heat off. If you keep adding people to your house, you will eventually create enough heat that the thermostat will probably turn the air-conditioner on. It's not a perfect analogy, but you can see where we are heading. Relatively few acts of deviance or unintended consequences will have little effect, but as they accumulate, the social system will respond and change just as your thermostat would.

Ambivalence generally refers to emotional or psychological attitudes that conflict with one another. **Sociological ambivalence,** on the other hand, refers to *"opposing normative tendencies in the social definition of a role"* (Merton, 1976, p. 12, emphasis original). It is important for us to see that the ambivalence here is structural, not individual—the ambivalence exists within the social structure. For example, medical doctors experience contradictory role expectations: they are expected to be emotionally detached in their professional relations with their patients, and at the same time they are expected to display compassion and concern for their patients. There is a way in which sociological ambivalence is an effect of modern social change: Ambivalence has "evolved to provide the flexibility of normatively acceptable behavior required dealing with changing states of a social relation" (Merton, 1976, p. 31).

Thus, in Merton's scheme of functionalism, social structures and systems are robust and complex. Merton argues that systems are not simply subject to functional consequences; rather, systems regularly experience dysfunctions, manifest and latent functions, and functional alternatives. With these ideas, Merton is proposing that functional analysis be open to the possibility of multiple consequences and focus on the net balance of outcomes. Merton also presents us with a picture of the individual social actor, with culture on one side and social structure on the other. These two structures form the salient environment for the person. That is, individuals set about doing their tasks within this environment that both restrains and enables. The different structural environments that individuals find themselves within produce differing rates and kinds of deviance and social ambivalence. Together, ambivalence, deviance, unanticipated consequences, and dysfunctions—all of which are intrinsic to social structure—accumulate to create social change.

Merton Summary

- Merton critiques Parsons's analytical approach to theory, arguing that scientific theory must accumulate over time and begin with testable middle-range theories. Once tested, these mid-range theories, in turn, can be brought together to form more general theories of social structure and action.

- Merton also wants to ground functionalism in the empirical world by getting rid of some of the more abstract assumptions underlying functional theorizing. In particular, Merton questions the functional unity of society, universal functionalism, and indispensability.

- Merton further establishes an empirical base for functionalism by arguing that there may be alternative outcomes to structural arrangements and individual social action. Merton proposes four concepts to help us think about different outcomes: functional alternatives, manifest and latent functions, dysfunctions, and unanticipated consequences.

- Merton also sensitizes us to more subtle and continuous social change. Merton sees quite a bit of social change building up through unanticipated consequences of social action, deviance, and sociological ambivalence. People rarely have full knowledge of the possible outcomes of their behaviors. If enough of the unanticipated consequences of behavior are similar, they will in the long run accrue enough presence or force to bring social change. Deviance also builds up over time; however, deviant behavior is more structurally based than unanticipated consequences. People have structured relations with the goals and means of any society. These relations result in four different types of deviance: innovation, ritualism, retreatism, and rebellion. As similar forms of deviance accumulate, they too will push for social change. Like deviance, sociological ambivalence is structured. If the behaviors that emerge from these structured positions become patterned, they will also create pressures within the social system for change.

Looking Ahead

As I mentioned in the introduction to this section, in many ways Talcott Parsons sets the tone for sociological theorizing in the twentieth century. You'll see as we work our way through the next few chapters that Parsons set the bar that had to be referenced in order to make theoretical statements. Although Parsons's scheme is brilliant and powerful, this preeminent status is also due to the kinds of questions he asked, specifically about social order, the relations among diverse social structures, and the process of institutionalization. These are among the enduring questions of sociology, and we'll see them referenced time and again as we journey through the various perspectives.

Ironically, these Parsonian concerns are in many ways at the heart of the postmodern critique (Chapter 14). As Parsons argues, structural differentiation causes individuation and institutional strains. Postmodernists see social and structural differentiation as having reached a threshold point past which fragmentation rather than integration, and cultural pluralism rather than cohesion, dominate. Functionalists, on the other hand, would point to the fact that typical postmodern societies, such as the United States, continue to exist and operate and must therefore be functionally integrated in some way. In the hands of someone like Parsons or Merton, then, the continued existence of a society under conditions of postmodernity becomes an intriguing empirical question.

In addition to this more general influence, Parsons is also the bedrock upon which neo-functionalism is built. The prefix (neo) simply means new or recent. According to Jeffrey C. Alexander (1985), neo-functionalism "indicates nothing so precise as a set of concepts, a method, a model, or an ideology. It indicates, rather, a tradition" (p. 9). What he means is that neo-functionalism is more of a perspective than a precise theory. It's a perspective that emphasizes certain issues: the interrelationship and distinctions among social parts, differentiation as a major force of social change, and equilibrium as a reference point of analysis rather than an assumption made about the system.

Functionalism and neo-functionalism, then, discourage us from looking at individual social phenomena as if they existed in a vacuum. Rather, this perspective helps us see that sociological inquiry ought to always consider the wider context and the relationships that might exist between one's research topic and the other elements in society; using the functionalist perspective forces us to take just such a point of view. In this vein, neo-functionalism also maintains Parsons's distinctions among the personality, culture, and social systems. Parsonian theory sees the individual, culture, and society as intertwined and influencing one another. The relations and interactions among and between these subsystems are vital for understanding society and social change—the tensions produced by their interpenetration are a source of change and control.

We will consider the concepts and theories of a man whose work is generally referred to as neo-functionalist in Chapter 12: Niklas Luhmann. There are several reasons why I've put Luhmann in a chapter separate from Parsons and Merton, chief of which is that Luhmann emphasizes systems in a way that Parsons does not. While Parsons looked at society in system terms, Luhmann takes the idea of a system much more seriously and, in the end, this makes his theory decidedly different. Another reason I've held off on Luhmann is that he gives us an interesting perspective on globalization and modernity—issues more indicative of contemporary theory.

Building Your Theory Toolbox

Knowing Functionalism

After reading and understanding this chapter, you should be able to define the following terms theoretically and explain their theoretical importance to functional theorizing:

Organismic analogy, evolutionary model, function, functionalism, voluntaristic action, action theory, the unit act, modes or orientation, adaptation, goal attainment, integration, latent pattern maintenance, socialization, cybernetic hierarchy of control, equilibrium, cultural strain, alienative motivational elements, middle-range theory, functional alternatives, manifest and latent functions, dysfunctions, unanticipated consequences, conformity, innovation, ritualism, retreatism, rebellion, sociological ambivalence

After reading and understanding this chapter, you should be able to

- Define functionalism

- Explain how social systems are formed through modes of orientation and types of action, and through roles, norms, and status positions

- Describe a system's functional requisites and interstructural relations using Parsons's AGIL analytical scheme

- Explain how the cybernetic hierarchy of control works and its importance for understanding how society functions

- Discuss the process of social change from the beginnings of a social movement to the ordering of the new social system (equilibrium)

- Explain what middle-range theories are and how they fit into the overall enterprise of theory building

- Discuss Merton's critiques of functionalism and his proposed alternatives

- Use Merton's structural theory of deviance to discuss how deviance occurs in a society

- Compare and contrast Parsons's and Merton's theories of social change

Learning More: Primary Sources

- Talcott Parsons:

  Parsons, T. (1937). *The structure of social action*. New York: McGraw-Hill. (Parsons's magnum opus)
  Parsons, T. (1964). *Social structure and personality*. New York: Free Press.
  Parsons, T. (1966). *Societies*. Englewood Cliffs, NJ: Prentice Hall. (Perhaps the most easily understood of Parsons's writings)

- Robert K. Merton:

  Merton, R. K. (1949). *On theoretical sociology*. New York: Free Press.
  Merton, R. K. (1957). *Social theory and social structure* (Rev. ed.). New York: Free Press.
  Merton, R. K. (1976). *Sociological ambivalence and other essays*. New York: Free Press.

Learning More: Secondary Sources

To read more about Talcott Parsons, I would recommend the following:

> Holton, R. J., & Turner, B. S. (1989). *Talcott Parsons (Key Sociologists)*. Chichester, UK: Ellis Horwood.
> Lidz, V. (2000). Talcott Parsons. In G. Ritzer (Ed.), *The Blackwell companion to major social theorists*. Oxford, UK: Blackwell.
> Robertson, R., & Turner, B. S. (Eds.). (1991). *Talcott Parsons: Theorist of modernity*. London: Sage.

For Robert K. Merton, read the following:

> Sztompka, P. (1986). *Robert K. Merton: An intellectual profile*. New York: St. Martin's Press.

Theory You Can Use (Seeing Your World Differently)

- Parsons's primary point of view is that he sees things as a system. Recall what makes a system a system and analyze this society in terms of a system. Is this society a system? If so, in what ways? If not, in what ways does it not meet the criteria? Let's take it down a level: analyze the university you attend in terms of system qualities. Is it a system? What about your classroom? Is it a system in Parsonian terms? Can you analyze your friendship network in terms of systems? What about you as a person? Do you exist as a system? If all these are systems, how are they linked together?

- Do you think that we can understand globalization from a systems perspective? If so, name at least five different ways that Parsons's theory could be used on a global basis.

- Remembering Parsons's idea of cultural strain, take a look at the society in which you live. Is it ripe for cultural strain? If so, why? What kinds of cultural strain can you identify? From a Parsonian approach, what are the effects we might expect?

- Use Parsons's unit act analytical scheme (see Figure 6.1) to explain your behaviors at school today. Take the scheme and use it in at least five different settings (such as school, home, shopping mall, crosswalk, beach, and so on). How does his scheme hold up? Were you able to analyze all of the behaviors equally well?

- Recalling Parsons's idea of generalized media of exchange, I'd like for you to choose two institutions. What generalized media of exchange do you think exist between these two institutions? How would you go about determining if the media you propose are actually at work?

- Pick two different social institutions, such as religion and the economy. Analyze each institution using Merton's ideas of functional alternatives, manifest and latent functions, dysfunctions, and unanticipated consequences.

- Using a copy of today's newspaper (local or national), find all the articles that cover some form of deviance. Use Merton's theory of deviance to categorize and understand what is going on.

CHAPTER 7

Conflict and Critical Theories

Part I: Conflict Theory: Lewis Coser (1913–2003)

Ralf Dahrendorf (1929–)

Randall Collins (1941–)

Seeing Further: Normal Conflict

What do an argument, the Enron case, bidding on eBay, the civil rights movement, and the U. S. invasion of Iraq have in common? They are all forms of conflict with various levels of intensity and violence. We may only think of war or arguments as conflict, but what the theorists in this first part of the chapter want to point out is that society is rife with conflict—conflict is a general social form that isn't limited to just overtly violent situations. More than that, conflict doesn't necessarily rip society apart. In fact, it might be one of the most important ways that society holds itself together.

Conflict theory has a long history in sociology. Without question, Karl Marx's work in the early to mid-1800s formed the initial statements of this perspective. As you know, Marx was centrally concerned with class and the dialectics of capitalism. He argued that capitalism would produce its own gravediggers by creating the conditions under which class consciousness and a failing economy would come into existence. In this juncture between structure and class-based group experience, the working class revolution would take place.

In the early twentieth century, Max Weber formulated a response to Marx's theory. Weber saw that conflict didn't overwhelmingly involve the economy, but that the state and economy together set up conditions for conflict. Of central importance to Weber's scheme is the notion of legitimation. All systems of oppression must be legitimated in order to function. Thus, legitimation is one of the critical issues in the idea of conflict. Weber also saw that class is more complex than Marx initially supposed, and that there are other factors that contribute to social inequality, most notably status and party (or power).

Since that time, a number of efforts have combined different elements from one or both of these theorists to understand conflict. In this chapter, we will consider three of those efforts. Our first theorist is Lewis Coser. Coser's work is interesting for two reasons. First, he intentionally draws the majority of his theoretical ideas from Georg Simmel rather than Marx or Weber. Coser uses Marx and Weber now and then to frame or elaborate upon what Simmel has to say, but by and large Coser (1956) presents "a number of basic propositions which have been distilled from theories of social conflict, in particular from the theories of Georg Simmel" (p. 8). Keep this in mind as we talk about Coser's theory: we could easily substitute Simmel's name for Coser's.

The second reason Coser is remarkable is that he is the first to consider the functional consequences of conflict—other than Simmel, that is. Before Simmel, conflict had been understood as a source of social change and disintegration. Simmel was the first to acknowledge that conflict is a natural and necessary part of society; Coser brought Simmel's idea to mainstream sociology, at least in America. From that point on, sociologists have had to acknowledge that

> groups require disharmony as well as harmony, dissociation as well as association; and conflicts within them are by no means altogether disruptive factors. . . . Far from being necessarily dysfunctional, a certain degree of conflict is an essential element in group formation and the persistence of group life. (Coser, 1956, p. 31)

In terms of the history of social thought and the layout of this book, it is interesting to note that Coser (1956) was motivated to consider the functional consequences of conflict to address a deficiency in Talcott Parsons's theory: "Parsons considers conflict primarily a 'disease'" (p. 21). In the same vein, it is worthy of note that Coser was a student of Merton's.

Our second theorist is Ralf Dahrendorf. He clearly blends elements from Marx and Weber and he sprinkles in elements from Coser to present a new understanding of conflict in society. From Marx he takes the idea of dialectical change: "social structures . . . are capable of producing within themselves the elements of their supersession and change" (Dahrendorf, 1957/1959, p. viii). If you don't recall Marx's use of the dialectic, I encourage you to look back at Chapter 1.

Dahrendorf also uses Marx's notion of political interests stemming from bipolarized social positions. Remember that Marx argued that capitalism contains only two classes that really matter: the owners and the workers. These two positions are inherently antagonistic and by their nature dictate different political interests; that is, all workers have the same political interests as do all owners. From Weber, Dahrendorf takes the idea of power and authority. Rather than seeing class as the central characteristic of modern society, Dahrendorf claims that *power* is the one unavoidable feature of all social relations. In light of the theorists covered in the previous chapter, it's worth noting that Dahrendorf (1957/1959) regards Merton's theories of the middle range as "the immediate task of sociological research" (p. x), and he sees his own theory as a necessary corrective of Parsons's "equilibrium approach."

On the other hand, our third conflict theorist, Randall Collins, is much less concerned with orienting his work around Parsons's project. Rather, Collins (1975) draws on the work of Weber, Durkheim, and Goffman to argue that symbolic goods and emotional solidarity are among the "main weapons used in conflict" (p. 59). This micro-level orientation is a unique and powerful addition to the conflict perspective. Most other conflict theories are oriented toward the macro level. Stratification is generally understood as operating through oppressive structures that limit access and choices (the idea of the "glass ceiling" is a good example), and power is conceived of as working coercively through the control of material resources and methods of social control. Collins also attunes us to a different level of analysis than either Coser or Dahrendorf—the global level of geopolitics where political conflicts are analyzed within the context of history and geography.

Defining Conflict Theory

In general, conflict theory seeks to scientifically explain the general contours of conflict in society: how conflict starts and varies, and the effects it brings. The central concerns of conflict theory are the unequal distribution of scarce resources and power. What these resources are might be different for each theorist, but conflict theorists usually work with Weber's three systems of stratification: class, status, and power. Conflict theorists generally see power as the central feature of society, rather than thinking of society as held together by collective agreement concerning a cohesive set of cultural standards, as functionalists do. Where power is located and who uses it (and who doesn't) are thus fundamental to conflict theory. In this way of thinking about things, power isn't necessarily bad: it is a primary factor that guides society and social relations.

Lewis Coser: The Functional Consequences of Conflict

Photo: © Reprinted with permission of the American Sociological Association.

The Essential Coser

Biography

Lewis Coser was born in Berlin, Germany, in 1913. His family moved to Paris in 1933 where he studied literature and sociology at the Sorbonne. Because of his German heritage, Coser was arrested and interned by the French government near the beginning of WWII. He later was able to get political asylum in the

United States and arrived in New York in 1941. Coser did his Ph.D. work at Columbia University, where he studied under Robert K. Merton. His dissertation, *The Functions of Social Conflict*, took conflict theory in a new direction and was later named as one of the best-selling sociology books of the twentieth century by the journal *Contemporary Sociology*. Coser also authored *Masters of Sociological Thought*, which became one of the most influential sociological theory books in the English language. In addition, Coser established the Department of Sociology at Brandeis University; founded *Dissent* magazine; served as president of the American Sociological Association (1975), the Society for the Study of Social Problems, and the Easter Sociological Association (1983); and is honored annually through the American Sociological Association's Lewis A. Coser Award for Theoretical Agenda-Setting. Coser died in July of 2003.

Passionate Curiosity

James B. Rule (2003), writing in memoriam for *Dissent* magazine, said of Coser,

> he always considered himself an intellectual first and a sociologist second. His aim was always to make some sort of comprehensive sense of the human condition—a sense of the best that social life could offer and a hardheaded look at the worst things human beings could do to one another, a vision of possibilities of change for the better and an assessment of the forces weighing for and against those possibilities.

Keys to Knowing

Crosscutting influences, absolute deprivation, relative deprivation, rational and transcendent goals, functional consequences of conflict, internal and external conflict, types of internal conflict, network density, group boundaries, internal solidarity, coalitions

Concepts and Theory: Variation in Conflict

Coser argues that conflict is instinctual for us, so we find it everywhere in human society. There is the conflict of war, but there is also the conflict that we find in our daily lives and relationships. But Coser also argues that conflict is different for humans than for other animals in that our conflicts can be goal related. There is generally something that we are trying to achieve through conflict, and there are different possible ways of reaching our goal. The existence of the possibility of different paths opens up opportunities for negotiation and different types and levels of conflict. Because Coser sees conflict as a normal and functional part of human life, he can talk about its variation in ways that others missed, such as the level of violence and functional consequences.

Basic Sources of Conflict

First, we want to consider what brings on social conflict in the first place. As I pointed out in the definition of conflict theory, most social conflict is based on the unequal distribution of scarce resources. Weber identified those resources for us as class, status, and power. Weber, as well as Simmel, also pointed out the importance of the *crosscutting influences* that originate with the different structures of inequality. For example, a working class black person may not share the same political interests as a working class white person. The different status positions of these two people may cut across their similar class interests. Thus, what becomes important as a source of social conflict is the covariance of these three systems of stratification. If the public perceives that the same group controls access to all three resources, it is likely that the legitimacy of the system will be questioned because people perceive that their social mobility is hampered.

The other general source of conflict comes from Marx. Marx's concern was with a group's sense of deprivation caused by class. This sense of deprivation is what leads a group to class consciousness and produces conflict and social change. Marx was primarily concerned with explaining the structural changes or processes that would bring the working class to this realization, such things as rising levels of education and worker concentration that are both structurally demanded by capitalism.

Contemporary conflict theory has modified the idea of deprivation by noting that it is the shift from absolute to relative deprivation that is significant in producing this kind of critical awareness. *Absolute deprivation* refers to the condition of being destitute, living well below the poverty line where life is dictated by uncertainty over the essentials of life (food, shelter, and clothing). People in such a condition have neither the resources nor the willpower to become involved in conflict and social change.

Relative deprivation, however, refers to a sense of being underprivileged relative to some other person or group. The basics of life aren't in question here; it's simply the sense that others are doing better and that we are losing out on something. These people and groups have the emotional and material resources to become involved in conflict and social change. But it isn't relative deprivation itself that motivates people; it is the shift from absolute to relative deprivation that may spark a powder keg of revolt. People who are upwardly mobile in this way have the available resources, and they may experience a sense of loss or deprivation if the economic structural changes can't keep pace with their rising expectations.

Predicting the Level of Violence

Simmel and Coser move us past these basic premises to consider the ways in which conflict can fluctuate. One of the more important ways that conflict can vary is by its level of violence. If people perceive conflict as a means to achieving clearly expressed *rational goals,* then conflict will tend to be less violent. A simple exchange is a good example. Because of the tension present in exchanges, conflict is likely, but it is a low-level conflict in terms of violence. People engage in exchange in order to achieve a goal, and that desired end directs most other factors. Another example is a worker strike. Workers generally go on strike to achieve clearly articulated goals and the strikers usually do not want the struggle to become violent—the violence can detract from achieving their goals (though strikes will become violent under certain conditions). The passive resistance movements of the sixties and early seventies are other examples. We can think of these kinds of encounters as the strategic use of conflict.

However, conflict can be violent, and Coser gives us two factors that can produce violent conflict: *emotional involvement* and *transcendent goals.* In order to become violent, people must be emotionally engaged. Durkheim saw that group interaction could increase emotional involvements and create moral boundaries around group values and goals. He didn't apply this to conflict, but Coser does. The more involved we are with a group, the greater is our emotional involvement and the greater the likelihood of violent conflict if our group is threatened.

Conflict will also tend to have greater levels of violence when the goals of a group are seen to be transcendent. As long as the efforts of a group are understood to be directed toward everyday concerns, people will tend to moderate their emotional involvement and thus keep conflict at a rational level. If, on the other hand, we see the goals of our group as being greater than the group and the concerns of daily life, then conflict is more likely to be violent. For example, when the United States goes to war, the reasons are never expressed by our government in mundane terms. We did not say that we fought the First Gulf War in order to protect our oil interests; we fought the war in order to defeat oppression, preserve freedom, and protect human rights. Anytime violence is deemed necessary by a government, the reasons are couched in moral terms (capitalists might say they fight for individual freedoms; communists would say they fight for social responsibility and the dignity of the collective). The existence of transcendent goals is why the Right to Life side of the abortion conflict tends to exhibit more violence than advocates of choice— their goals are more easily linked to transcendent issues and can thus be seen as God-ordained.

Concepts and Theory:
The Integrating Forces of Conflict

Coser makes the case for two kinds of *functional consequences of conflict:* conflict that occurs within a group and conflict that occurs outside the group. An example of internal conflict is the tension that can exist between indigenous populations or first nations and the national government. Notice that this internal conflict is actually between or among groups that function within the same social system. Examples of external group conflicts are the wars in which a nation may involve itself. When considering the consequences for internal group conflict, Coser is concerned with low-level and more frequent conflict. When explaining the consequences for external conflict, he is thinking about more violent conflict.

Internal Conflict

Internal conflict in the larger social system, as between different groups within the United States, releases hostilities, creates norms for dealing with conflict, and develops lines of authority and judiciary systems. Remember that Coser sees conflict as instinctual for humans. Thus, a society must always contend with the psychological need of individuals to engage in conflict. Coser appears to argue that this need can build up over time and become explosive. Low-level, frequent conflict tends to *release hostilities* and thus keep conflict from building and becoming disintegrative for the system.

This kind of conflict also creates pressures for society to produce *norms governing conflict.* For example, most of the formal norms (laws) governing labor in Western capitalist countries came about because of the conflict between labor and management. We can see this same dynamic operating at the dyad level as well. For example, when a couple in a long-term relationship experiences repeated episodes of conflict, such as arguing, they will attempt to come up with norms for handling the tension in a way that preserves the integrity of the relationship. The same is true for the social system, but the social system will go a step further and develop formal authorities and systems of judgment to handle conflict. Thus, frequent, low-level conflict creates moral and social structures that facilitate social integration.

Coser also notes that not every internal conflict will be functional. It depends on the types of conflict and social structure that are involved. In Coser's theory, there are two basic *types of internal conflict:* those that threaten or contradict the fundamental assumptions of the group relationship and those that don't. Every group is based on certain beliefs regarding what the group is about. Let's take marriage as an example of a group. For many people, a basic assumption undergirding marriage is sexual fidelity. A husband and wife may argue about many things—such as finances, chores, toilet seats, and tubes of toothpaste—but chances are good that none of these will be a threat to the stability of the "group" (dyad) because they don't contradict a basic assumption that provides the basis of the group in the first place. Adultery, on the other hand, may very well put the marriage in jeopardy because it goes against one of the primary defining features of the group. Conflict over such things as household chores may prove to be functional in the long run for the marriage, while adultery may be dysfunctional and lead to the breakup of the group.

However, I want you to notice something very important here: In Coser's way of thinking about things, adultery won't break a marriage up because it is morally wrong. Whether the relationship will survive depends on the couple's basic assumptions as to its reasons for existence. A couple may have an "open marriage" based on the assumption that people are naturally attracted to other people and sexual flings are to be expected. In such a case, outside sexual relations will probably not break the group apart. Couples within such marriages may experience tension or fight about one another's sexual exploits—and research indicates that they often do—but such conflict will tend to be functional for the marriage because of its basic assumptions. Note also that conflict over household chores may indeed be dysfunctional if the underlying assumption of the marriage is egalitarianism, but the actual division of labor in the house occurs along stereotypical gender lines.

The *group structure* will also help determine whether or not a conflict is functional. As Coser (1956) explains, "social structures differ in the way in which they allow expression to antagonistic claims" (p. 152). To talk about this issue, let's make a distinction based on network density. *Network density* speaks of how often a group gets together, the longevity of the group, and the demands of the group in terms of personal involvement. Groups whose members interact frequently over long periods of time and have high levels of personal and personality involvement have *high network density.* Such groups will tend to suppress or discourage conflict. If conflict does erupt in such a group, it will tend to be very intense for two reasons. First, the group will likely have built up unresolved grievances and unreleased hostilities. Once unfettered, these pent-up issues and emotions will tend to push the original conflict over the top. Second, the kind of total personal involvement these groups have makes the mobilization of all emotions that much easier. On the other hand, groups whose members interact less frequently and that demand less involvement—those with *low network density*—will be more likely to experience the functional benefits of conflict.

External Conflict

The different groups involved in conflict also experience functional results, especially when the conflict is more violent. As a group experiences external conflict, the boundaries surrounding the group become stronger, the members of the group experience greater solidarity, power is exercised more efficiently, and the group tends to form coalitions with other groups (the more violent the conflict is, the more intensified are these effects). In order for any group to exist, it must include some people and exclude others. This inclusion/exclusion process involves producing and regulating different behaviors, ways of feeling and thinking, cultural symbols, and so forth. These differences constitute a *group boundary* that clearly demarcates those who belong from those who do not.

As a group experiences conflict, the boundaries surrounding the group become stronger and better guarded. For example, during WWII the United States incarcerated those Americans of Japanese descent. Today we may look back at that incident with shame, but at the time it made the United States stronger as a collective; it more clearly demarcated "us" from "them," which is a necessary function for any group to exist. Conflict makes this function more robust: "conflict sets boundaries between groups within a social system by strengthening group consciousness and awareness of separateness, thus establishing the identity of groups within the system" (Coser, 1956, p. 34).

Along with stronger external boundaries, conflict enables the group to also experience higher levels of *internal solidarity.* When a group engages in conflict, the members will tend to feel a greater sense of camaraderie than during peaceful times. They will see themselves as more alike, more part of the same family, existing for the same reason. Group-specific behaviors and symbols will be more closely guarded and celebrated. Group rituals will be engaged in more often and with greater fervency, thus producing greater emotional ties between members and creating a sense of sacredness about the group.

In addition, a group experiencing conflict will tend to produce a more *centralized power structure*. A centralized government is more efficient in terms of response time to danger, regulating internal stresses and needs, negotiating external relations, and so on. Violent conflict also tends to produce *coalitions* with previously neutral parties. Again, WWII is a clear example. The story of WWII is one of increasing violence with more and more parties being drawn in. Violent conflict produces alliances that would have previously been thought unlikely, such as the United States being allied with Russia.

> Coalition . . . permits the coming together of elements that . . . would resist other forms of unification. Although it is the most unstable form of socialization, it has the distinct advantage of providing some unification where unification might otherwise not be possible. (Coser, 1956, p. 143)

Coser Summary

- Contrary to the claims of most previous theorists, Coser argues that conflict can have integrating as well as disintegrating effects. Conflict functions differently whether it is between unrelated groups (external) or inside a group, between factions (internal).

- For internal conflict, the question of functionality hinges on the conflict being less violent and more frequent, not threatening the basic assumptions of the group at large, and the group having low interactional network density. Under these conditions, internal conflict will produce the following functional consequences: conflicts will serve to release pent-up hostilities, create norms regulating conflict, and develop clear lines of authority and jurisdiction (especially around the issues that conflict develops).

- External conflict that is more violent will tend to have the following functional consequences: stronger group boundaries, higher social solidarity, and more efficient use of power and authority. Conflict violence will tend to increase in the presence of high levels of emotional involvement and transcendent goals.

Ralf Dahrendorf: Power and Dialectical Change

We move now to Ralf Dahrendorf's theory of power and dialectical change. Like Coser, Dahrendorf sees conflict as universally present in all human relations. But Dahrendorf doesn't see the inevitability of conflict as part of human nature; he sees it, rather, as a normal part of how we structure society and create social order. In this sense, Dahrendorf is concerned with the same issue as Talcott Parsons: How is social order achieved? However, rather than assuming collective agreement about norms, values, and social positions, as Parsons does, Dahrendorf argues that it is *power* that both defines and enforces the guiding principles of society. Dahrendorf also follows Coser in talking about the level of violence and its effects, but Dahrendorf adds a further variable: conflict intensity.

Photo: Hulton Archives/Getty Images.

The Essential Dahrendorf

Biography

Ralf Dahrendorf was born in Hamburg, Germany, on May 1, 1929. His father was a Social Democratic politician and member of the German Parliament who was arrested and imprisoned by the Nazis during WWII. The younger Dahrendorf was arrested as well, fortuitously escaping death by only a few days. His father continued in politics after WWII in the Soviet-held portion of Germany, but was again arrested, this time by the Soviets. He eventually escaped and fled with Ralf to England. Young Dahrendorf later returned to Germany to study at the University of Hamburg, where he received his first

Ph.D. in philosophy; he earned his second Ph.D. (sociology) in England at the London School of Economics. Dahrendorf taught sociology at the universities of Hamburg, Tübingen, and Konstanz between 1957 and 1969. In 1969, Dahrendorf turned to politics and became a member of the German Parliament. In 1970, he was appointed a commissioner in the European Commission in Brussels. From 1974 to 1984, Dahrendorf was the director of the London School of Economics. In 1988, Dahrendorf became a British citizen, and in 1993 he was given life peerage and was named Baron Dahrendorf of Clare Market in the City of Westminster by Queen Elizabeth II. Sir Dahrendorf is currently a member of the House of Lords.

Passionate Curiosity

In describing his own intellectual search, Dahrendorf (1989) says that it is

> my firm belief that the regulation of conflict is the secret of liberty in liberal democracy. That if we don't manage to regulate conflict, if we try to ignore it, or if we try to create a world of ultimate harmony, we are quite likely to end up with worse conflicts than if we accept the fact that people have different interests and different aspirations, and devise institutions in which it is possible for people to express these differences, which is what democracy, in my view, is about. Democracy, in other words, is not about the emergence of some unified view from "the people," but it's about organizing conflict and living with conflict.

Keys to Knowing

Power, authority, imperatively coordinated associations, Hobbesian problem of social order, class, quasi-groups, interest groups, technical conditions, political conditions, social conditions, conflict violence and intensity

Concepts and Theory: Power and Group Interests

Power

It comes to this: dwarf-throwing contests,
dwarfs for centuries given away
as gifts, and the dwarf-jokes
at which we laugh in our big, proper bodies.
And people so fat they can't
scratch their toes, so fat
you have to cut away whole sides of their homes
to get them to the morgue.
Don't we snicker, even as the paramedics work?

And imagine the small political base
of a fat dwarf. Nothing to stop us
from slapping our knees, rolling on the floor.
Let's apologize to all of them, Roberta said
at the spirited dinner table. But by then
we could hardly contain ourselves.

—Stephen Dunn (1996, p. 61)*

Power

Power is an uneasy word, a word we don't like to acknowledge in proper company. Perhaps we may even shy away from it in improper company, because to speak it is to make it crass. It is certainly a word that social scientists are uncomfortable yet obsessed with. Social scientists understand that power makes the human world go round, but they have a devil of a time defining it or determining where it exists. One of the reasons it is hard to define is that it is present in every social situation.

Who has power, where is it located, and how is it exercised? Those questions have proven themselves to be quite difficult for social scientists to answer. Some theorists see power as an element of social structure—something attached to a position within the structure, such as the power that comes with being the president of the United States. In this scheme, power is something that a person can possess and use (see Janet Chafetz, Chapter 8).

Other theorists define power as an element of exchange (see Chapter 10). Others see power more in terms of influence. This is a more general way in which to think of power, because many types of social relationships and people can exercise influence. Still other thinkers, as we will see when we get to Michel Foucault (Chapter 14), define power as insidiously invested in text, knowledge, and discourse (see also Dorothy E. Smith, Chapter 13). I want to encourage you to pay close attention to the way our theorists speak of power and how it is used in society and social relations. It's an extremely important social factor and one that is multifaceted in the ways it is used.

For his part, Dahrendorf (1957/1959), here quoting Weber, defines **power** as "the probability that one actor within a social relationship will be in a position to carry out his own will despite resistance, regardless of the basis on which this probability rests" (p. 166). Dahrendorf also makes the distinction, along with Weber, between power and authority. Power is something that can be exercised at any moment in all social relations and depends mostly on the personalities of the individuals involved. Because of its universal characteristic, Dahrendorf calls power "factual": it is a fact of human life.

Power can be based on such different sources as persuasion and brute force. If someone has a gun pointed at your head, chances are good that the person has

---

the power in the encounter; that is, if he or she is willing to use it and you're afraid of dying, then chances are good you'll do what the person says—those individual features are where personality comes in. Persuasion works subtly as we are drawn in by the personal magnetism of the other person. Persuasion can also be based on skills: if someone knows how interactions work and knows social psychology, then she or he can manipulate those factors and achieve power in the interaction. Again, a specific personality is involved—knowing how to manipulate people and actually doing it are two different things.

However, like Weber, Dahrendorf is more interested in authority than this kind of factual power. Authority is a form of power, of course, but it is legitimate power. It is power that is "always associated with social positions or roles" (Dahrendorf, 1957/1959, p. 166). *Authority* is part of social organization, not individual personality. Please note where Dahrendorf locates authority—the legitimated use of power is found in the status positions, roles, and norms of organizations. Obvious examples are your professors, the police, your boss at work, and so on. Because of its organizational embeddedness, Dahrendorf refers to authoritative social relations as **imperatively coordinated associations (ICAs)**. I know that sounds like a complex idea, but it actually isn't. If something is imperative, it is binding and compulsory; you must do it. So the term simply says that social relations are managed through legitimated power (authority). While the term is straightforward, it is also important.

As I mentioned before, Dahrendorf positions himself against Parsons, and here is where we can see the differences that he wants to accentuate. Dahrendorf (1968) makes the distinction between the "equilibrium approach" to social order and the "constraint approach" (pp. 139–140). Parsons is concerned with what is commonly called the *Hobbesian problem of social order*, after the philosopher Thomas Hobbes. Hobbes felt that, apart from social enforcement, some kind of glue binding people together, society would disintegrate into continual chaos and confrontation.

The problem, then, is to explain how selfishly motivated actors create social order. If all you care about is yourself, why would you cooperate with other people to achieve goals you don't care about? One solution to the problem is found in exchange theory (Chapter 10); another prominent idea is proposed by both Durkheim and Parsons. Functionalists argue for the equilibrium approach to the problem of social order: society is produced as individuals are constrained and directed through a cohesive set of norms, values, and beliefs. For Durkheim, this took the form of a moral collective consciousness that imposes its will on the individual members of the group. You'll remember that, for Parsons, the solution is found in modes of orientation, commonly held cultural belief systems, expressive symbols, value orientations, and recognizable types of action.

Dahrendorf (1968) recognizes that "continuity is without a doubt one of the fundamental puzzles of social life" but argues that social order is the result of constraint rather than some consensus around social beliefs (pp. 139–140). In the constraint approach, the norms and values of society are established and imposed through authoritative power. Be careful to see the distinction that's being made. In the equilibrium model, the actions of individuals are organized through a collectively held and agreed-upon set of values, roles or types of action, expressive symbols, and so on. In this Durkheim–Parsons model, these cultural elements hold sway because they are functional and/or they have moral force. These elements produce an equilibrium or balance between individual desires and social needs.

Dahrendorf, however, points out that there is an assumed element of power in the equilibrium model. By definition, "a norm is a cultural rule that associates people's behavior or appearance with rewards or punishments" (Johnson, 2000, p. 209). Not all behaviors are *normative*—that is, not all are governed by a norm or standard. To bring out this point, let's compare normal (in the usual sense) and normative. Some behaviors can be normal (or not) and yet not be guided by a norm. For example, I usually wear jeans, T-shirts, and Chuck Taylor shoes to teach in. That's not normal attire for a professor at my school, but I'm not breaking a norm in dressing like that. There are no sanctions involved—I don't get rewarded or punished. I'm sure you see Dahrendorf's point: norms always presume an element of power in that they are negatively or positively enforced.

Dahrendorf agrees with Durkheim and Parsons that society is created through roles, norms, and values, but he argues that they work through power rather than collective consensus. Here is where we can see the primary distinction between the functional and conflict theory approaches: Functionalists assume some kind of cultural agreement and don't see power as a central social factor; in contrast, conflict theorists argue that power is the central feature of society. Further, as a conflict theorist, Dahrendorf (1968) sees that the substance of social roles, norms, status positions, values, and so forth "may well be explained in terms of the interests of the powerful" (p. 140). Like Marx, Dahrendorf argues that the culture of any society reflects the interests of the powerful elite and not the political interests of the middle or lower classes.

It is also important to note that Dahrendorf sees **class** as related more to power than to money or occupation. Both of those might be important, but the reason for this is that they contribute to an individual's power within an ICA. Thus, for Dahrendorf (1957/1959), classes "are social conflict groups the determinant . . . of which can be found in the participation in or exclusion from the exercise of authority within any imperatively coordinated association" (p. 138). Keep this distinction in mind. It implies that Dahrendorf's concern with conflict is more narrowly defined than is Coser's. Coser is interested in explaining *any* internal and external conflict, while Dahrendorf's main interest is internal *class* conflict.

Latent and Manifest Interests

Like Marx, Dahrendorf sees the interests of power and class in dichotomous terms: you either can wield legitimated power or you can't. Now that I've said that, I need to qualify it. Remember that Dahrendorf calls the social relationships organized around legitimated power imperatively coordinated associations. One of the ideas implied in the term is that social relations are embedded within a hierarchy of authority. What this means is that most people are sandwiched in between power relations. That is, they exercise power over some and are themselves subject to the authority of those above them. However, this idea also points out that embedded within this hierarchy of power are dichotomous sets of interests.

For example, let's say you are a manager at a local eatery that is part of a restaurant chain. As manager, you will have a number of employees over whom you have authority and exercise power. You will share that power with other shift or section managers. In the restaurant, then, there are two groups with different power interests: a group of managers and a group of employees. At the same time, you have regional and corporate managers over you. This part of the organizational structure sets up additional dichotomous power interests. In this case, you are the underling and your bosses exercise power over you and others. If you stop and think about it, you'll see what Dahrendorf wants us to see: Society is set up and managed through imperatively coordinated associations. Society is a tapestry that is woven together by different sets of power interests.

Okay, social relationships are coordinated through authority and power is everywhere. What's the big deal? What else does Dahrendorf want us to see? There's an important distinction and significant question that Dahrendorf wants us to become aware of. Using two terms from Merton, Dahrendorf argues that everyone is involved in positions and groups with latent power interests. People with these similar interests are called quasi-groups. *Quasi-groups* "consist of incumbents of roles endowed with like expectations of interests" and represent "recruiting fields" for the formation of real *interest groups*. Interest groups, Dahrendorf tells us, "are the real agents of group conflict" (Dahrendorf, 1957/1959, p. 180). Everybody is part of various quasi-groups. For example, you and your fellow students form a loose aggregate of interests opposed to the professors at your university. Here's the significant question that Dahrendorf wants us to consider: How do latent interests become manifest interests? In other words, what are the social factors that move an aggregate from a quasi-group to an interest group?

Concepts and Theory: Conflict Groups and Social Change

Conditions of Conflict Group Formation

Before we get into these conditions, let me reemphasize an important sociological point. Every single one of us maintains different positions within social aggregates. An aggregate is simply "a mass or body of units or parts somewhat loosely associated with one another" (Merriam-Webster, 2002). For example, you have an economic class position; perhaps you're working or middle class. Yet, while you share that position with a vast number of others, you may not experience any sense of group identity or shared interests. When, why, and how these aggregates actually form into social groups is a significant sociological question. As an illustration, ask yourself what would have to happen for you and your fellow students to become an active social group that would rise up against the authority of your professors or campus administrators? More significantly, what are the conditions under which disenfranchised groups such as gays and lesbians (in the United States) would challenge the existing power arrangements?

Dahrendorf gives us three sets of conditions that must be met for a group to become active in conflict: technical, political, and social conditions. The *technical conditions* are those things without which a group simply can't function. They are the things that actually define a social group as compared to an aggregate. The technical conditions include members, ideas or ideologies (what Dahrendorf calls a "charter"), and norms. The members that Dahrendorf has in mind are the people who are active in the organization of the group. For an illustration, we can think of a Christian church. As any pastor knows, within a congregation there are active and inactive members. There are the people who actually make the church work by teaching Sunday school or organizing bake sales; and then there are the people who show up once or twice a week and simply attend. We can see the same thing in political parties: There are those who are active year in and year out and there are those who simply vote. It's the workers or "leading group" that Dahrendorf has in mind as members.

For a collective to function as a group, there also has to be a defining set of ideas, or an ideology. These ideas must be distinct enough from the ruling party to set the conflict group apart. For example, for the students at your school to become an interest group, there would have to be a set of ideas and values that are different from the ones the administration and faculty hold. Just such an ideology was present during the free speech movement at the University of California at Berkeley during the sixties. A friend of mine taught his first introduction to sociology class at Berkeley during this time. He walked in on the first day of class and handed out his syllabus. In response, the students, all 300 of them, got up and walked out. Why? The students believed that they should have had input in making up the syllabus— a value that most professors don't hold. (My friend, by the way, invited them back to collectively negotiate a syllabus.)

A group also requires norms. Groups are unruly things. Without norms, people tend to go off in their own direction either by mistake or intention. There must be some social mechanism that acts like a shepherd dog, nipping at the heels of the sheep to bring them back to the flock. So important are norms to human existence, Durkheim argued that people would commit suicide if there were no clear norms to guide behavior (anomic suicide). Norms are particularly important for interest groups involved in conflict. Conflict demands a united stand from the interest group, and norms help preserve that solidarity. Note also that the existence of norms implies a power hierarchy within the interest group itself—a leadership cadre.

The *political conditions* refer specifically to the ability to meet and organize. This is fairly obvious but is nonetheless important. Using our student revolt example, let's say that your university administration got wind of student unrest. Now, where is the most logical and the easiest place for a group of students to meet? The college campus would be the best place; many students live there and perhaps have limited transportation, and the campus is also the place that every student knows. However, the administration controls access to all campus facilities and could forbid students to gather, especially if they knew that the students were fomenting a revolt.

The administration could further hamper meetings through the way the campus is built. I attended a school that was building a student center while I was there. Everybody was excited, and we students were looking forward to having all the amenities that come with such a facility, such as greater choices in food (we would be getting Burger King, Kentucky Fried Chicken, Pizza Hut, and assorted other options) and a movie theater. What most of us didn't realize at the time was that the university had had plans long before to build a student center, but those plans got scrapped. Why? The original center was supposed to be built in 1964, right in the middle of the civil rights and free speech movements. The university didn't build the center then because they didn't want to provide the students with an opportunity to gather together. The center was eventually built during the latter part of the 1980s, when students seemed most content with capitalist enterprise. Now, move this illustration out to general society and you'll see the importance of these political conditions: governments can clearly either hamper or allow interest groups to develop.

Social conditions of organization must also be met. There are two elements here: communication and structural patterns of recruitment. Obviously, the more people (quasi-groups) are able to communicate, the more likely they will form a social group (interest group). A group's ability to communicate is of course central to Marx's view of class consciousness. Dahrendorf (1957/1959) brings it into his theory with updates: "In advanced industrial societies this condition may be assumed to be generally given" (p. 187).

Marx of course was aware of some communication technologies, such as printing and newspapers, but still saw that bringing people together in physical proximity was necessary for communication. Dahrendorf, writing in the 1950s, saw even more technological development than did Marx, and you and I have seen this condition fully blossom with the advent of computer technologies and the Internet. Communication is thus a given in modern society. But hold onto this idea of non–face-to-face communication until we get to Randall Collins; he's going to give us a caveat to Dahrendorf's assumed level of communication.

The second part of Dahrendorf's social conditions also sets a limit on communication. The social connections that people make must be structurally predictable for an interest group to develop. Let's use Internet communication as an example. When email and the Internet first began, there were few mechanisms that patterned the way people got in touch with one another. People would email their friends or business acquaintances, and in that sense computer technologies only enhanced already established social connections. But with the advent of search engines like Google and Web sites like Yahoo, there are now structural features of the Internet that can more predictably bring people together.

For instance, I just opened the Yahoo homepage. Under "Groups" is listed "From Trash to Treasure; React locally, impact globally." If I'm concerned about ecological issues, then my communication with other like-minded people is now facilitated by the structure of the Internet. However, my accessing the Yahoo homepage is not structured. Whether or not you or I use Yahoo and see the discussion group is based on "peculiar, structurally random personal circumstances," which "appear generally unsuited for the organization of conflict groups" (Dahrendorf, 1957/1959, p. 187). Thus, while parts of these social conditions appear to be structured, others are not. The thing I want you to see here is that this condition is highly variable, even though we are living in a technologically advanced society.

Social Change

According to Dahrendorf, conflict will vary by its level of intensity and violence. *Conflict intensity* refers to the amount of costs and involvement. The cost of conflict is rather intuitive; it refers to the money, life, material, and infrastructure that are lost due to conflict. Involvement refers to the level of importance the people in the conflict attach to the group and its issues. We can think of this involvement as varying on a continuum from the level that a game of checkers requires to that of a front-line soldier. Checkers only requires a small portion of a person's personality and energy, while participating in a war where life and death are at stake will engulf an individual's entire psyche. For Dahrendorf, *conflict violence* refers to how conflict is manifested and is basically measured by the kinds of weapons used. Peaceful demonstrations are conflictual but exhibit an extremely low level of violence, while riots are far more violent.

While violence and intensity can go together, as in a nuclear war, they don't necessarily covary, and they tend to influence social change in different directions. More intense conflicts will tend to generate more profound social changes. We can think of the life of Mahatma Gandhi as an example of conflict with a high level of intensity but no violence. Gandhi is also a good example of the profound social changes that intense conflict can engender. Not only was he centrally responsible for major structural changes in Indian society, he has also had a profound and lasting impact worldwide.

On the other hand, the violence of a conflict will influence how quickly the changes occur. We can think of the recent invasion of Iraq by the U.S. military as an example of violent conflict and rapid social change. The United States invaded Iraq on March 20, 2003. On April 9, 2003, Baghdad fell to the U.S.-led military forces. On that day, U.S. marines pulled down the 20-foot-tall statue of Saddam Hussein, thus symbolically ending his regime. An interim Iraqi government was appointed in 2004 and elections for a permanent government occurred in 2005. How deep these structural changes go remains to be seen, yet there is little doubt that the rapidness of the changes is due to the level of violence the United States government was willing to employ.

Important note: the two examples I've just given are somewhat outside the scope of Dahrendorf interests. Remember that Dahrendorf is concerned primarily with explaining class conflict within a society. The reason I used those examples is that they clearly point out the differences between the violence and the intensity of conflict. Often class conflict, especially over longer periods of time, involves both intensity and violence and thus they are difficult to empirically disengage from one another. A good example of these factors is the civil rights movement in the United States. I invite you to check out a civil rights timeline by using your favorite Internet search engine; be sure to use a timeline that goes back at least to 1954. Think about the types of conflict, whether intense or violent, and the kinds of social changes occurring.

Level of Violence

Within a society, the violence of class conflict, as defined by Dahrendorf, is related to three distinct groups of social factors: (1) the technical, political, and social conditions of organization; (2) the effective regulation of conflict within a society; and (3) the level of relative deprivation. Violence is negatively related to the three conditions of organization. In other words, the more a group has met the technical, political, and social conditions of organization, the less likely it is that the conflict will be violent. Remember, we saw this idea in a more basic form with Coser. While some level of organization is necessary for a group to move from quasi- to interest group, the better organized a group is, the more likely it is to have rational goals and to seek reasonable means to achieve those goals.

The violence of a conflict is also negatively related to the presence of legitimate ways of regulating conflict. In other words, the greater the level of formal or informal norms regulating conflict, the greater the probability that both parties will use the norms or judicial paths to resolve the conflict. However, this factor is influenced by two others. In order for the two interested parties to use legitimate roads of conflict resolution, they must recognize the fundamental justice of the cause involved (even if they don't agree on the outcome), and both parties need to be well-organized. In addition, the possibility of violent conflict is positively related to a sense of *relative deprivation*. We reviewed this idea with Coser, but here Dahrendorf is specifying the concept more and linking it explicitly to the level of violence.

Level of Intensity

Within a social system, the level of *conflict intensity* is related to the technical, political, and social conditions of organization; the level of social mobility; and to the way in which power and other scarce resources are distributed in society. Notice that both violence and intensity are related to group organization and the relationship in both cases is negative. The violence and intensity of conflict will tend to go down as groups are better organized—again, for the same reason: better organization means more rational action.

With Coser, we saw that people will begin to question the legitimacy of the distribution of scarce resources as the desired goods and social positions tend to all go to the same class. Here, Dahrendorf is being more specific and is linking this issue with conflict intensity. The relationship is positive: the more society's scarce resources are bestowed upon a single social category, the greater will be the intensity of the conflict. In this case, the interest groups will see the goals of conflict as more significant and worth more involvement and cost. Finally, the intensity of a conflict is negatively related to social mobility. If an ICA (imperatively coordinated association) sees its ability to achieve society's highly valued goods and positions systematically hampered, then chances are good the group members will see the conflict as worth investing more of themselves in and possibly sustaining greater costs.

In Table 7.1, I've listed the various propositions that Coser and Dahrendorf give us concerning the varying levels of conflict violence and intensity. As you can see, the level of violence tends to go up with increasing levels of emotional involvement, the presence of transcendent goals, and a sense of change from absolute to relative deprivation. Conversely, the likelihood of violence in conflict tends to go down when the interest groups meet the technical, social, and political conditions of organization (class organization); when they have explicitly stated rational goals; and when there are norms and legal channels available for resolving conflict. As the violence of conflict increases, we can expect social changes to come rapidly and we can anticipate groups to experience stronger boundaries, solidarity, and more efficient control and authority. Only Dahrendorf comments on conflict intensity, and he argues that decreasing class organization and social mobility and increasing covariance of authority and rewards will tend to produce higher levels of intensity, which in turn will produce more profound structural changes.

Table 7.1   Coser and Dahrendorf's Propositions of Conflict Violence and Intensity

Propositions Concerning the Level of Conflict Violence	
↑ Emotional Involvement	↑ Violence
↑ Transcendent Goals	↑ Violence
↑ Sense of Absolute to Relative Deprivation	↑ Violence
↑ Class Organization	↓ Violence
↑ Explicitly Stated Rational Goals	↓ Violence
↑ Normative Regulation of Conflict	↓ Violence
Possible functional effects: greater rapidness of change; stronger group boundaries; greater group solidarity; centralization of power	
Propositions Concerning the Level of Conflict Intensity	
↓ Class Organization	↑ Intensity
↓ Social Mobility	↑ Intensity
↑ Association of Authority and Rewards	↑ Intensity
Possible effects: more profound structural changes	

Regardless of how fast or how dramatically societies change, the changes must be institutionalized. We saw this idea with Parsons. For Dahrendorf (1957/1959), institutionalization occurs within structural changes "involving the personnel of positions of domination in imperatively coordinated associations" (p. 231). What you should notice about this statement is that social change involves changing personnel in ICAs. Remember that ICAs are how Dahrendorf characterizes the basic structure of society. The roles, norms, and values of any social group are enforced through the legitimated power relations found in ICAs. Every ICA contains quasi-groups that are differentiated around the issue of power. ICAs move from quasi-group status to interest groups, and concerns of power move from latent to manifest, as these groups meet the technical, political, and social conditions of group organization. This conflict then brings different levels and rates of change based on its intensity and violence. These changes occur in the structure of ICAs, with different people enforcing different sets of roles, norms, and values, which, in turn, sets up new configurations of power and ICAs. Then this power dialectic starts all over again.

Dahrendorf Summary

- Dahrendorf argues that underlying all social order are imperatively coordinated associations (ICA). ICAs are organizational groups based on differential power relations. These ICAs set up latent power interests between those who have it and those who don't. These interests will tend to become manifest when a group meets the technical, political, and social conditions of group organization. Conflict generated between interest groups varies by intensity and violence.

- The intensity of conflict is a negative function of group organization and social mobility, and a positive function of association among the scarce resources within a society. The more intense conflicts are, the more profound are the structural changes.

- The violence of conflict is a negative function of the conditions of group organization and already existing legitimate ways of resolving conflict, and a positive function of relative deprivation. The more violent is the conflict, the quicker structural change occurs.

- Social change involves shifts in the personnel of ICAs. The new personnel impose their own hierarchy of status positions, roles, norms, and values, which sets up another grouping of ICAs and latent power interests.

Randall Collins: Emotion and the World in Conflict

Randall Collins takes us in a different direction from either Coser or Dahrendorf. First, Collins's work of synthesis is broader and more robust. As I've already mentioned, Collins draws not only from the classical conflict theorists, he also uses Durkheim and Erving Goffman (Chapter 9). The inclusion of Durkheim is extremely important. Using Durkheim allows Collins to consider the use of emotion and ritual in conflict. As you'll see, these are important contributions to our understanding of conflict. In talking about Collins's theory, I'm not going to review what Durkheim said about rituals and emotion. So, be sure to bring the information you learned in Chapter 3 into your thinking here. If you need to, please review Durkheim's theory of ritual.

But more than adding new ideas, the scope of Collins's project is much wider. In 1975 Collins published *Conflict Sociology*. His goal in the book was to draw together all that sociologists had learned about conflict and to scientifically state the theories in formal propositions and hypotheses. The end result is a book that contains hundreds of such statements.

Without a doubt, his book represents the most systematic effort ever undertaken to scientifically explain conflict, even to this day. Then, in 1993, Collins reduced the hundreds of theoretical statements from his 1975 work to just "four main points of conflict theory" (1993a, p. 289). Anytime a theorist does something like this, the end statement is theoretically powerful. In essence, what Collins is saying is that most of what we know about conflict can be boiled down to these four points. Collins also takes us further because he considers more macro-level, long-range issues of conflict in a new theoretical domain called "geopolitical theory."

Photo: © Courtesy of Randall Collins.

The Essential Collins
Concepts and Theory: Four Main Points in Conflict Sociology
    Scarce Resources and Mobilization
    The Propagation and End of Conflict
Concepts and Theory: Geopolitics
    The Role of the State
    Geopolitical Dynamics
    The Demise of Soviet Russia
Collins Summary

The Essential Collins

Biography

Randall Collins was born in Knoxville, Tennessee, on July 29, 1941. His father was part of military intelligence during WWII and then a member of the state department. Collins thus spent a good deal of his early years in Europe. As a teenager, Collins was sent to a New England prep school, afterward studying at Harvard and the University of California, Berkeley, where he encountered the work of Herbert Blumer and Erving Goffman, both professors at Berkeley at the time. Collins completed his Ph.D. at Berkeley in 1969. He has spent time teaching at a number of universities, such as the University of Virginia and the Universities of California at Riverside and San Diego, and has held a number of visiting professorships at Chicago, Harvard, Cambridge, and at various universities in Europe, Japan, and China. He is currently at the University of Pennsylvania.

Passionate Curiosity

Collins has enormous breadth, but seems focused on understanding how conflict and stratification work through face-to-face ritualized interactions. Specifically, his passion is to understand how societies are produced, held together, and destroyed through emotionally rather than rationally motivated behaviors.

Keys to Knowing

Conflict mobilization, material and emotional resources, resource mobilization, ritualized exchange of atrocities, bureaucratization of conflict, ritual solidarity, geopolitical theory, the state, state legitimacy, heartland advantage, marchland advantage, overexpansion

---

Concepts and Theory: Four Main Points in Conflict Sociology

Scarce Resources and Mobilization

Point One: The unequal distribution of each scarce resource produces potential conflict between those who control it and those who don't. Dahrendorf argues that there is one primary resource in society: power. Randall Collins, on the other hand, follows the basic outline that Weber gave us of the three different types of scarce resources: *economic resources*, which may be broadly understood as all material conditions; *power resources*, which are best understood as social positions within control or organizational networks; and status or *cultural resources*, which Collins understands as control over the rituals that produce solidarity and group symbols.

Notice that Collins expands and generalizes two of these resources. Both Marx and Weber saw economic resources in terms of class position; Collins, however, argues that economic resources ought to be seen as encompassing a much broader spectrum of issues—control over any material resources. These may come to us as a consequence of class, but they also may accrue to a person working in an underground social movement through thievery or other illegal means.

Point Two: Potential conflicts become actual conflicts to the degree that opposing groups become mobilized. There are at least two main areas of *resource mobilization:* The first area involves emotional, moral, and symbolic mobilization. The prime ingredient here is collective rituals. This is one of Collins's main contributions to conflict theory. Groups don't simply need material goods to wage a battle; there are also clear emotional and symbolic goods used in conflict. As Durkheim (1912/1995) says, "we become capable to feelings and conduct of which we are incapable when left to our individual resources" (p. 212). Collins uses Durkheim's theory of ritual performance to explain symbolic mobilization. In general, the more a group is able to physically gather together, create boundaries for ritual practice, share a common focus of attention, and have a common emotional mood, the more group members will

1. Have a strong and explicit sense of group identity

2. Have a worldview that polarizes the world into two camps (in-group and out-group)

3. Be able to perceive their beliefs as morally right

4. Be charged up with the necessary emotional energy to make sacrifices for the group and cause

The second main area for mobilization concerns the material resources for organizing. Material mobilization includes such things as communication and transportation technologies, material and monetary supplies to sustain the members while in conflict, weapons (if the conflict is military), and sheer numbers of people. While this area is pretty obvious, the ability to *mobilize* material resources is a key issue in geopolitical theory.

There are a couple of corollaries or consequences that follow these propositions. If there are two areas of mobilization, then there are two ways in which a party can win or lose a conflict. The first has to do with material resources, which get used up during conflicts. People die; weapons are spent; communication and transportation technologies are used up, break down, or are destroyed; and so on. A conflict outcome, then, is dependent not only upon who has the greatest resources at the beginning of a war, but also upon who can replenish those supplies.

A group can also win by generating higher levels of ritual solidarity as compared to their enemies. Collins gives the example of Martin Luther King Jr. King obviously had fewer material resources than the ruling establishment, but the civil rights movement was able to create higher levels of ritualized energy and was able to generate broad-based symbolic, moral appeal. Of course, a group can also lose the conflict if its members are unable to renew the necessary emotional energies. Emotional energy and all the things that go with it—motivation, feelings of morality, righteous indignation, willingness to sacrifice, group identity, and so on—thus have a decay factor.

Symbols and ideas aren't themselves sacred or moral, nor do they actually "carry" sacredness or morality; they only act as prompts to evoke these emotions in people. It is necessary, then, to renew the collective effervescence associated with the symbol, moral, or group identity. If collective rituals aren't continually performed, people will become discouraged, lose their motivation, entertain alternatives views of meaning and reality, and become incapable of making the necessary sacrifices.

The Propagation and End of Conflict

Point Three: Conflict engenders subsequent conflict. In order to activate a potential conflict, parties must have some sense of moral rightness. Groups have a difficult time waging war simply on utilitarian grounds. They have to have some sense of moral superiority, some reason that extends beyond the control of oil or other material good. As a result, conflicts that are highly mobilized tend to have parties that engage in the *ritualized exchange of atrocities.* Collins calls this the negative face of social solidarity. This is a somewhat difficult subject to illustrate, because if you hold to or believe in one side in a conflict, its definition of atrocities or terrorism will seem morally right. The trick is to see and understand that there has never been a group that has entered into a conflict knowing or feeling that they are wrong. For instance, the people who flew the airplanes into the World Trade Center felt morally justified in doing so.

We can think of many, many examples from around the world, such as the Croats and Serbs and the Irish Catholics and Protestants. And the history of the United States is filled with such illustrations. For example, there is still a debate concerning the reasons and justifiability of the use of nuclear weapons during WWII. Whatever side of the debate people take, it is undeniable that retribution was and is part of the justification. As President Truman (1945a, 1945b) said,

> The Japanese began the war from the air at Pearl Harbor. They have been repaid many fold. And the end is not yet. With this bomb we have now added a new and revolutionary increase in destruction. . . . Having found the bomb we have used it. We have used it against those who attacked us without warning at Pearl Harbor, against those who have starved and beaten and executed American prisoners of war, against those who have abandoned all pretense of obeying international laws of warfare.

In addition to satiating righteous indignation and affirming social solidarity, ritualized retributions are used to garner support. We can see this clearly in the United States' use of the attacks of September 11, Israel's use of the holocaust, the antiabortionists' conceptualization of abortion as murder, and the various civil rights groups' use of past atrocities. Atrocities thus become a symbolic resource that can be used to sway public opinion and create coalitions.

Point Four: Conflicts diminish as resources for mobilization are used up. Just as there are two main areas of conflict mobilization, there are two fronts where demobilization occurs. For intense conflicts, emotional resources tend to be important in the short run, but in the long run, material resources are the key factors. Many times the outcome of a war is determined by the relative balance of resources. Randall Collins gives us two corollaries. The first is that milder or sporadic forms of conflict tend to go on for longer periods of time than more intense ones. Fewer resources are used and they are more easily renewed. This is one reason why terrorism and guerilla warfare tend to go on almost indefinitely. Civil rights and relatively peaceful political movements can be carried out for extended periods as well.

The second corollary Collins gives us is that relatively mild forms of conflict tend to deescalate due to the *bureaucratization of conflict.* Bureaucracies are quite good at co-optation. To co-opt means to take something in and make it one's own or make it part of the group, which on the surface might sound like a good thing. But because bureaucracies are value and emotion free, there is a tendency to downplay differences and render them impotent. For example, one of the things that our society has done with race and gender movements is to give them official status in the university. One can now get a degree in race or gender relations. Inequality is something we now study, rather than it being the focus of social movements. In this sense, these movements have been co-opted. "This is one of the unwelcome lessons of the sociology of conflict. The result of conflict is never the utopia envisioned in the moments of intense ideological mobilization; there are hard-won gains, usually embedded in an expanded bureaucratic shell" (Collins, 1993a, p. 296).

The second front where conflicts may be lost is *deescalation of ritual solidarity.* A conflict group must periodically gather to renew or create the emotional energy necessary to sustain a fight. One of the interesting things this implies is that the intensity of conflicts will vary by focus of attention. Conflict that is multifocused will tend not to be able to generate high levels of emotional energy. The conflict over civil rights in the United States is just such a case. The civil rights movement today has splintered because the idea of civil rights isn't held by everyone involved as a universal moral. That is, the groups involved don't focus on civil rights per se; they focus on civil rights for their group. For example, there are those working for the equal rights of African Americans who would deny those same rights to homosexuals.

Concepts and Theory: Geopolitics

There are two things that I want to point out before we consider geopolitical theory. The first is that geopolitical processes happen over the long run. These forces take time to build up and aren't readily apparent, especially to most of us living in the United States. In this country, we have difficulty thinking in the long term. We are focused on the individual and immediate gratification, and even the economic planning that is done is oriented toward short-term portfolio management. Geopolitical theory is sociology over the long term. It explains how nations grow and die. The processes and dynamics can't be seen by just looking at our daily concerns. We have to rise above ourselves and look historically.

The second thing I want to point out is that geopolitical theory focuses on the state rather than the economy. Generally speaking, world-systems theory, like that of Immanuel Wallerstein (Chapter 12), focuses on the economy. Collins understands the world system in more Weberian terms, where the nation-state is the key actor on the world stage. As mentioned earlier, nation-states are relatively recent inventions. Up until the sixteenth century, the world was not organized in terms of nation-states. People were generally organized ethnically with fairly fluid territorial limits, as with feudalism. Feudalistic states were based on land stewardship established through the relation of lord to vassal. Its chief characteristics were homage, the service of tenants under arms and in court, wardship, and forfeiture. A nation-state, on the other hand, is a collective that occupies a specific territory, shares a common history and identity, is based on free labor, and sees its members as sharing a common fate.

The Role of the State

In Weberian terms, *the state* is defined as an entity that exercises a monopoly over the legitimate use of force within and because of a specific geographic territory. First and foremost, nation-states have a monopoly on force. In fact, one of the main impetuses behind the nation is the ability to regularly tax people for the purpose of creating a standing army. Previously, armies were occasional things that were gathered to fight specific wars. A standing army is one that is continually on standby; it is ready to fight at a moment's notice.

Notice that nation-states are organized around the legitimate use of power. Thinking about power in terms of legitimacy brings in cultural and ritual elements. If power is defined as the ability to get people to do what you want, then legitimacy is defined in terms of the *willingness* of people to do what you want. In order for any system of domination to work, people must believe in it. As we saw in Weber's theory, to maintain a system of domination not based on legitimacy costs a great deal in terms of technology, money, and peoplepower. In addition, people generally respond in the long run to the use of coercion by either rebelling or giving up—the end result is thus contrary to the desired goal. Authority and legitimacy, on the other hand, imply the ability to require performance that is based upon the performer's belief in the rightness of the system.

With nation-states, there is an interesting relationship between force and legitimacy. According to Randall Collins (1986c), this legitimacy is a special kind of emotion: it's "the emotion that individuals feel when facing the threat of death in the company of others" (p. 156). Legitimacy isn't something that is the direct result of socialization, though it plays a part. Rather, legitimacy is active; it ebbs and flows and is stronger at some times than at others—people feel more or less patriotic depending on a number of factors, most notably ritual performance.

The governments of nation-states are painfully aware of the active nature of legitimacy. Legitimacy provides the government's right to rule. Though also associated with economic prosperity and mass education, nationalism—the nation-state's particular kind of legitimacy—is dependent upon a common feeling that is most strongly associated with ritualized interactions performed in response to perceived threat. This threat can be internal, as in the case of minority group uprisings, crime, and deviance, but it is most strongly associated with externally produced threat and shock. You will notice that *state legitimacy* comes up again in the next section on critical theory, but from a different perspective.

The other defining feature of the nation-state is the control of a specific geographic territory. One of the reasons that a standing army originally came about was to defend a specific territory. As humans first became settled due to agriculture, it became increasingly necessary to defend the territory and internally organize a population that was growing in both size and diversity. The geographic contours of this territory are extremely important for Collins. Collins (1987) argues that the idea of property "upholds the macroworld as a social structure" (p. 204). The reason behind this is that property is the fundamental backdrop against which all interaction rituals are produced. Further, geographic space is not simply the arena in which interactions take place; it is one of the fundamental elements over which people struggle for control, thus making space a strong ritual focus of attention. Thus, on one level, the explicitness and increased size of the territories associated with nation-states have important implications for the production of interaction ritual chains and macro-level phenomena in general.

Geopolitical Dynamics

Territory is also important because specific geopolitical issues are linked to it. All forms of political organization come and go, including nation-states. Nations are born and nations die. A sociological study in the long run ought to explain—and predict, if it is scientific—the life course of a nation. The geopolitical factors that predict and explain the rise and fall of nations are linked to territory. There are two territorial factors: heartland and marchland advantages. *Heartland advantage* is defined in terms of the size of the territory, which is linked to the level of natural resources and population size. The logic here is simple. Larger and wealthier territories can sustain larger populations that in turn provide the necessary tax base and manpower for a large military. Larger nations can have larger armies and will defeat smaller nations and armies. *Marchland advantage* is defined in terms of a nation's borders: nation-states with fewer enemies on their immediate borders will be stronger than other nations with more enemies nearby but a similar heartland advantage. Marchland nations are geographically peripheral; they are not centered in the midst of other nations.

Taken together, we can see that larger, more powerful states have a cumulative resource advantage: nations with both heartland and marchland advantage will tend to grow cumulatively over time, and the neighbors of such nations will tend to diminish. Eventually, as smaller nations are annexed, larger nations confront one another in a "showdown" war, unless a natural barrier exists (such as an ocean). Natural barriers form a buffer between powerful states and will bring a stable balance of power. On the other hand, nations that are geographically central and have multisided borders will tend to experience internal political schisms and conflict that can lead to long-term fragmentation.

The key to geopolitical theory and the demise of heartland/marchland nations is *overexpansion*. A nation can overextend itself materially and culturally. One of the important features of warfare is the cost involved with keeping an army supplied. The further away an army has to go to fight, the greater are the costs involved in transporting goods and services to it. This issue becomes important as the size of the army increases past the point where it can forage or live off the land. A critical point is reached when a nation tries to support an army that is more than one heartland away (if there is another nation or more in between the two warring factions). A nation-state can also overextend itself culturally. Remember that legitimacy is a cultural good. The legitimacy of a nation is strained the farther away it moves from its ethnic base. In other words, there is an increase in the number and extent of tension points the more a nation increases its social diversity. There are more areas of potential disagreement within a diverse population than among a homogeneous population, especially if the other ethnic groups are brought into society through warfare or other measures of forced annexation.

The Demise of Soviet Russia

Randall Collins gives us an example of these geopolitical forces in the case of the USSR. On Christmas day in 1991, the Union of Soviet Socialist Republics officially collapsed. Five years prior, Collins (1986c, pp. 186–209) published a book with a chapter entitled "The Future Decline of the Russian Empire." Collins's prediction of the fall of the USSR was based on geopolitical theory. The historical expansion of Russia illustrates these principles of geopolitical theory.

The expansion began with Moscow in the late fourteenth century, a small state with a marchland advantage. Fighting fragmented rivals, Moscow made slow cumulative growth. By 1520, Moscow had annexed all of ethnic Russia. By the late 1700s, Russia had expanded across Siberia and the Southern Steppes and was a strong military power in Europe. Russia further expanded by taking advantage of Napoleon's wars, the fall of the Ottoman Empire, and China's prolonged civil wars—this further expansion was based on geopolitical factors. In the end, the USSR was the largest country on the globe, consisting of 15 soviet socialist republics whose territories reached from the Baltic and Black Seas to the Pacific Ocean, an area of 8,649,512 square miles, 11 time zones, and, most importantly, that shared common boundaries with six European and six Asian countries.

Thinking in terms of geopolitical issues, the problems that faced the USSR are obvious. The nation was overextended both culturally and economically. It no longer held heartland advantage: in terms of total population, the enemies of the USSR outnumbered them 3.5 to 1; and in terms of economic resources, it was

4.6 to 1. In addition, because of its successful expansion, the USSR no longer had a marchland advantage. It had done away with all weak buffer states and only faced powerful enemy nations in all directions. Further, the USSR had to exert military control over its Eastern European satellites, which were two and three times removed from the heartland. All told, it had to defend borders totalling 58,000 kilometers, or over 36,000 miles. What's more, the USSR contained at least 120 different ethnic groups. As Collins (1986c) projected, "if Russia has shifted from a marchland to an interior position, it may be expected that in the long-term future Russia will fragment into successively smaller states" (p. 196).

Collins Summary

- According to Collins, in order for conflict to become overt, people must become mobilized through the material resources for organizing, and they must be emotionally motivated and sustained, feel moral justification, and be symbolically focused and united. Once conflict begins, it tends to reproduce itself through a ritualized exchange of atrocities. The back and forth exchange of atrocities reproduces and boosts emotional motivation and moral justification, and it creates further representative symbols for additional ritual performances. After a time, conflicts are won or lost primarily as the two different kinds of resources are gained or lost.

- Nation-states are based on the legitimate use of force and territorial boundaries. Legitimacy is a product of ritual performance. The rituals that produce nationalism, the nation-state's specific form of legitimacy, occur most frequently in response to the perception of threat. Threat can come from outside, as from other nations, or inside, as from social movements. Because nationalism, as with all forms of emotional energy, has a natural decay factor, it is in the government's best interest to keep the perception of threat somewhat high.

- The other defining feature of nation-states is territory, and territory, like legitimacy, carries its own set of influences, specifically heartland and marchland advantages. Heartland advantages concern material resources: natural resources, population size, and tax base. Marchland advantage is an effect of national boundaries and the number and distance from enemy territories. The key variable in geopolitical theory is overexpansion, a condition where a nation overextends its reach materially (supporting armies too far from the heartland) and culturally (controlling too diverse a population).

Part II: Jürgen Habermas: Critical Theory and Modernity

Photo: © Corbis.

The Essential Habermas
Seeing Further: The Modern Hope of Equality
    Defining Critical Theory
Concepts and Theory: Capitalism and Legitimation
    Liberal Capitalism and the Hope of Modernity
    Organized Capitalism and the Legitimation Crisis
Concepts and Theory: The Colonization of Democracy
    Colonization of the Life-world
    Colonization of the Public Sphere
Concepts and Theory: Communicative Action and Civil Society
    Ideal Speech Communities
    Civil Society
Habermas Summary

The Essential Habermas

Biography

Jürgen Habermas was born on June 18, 1929, in Düsseldorf, Germany. His teen years were spent under Nazi control, which undoubtedly gave Habermas his drive for freedom and democracy. His educational background is primarily in philosophy, but also includes German literature, history, and psychology. In 1956,

Habermas took a position as Theodor Adorno's assistant at the Institute of Social Research in Frankfurt, which began his formal association with the Frankfurt School of critical thought. In 1961, Habermas took a professorship at the University of Heidelberg, but returned to Frankfurt in 1964 as a professor of philosophy and sociology. From 1971 to 1981, he worked as the director of the Max Planck Institute, where he began to formalize his theory of communicative action. In 1982, Habermas returned to the institute in Frankfurt where he remained until his retirement in 1994.

Passionate Curiosity

Born out of the political oppression of Nazi Germany, Habermas was driven to produce a social theory of ethics that would not be based on political or economic power and would be universally inclusive. He is a critical theorist who sees humankind's hope of rational existence within the inherent processes of communication.

Keys to Knowing

Critical theory; analytical, interpretive, and critical forms of knowledge; liberal capitalism; life-world; public sphere; organized capitalism; legitimation crisis; colonization of the life-world; instrumental and value rationality; colonization of the public sphere; communicative action; ideal speech community; civil society.

Seeing Further: The Modern Hope of Equality

We have covered quite a bit of ground thus far in this chapter. I've introduced you to the ideas of three theorists whose collective work spans over 50 years and synthesizes the ideas of many of our classical theorists. These three thinkers also have some fundamental issues in common. The most notable is that each one of them is guided by the ideals of social science and theory cumulation. They see conflict as a universal attribute of society and are interested in explicating the general principles that govern conflict. Each one, then, gives us a sociological theory.

In the next section of this chapter, things change. One thing you will probably notice is that the "voice" changes. It's kind of like the differences between writing a mechanic's manual and writing prose—a different voice is required for each. The same is true with different kinds of theory. In this case, we've been talking about positive and negative relationships and not talking at all about context. That's because positivistic theories—the kind espoused by social science—see universal social laws at work in every circumstance. So, it matters little if we are looking at conflict in a traditional or modern society; there are certain laws that govern and explain conflict in all situations. That's what sociological theory tries to get at.

In this section, we will have our first taste of critical social theory; watch how the voice changes. Critical theory sees context as vitally important. Thus, what you'll find here is a relatively detailed explanation of our historical milieu. We're going to talk about such things as feudalism, nation-states, anti-trust legislation, Keynesian economics, democracy, and so on. You'll also see that the voice becomes moralizing. Critical theory is a perspective that knowingly embraces its values and ethics.

Social theory becomes more prominent as we move through the book, especially in the last two chapters, Identity Politics and Post-Theories. So, why introduce critical, social theory here? I have two reasons. First, putting these two approaches side by side lets you clearly see the differences—and it's important to see them. For example, in Chapter 8 we will see a sociological explanation of gender inequality; and in Chapter 13 we will see a social theory about gender. They are two very different things, with different voices and diverse goals. Introducing you to conflict and critical theories here makes plain what would perhaps be missed in chapters that are several hundred pages apart.

The second reason is that conflict and critical theory are related—they both come from the same source: Karl Marx. As we've seen, conflict theory began with Karl Marx and was significantly modified by Max Weber. Marx focused on the dynamics surrounding class, while Weber argued that the crosscutting influences of class, status, and power significantly impact conflict and change in society. Weber also introduced a key element in stratification: legitimacy. But there is a further distinction between Marx and Weber. While Weber was disheartened and had grave concerns about modern life, especially as related to bureaucracies and rationalization, he did not have the critical, revolutionary edge that Marx did. As a result, Marx has had a unique influence on contemporary social theory.

Marx spawned two distinct theoretical approaches. One approach focuses on conflict and class as general features of society. The intent with this more sociological approach is to analytically describe and explain conflict as a normal part of society. We've seen this emphasis in our three conflict theorists. We'll also see it explicitly applied to the issues of race and gender in Chapter 8.

The other approach that Marx inspires is more critical and is focused on emancipatory politics. Rather than maintaining analytical distance, the intent here is to expose the oppressive elements in any social system, most specifically capitalism in modernity. In this case, the capitalist system is understood in terms of all the connections and effects capitalism has on the state, education, mass media, and society at large. Critical theory, then, asks us to see all of society and all the hidden ways in which human beings are devalued, oppressed, and prevented from expressing themselves fully. In addition, critical theory holds onto the hope of modernity: improving the human condition through reason.

But, as we will see further into this book, modernity has its detractors. For example, in Chapter 11 Anthony Giddens (1990) argues that modernity has become like a runaway train, with no one at the helm; he asks, "Why has the generalising of 'sweet reason' not produced a world subject to our prediction and control?" (p. 151). The modern world is perhaps "not one in which the sureties of tradition and habit have been replaced by the certitude of rational knowledge" (Giddens, 1991, p. 3).

In response to such critiques, Erich Fromm (1955) pointed the way for critical theory when he said, "But all these facts are not strong enough to destroy faith in man's reason, good will and sanity. As long as we can think of other alternatives, we are not lost; as long as we can consult together and plan together, we can hope" (p. 363). In many ways, critical theory points to a way to fulfill the social task of modernity. Can we take control of society and move it to become better, more humane, and truly free? Can reason prevail in the face of the alienating forces of modernity? Our critical theorist in this section, Jürgen Habermas, thinks so, and gives us theoretical reasons for our doing the same.

Defining Critical Theory

In one sense, there are many kinds of critical theory, such as feminism, post-modernism, critical race theory, queer theory, and so on. But the term critical theory is also used in a very specific sense: to refer to the Frankfurt Institute for Social Research. Briefly, the Frankfurt School began in the early 1920s at the University of Frankfurt in Germany. It was formed by a tight group of radical intellectuals and, ironically, financed by Felix Weil, the son of a wealthy German merchant (Karl Marx's work was partially financed by Friedrich Engels, son of a wealthy textile baron). Weil's goal was to create "an institutionalization of Marxist discussion beyond the confines both of middle-class academia and the ideological narrow-mindedness of the Communist Party" (Wiggershaus, 1986/1995, p. 16). As the Nazis gained control in Germany, the Frankfurt School moved first to Switzerland in 1935 and eventually to California. In 1953, the school was able to move back to its home university in Frankfurt. The various leaders and scholars associated with the school include Theodore Adorno, Max Horkheimer, Herbert Marcuse, Erich Fromm, and Jürgen Habermas.

In general, the Frankfurt School elaborated and synthesized ideas from Karl Marx, Max Weber, and Sigmund Freud, and focused on the social production of knowledge and its relationship to human consciousness. This kind of Marxism focuses on Marx's indebtedness to Georg Wilhelm Hegel. Marx basically inverted Hegel's argument from an emphasis on ideas to material relations in the economy. The Frankfurt School reintroduced Hegel's concern with ideas and culture but kept Marx's critical evaluation of capitalism and the state. Thus, like Marx, the Frankfurt School focuses on ideology, but, unlike Marx, critical theory sees ideological production as linked to culture and knowledge rather than class and the material relations of production. Ideology, according to these theorists, is more broadly based and insidious than Marx supposed.

Max Horkheimer became the director of the Frankfurt School in 1930 and continued in that position until 1958. Horkheimer criticized the contemporary Western belief that positivistic science was the instrument that would bring about necessary changes, arguing instead that the questions that occupy the social sciences simply reflect and reinforce the existing social and political orders. Horkheimer believed that the kind of instrumental reasoning or rationality that is associated with science is oriented only toward control and exploitation, whether the subject is the atom or human beings. Science is thus intrinsically oppressive, and a different kind of perspective is needed to create knowledge about people.

Jürgen Habermas, the director of the institute from 1963 to the early 1990s, picked up Horkheimer's theme. He argues that there are three kinds of knowledge and interests: empirical, *analytic knowledge* that is interested in the technical control of the environment (science); hermeneutic or *interpretive knowledge* that is interested in understanding one another and working together; and *critical knowledge* that is interested in emancipation. Because scientific knowledge seeks to explain the dynamic processes found within a given phenomenon, science is historically bound. That is, it only sees things as they currently exist. That being the case, scientific knowledge of human institutions and behaviors can only describe and thus reinforce existing political arrangements (since society is taken "as is"). As such, science in sociology is ideological.

Critical theory, on the other hand, situates itself outside the historical, normative social relations and thus isn't susceptible to the same limitations as science. Truly important social questions must be addressed from outside science and the historical confines of present-day experience. The intent of critical knowledge is to get rid of the distortions, misrepresentations, and political values found in our knowledge and speech. This ideology of course comes from the bourgeoisie: "The ideas of the ruling class are in every epoch the ruling ideas" (Marx, 1932/1978b, p. 172). Under capitalism, then, the working class participates in their own oppression by holding and believing in ideas produced by the ruling elite.

Now, think about this: According to Marx, the ideas that we hold and see the world through are based on alienation and ideology. Our minds have a false consciousness about them. If this is the case, how can we think outside the box that capitalism has given us? How can we think outside our own thoughts (which are, in fact, ideological thoughts)? If we are caught up in false consciousness, how will we ever become truly aware? The answer is praxis. Initially, *praxis* is an attitude of the mind—it is founded on the desire for equality and freedom—but, more importantly, praxis is practice. It is practice that is aimed at changing both the world and the individual. Praxis, or critical consciousness, is thus a penetrating, reflexive examination of self and current class conditions. Praxis is the practice through which the social world can be changed, beginning first with the self and then extending to others. In the end, then, real critical theory comes through the process of emancipatory work and analytical thought.

In brief, then, the main points of critical theory are these:

• A critique of positivism and the idea that knowledge can be value-free. With humans, all knowledge is based on and reflects values. Positivistic knowledge through social science is simply blinded to its own biases. The supreme use of reason is emancipation, bringing equality and freedom to all humankind. Thus, the "truest" form of knowing is critical theory aimed at this end. "The philosophers have only *interpreted* the world, in various ways; the point, however, is to *change* it" (Marx, 1888/1978d, p. 145, emphasis original).

• An emphasis on the relationship between history and society on the one hand and social position and knowledge on the other. To understand any society, you must first understand its historical path and the structural arrangements within it.

The knowledge and political interests that people hold within any society are based upon a person's social position within a historically specific social context. Critical theory, then, "has as its object human beings as producers of their own historical form of life" (Horkheimer, 1993, p. 21).

- Praxis/practice. "Man must prove the truth, that is, the reality and power, the this-sideness of his thinking in practice. The dispute over the reality or non-reality of thinking which is isolated from practice is a purely *scholastic* question" (Marx, 1888/1978d, p. 144, emphasis original).

- Participatory democracy. The vehicle for social change is participatory democracy guided by critical and rational thought. Most of what passes as "democracy" is nothing but oligarchy by the powerful elite. This rule by few produces a false totality, which, in truth, is antagonistic and held together by oppressive power. In contrast, the intent of social democracy is the intellectual and practical involvement of all its citizens.

- A utopian vision. Being critical of society isn't enough. There must be a sense (and consensus) of where society ought to be heading.

Concepts and Theory: Capitalism and Legitimation

Drawing on Karl Marx's theory of capitalism, Max Weber's ideas of the state and legitimation, Edmund Husserl's notion of the life-world, and Talcott Parsons's view of social systems, Habermas gives us a model of social evolution and modernity (our current historical context). By now, you should be familiar enough with Marx, Weber, and Parsons's ideas. So, let me just briefly introduce you to the idea of the life-world. The concept of **life-world** originally came from the philosopher Edmund Husserl. Habermas uses the idea to refer to the individual's everyday life—the world as it is experienced immediately by the person, a world built upon culture and social relations, and thus filled with historically and socially specific meanings. The purpose of the life-world is to facilitate communication: to provide a common set of goals, practices, values, languages, and so on that allow people to interact, to continually weave their meanings, practices, and goals into a shared fabric of life. Hold onto this idea of the life-world; it's important and we will come back to it shortly.

Liberal Capitalism and the Hope of Modernity

Drawing from Marx and Weber, Habermas argues that there have been two phases of capitalism: liberal capitalism and organized capitalism. Each phase is defined by the changing relationship between capitalism and the state. In *liberal capitalism,* the state has little involvement with the economy. Capitalism is thus able to function without constraint. Liberal capitalism occurred during the beginning phases of Western capitalism and the nation-state.

Capitalism and the nation-state came into existence as part of sweeping changes that redefined Western Europe and eventually the world. Though beginning much earlier, these changes coalesced in the seventeenth and eighteenth centuries. Prior to this time, the primary form of government in Europe was feudalism, brought to Europe by the Normans in 1066. Feudalism is based on land tenure and personal relationships. These relationships, and thus the land, were organized around the monarchy with a clear social division between royalty and peasants. Thus, the life-world of the everyday person in feudal Europe was one where personal obligations and one's relationship to the land were paramount. The everyday person was keenly aware of her or his obligations to the lord of the land (the origin of the word "land-lord"). This was seen as a kind of familial relationship, and fidelity was its chief goal. Notice something important here: people under feudalism were subjects of the monarchy, not citizens.

Capitalism came about out of an institutional field that included the state, Protestantism, and the Industrial Revolution. The nation-state was needed to provide the necessary uniform money system and strong legal codes concerning private property; the Protestant Reformation created a culture with strong values centered on individualism and the work ethic; and the Industrial Revolution gave to capitalism the level of exploitation it needed.

Habermas argues that together the nation-state and capitalism depoliticized class relations, proposed equality based on market competition, and contributed strongly to the emergence of the public sphere. In contrast to Dahrendorf, Habermas uses the term "class" in its traditional Marxian sense. In this sense, the term first came into the English language in the seventeenth century (see Williams, 1985, pp. 60–69). At that time, it had reference mainly to education; our use of *classic* and *classical* to refer to authoritative works of study came from this application. The true modern use of the term class came into existence between 1770 and 1840—a time period that corresponds to the Industrial Revolution as well as the French and American political revolutions.

Almost everything about society changed during this time, in particular the ideas of individual rights and accountability and the primacy of the economic system. The modern word class, then, carries with it the ideas that the individual's position is a product of the social system and that social position is made rather than inherited.

> What was changing consciousness was not only increased individual mobility, which could be largely contained within the older terms, but the new sense of a society or a particular social system which actually created social division, including new kinds of divisions. (Williams, 1985, p. 62)

Thus, according to Habermas, class is no longer a *political* issue, it is an *economic* one—class relations are no longer seen in terms of personal relations and family connections, but rather as the result of free market competition. Under capitalism and the civil liberties brought by the nation-state, all members of society are seen equally as citizens and economic competitors. Any differences among members in society are thus believed to come from economic competition and market forces, rather than birthright and personal relationships. Clearly, liberal capitalism brought momentous changes to the life-world: it became a world defined by democratic freedoms and responsibilities. Social relationships were no longer familial but rather legal and rational. The chief goal for the person in this life-world was full democratic participation. According to Habermas, the mechanism for this full participation is the public sphere.

The combination of the ideals of the Enlightenment, the transformation of government from feudalism to nation-state democracy, and the rise of capitalism created something that had never before existed: the public sphere. The **public sphere** is a space for democratic, public debate. Under feudalism, subjects could obviously complain about the monarchy and their way of life, and no doubt they did. But grumbling about a situation over which one has no control is vastly different from debating political points over which one is expected to exercise control. Remember, this was the first time Europe or the Americas had citizens, with rights and civic responsibilities; there was robust belief and hope in this new person, the citizen. The ideals of the Enlightenment indicated that this citizenry would be informed and completely engaged in the democratic process, and the public sphere is the place where this strong democracy could take place.

Habermas sees the public sphere as existing between a set of cultural institutions and practices on the one hand and state power on the other. The function of the public sphere is to mediate the concerns of private citizens and state interests. There are two principles of this public sphere: access to unlimited information and equal participation. The public sphere thus consists of cultural organizations such as journals and newspapers that distribute information to the people, and it contains both political and commercial organizations where public discussion can take place, such as public assemblies, coffee shops, pubs, political clubs, and so forth. The goal of this public sphere is pragmatic consensus.

Thus, during liberal capitalism, the relationship between the state and capitalism can best be characterized as *laissez-faire,* which is French for "allow to do." The assumption undergirding this policy was that the individual will contribute most successfully to the good of the whole if left to her or his own aspirations. The place of government, then, should be as far away from capitalism as possible. In this way of thinking, capitalism represents the mechanism of equality, the place where the best are defined through successful competition rather than by family ties. During liberal capitalism, then, it was felt that the marketplace of capitalism had to be completely free from any interference so that the most successful could rise to the top. In this sense, faith in the "invisible hand" of market dynamics corresponded to the evolutionist belief in survival of the fittest and natural selection.

Organized Capitalism and the Legitimation Crisis

Such was the ideal world of capitalism and democracy coming out of the Enlightenment. The central orienting belief was progress; humankind was set free from the feudalistic bonds of monarchical government and each individual would stand or fall based on his or her own efforts. In addition to economic pursuit, these efforts were to be focused on full democratic participation. Each citizen was to be fully and constantly immersed in education—education that came not only from schools but also through the public sphere. The hope of modernity was thus invested in each citizen and that person's full participation—people believed that rational discourse would lead to decisions made by reason and guided by egalitarianism.

Two economic issues changed the relationship between the economy and the state, which, in turn, had dramatic impacts on the life-world and public sphere. First, rather than producing equal competitors on an even playing field, free markets tend to create monopolies. Thus, by the end of the nineteenth and beginning of the twentieth century, the United States' economy was essentially run by an elite group of businessmen who came to be called "robber barons." Perhaps the attitude of these capitalists is best captured by the phrase attributed to William H. Vanderbilt, a railroad tycoon: "The public be damned."

These men emphasized efficiency through "Taylorism" (named after Frederick Taylor, the creator of scientific management) and economies of scale. The result was large-scale domination of markets. These monopolies weren't restricted to the market; they extended to "vertical integration" as well. With vertical integration, a company controls before-and-after manufacture supply lines. One example is Standard Oil, who at this time dominated the market, owned wells and refineries, and controlled the railroad system that moved its product to market.

The response of the U.S. government to widespread monopolization was to enact antitrust laws. The first legislation of this type in the United States was the Sherman Antitrust Act of 1890. In part the act reads,

> Every contract, combination in the form of trust or otherwise, or conspiracy, in restraint of trade or commerce among the several States, or with foreign nations, is declared to be illegal. . . . Every person who shall monopolize, or attempt to monopolize, or combine or conspire with any other person or persons, to monopolize any part of the trade or commerce among the several States, or with foreign nations, shall be deemed guilty of a felony.

However, capitalists fought the act on constitutional grounds and the Supreme Court prevented the government from applying the law for a number of years. Eventually the Court decided for the government in 1904, and the Antitrust Act was used powerfully by both Presidents Theodore Roosevelt and William Taft. This regulatory power of the U.S. government was further extended under Woodrow Wilson's administration and the passing of the Clayton Antitrust Act in 1914.

The second economic issue that modified the economy's relationship with the state was economic fluctuations. As Karl Marx had indicated, capitalist economies are subject to periodic oscillations, with downturns becoming more and more harsh. By the late 1920s, the capitalist economic system went into severe decline, creating worldwide depression in the decade of the thirties. What came to be called "classic economics" fell out of favor and a myriad of competitors clamored to take its place. Eventually the ideas of John Maynard Keynes took hold and were explicated in his 1936 book, *The General Theory of Employment, Interest and Money.* His main idea was simple, and reminiscent of Marx: capitalism tends toward over-production—the capacity of the system to produce and transport products is greater than the demand. Keynes's theory countered the then-popular belief in the invisible hand of the market and argued that active government spending and management of the economy would reduce the power and magnitude of the business cycle.

Keynes's ideas initially influenced Franklin D. Roosevelt's belief that insufficient demand produced the depression, and after WWII Keynes's ideas were generally accepted. Governments began to keep statistics about the economy, expanded their control of capitalism, and increased spending in order to keep demand up. This new approach continued through the 1950s and 1960s. While the economic problems of the 1970s cast doubt upon Keynesian economics, new economic policies have continued to include some level of government spending and economic manipulation.

Thus, due to the tendency of completely free markets to produce monopolies and periodic fluctuations, the state became much more involved in the control of the economy. *Organized capitalism,* then, is a kind of capitalism where economic practices are controlled, governed, or organized by the state. According to Habermas, the change from liberal to organized capitalism, along with the general dynamics of capitalism (such as commodification, market expansion, advertising, and so on), have had three major effects.

First, there has been a shift in the kind and arena of crises. As we've seen, liberal capitalism suffered from economic crises. Under organized capitalism, however, the economy is managed by the state to one degree or another. This shift means that the crisis, when it hits, is a crisis for the state rather than the economy. It is specifically a legitimation crisis for the state and for people's belief in rationality.

There are two things going on here in the relationship between the state and the economy: the state is attempting to organize capitalism and it is employing scientific knowledge to do so. Together, these issues create *crises of legitimation and rationality* rather than simply economic disasters. Nevertheless, Habermas argues that the economy is the core problem: capitalism has an intrinsic set of issues that continually create economic crises. However, due to the state's attempts to govern the economy, what the population experiences are ineffectual and disjointed responses from the state rather than economic crises. More significantly, in attempting to solve economic and social problems, the state increasingly depends upon scientific knowledge and technical control. This reliance on technical control changes the character of the problems from social or economic issues to technical ones (recall Habermas's three types of knowledge).

Concepts and Theory: The Colonization of Democracy

The other two important effects concern the life-world and the public sphere. In our discussion of legitimation and rationality crises, we can begin to see the changes in the life-world. The life-world of liberal capitalism was constructed out of a culture that believed in progress through science and reason. In this life-world, the person was expected to be actively involved in the democratic process. However, the general malaise that grows out of the crisis of legitimation reduces people's motivation and the meaning they attach to social life.

Colonization of the Life-World

In addition, according to Habermas, the life-world is becoming increasingly colonized by the political and economic systems. To understand what Habermas means, we have to step back a little. As I've already noted, Habermas gives us a theory that involves social evolution. In general, social evolutionists argue that society progresses by becoming more complex: structures and systems differentiate and become more specialized. The evolutionary argument is that this specialization and complexity produce a system that is more adaptable and better able to survive in a changing environment.

One of the problems that comes up in differentiated systems concerns coordination and control, or what Habermas refers to as "steering." This is one of the problems that Talcott Parsons identified. The issue is trying to guide social structures that have different values, roles, status positions, languages, and so forth. Differentiated social structures tend to go off in their own direction. We have seen that Parsons felt this problem was solved through generalized media of exchange. The idea of media is important, so let's consider it again for a minute. Merriam-Webster (2002) defines *medium* (media is plural) as "something through or by which something is accomplished, conveyed, or carried on." For example, language is a form of media: it's the principal medium through which communication is organized and carried out. Different social institutions or structures use different media. In education, for instance, it's knowledge, and in government, it's power. These are the instruments or media through which education and government are able to perform their functions.

For Parsons, the solution to the problem of social integration and steering is for the different social subsystems to create media that are general or abstract enough that all other institutions could use them as means of exchange. We can think about this like boundary crossings. Visualize a boundary between different social structures or subsystems, such as the economy and education. How can the boundary between economy and education be crossed? Or, using a different analogy, how can the economy and education talk to each other when they have different languages and values?

Habermas is specifically concerned with the boundaries between the life-world and the state and economy. In Habermas's terms, Parsons basically argues that the state and economy use power and money respectively as media of exchange with the life-world. If you think about this for a moment, it seems to make sense: You exist in your life-world, so what does the economy have that you want? You might start a list of all the cars, houses, and other commodities that you want, but what do they all boil down to? Money. And how does the economy entice you to leave your life-world and go to work? Money. So, money is the medium of exchange between the life-world and the economy. The same logic holds for the boundary between the life-world and the state: power is what the state has and what induces us to interact with the state. However, Habermas (1981/1987) sees a problem:

> I want to argue against this—that in the areas of life that primarily fulfill functions of cultural reproduction, social integration, and socialization, mutual understanding cannot be replaced by media as the mechanism for coordinating action—that is, it *cannot be technicized*—though it can be expanded by technologies of communication and organizationally mediated—that is, it can be *rationalized*. (p. 267, emphasis original)

Habermas (1981/1987) is arguing that there is something intrinsic about the life-world that cannot be reduced to media, such as money and power, "without sociopathological consequences" (p. 267). Let me give you an easy example from a different issue: the sex act. Most people would agree that you cannot "technicize" this behavior using the medium of money without fundamentally changing the nature of it; there is a clear distinction between making love with your significant partner and having sex with a prostitute. Habermas is making the same kind of argument about humanity and communication in general. For him, the sphere of mutual understanding—the life-world—cannot be reduced to power and money without essentially changing it.

Yet Habermas isn't arguing that Parsons made a theoretical mistake. Parsons saw himself as an empiricist and merely sought to describe the social world. So in this sense, Parsons was right: there is something going on in modernity that tries to mediate the life-world. Habermas takes this idea from Parsons and argues that in imposing their media on the life-world, the state and economy are fundamentally changing it. The life-world, by definition, cannot be mediated through money or power without deeply altering it.

According to Habermas, the life-world is naturally achieved through consensus. This is basically the same thing that Mead and the symbolic interactionists argue. Remember, interactions emerge and are achieved by individuals consciously and unconsciously negotiating meaning and action in face-to-face encounters. This negotiation, or consensus building, occurs chiefly through speech. Thus, using money or power fundamentally changes the life-world. In Habermas's (1981/1987) words, it is colonized: "the *mediatization* of the life-world assumes the form of a *colonization*" (p. 196, emphasis original).

This idea of the **colonization of the life-world** is perhaps one of Habermas's best-known and most provocative concepts. Using Merriam-Webster (2002) again, a colony is "a body of people settled in a new territory, foreign and often distant, retaining ties with their motherland or parent state . . . as a means of facilitating established occupation and [governance] by the parent state." Habermas is arguing that the modern state and economic system (capitalism) have imposed their media upon the life-world. In this sense, money and power act just like a colony—they are means through which these distant social structures seek to occupy and dominate the local life-world of people.

Habermas (1981/1987, p. 356) argues that four factors in organized capitalism set the stage for the colonization of the life-world. First, the life-world is differentiated from the social systems. Historically, there was a closer association between the life-world and society; in fact, in the earliest societies they were *coextensive;* in other words, they overlapped to the degree that they were synonymous. As society increases in differentiation and complexity, the life-world becomes "decoupled" from institutional spheres.

Second, the boundaries between the life-world and the different social subsystems become regulated through differentiated roles. Keep in mind that social roles are scripts for behavior. In traditional societies, most social roles were related to the family. So, for example, the eldest male would be the high priest—the family and religious positions would be filled and scripted by the same role. This kind of role homogeneity made the relationship between the life-world and society relatively nonproblematic, and, more importantly, it served to connect the two spheres.

Third, the rewards for workers in organized capitalism in terms of leisure time and expendable cash offset the demands of bureaucratic domination. "Wherever bourgeois law visibly underwrites the demands of the lifeworld against bureaucratic domination, it loses the ambivalence of realizing freedom at the cost of destructive side effects" (Habermas, 1981/1987, p. 361). And fourth, the state provides comprehensive welfare. Worker protection laws, social security, and so forth reduce the impact of exploitation and create a culture of entitlement where legal subjects pursue their individual interests and the "privatized hopes for self-actualization and self-determination are primarily located . . . in the roles of consumer and client" (Habermas, 1981/1987, p. 350).

For simplicity's sake, we can group the first two and last two items together. The first two factors are generally concerned with the effects of complex social environments. The more complex the social environment, due to structural differentiation, the greater will be the number and diversity of cultures and roles with which any individual will have to contend. This in turn dismantles the connections among the elements that comprise the life-world: culture, society, and personality.

The second two factors concern the effects of the state's position under organized capitalism. Under organized capitalism, the state protects the capitalist system, the capitalists, and the workers. In doing so, the state mitigates some of the issues that would otherwise produce social conflict and change. But perhaps more importantly, the state further individualizes the person. The roles of consumer and client, both associated with a climate of entitlement, overshadow the role of democratic citizen.

As a result of these factors, everything in organized capitalism that informs the life-world, such as culture and social positions, comes to be defined or at least influenced by money and power. Money and power have a certain logic or rationality to them. Weber (Chapter 2) talked about four distinct forms of rationality, two of which are pertinent here: instrumental and value rationality. As a reminder, *instrumental-rational action* is related to means-and-ends calculation and *value-rational action* is defined as behavior that is motivated by a person's values or morals.

Value rationality is specifically tied to the life-world and instrumental rationality to the state and economy. Thus, a good deal of what happens when the life-world is colonized is the ever-increasing intrusion of instrumental rationality and the emptying of value rationality from the social system. The result is that "systemic mechanisms—for example, money—steer a social intercourse that has been largely disconnected from norms and values. . . . [And] norm-conformative attitudes and identity-forming social memberships are neither necessary nor possible" (Habermas, 1981/1987, p. 154).

In turn, people in this kind of modern social system come to value money and power, which are seen as the principal means of success and happiness. Money is used to purchase commodities that are in turn used to construct identities and impress other people. Rather than being a humanistic value, respect becomes something demanded rather than given—a ploy of power rather than a place of honor.

To see the significance of this, let's recall the ideal of the life-world of modernity. When the life-world changed in the move from traditional to modern society, it took on new priorities and importance. The life-world was ideally to be dominated by democratic freedoms and responsibilities and occupied by citizens fully engaged in reasoning out the ways to fulfill the goals of the Enlightenment—progress and equality—through communication and consensus building. As Habermas (1981/1987) says, "the burden of social integration [shifts] more and more from religiously anchored consensus to processes of consensus formation in language" (p. 180).

As you can see, using money or power as steering media in the life-world is the antithesis of open communication and consensus building. One of the results of this situation is that the life-world decouples from or becomes incidental to the social system, in terms of its integrative capacities. A life-world colonized by money and power cannot build consensus through reasoning and communication, and people in this kind of life-world lose their sense of responsibility to the democratic ideals of the Enlightenment.

Colonization of the Public Sphere

This process is further aggravated by developments in the public sphere. As we've seen, the public sphere and its citizens came into existence with the advent of modernity. Citizens "are endowed by their Creator with certain unalienable Rights." This phrasing in the U.S. Declaration of Independence is interesting because it implies that these rights are moral rather than simply legal. There is a moral obligation to these rights that expresses itself in certain responsibilities:

[W]henever any Form of Government becomes destructive of these ends, it is the Right of the People to alter or to abolish it, and to institute new Government, laying its foundation on such principles and organizing its powers in such form, as to them shall seem most likely to effect their Safety and Happiness.

Thus, the most immediate place for involvement of citizens is the public sphere. In that space between power on the one hand and free information on the other, citizens are meant to engage in communication and consensus formation. It is in that space that discussion and decisions about any "form of government" are to be made. However, the public sphere has been colonized in much the same way as the life-world. Specifically, the public sphere, which began in the eighteenth century with the growth of independent news sources and active places of public debate, transformed into something quite different in the twentieth century. It became the place of public opinion—something that is measured through polls, used by politicians, and influenced by a mass media of entertainment.

There are two keys here. First, public opinion is something that is manufactured through social science. It's a statistic, not a public forum or debate that results in consensus. Recall what we saw earlier about how Habermas views the knowledge of science, even social science: its specific purpose is to control. Transforming consensus in the public sphere into a statistic makes controlling public sentiment much easier for politicians, both subjectively and objectively.

The second key issue I want us to see is the shift in news sources. Most of the venues through which we obtain our news and information today are motivated by profit. In other words, public news sources aren't primarily concerned with creating a democratic citizenry or with making available information that is socially significant. As such, information that is given out is packaged as entertainment most of the time. In a society like the United States, the consumers of mass media are more infatuated with "wicked weather" than they are concerned about the state of the homeless.

Concepts and Theory: Communicative Action and Civil Society

When we began this discussion, I mentioned that Habermas still holds out the promise of modernity. This hope is anchored in two arenas: speech communities and civil society. Both of these are rather straightforward proposals, though achieving them is difficult under the conditions created by organized capitalism, where the possibility and horizon of moral discourse is stunted.

Ideal Speech Communities

Let's talk first about **ideal speech communities.** These communities or situations are the basis for ethical reasoning and occur under certain guidelines to communication. Before we get to those guidelines, we need to consider what Habermas calls *communicative action:* action with the intent to communicate. Habermas makes the point that all social action is based on communication. However, to understand Habermas's intent, it might be beneficial to consider something that looks like social communication but isn't. We can call this "strategic speech." Strategic speech is associated with instrumental rationality and it is thus endemic within the life-world of organized capitalism as well as the social system.

In this kind of talk, the goal is not to reach consensus or understanding, but rather for the speaker to achieve his or her own personal ends. For example, the stereotypical salesperson or "closer" of a deal isn't trying to reach consensus; she or he is trying to sell something (a more immediate example is the student explaining why he or she missed the test). In strategic talk, speech isn't being practiced simply as communication; communication is being *used* to achieve egocentric ends, which is contrary to the function of communication: "Reaching understanding is the inherent telos [ultimate end] of human speech. Naturally, speech and understanding are not related to one another as means to end" (Habermas, 1981/1984, p. 287).

Communicative action within an ideal speech situation is based upon some important assumptions. As we are reviewing these assumptions, keep in mind that Habermas is making the argument that communication itself holds the key and power to reasoned existence and emancipatory politics. Communication has intrinsic properties that form the basis of human connection and understanding. Habermas points out that every time we simply talk with someone, in every natural speech act, we assume that communication is possible. We also assume that it is possible to share intersubjective states. These two assumptions sound similar but are a bit different. Communication simply involves your assuming that your friend can understand the words you are saying. Sharing intersubjective states is deeper than this. With *intersubjectivity,* we assume that others can share a significant part of our inner world—our feelings, thoughts, convictions, and experiences.

A third assumption we make in speech acts is that there is a truth that exists apart from the individual speaker. In this part of speech, we are making *validity claims.* We claim that what we are saying has the strength of truth or rightness. This is an extremely important point for Habermas and forms the basis of discourse ethics and universal norms. All true communication is built upon and contains claims to validity, which inherently call for reason and reflection. Further, these claims assume validity is possible; that truth or rightness can exist independent of the individual, which implies the possibility of universal norms or morals; and that validity claims can be criticized, which implies that they are in some sense active and accountable to reason. Validity claims also facilitate intersubjectivity in that they create expectations in both parties. The speaker is expected to be responsible for the reasonableness of her or his statement, and the hearer is expected to accept or reject the validity of the statement and provide a reasonable basis for either.

These assumptions are basic to speech: we assume that we can communicate; we assume we can share intersubjective worlds; and we assume that valid statements are possible. What Habermas draws out from these basic assumptions of speech is that it is feasible to reasonably decide on collective action. This is a simple but profound point: intrinsic to the way humans communicate is the hope of decisive collective action. It is possible for humanity to use talk in order to build consensus and make reasoned decisions about social action. This is both the promise and hope of modernity and the Enlightenment.

Ethical reason and substantive rationality are thus intrinsic to speech, but it isn't enough in terms of making a difference in organized capitalism. As with all critical theorists, Habermas has a praxis component. Praxis for Habermas is centered in communication and the creation of ideal speech situations. Here, communication is a skill, one that as democratic citizens we need to cultivate in order to participate in the civil society. As we consider these points of the ideal speech community, notice how many of them have to do more with listening than with speaking. In an ideal speech situation,

- Every person who is competent to speak and act is allowed to partake in the conversation—full equality is granted and each person is seen as an equal source of legitimate or valid statements
- There is no sense of coercion; consensus is not forced; and there is no recourse to objective standings such as status, money, or power
- Anyone can introduce any topic; anyone can disagree with or question any topic; everyone is allowed to express opinions and feelings about all topics
- Each person strives to keep his or her speech free from ideology

Let me point out that this is an ideal against which all speech acts can be compared, and toward which all democratic communication must strive. The closer a community's speech comes to this ideal, the greater is the possibility of consensus and reasoned action.

If we assume that the human species maintains itself through the socially coordinated activities of its members and that this coordination has to be established through communication . . . then the reproduction of the species also requires satisfying the conditions of a rationality that is inherent in communicative action. (Habermas, 1981/1984, p. 397)

Civil Society

Ideal speech communities are based upon and give rise to civil society. **Civil society** for Habermas is made up of voluntary associations, organizations, and social movements that are in touch with issues that evolve out of communicative action in the public sphere. In principle, civil society is independent of any social system, such as the state, the market, capitalism in general, family, or religion. Civil society, then, functions as a midpoint between the public sphere and social interactions. The elements of civil society provide a way through which the concerns developed in a robust speech community get expressed to society at large. One of the more important things civil society does is continually challenge political and cultural organizations in order to keep intact the freedoms of speech, assembly, and press that are constitutionally guaranteed. Examples of elements of civil society include professional organizations, unions, charities, women's organizations, advocacy groups, and so on.

Habermas gives us several conditions that must be met for a robust civil society to evolve and exist.

- It must develop within the context of liberal political culture, one that emphasizes equality for all, and an active and integrated life-world.

- Within the boundaries of the public sphere, men and women may obtain influence based on persuasion but cannot obtain political power.
- A civil society can exist only within a social system where the state's power is limited. The state in no way occupies the position of the social actor designed to bring all society under control. The state's power must be limited and political steering must by indirect and leave intact the internal operations of the institution or subsystem.

Overall, Habermas rekindles the social vision that was at the heart of modernity's birth. Modernity began in the fervor of the Enlightenment and held the hope that humanity could be the master of its own fate. There were two primary branches of this movement, one contained in science and the other in democratic society. In many ways, science has proven its worth through the massive technological developments that have occurred over the past 200 years or so.

However, Habermas argues that the hope of democracy has run aground on the rocks of organized capitalism. In communicative action and civil society, he points the way to a fully involved citizenry reasoning out and charting their own course. But what Habermas gives us is an ideal—not in the sense of fantasy, but in the sense of an exemplar vision. In his theory, it is the goal toward which societies and citizens must strive if they are to fulfill the promise of modernity. Habermas, then, lays before us a challenge, "the big question of whether we could have had, or can now have, modernity without the less attractive features of capitalism and the bureaucratic nation-state" (Outhwaite, 2003, p. 231).

Habermas Summary

- Habermas's theory of modernity is in the tradition of the Frankfurt School of critical theory. His intent is to critique the current arrangements of capitalism and the state, while at the same time reestablishing the hope of the Enlightenment, that it is possible for human beings to guide their collective life through reason.

- Habermas argues that modernity has thus far been characterized by two forms of capitalism: liberal and organized. The principal difference between these two forms is the degree of state involvement. Under liberal capitalism, the relationship between the state and capitalism was one of *laissez-faire*. The state practiced a hands-off policy in the belief that the invisible hand of market competition would draw out the best in people and would result in true equality based on individual effort. However, *laissez-faire* capitalism produced two counter-results: the tendency toward monopolization and significant economic fluctuations due to overproduction. Both unanticipated results prompted greater state involvement and oversight of the capitalist system.

- Organized capitalism is characterized by active government spending and management of the economy. This involvement of the state in capitalism facilitates three distinct results, all of which weaken the possibility of achieving the social promise of modernity:

1. A crisis of legitimation and rationality. Because the state is now involved in managing the economy, fluctuations, downturns, and other economic ills are perceived as problems with the state rather than the economy. When they occur, these problems threaten the legitimacy of the state in general. In addition, because the state uses social scientific methods to forecast and control economies, belief in rationality is put in jeopardy. These crises in turn reduce the levels of meaning and motivation felt by the citizenry.

2. The colonization of the life-world. The life-world is colonized by the state and economy, as the media of power and money replace communication and consensus as the chief values of the life-world.

3. The reduction of the public sphere to one of public opinion. This occurs principally as the media have shifted from information to entertainment value and as the state makes use of social scientific methods to measure and then control public opinion.

• However, Habermas argues that the hope of social progress and equality can be embraced once again through communicative action and a robust civil society. Communication is based upon several assumptions, the most important of which concern validity claims—these inherently call for reason and reflection. Together, such assumptions lead Habermas to conclude that the process of communication itself gives us warrant to believe it is possible to reach consensus and rationally guide our collective lives.

• Communicative action is also a practice. True communicative action occurs when full equality is granted and each person is seen as an equal source of legitimate or valid statements; objective standings such as status, money, or power are not used in any way to persuade members; all topics may be introduced; and each person strives to keep her or his speech free from ideology.

• Communicative action results in and is based upon a robust civil society. Civil society is made up of mid-level voluntary associations, organizations, and social movements. Such organizations grow out of educated, rational, and critical communicative actions and become the medium through which the public sphere is revitalized. A civil society is most likely to develop under the following conditions: A liberal political culture is present that emphasizes education, communication, and equality; men and women are prevented from obtaining or using power in the public sphere; the state's power is limited.

Looking Ahead

In Chapter 6, you were introduced to the idea of neo-functionalism. There is also a field referred to as neo-Marxism. In fact, so profound is Marx's influence on contemporary theory, I've already "looked ahead" twice concerning the impact of his theories—the first time at the end of Chapter 1 and again in this chapter at the crossover point between critical and conflict theories. As a result, I don't want to spend too much time looking ahead here.

Nevertheless, I do want to mention two other places where Marx shows up: with Jean Baudrillard (Chapter 14) and Dorothy E. Smith (Chapter 13). To one degree or another, both of these thinkers are concerned with consciousness. Though Baudrillard began as a Marxist, the disappointing results of the 1968 student uprisings made him question economic Marxism. Baudrillard ends up viewing Marxism as actually substantiating certain aspects of the capitalist mentality. In particular, Marx's theory of species-being argues that human beings are by nature economic producers, and that true consciousness comes through being intimately involved in the production process. The product of this kind of species production acts like a mirror that reflects human nature back to people.

Baudrillard maintains that in making this kind of argument, Marx is giving ultimate legitimation to the entire capitalist scheme of production. Marx reduced humankind to economic producers, just the way capitalism does. Baudrillard (1973/1975) argues that "in order to find a realm beyond economic value (which is in fact the only revolutionary perspective), then the *mirror of production* in which all Western metaphysics is reflected, must be broken" (p. 47, emphasis original). One of the ways that Baudrillard breaks this mirror is through his notion of symbolic exchange.

In her theorizing, Dorothy E. Smith proposes a "new materialism." Recall that Marx's materialism is the basis of his understanding of human nature and consciousness. Marx used this argument to see how people are alienated from their inner being; he was specifically concerned with seeing how money and commodities come to stand between humanity and their species-consciousness. Smith's new materialism shifts the focus away from production to knowledge. Like Weber, Smith sees the modern era as one dominated by bureaucratic institutions that run on objective and rationalized knowledge. What comes to stand in the way of true consciousness, then, are texts and facts. Smith opens our eyes to see that texts and facts don't exist by themselves as some pure expression of "the way things are." Rather, like Marx, Smith argues that texts and facts are, in fact, expressions of the relations of ruling.

Building Your Theory Toolbox

Knowing Conflict Theory

After reading and understanding this chapter, you should be able to define the following terms theoretically and explain their theoretical importance to conflict theory:

Cross-cutting influences; absolute deprivation; relative deprivation; rational and transcendent goals; functional consequences of conflict; internal and external conflict; types of internal conflict; network density; group boundaries; internal solidarity; coalitions; power and authority; imperatively coordinated associations; Hobbesian problem of social order; class; quasi-groups; interest groups; technical conditions; political conditions; social conditions; conflict violence and intensity; conflict mobilization;

material and emotional resources; resource mobilization; ritualized exchange of atrocities; bureaucratization of conflict; ritual solidarity; geopolitical theory; state legitimacy; heartland advantage; marchland advantage; overexpansion; critical theory; analytical, interpretive, and critical forms of knowledge; liberal capitalism; life-world; public sphere; organized capitalism; legitimation crisis; colonization of the life-world; instrumental and value rationality; colonization of the public sphere; communicative action; ideal speech community; civil society

After reading and understanding this chapter, you should be able to

- Explicate the defining features of conflict theory

- Explain the two basic sources of conflict

- Identify the factors that predict the level of violence in conflict and how the level of violence affects the results of conflict

- Explain the social factors that influence the level of conflict intensity and how the level of intensity affects the results of conflict

- Describe the functional consequences for groups experiencing internal and external conflict

- Discuss the social factors that influence the functionality of conflict for internal conflict

- Explain the social factors that move an aggregate of people from quasi-group to interest group status

- Describe the place that resource mobilization and depletion play in determining the outcome of a conflict

- Analyze how conflict engenders further conflict, paying special attention to the place that the ritualized exchange of atrocities has in the process

- Evaluate how bureaucracies influence conflict

- Explain the role the state plays in geopolitical theory, giving specific attention to heartland and marchland advantages and overexpansion

- Analyze the downfall of the Soviet Union in geopolitical terms

- Define the Frankfurt School's critical theory and explain its view of knowledge and culture

- Explain praxis and how it is associated with critical knowledge

- Explain the differences between liberal and organized capitalism, paying particular attention to the changing relations between the state and economy

- Define the life-world and its purpose, and explain how it became colonized

- Define the public sphere and explain how it came about, its purpose, and its colonization

- Define communicative action and explain how it forms the basis of value-rational action

- Define ideal speech situations (or communities) and explain how they give rise to civil society

- Define civil society, explain the conditions under which it can survive, and discuss how it is important to a democratic society

Learning More: Primary Sources

- Lewis Coser:

  Coser, L. (1956). *The functions of social conflict.* New York: Free Press.

- Ralf Dahrendorf:

  Dahrendorf, R. (1959). *Class and class conflict in industrial society.* Palo Alto, CA: Stanford University Press.

- Randall Collins:

  Collins, R. (1975). *Conflict sociology.* New York: Academic Press.
  Collins, R. (1993). What does conflict theory predict about America's future? *Sociological Perspectives, 36,* 289–313.

- Jürgen Habermas:

  Habermas, J. (1987). *The theory of communicative action, Vol. 1: Reason and the rationalization of society.* Boston: Beacon Press, 1984; *Vol. 2: Lifeworld and system: A critique of functionalist reason.* Boston: Beacon Press.
  Habermas, J. (1990). *The philosophical discourse of modernity: Twelve lectures.* Boston: MIT Press.
  Habermas, J. (1991). *The structural transformation of the public sphere: An inquiry into a category of bourgeois society.* Boston: MIT Press.

Learning More: Secondary Sources

  Braaten, J. (1991). *Habermas's critical theory of society.* New York: SUNY Press.
  Calhoun, C. (Ed.). (1993). *Habermas and the public sphere.* Boston: MIT Press.
  Outhwaite, W. (1995). *Habermas: A critical introduction.* Palo Alto, CA: Stanford University Press.

Theory You Can Use (Seeing Your World Differently)

- Consult a daily national newspaper for one week. How many of the reported events would you say qualify as conflictual? Pick two of the most interesting and analyze each using the various ideas we covered with Coser, Dahrendorf, and Collins. Which ideas seem most important? Why? Based on what you've been able to glean from the news, can you make any predictions about the intensity, violence, or consequences of the conflicts? Explain how you used theory to predict the results.

- For a larger project, analyze the U.S. civil rights movement, from the late 1950s through to present times, using our three conflict theorists. Which theories seem to explain the most?

- Using your favorite Internet search engine, look up "participatory democracy." How would Habermas's ideal speech community fit this model? Does the Internet provide greater possibilities for ideal speech situations to develop? How would Internet communities be linked to civil society?

- Habermas argues that your life-world has been colonized. Using Habermas's ideal of the life-world in modernity, come up with a list of six ways that your life-world has been colonized (three from the general effects of complex social environments and three from the state's work under capitalism).

- Racial, ethnic, gender, sexual identity, and religious groups have all been and are being disenfranchised in modern society. How does the ideal speech situation "enfranchise" these groups? In other words, how does the ideal speech situation do away with the possibility of disenfranchised groups?

- What social group do you belong to that most nearly approximates the ideal speech community?

- How can you begin your own praxis?

Further Explorations—Web Links

- Collins is one of the few contemporary theorists for whom we have a biographical statement that links his life experiences with his theorizing. See Alair Maclean and James Yocom, *Interview With Randall Collins, 2000,* http://www.ssc.wisc.edu/theory@madison/papers/ivwCollins.pdf

- There's also an interview with Dahrendorf that was given in 1989. In it he talks about his ideas and how his past influenced his theorizing and political involvement: http://globetrotter.berkeley.edu/Elberg/Dahrendorf/dahrendorf0.html

- There are several good Web sites for Habermas. Check out the following site that contains a collection of Habermas Web links: http://www.habermasonline.org/

- See also the various sites for critical theory:

  http://home.case.edu/~ngb2/Pages/Intro.html

  http://www.cla.purdue.edu/academic/engl/theory/

  http://sun3.lib.uci.edu/~scctr/online.html

CHAPTER 8

Structures of Inequality— Race and Gender

William Julius Wilson (1935–)

Janet Chafetz (1942–)

I want you to think for a moment about how you made your way into the room you are in right now. Unless the room is standing all by itself out in the middle of a field with no roads in or out, you probably got to it by driving on roads, traveling on paths, and walking through hallways and doorways. Okay, now imagine that you wanted to get into this room using something besides the door or the window. It would be difficult, wouldn't it? Walking through walls is no easy task. The point of all this is that you used different kinds of structures to get around (roads, walls, floors, doors, and so on). Those structures helped you accomplish your goal of getting to this room (which is itself a structure), but they also restricted and guided your options. The same is true about social structures.

Like physical structures, *social structures* enable and simultaneously restrain human action. With Durkheim and Parsons, we saw how social structures enable us to interact socially in patterned and predictable ways. In the first part of Chapter 7, we saw some of the general ways in which social structures restrict people by creating patterned inequalities around class and power. We also saw that those restrictions produce patterned and somewhat predictable conflicts in society. In this chapter, we are going to narrow our gaze to consider two types of status groups and the inequality that surrounds them. Our two theorists in this chapter want us to see that race and gender matter. Not only is the social esteem or status for people of color and women lower, but race and gender influence how class and power are distributed as well.

Another characteristic of social structures that is like physical structures is that the *structural* parts are the *connections* among the units. For example, the substructure of a house isn't created simply by piling 2 × 6s in a haphazard manner. It's putting them together in a specific way that creates the structure. Another way to understand the principle of structure is to compare it to an organizational chart. There are positions on the chart, such as vice president or secretary, and there are expectations associated with the positions (such as making decisions or typing). The expectations, of course, are based on the relationships among and between the nodes or positions on the chart.

Your social status position is like your place in the organizational chart of society. You have several status positions, such as student, son or daughter, friend, and so on. Those structured positions tell people what to expect from you—there are roles (behavioral expectations) associated with each position. These roles exist by virtue of the connections to other positions. For example, because you are related differently to other positions, your role as a daughter is different from your role as a worker. Further, you act like a student or you act like a young person not because you *are* a student or young, but because that's what is expected of that status position in the country where you live.

Keep these two things in mind as we travel through this chapter: social structures restrict as well as enable, and social structures establish relationships among people. Putting these two ideas together, we can see that most of the social relations you have are established because of various social structures. This idea is very important in William Julius Wilson's theory of race relations—the nature of black–white contact is strongly influenced by economic and political social structures. Social structures also restrict you from certain relations, positions, and life course opportunities. We'll see this idea clearly brought out in Janet Chafetz's theory of gender inequality.

William Julius Wilson: Race and Class

Photo: © Courtesy of Martha Stewart.

The Essential Wilson

Biography

William Julius Wilson was born on December 20, 1935, in Derry Township, Pennsylvania. Wilson attended Wilberforce and Bowling Green Universities before completing his Ph.D. in sociology at Washington State University in 1966. His first professorship was at the University of Massachusetts at Amherst, and he joined the faculty of the University of Chicago in 1972. There he held the position of professor and was the director of the Center for the Study of Urban Inequality. He moved to Harvard in 1996, where he currently holds the Lewis P. and Linda L. Geyser University Professorship. Wilson has received many top honors in his career, including the 1998 National Medal of Science (the highest such honor given in the United States), and he was named by *Time* magazine as one of America's 25 Most Influential People in 1996. He has authored several groundbreaking and significant books including *The Declining Significance of Race, The Truly Disadvantaged,* and *When Work Disappears: The World of the New Urban Poor.*

Passionate Curiosity

Wilson is driven to discover the structural causes of poverty that exist in the United States, especially African American poverty. His work is based on solid and detailed analysis of statistical data, which he uses to argue for policy changes that would alleviate needless destitution.

Keys to Knowing

Marxist elite theory, race relations, exploitation, state-capitalist collusion, dark nations, split labor market, racial-caste system, Jim Crow segregation, urbanization, postindustrial, tight labor market

Seeing Further: Structured Racial Inequality

- The Tuskegee study: In 1932, the U.S. Public Health Service began a study of untreated syphilis in black men. A total of 399 men were enlisted for the study, drawn from participants in free medical clinics throughout Macon County, Alabama. The men were never told they had syphilis, just that they had "bad blood"; they were never given treatment (even after the advent of penicillin in 1947); as men died, went blind, or went insane, treatment was still withheld; the study discontinued after becoming public in 1972—but by that time 128 men had died of syphilis or related complications, at least 40 wives had been infected, and 19 children were infected at birth.

- The proportion of males in prison: For whites, the proportion is 462 per 100,000 in the population; for blacks, it is 3,500 per 100,000 in the population.

- Nearly twice as many blacks as whites are convicted of drug offenses, even though it is estimated that there are five times more white users than black.

- Of youths charged with drug use, blacks are 48 times more likely than whites to receive a prison sentence.

- Of all racial groups, blacks have the lowest life expectancy; highest infant mortality; highest rate of most cancers, diabetes, heart disease, high blood pressure, new cases of HIV and AIDS; and the highest death rate from treatable diseases, gunshots, and drug/alcohol-related illness or incidents.

- Average annual income for black households: $30,436; for whites: $44,232.

- Black unemployment is twice as high as that for whites.

- An estimated 48% of blacks in the United States own their home; 74% of whites do.

- The National Cancer Institute spends a mere 1% of its budget on studies of ethnic groups, even though black women are 50% more likely than white women to get breast cancer before the age of 35 and 50% more likely to die before 50.

- EPA study: An estimated 9 out of 10 of the major sources of industrial pollution are in predominately black neighborhoods. The town of Carville, Louisiana, is 70% black and its industries produce 353 pounds of toxic material per person per year—the Louisiana state average is 105 pounds.

There is no doubt that race matters and significantly impacts the life chances, as well as the quality of life, of millions of people of color. But how does it work? William Julius Wilson gives us a unique and complex way of seeing racial inequality. Obviously, his concern is the oppression of black people in the United States, but to simply see the above inequalities as a product of racism is too simplistic. Wilson argues that racism as such requires a particular kind of structural configuration among the state, the economy, and social relations. This triangle of relations comes from Marx. You'll remember that Marx argued that the means of production determines the relations of production. In other words, different economies create different kinds of social relationships. In Marx's scheme, the primary social relations are found in the two classes of capitalism: workers and owners. As you may also recall, Marx argued that the state and the economic elite are in collusion and work in harmony to oppress the working class.

Wilson adds two things to Marx's theory. First, he adds race. Wilson wants us to see that within the relations of production are *race relations*. It's this issue of black–white contact and the accompanying ideology, which together define race relations, that is Wilson's theoretical focus. In general he asks, do relations between blacks and whites change in response to structural changes? The second idea Wilson adds is that the relationship between the state and the economy can vary. So, historically speaking, the economic elite may strongly influence government at one point in time; but at another time the influence of capitalists on the state may be very weak.

In the end, Wilson argues that racism as a causal factor in determining the overall condition of blacks in the United States has been declining in significance since the Civil War. To document the causal force of racism, Wilson tracks the changing relationships between the economy and the state. We'll see that to be effective, racism needs a particular kind of institutional configuration. In its absence, other factors become important for influencing the position of blacks in the United States—in particular, a split labor market and class. As we discuss Wilson's theory, keep in mind that he is not arguing that racism is no longer present or important. Racism is still a problem, but Wilson is pointing out that because of shifts in the relationship between the economy and the state, the *relative importance* of race has declined over time and has been replaced by class-based issues.

Concepts and Theory: Three Theories of Race Relations

Orthodox Marxist Theory

Generally speaking, there have been two main approaches to understanding race relations in America: Marxist economic elite and split labor market theories. In the main, *Marxist elite theory* argues that capitalism depends on and results in exploitation and *state–capitalist collusion*. We went over exploitation in the chapters on Marx and Du Bois, so I won't go into detail here. However, there are a couple of things to keep in mind from Marx's argument. First, capitalist profits are based on some form and degree of **exploitation.** Capitalists must pay workers less than they earn (in terms of actual production) in order to make a profit. Marx saw two basic forms of exploitation: absolute (increasing the number of hours worked) and relative (using machines to increase the worker's output).

The second thing to remember from Marx is that capitalists *need* exploitation. Because capitalists need workers to exploit, workers have a certain amount of power in their relations with owners. This dependency is the basis of the power behind worker demands and strikes, and it is a primary factor behind all the labor and work safety laws passed since capitalism began. Once gained, this power can be used in a variety of ways, sometimes influencing the state, sometimes the capitalists themselves, and at times other workers in the job market (as we'll see with split labor market theory).

While Marx saw two forms of exploitation, Du Bois showed us another: race. You'll recall that Du Bois (1920/1996b) argued that capitalism succeeded as a world economy because capitalists had the "dark nations" to exploit: The "degrading of men by men is as old as mankind. . . . It has been left, however, to Europe and to modern days to discover the eternal world-wide mark of meanness,—color!" (p. 504). Using race as a key in exploitation gave capitalists a group of workers they could exploit without limit, a group that had no power. With the invention of chattel slavery, capitalism changed blacks into commodities, and commodities have no power over their owners whatsoever. But notice that this type of exploitation is based on cooperation with the state. The state must support the capitalists' claim that certain racial groups have no civil rights.

This Marxian way of understanding race relations, then, argues that it is in the best interest of the bourgeoisie or elite to propagate racist ideologies and practices of discrimination:

> [T]he capitalist class benefit not only because they have created a reserved army of labor that is not united against them and the appropriation of surplus from the black labor force is greater than the exploitation rate of the white labor force, but also because they can counteract ambitious claims of the white labor force for higher wages either by threatening to increase the average wage rate of black workers or by replacing segments of the white labor force with segments of the black labor force. (Wilson, 1980, p. 5)

Split Labor Market Theory

The *split labor market theory* also sees race being used as a dividing technique, but the causes are different. This theory assumes that, in general, business would support a free and open market where all laborers compete against one another regardless of race. This kind of competition would result in an overall higher level of exploitation. In addition, split labor theory proposes three key classes, rather than the two of orthodox Marxism: business, higher-paid labor, and cheaper labor. Overall, these theoretical moves shift the dynamics of race relations from the bourgeois to the higher-paid working class. In other words, rather than race being used by capitalists, it is seen as being used by the higher-paid working class to preserve their own interests—this is part of the power that comes to the working class as a result of the capitalists' need for exploitation.

The emphasis in this theory is on how the market for labor splits and who benefits. The labor market refers to any collective of workers vying for the same or similar positions within a capitalistic economy. A labor market splits when there are two or more social groups whose price for the same work is different, one being cheaper than the other. The price difference is primarily based on dissimilar resource levels, determined by economic and political resources and the availability of information. That is, if there is an ethnic or racial group within a labor market whose standard of living is significantly lower, who lacks the ability to politically organize, and who is less informed about labor market conditions, then the labor market will split.

The important thing to see here is that when a market splits, it is more beneficial to higher-paid workers than to business owners. According to this theory, free and open competition would displace the higher-paid workforce and result in lower wages and higher exploitation generally. "Only under duress does business yield to labor aristocracy" (Bonacich, 1972, p. 557). Race, then, becomes a tool of the higher-paid working class to preserve their own economic interests. The white higher-paid working class promotes racist ideologies and discriminatory practices in order to monopolize skilled labor and management positions, prevent blacks from obtaining necessary skills and education, and deny blacks political resources.

Wilson's Class–State Theory

One of the problems with both of these approaches is that they are ahistorical. That is, neither classic Marxism nor split labor market theory indicates that historic changes in the system of production will influence the theory one way or another. In fact, Marxian state theory isn't comprehensive enough to explain race relations in the United States from their inception to the present, and neither is split labor market theory. Wilson, however, does see the theories as historically specific: each provides crucial insights at different historical moments.

In addition to recognizing historically specific differences, Wilson's approach adds another variation to the equation: the state. In classic Marxian theory, the state is seen as a tool or extension of the business elite: "the Bourgeoisie has at last . . . conquered for itself, in the modern representative State, exclusive political sway. The executive of the modern State is but a committee for managing the common affairs of the whole bourgeoisie" (Marx & Engels, 1848/1978, p. 475). Wilson, however, modifies this approach and argues that the relationship between the state and the economic elite is not a given. What this means is that the state can be influenced by either the elite or the workers and, more importantly, the state can act autonomously.

In Figure 8.1, I've given you a picture of what Wilson is talking about. The figure simply says that the economic, political, and social relations structures mutually influence and create one another. The important thing for us to see and keep in mind is that our social relationships are structured by the present political and economic institutions. Most of us don't usually think of our relationships as being structured, but they are. Stop and think for a moment how often you interact with someone who is extremely wealthy or very poor. If you're like me, you fall somewhere in the middle class range and thus most of your contacts and relationships are middle class as well.

The same is true for race. The likelihood of your coming in contact with someone from a different racial group is based on how the different populations are dispersed through society, and those dispersion patterns are heavily influenced by economic and political concerns. But Wilson wants us to see something more significant than simple contact. Polity and economy also influence how we view and act toward other races. A simple example is to compare how the races interact today with how those interactions were structured prior to the U.S. Civil War. Here's the bottom line: how we relate to other races, specifically how blacks and whites relate, is shaped by the current relations between government and the economy.

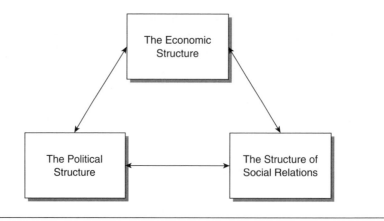

Figure 8.1   Wilson's Class–State Relations

Concepts and Theory: Three Periods of American Race Relations

Wilson argues that none of these individual theories is adequate to explain the place and use of race in capitalism over time. However, each yields better explanatory power under certain historical conditions. To show this, Wilson delineates three distinct phases of black–white relations in the United States: the period of plantation economy and racial-caste oppression (pre– and early post–Civil War America); the era of industrial expansion and class-based racial oppression (the latter part of the nineteenth century through the New Deal of the 1930s); and the time of progressive transition from racial inequalities to class inequalities (the current period beginning after WWII and taking shape during the 1960s and 1970s).

The Plantation Economy: Racial-Caste Oppression

Wilson argues that orthodox Marxian theory is much better at explaining the *racial-caste system* that worked under the American plantation economy, at least in the South. This type of non-manufacturing agrarian society is characterized by a simple division of labor and a small aristocracy. In such a society, there is very little if any job market competition. During this period in the United States, working whites were either craftsmen or serfs, and blacks were generally held as slaves. In addition, in a plantation economy there is a vast distance between the upper and lower classes. Because of this distance, there is little contact between classes, and what contact does happen is highly ritualized and subject to strong social norms of manners and etiquette. These conditions result in little to no class conflict, with the white workers having "little opportunity to challenge the control of the aristocracy" (Wilson, 1980, p. 13).

In such systems of production, the aristocracy dominates both economic and political life. In the United States, the white, landed elite were able to secure laws and policies extremely favorable to their economic interests, and they were able to propagate a ruling ideology concerning the differences between the races. As in classic Marxism, the system of production and the state formed a mutually reinforcing cycle. As a result, "the system of slavery severely restricted black vertical and horizontal mobility" (Wilson, 1980, p. 24) and race relations with elite whites took the form of paternalism.

This is not to say, however, that a split labor market did not exist—it simply was not crucial during this time in determining race relations. In the South, split labor markets existed between white workers and slaves. The result was that "the more frequent the contact between black slaves and white workers . . . the more the wages of white workers were depressed" (Wilson, 1980, p. 42). However, the white workers lacked the political and economic power to do much about this confrontation and thus were unable to split the labor market.

In the North, the labor market was noticeably split along racial lines. In the northern case, the system of production was more diversified with a more prominent manufacturing sector. This difference meant that white workers in the North were much more involved in the economic system than were their southern counterparts. Thus, white higher-paid workers had political clout and were able to preserve their interests against black competition. However, the issue for white workers in the North wasn't so much the presence of black workers in the northern job market as it was the fear of a large migration of black workers from the South. It is important to note that the greater political power of white higher-wage earners actually resulted in higher levels of race antagonism and strained race relations in the North than were actually experienced in the South.

Post Civil War to New Deal Politics: The Split Labor Market

After the U. S. Civil War, the industrialization of the U.S. economy grew quickly and the southern economy in particular expanded rapidly. In addition, the Thirteenth and Fourteenth Amendments to the Constitution abolished slavery and granted civil rights to the black population. As a result, from the latter part of the nineteenth century through the 1930s, there were massive changes in the system of production and race relations. This period marks a shift from race relations based on a paternal racial-caste system to a more class-based labor market.

In the South, economic expansion greatly increased the political power of the white working class. Blacks were freed but had very little economic or political power. White workers, then, attempted to control the newly available skilled and unskilled positions—the classic split labor market condition. The outcome was an elaborate system of *Jim Crow segregation* that was reinforced with a strong ideology of biological racism.

The name "Jim Crow" doesn't refer to an actual person but to the stereotypical characterization of blacks in minstrel shows at the time. The idea of Jim

Crow segregation references the laws that many states enacted after the Civil War in order to control blacks and preserve white privilege. Wilson's point here is that these laws did not directly benefit the capitalist elite. It would have been much more beneficial in terms of exploitation for the labor market to be open equally to blacks and whites. The effect would have been a generally higher level of exploitation for capitalists. Rather, Jim Crow segregation benefited the higher-paid white working class by keeping blacks out of the competition for jobs, especially in the South.

In addition, the initial support from the white elite that blacks enjoyed disappeared once the politicians began to worry about the black vote. To alleviate their uncertainty, white politicians effectively eliminated the black vote through such prerequisites as poll taxes, literacy tests, and property qualifications:

> The almost total subordination of blacks in the South was clearly related to the disintegration of the paternalistic bond between Negroes and the southern economic elite, because this disintegration cleared the path for what ultimately resulted in a united white segregation movement. (Wilson, 1980, p. 83)

The North experienced a different configuration. Due to high levels of migration of blacks from the South and high immigration rates of European whites, blacks most often entered the job market as strikebreakers. White workers would strike for better wages or working conditions and management would bring in black workers to keep production going. In some cases, management would preempt a strike by hiring black workers on permanently. This move obviously created high tension between black and white workers, which culminated in a number of race riots in 1917 and 1919.

The Great Depression of the 1930s shifted things considerably for both black and white workers in the North. During the Depression, there was a strong, positive movement toward unionizing. The unions themselves began to recruit black workers. As a result, black antagonism toward the unions was reduced; black and white workers saw themselves as united in their stand for economic reforms; and the practice of employers using blacks as strikebreakers was eliminated.

WWII and Beyond: Class-Based Racial Inequality

Notice the shifts in the sources of oppression. During the plantation economy, oppression came out of collusion between the state and the white bourgeoisie. However, during the second era of race relations in the United States, a split labor market developed where the bourgeoisie wanted to give blacks economic freedoms in order to use them to keep white workers' wages down. Thus, while race was still an issue, it originated with white workers rather than collusion between capitalists and the state. Also notice that class became increasingly important over time. Labor markets are primarily split over class, not race. Race, then, becomes a marker for class antagonisms rather than for racism itself.

Wilson argues that the role of the state continued to change from the classic Marxian model. WWII brought a ban on discrimination in defense and government agencies. This move also provided for on-the-job training for blacks. Black workforce participation continued to expand under the equal employment legislation of the 1950s and 1960s and growing affirmative action programs. These changes obviously didn't come as a result of the government's desire for equality, but in response to civil rights movements, which also boasted black political involvement. But regardless of the source, the state took successive steps to address black inequality.

In addition, the trend toward industrialization that began in the North prior to the Civil War expanded geographically and exponentially from the 1940s onward. This facilitated a shift in the black population toward **urbanization**, away from rural, agricultural settings and low-paying farm jobs and to cities and industries with better-paying jobs. A large black population thus began to develop in urban centers, which, in turn, prompted the growth of black business owners and black professionals oriented toward serving the needs of the growing black community.

As a result of affirmative action and these economic and population shifts, more and more businesses were seeking black employees. For example, during the 10-year period between 1960 and 1970, the average number of corporate recruitment visits to traditionally black colleges jumped from 4 to 297; in some southern colleges, the number rose from zero to 600 corporate visits. During this time, there was also a jump in the percentage of blacks working in government jobs, rising from 13% to almost 22%, and the overall percentage of black males in white-collar positions rose from 16% to 24% (Wilson, 1980, pp. 88–109).

However, the U.S. began to noticeably shift toward a *postindustrial* economy beginning in the 1970s. This move away from manufacturing and toward service- and knowledge-based goods brought the decentralization of U.S. businesses, further expansion in government and corporate sectors, and demographic shifts from urban to suburban settings (sometimes referred to as "white flight"). These economic and population changes created a situation in which city tax resources either declined or increased at slower than necessary rates. At the same time, and due to the same social factors, cities experienced a sharp increase in expenditures. This situation obviously creates problems for municipal services, public assistance, and urban schools.

The picture that Wilson gives us of the time following WWII involves two push–pull forces. On the one hand, political and economic opportunities for blacks increased dramatically. Through the 1930s, 1940s, and 1950s, the black working class experienced increasing opportunities and urbanization, which at the time was a positive move. On the other hand, from the 1960s on, there was the decentralization of American business, decreases in manufacturing and increases in government and corporate jobs, and white flight from urban to suburban settings. These overlapping yet opposing forces fragmented the black labor force and resulted in "vastly different mobility opportunities for different groups in the black population" (Wilson, 1980, p. 121).

Those African Americans who were already moving toward the middle class were poised to take advantage of the economic and political shifts. They continued to experience upward mobility and "unprecedented job opportunities in the corporate and government sectors" (Wilson, 1980, p. 121). These middle class blacks, like their white counterparts, have been able to move to more affluent neighborhoods. The other segment of the black labor force, however, has become locked into the cycle of inner city problems: declining city revenues in the face of increasing social needs. These are "the relatively poorly trained blacks of the inner city, including the grow- ing number of younger blacks emerging from inferior ghetto schools" (Wilson, 1980, p. 121) who are locked into low-paying jobs with high turnover rates and little hope of advancement.

Wilson wants us to see that race, as it was used in previous times, is a declining factor in predicting the economic and political success of blacks in the United States. Again, this is a proportional evaluation. Race still matters, but class distinc- tions within the black population have greater impact on black opportunities than does race itself. Prior to the mid-1960s, studies indicated that "race was so much of a dominant factor that very little of black economic achievement was determined by class background" (Wilson, 1980, p. 167). Since that time, those blacks already "in the system" have continued to experience occupational and salary gains. "For those blacks who are not in the system, however, who have not entered the main- stream of the American labor market, the severe problems of low income, unem- ployment, underemployment, and the decline in labor-force participation remain" (Wilson, 1980, p. 171). It is thus the class positions prior to the sixties and seventies that currently oppress poor and working class blacks rather than race itself.

In addition, Wilson argues that the character of racial strife has changed. Previous to this time period, racial tensions focused on the economy. From the Civil War through the civil rights era, racial tensions revolved around granting blacks equal access to economic opportunities. Since the segmentation of the black labor pool, racial tensions have shifted to the sociopolitical order. The actors are the same—blacks and the white working class—"but the issues now have more to do with racial control of residential areas, schools, municipal political systems, and recreational areas than with the control of jobs" (Wilson, 1980, p. 121).

Policy Implications

Wilson's basic conclusion is that concern for racial equality needs to move from a focus on race to a focus on class. In other words, policies pointed at creating a tight labor market will go further toward improving the overall condition of blacks in the United States today than will laws aimed at ending racial discrimination. The dif- ferences between a tight and a slack labor market basically revolve around the level of employment. In a *tight labor market,* there are a high number of job openings relative to the labor pool. In other words, those who want to work *can* work. High employment/low unemployment rates mean that it is a worker's market: workers have choices of positions; wages are high; and unemployment, when it does occur, is relatively short. On the other hand, a slack labor market means that there are more workers than positions (unemployment is high). This is a capitalist's market:

business owners have their choice of workers, wages are low, and unemployment is chronic.

Unfortunately, in recent times the United States has moved away from "using public policy as a means to fight social inequality" and instead has placed emphasis "on personal responsibility, not inequalities in the larger society" (Wilson, 1996–1997, pp. 569–570). For example, in the year 2000, low-income programs made up 21% of the federal budget but constituted 67% of the spending cuts that year. In order to combat this trend, Wilson is acting to galvanize both private and public support for creating national performance standards for schools. Wilson specifically cites other capitalist democracies that have national policies that emphasize critical and higher-order thinking skills. Along with this emphasis on national standards, Wilson argues that national policy should provide the kind of support needed by inner city and disadvantaged neighborhood schools to meet such standards. Currently, the unequal funding of schools produces what Jonathon Kozol (1991) refers to as "savage inequalities."

Wilson also advocates improving the family support system in the United States. Currently, the United States is the only modernized country that does not provide universal preschool, child support, and parental leave programs—much-needed support structures in the face of changes in the family and the social structures surrounding it. Along with this support for the changing family structure, Wilson argues that the United States needs to do a better job at linking families, schools, and work. Currently, U.S. firms take five years longer than other developed nations to hire high school graduates. Where in Germany and Japan students are hired directly out of high school, typically the larger firms in the United States don't hire high school graduates until they have reached their mid-twenties.

Unfortunately, we don't have the time or space to review all of Wilson's proposals. Wilson has fully documented his policy concerns in *When Work Disappears: The World of the New Urban Poor*, and I encourage you to read it. But from what was just discussed, you can see that Wilson has shifted the discourse concerning the plight of black Americans from one that focuses specifically on race to one that emphasizes class. I want to point out that Wilson's conclusions are based on his ability to use different theories in diverse contexts. Explanations of social injustice aren't as simple as most of us think, and this is becoming truer the more our society and economy become globalized. Having a number of theories at our disposal and possessing the flexibility of mind to use them creatively will go a long way in enabling us to see, explain, and impact the social world around us.

Wilson shows us that our very best efforts at ending racial inequality should be vitally concerned with and aimed at improving the class opportunities of Americans as a whole, especially as we move into a postindustrial, globalized economy. While these efforts would alleviate the economic suffering of many people,

> Their most important contribution would be their effect on the children of the ghetto, who would be able to anticipate a future of economic mobility and share the hopes and aspirations that so many of their fellow citizens experience as part of the American way of life. (Wilson, 1997, p. 238)

Wilson Summary

- Wilson considers three different theories in explaining race relations in the United States: Marxist elite theory, which sees complicity between elite capitalists and the state; split labor market theory, which calls attention to the different resource levels of diverse class positions; and Wilson's own class–state theory. Wilson's theory argues that economic and race relations are based on changing configurations among the state, economy, and class relations. Further, Wilson argues that racism requires a specific kind of relationship between economic elite and polity that was most purely found in the plantation South.

- Using these three different theories, Wilson examines three periods of American race relations: pre– and early post–Civil War, the latter 1800s to the 1930s, and the period from WWII on. The race relations within each of these periods is best explained by different theories: Marxist elite theory best explains the racism prevalent in pre– and post–Civil War America; split labor market theory explains the class growth period up until the Great Depression of the 1930s; and Wilson's class-state theory describes current race relations. Wilson's conclusion is that over time there has been a declining significance of race in explaining the position of African Americans.

- Based on his analysis, Wilson proposes a number of policy changes aimed at improving the overall welfare of the American worker, which, in turn, will improve the conditions of African Americans.

Looking Ahead

As we've seen, Wilson argues that race still matters, though it's importance is declining in the face of class inequality. Later, in Chapter 13, we'll see that Cornel West has a different way of understanding race matters. For West, the issue isn't so much the *importance* of race, but, rather, the *matters* of race: the cultural and political ramifications of living as an African American in the United States. While he would certainly agree with Wilson's analysis of class, West sees the upward mobility of some blacks and the split between black classes more in terms of market forces and culture rather than social structure and class.

Most segments of American society have now been identified as target markets in capitalist consumerism. But saturation of the African American market has led to unique effects on the experience of black identity. It is these with which West is most concerned. West asks, how have the changes in class and marketability significantly altered black subjectivity and identity? Being concerned with culture and political identity leads West to conclude that race continues to matter profoundly. One further note on West: he clearly sees race and democracy as related. True democracy cannot exist in the face of racism. As West (1999) says, "To be American is to raise perennially the frightening democratic question: What does the public interest have to do with the most vulnerable and disadvantaged in our society?" (p. xix).

Our other theorist on race in Chapter 13, Patricia Hill Collins, argues that it is inadequate to view race in isolation. Rather than seeing class as more important, as Wilson does, Collins argues that class and race work together. Because people occupy more than one status position, they stand at a place of "intersectionality" where race, class, gender, sexuality, religion, age, and so on come together. Collectively, these systems of inequality form a "matrix of domination" in society, which captures the overall organization of power relations. Thus, Collins gives us a very nuanced look at how power, oppression, and identity come together.

Janet Chafetz: Gender Equity

Photo: © Courtesy of Henry Chafetz.

The Essential Chafetz

Biography

Janet Saltzman Chafetz received her B.A. in history from Cornell University and her M.A. in history from the University of Connecticut. While at the University of Connecticut, she began graduate studies in sociology and completed her Ph.D. at the University of Texas at Austin. Chafetz served as president of Sociologists for Women in Society (SWS), 1984–1986, and as chairperson of the American Sociological Society (ASA) Theory Section, 1998–1999. Chafetz was also honored as the first invited lecturer for the Cheryl Allyn Miller Endowed Lectureship Series, sponsored by SWS; her book *Gender Equity* won the American Educational Studies Association Critic's Choice Panel Award (1990) and was selected by *Choice Magazine* for their list, Outstanding Academic Books (1990–1991). Chafetz is currently professor of sociology at the University of Houston and has recently been working on issues in immigrant and transnational religion.

Passionate Curiosity

Chafetz is a positivist seeking to understand and facilitate change in the system of gender inequality. She approaches gender inequality the way a scientist would approach cancer. In order to eradicate cancer, the laboratory researcher first has to understand how it works. Understanding how it works can then lead to targeted methods of cure rather than trial-and-error shots in the dark. As Chafetz (1990) says, "in practical terms, a better understanding of how change occurs . . . could contribute to the development by activists of better strategies to produce change" (p. 100).

Keys to Knowing

Unpaid labor force, women's workforce participation, organizational personality, social exchange and micro power, gender definitions, intrapsychic structures, social learning theory, engenderment, unintentional and intentional changes in the structure of gender inequality

---

Seeing Further: Structured Gender Inequality

Throughout history, the social category of gender has been and continues to be the most fundamental way in which distinctions are made among people. As such, gender as a social category influences almost every form of discrimination. Take the obvious example of class. According to the Business and Professional Women's Foundation (2004), women generally get paid 76% of what a man makes for the same job with the same education, which means that women are underpaid by about half a million dollars over a lifetime. Interestingly, this discrepancy increases for some of the best-paying and most powerful occupations. Female physicians, for instance, earn about 68% of what comparable male doctors do.

Women also experience what is referred to as "the glass ceiling," an invisible wall that stops women from advancing up corporate and political ladders past a certain point. Thus, men hold most of the top company positions in the United States: in 2002, there were only six female chief executive officers in the Fortune 500. Moreover, men hold most of the political positions: in the 108th Congress (2003–2004), women held only 14% of the seats in both the U.S. Senate and the House of Representatives.

The list of the social consequences of gender is almost endless: women are more likely to have been sexually molested than men; women suffer far more domestic violence than men; women have less decision-making power in organizations and relationships; women typically receive less education than men; women control fewer conversations than men; women are more supportive in conversations than men; the sexual expectations for women are oppressive when compared to men; women are more likely to be hassled in public than are men; and on and on.

Chafetz's central concern is to explain the structural conditions that create these indices of inequality. One of the delights in working with the theories of someone like Chafetz is the amount and diversity of theoretical perspectives and principles she brings together. In explaining gender inequality, Chafetz brings together the work of Karl Marx, Rosabeth Kanter, Nancy Chodorow, Albert Bandura, Erving Goffman, and Ralf Dahrendorf, as well as elements from exchange theory and social movements theory—and Chafetz manages to make all those diverse theories fairly easy to understand. Because she brings in so many different viewpoints, with Chafetz, it will be much easier for us to review her perspective as we go through her theory.

To begin, Chafetz assumes gender inequality. She isn't so much interested in explaining how it came about as she is in explicating how gender systems of inequality are stabilized and maintained—and she is quite inclusive in her explanation. She considers all three levels of analysis—macro, meso, and micro—and she argues that gender stratification is maintained through voluntaristic as well as coercive means. Chafetz argues that coercion usually functions as a background feature that helps to legitimate gender inequality. Both men and women generally conform to gender expectations and coercion isn't necessary. Coercion usually only comes to the fore during times of change or uncertainty.

Concepts and Theory: Coercive Gender Structures

Macro-Level Coercive Structures

Chafetz draws from specific theories to explain how the coercive and voluntaristic features work. The coercive theories tend to correspond to the three levels of analysis, so that there is a particular theory that is used at each level. Her primary orientation for the macro-level structural features of gender stratification comes from Marxian feminist theory. The basic orientation here is that patriarchy and capitalism work together to maintain the oppression of women, and the central dynamic in the theory is women's workforce participation.

Marx argued that social structure sets the conditions of social intercourse. In this sense, society itself has an objective existence and it thus strongly influences human behavior. More than that, social change occurs because of structural change. In Marx's theory, the economic structure itself contains dynamics that push history along. In general, Chafetz draws from Marx's emphasis on the economy as the most important site for social stability and change. She also explores his ideas about the way capitalism works.

Capitalism requires a group that controls the means of production as well as a group that is exploited. This basic social relationship is what allows capitalists to create profit. Patriarchy provides both: men who control the means of production and profit and women who provide cheap and often free labor. That latter part is particularly important. Much of what women do in our society is done for free. No wages are paid for the wife's domestic labor—this work constitutes the *unpaid labor force* of capitalism. Without this labor, capitalism would crumble. Paying women for caring for children and domestic work would significantly reduce profit margins and the capitalists' ability to accumulate capital. In addition, the man's ability to devote his time entirely to a job or career is dependent upon the woman's exploitation as a woman as well. When women are allowed in the workforce, they tend to be kept in menial positions or given lower wages for the same work as men.

Because of the importance of women's cheap and free labor to the capitalist system, elite males formulate and preach a patriarchic ideology that gives society a basis for believing in the rightness of women's primary call to childrearing and domestic labor. Elite men also use their structural power to disadvantage women's workforce participation. For example, in the United States, elite males have been able to systematically block most attempts at passing national comparable worth amendments or laws that would guarantee equal pay for equal work.

In brief, Chafetz argues that the greater the *workforce participation* of women, particularly in high-paying jobs, the less the structure of inequality is able to be maintained. The inverse is true as well: the less workforce involvement on the part of women, the greater the inequality on all levels. Thus, the type and level (macro) of their involvement in the workforce plays out at both the meso and micro levels. Though we've framed our discussion in terms of capitalism, Chafetz notes that ever since humans began to farm and herd animals, men have disproportionately controlled the means of production and its surplus. Throughout time, men have been slow to give up their economic positions.

Meso-Level Coercive Structures

To explicate the dynamics that sustain gender inequality at the meso level, Chafetz cites Rosabeth Kanter's (1977) work on organizations. Kanter gives us a social-psychological argument in which the structural position of the person influences her or his psychological states and behaviors. Kanter points to three factors related to occupational position that influence work and gender in this way: the possibility of advancement, the power to achieve goals, and the relative number of a specific type of person within the position. Each of these factors in turn influences the individual's attitudes and work performance—or what we could call his or her *organizational personality.*

The Possibility of Advancement

Most positions in an organization fall within a specific career path for advancement. The path for a professor, for example, goes from assistant, to associate, to full professor. The position of dean doesn't fall within that path. A professor could aspire to become dean, but the person would have to change her or his career trajectory. Kanter argues that women typically occupy positions within an organization that have limited paths for advancement. We can think of the occupational path for women as constricted in two ways: (1) the opportunities for advancement in feminized occupations, such as administrative assistant or secretary, are limited by the nature of the position; and (2) women who are on a professional career path more often than not run into a glass ceiling that hinders their progress.

The Power to Achieve Goals

Positions also have different levels of power associated with them. Again, this is a feature of the location within the organization, not of the individual. As an example of organizational power, let's use the position of professor. If you were in my theory class, I would have you write theory journals. Most of my students do the journals because I tell them to. But my authority to assign work doesn't have anything to do with me; it's a quality of the position. If there were another person in my position, the students would do what he or she told them. Thus, the kind of power that Kanter is talking about is a feature of the organizational position and not the person. Again, we typically find that, across the labor market, women hold organizational positions with less power attached to them than do men. Clearly, there are some women who transcend this situation, but women who are in positions of power are typically seen as "tokens," because there are no similar others in those positions within the company (Kanter's third organizational variable).

Relative Numbers

One of the things within organizations that facilitates upward mobility is the relative number of a social type within a position. Imagine being a white male reporting to work on your first day. You're given a tour of the facilities. As you are introduced to different people in the company, you notice that almost all the positions of power are held by black men. Most of the offices are occupied by black men and most of the important decisions are made by black men. There are whites at this place of employment, but they almost all hold menial jobs. By and large they are the secretaries and assistants and frontline workers. Being white, how would you gauge your chances for advancement at such a firm?

Further, imagine that after the end of your first week, you notice that there are two or three whites who seem to hold important positions. Would the presence of a few whites in management change your perception? It isn't likely. Though we like to talk as if individuals are the only things that matter, in fact, humans respond more readily to social types than to individual figures. That's why the few minorities that do make it up the corporate ladder are seen as tokens—exceptions to the rule—rather than a hopeful sign that things are changing.

As I've pointed out, these qualities are attributes of the position more than the person. This issue is important for Kanter's argument because social contexts influence individuals and their attitudes and behaviors. Positions that are similar in the organization produce similar contexts for people that most importantly include power and opportunity. These contexts influence the way the incumbents—the people that occupy the position—think and act.

In situations where power is available and the gates in the organizational flow lines are open, people develop a sense of efficacy. They feel empowered to control their destiny within the organization and they behave in a "take charge" manner. The reverse is true as well. People in positions where there are few opportunities and where power is limited are much less certain about showing positively aggressive behaviors. They feel ineffectual and limited in what they can achieve within the organization.

These issues are true for any who occupy these different positions. Humans are intimately connected to their context. Our social environment always influences who we are and how we act. The problem in organizations is that women are systematically excluded from positions of power and opportunity. As a result, they experience and manifest a self that corresponds to the position, one that feels and behaves ineffectually and in a limited manner. Though these differences in behavior and attitude are linked to organizational position, people generally attribute them to the person. Thus, women who occupy positions that have little power or hope of advancement demonstrate powerlessness and passivity. These attitudes and behaviors are then used to reinforce negative stereotypes of gender and work, which, in turn, are used to reinforce gender inequality within the organization.

Micro-Level Coercive Structures

Before we begin this section, I want to point out that even though we are talking about the micro level, Chafetz is talking about structures that have the power of coercion. This is important to keep in mind because many micro-level theories are oriented toward choice and agency. Chafetz talks about those issues in the section of this chapter on voluntaristic gender inequality, but for now we are considering how gender inequality is a structural force at the micro level.

Chafetz uses exchange theory to explicate coercive processes at the micro level. *Exchange theory* argues that people gravitate toward equal exchanges. Both partners in an exchange need to feel they are getting as much as they are giving. If an exchange isn't balanced, if one of the participants has more resources than the other, the person who has less will balance the exchange by offering compliance and deference. According to exchange theory, this is the source of power in social relationships.

Exchange theory also makes a distinction between economic and social exchanges. Economic exchanges are governed by explicit agreements, often in the form of contracts. The particulars of the exchange are well-known in advance and there is a discernable end to the exchange. For example, if you are buying a car on credit, you know exactly how much you have to pay every month and when the payments will stop. You know when your debt is paid off. However, in social exchanges the terms of the exchange cannot be clearly stated or given in advance. Imagine a situation where a friend of yours invites you over for dinner and tells you when and how you will be expected to repay. Chances are, you wouldn't go to dinner because the other individual broke the norms that make an exchange social. Thus, social exchange is implicit rather than explicit, and it is never clear when a debt has been paid in full.

Chafetz argues that these two issues together create a coercive micro structure that perpetuates gender inequality for women. Because of their systematic exclusion from specific workforce participation, women typically come into intimate relationships with fewer resources than men in terms of power, status, and class. The imbalance is offset by the woman offering deference and compliance to the man. This arrangement gives the man power within the relationship. The man's power in this exchange relationship is insidious precisely because it is based on social exchange. As we've seen, social exchanges are characterized by implicit agreements rather than explicit ones, with no clear payoff date or marker.

While insidious, this *micro power* is variable. Generally speaking, the more the economic structure favors men in the division of labor, the greater will be a man's material resources relative to the woman's, and the greater will be his micro-level power. The inverse is also true: "The higher the ratio of women's material resource contribution to men's, the less the deference/compliance of wives to their husbands" (Chafetz, 1990, p. 48).

The greater power that men typically have is used in a variety of ways. One common way is in relation to household work. Much of the work around the home, especially in caring for the young, is dull, repetitive, and dirty. Men typically choose the kinds of tasks that they will do around the home, as well as the level of work they contribute. Thus, men usually do more of the occasional work rather than repetitive work, such as mowing the lawn or fixing the car rather than the daily tasks of doing dishes or changing diapers. Men can also use their power to decide whether or not women work out of the home and to influence what kinds of occupations their wives take. Because women in unbalanced resource relations bear the greater workload responsibility for the children and home, they are restricted to jobs that can provide flexible hours and close proximity to home and school.

Concepts and Theory: Voluntaristic Gender Inequality

As we noted before, gender inequality usually functions without coercion. This implies that women cooperate in their own oppression. More exactly, "people of both genders tend to make choices that conform to the dictates of the gender system status quo" (Chafetz, 1990, p. 64). Chafetz refers to this as voluntaristic action, but, as we'll see, some of these behaviors and attitudes are unthinkingly expressed and so aren't "voluntary" in the sense of chosen. The patterns that support gender inequality are latently maintained; they are quiet and hidden.

The reason for this kind of maintenance is simple and it is how society works in general: we simply believe in the culture that supports our social structures. Max Weber pointed this out many years ago. Every structural system is sustained through legitimation, whether it is a system of inequality or the most egalitarian organization imaginable, and legitimation provides the moral basis for power—it gives us reasons to believe in the right to rule. Part of the way legitimation works has to do with the place culture has in human existence: culture works for humans as instinct does for animals. Another important reason for the significance of legitimation is that coercive power is simply too expensive in terms of costs of surveillance and enforcement to use on anything but an occasional basis.

Thus, Chafetz argues that much of what sustains the system of gender inequality is voluntarism. Both men and women continue to freely make choices and display behaviors that are stereotypically gendered. There are three types of *gender definitions* that go into creating gendered voluntaristic action: gender ideology, norms, and stereotypes. These three types vary by the level of social consensus and the extent to which gender differences are assumed.

Chafetz draws on three theoretical traditions to explain these issues: Freudian psychodynamic theory; social learning theory; and theories of everyday life, including symbolic interactionism, ethnomethodology, and dramaturgy. It's interesting to note that Chafetz is still working with levels of analysis: psychodynamic theory addresses the inner structural core of the individual, social learning focuses on the behaviors of the person, and theories of everyday life explicate the interaction.

Intrapsychic Structures

Nancy Chodorow's (1978) work forms the basis of Chafetz's psychodynamic theory. Chodorow argues that men's and women's psyches are structured differently due to dissimilar childhood experiences. The principal difference is that the majority of parenting is given by the mother with an absent father. Before we talk too much about Chodorow's theory, we need to make sure we understand the idea of psychic structure, or, more specifically, intrapsychic structure.

Freud argues that the psychic energy of an individual gets divided up into three parts: the id, the ego, and the superego. These three areas exist as structures in the obdurate sense. They are hard and inflexible and produce boundaries between the different internal elements of the person. Thus, when we are talking about *intrapsychic structures,* we mean the parts of the inner person that are fixed and divided off from one another, like the parts of the brain. The "intra" part of intrapsychic structures refers to how these three parts are internally related to each other. Thus, Chodorow's—and Chafetz's—argument is that the internal workings of boys and girls are structurally different. However, the difference isn't present at birth but rather is produced through social experiences.

Both boys and girls grow up with their chief emotional attachment being with their mother. Girls are able to learn their gender identity from their mothers, but boys have to sever their emotional attachment to their mother in order to learn their gender identity. The problem, of course, is that the father has historically been absent. His principal orientation is to work, a situation that was exacerbated through industrialization and the shift of work from the agricultural home to the factory.

Girls' intrapsychic structure, then, is one that is built around consistency and relatedness—they don't have to break away to learn gender and their social network is organically based in their mother. Women, then, value relationships and are intrapsychically oriented toward feelings, caring, and nurturing. Boys have to separate from their mother in order to learn gender, but there isn't a clear model for them to attach to and emulate. More exactly, the model they have is "absence."

The male psyche, then, is one that is disconnected from others, values and understands individuality, is more comfortable with objective things than relational emotions, and has and values strong ego boundaries. According to Freud, this male psyche also develops a fear and hatred of women (misogyny)—as the boy tries to break away from his mother, she continues to parent because of the absent father. The boy unconsciously perceives her continued efforts to "mother" as attempts to smother his masculinity under an avalanche of femininity. He thus feels threatened and fights back against all that is feminine.

Remember, these differences between males and females are dissimilarities in the structure of their psyches. This is a much stronger statement than saying that boys and girls are socialized differently. Intrapsychic structures are at the core of each person, according to Freudian theory, and much of what happens at this level is unconscious. Chafetz isn't necessarily saying that men consciously fear or hate women—it's deeper and more subtle than that. It is at the core of their being, their psyche. But also keep in mind that these structures vary according to the kind of parenting configuration a child has. It's very possible today for a boy to be raised principally by the father while the mother works (though our economic structure makes this unlikely), or for a single father to raise a child. The intrapsychic structure of such a boy would be dramatically different from that of a boy who is raised in a situation where maleness is defined by absence and separation.

These intrapsychic differences play themselves out not only in male–female relationships, but also in the kinds of jobs men and women are drawn to. Generally, men are drawn to the kinds of positions that demand individualism, objectification, and control. Women, on the other hand, are frequently drawn to helping occupations where they can nurture and support. While there have certainly been changes in occupational distribution over the past 30 years, most of the stereotypically gendered fields continue to have disproportionate representation. Thus, for example, while there are more male primary school teachers and nurses today than there were 30 years ago, the majority continue to be women; likewise, while there are more female CEOs, construction workers, and politicians today, those fields continue to be dominated by men. The important point Chafetz is making is that the "personal preference" individuals feel to be in one kind of occupation rather than another is strongly informed by gendered intrapsychic structures. These personal choices, then, "voluntarily" perpetuate gender structures of inequality.

Social Learning

Chafetz also draws on socialization theories such as social learning to explain the voluntaristic choices men and women make. The important components of *social learning theory* come to us from Albert Bandura (1977). Bandura argues that learning through experimentation is costly and therefore people tend to learn through modeling. For example, it's much easier to learn that fire is hot by the way others act around it than by sticking your hand in it. Social learning occurs through four stages: attention, retention, motor reproduction, and motivation.

Children pay attention to those models of behavior that seem to be the most distinctive, prevalent, emotionally invested, or that have functional value. They retain those models through symbolic encoding and cognitive organization, as well as through rehearsing the behaviors symbolically and physically. Motor reproduction refers to acting out the behaviors in front of others. Further motivation to repeat behaviors comes through rewards and positive reinforcement. Children are discouraged from repeating inappropriate behaviors through punishment and negative reinforcement. Eventually, children negatively or positively reinforce their own behaviors—as adults we thus self-sanction most of our gendered behaviors.

Performing Gender

Theories of everyday life look at how people produce social order at the level of the interaction and manage their self-identities. Chafetz specifically draws on Erving Goffman's (1977) work on gender. Goffman (whom we will cover in Chapter 9) argues that selves are hidden and the only way others know the kind of self we are claiming is by the cues we send out. Others read these cues, attribute the kind of self that is claimed, and then form righteously imputed expectations. People expect us to live up to the social self we claim. If we don't, that part of our self will be discredited and stigmatized.

Because gender is arguably the very first categorization that we make of people—one of the first things we "see" about someone is whether the person claims to be male or female—gender is thus an extremely important part of impression management and self-validation. As such, Goffman (1959) would see gender as a form of idealized performance in that we "incorporate and exemplify the officially accredited values of the society" (p. 35). In other words, social norms, ideologies, and stereotypes are used more strongly in gendered performances than in most others. Goffman also sees idealization as referencing a part of the "sacred center of the common values of the society" (p. 36), and he sees this kind of performance as a ritual.

Thus, gender is an especially meaningful and risky performance that is produced in almost every situation. We tend to pay particular attention to the cues we give out about our gender and the cues others present. Goffman further points out that we look to the opposite gender to affirm our managed impression. Our gendered performances, then, are specifically targeted to the opposite sex and tend to be highly stereotypical. According to Chafetz (1990), "For men, this quest [for affirmation] entails demonstrations of strength and competence. However, for women it entails demonstrations of weakness, vulnerability, and ineptitude" (p. 26).

Concepts and Theory: Stratification Stability

In Figure 8.2, we have Chafetz's model of gender stability. In it we can see all of the theoretical issues we've discussed coming together. Chafetz argues that gender inequality is initially stabilized through the economic structure and the division of labor. A gendered division of labor gives men and women different resource levels. Beginning with the top part of the model, we can see that because men have superior material resources, women at the micro level must balance the exchange by offering deference and compliance. Wifely compliance means that women are either kept at home, and thus utterly dependent upon the man for material survival, or are allowed to work but must also carry the bulk of the domestic duties (double workday). Wives' absence or double workday feeds back to and reinforces the gendered economic division of labor.

Follow the other path coming from male micro-resource power to male micro-definitional power, which, in turn, leads to gender social definitions. Here Chafetz is arguing that men have the greatest influence on society's definitions of male and

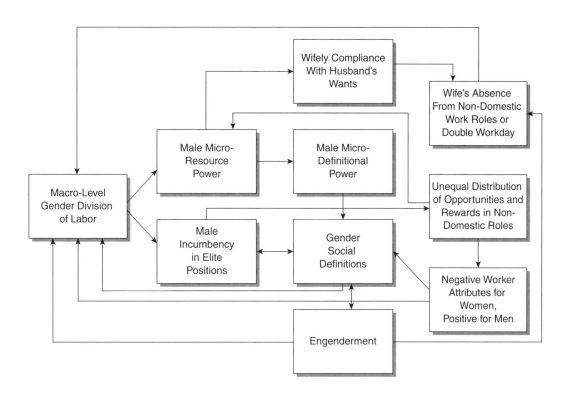

Figure 8.2   Gender Stratification Stability

Source: From Chafetz, J. S., *Gender Equity: An Integrated Theory of Stability and Change,* copyright © 1990, Sage Publications, Inc. Reprinted with permission.

female due to their greater control of resources. These definitions include stereotypes, norms, and ideologies that devalue women's work, legitimate unequal opportunities and rewards, and place higher value on masculine traits than on feminine ones (Chafetz, 1990, pp. 84–85).

This masculinized culture produces engenderment. *Engenderment* is Chafetz's term to talk about the internalization of gendered selves, identities, and behaviors. It specifically includes psychodynamic as well as social learning processes. Notice that gender social definitions are influenced by male incumbency in elite positions. The incumbency variable comes from the meso-level theories of organization. Thus, social definitions of gender are produced by male superior power at both the micro and meso levels.

As you can see from the model, engenderment feeds into wives' absence from the economy and the double workday. Chafetz (1990) is telling us that "a major component—conscious or not—of the feminine personality and self-identity is an orientation toward nurturant and caretaking roles . . ." and that "women, therefore, choose to place priority on family responsibilities, and, where financially possible, this priority often includes the choice to forgo other forms of work altogether" (p. 75). Engenderment also reciprocally influences the macro-level division of labor and definitions of gender.

The other main effect coming from the macro division of labor is male incumbency. The elite of any group are more likely to give positions of power to group members than to outsiders. This tendency toward like members is particularly strong during times of uncertainty or threat. In addition, given the current social definitions of gender, "elites are likely to believe that women lack the personal traits required to fill positions of responsibility" (Chafetz, 1990, p. 53). Taken together, the initial gender inequality in the division of labor with men holding the positions of greater power plus gendered stereotypes of work expectations imply that there will be a strong tendency for men to guard and keep the positions of power for other men. This incumbency, in turn, creates an unequal distribution of opportunities and rewards, which helps create male resource supremacy at the micro level as well as produce negative worker attributes for women and positive ones for men at the meso level of organizations. Note that, though not indicated on the model, all these relationships are positive (vary in the same direction) and thus work to structurally maintain gender stratification.

Concepts and Theory: Changing Gender Inequality

To begin this section, we should note that Chafetz argues that structural rather than cultural changes are necessary to bring gender equality. As we see throughout this book, sociologists give different weights to culture and structure. Some argue that culture is an extremely important and independent variable within society; others claim that culture simply reinforces structure and that structure is the most important feature of society. Chafetz falls into the latter camp. While voluntaristic processes, which are associated in one way or another with culture, are the key way gender inequality is sustained, "substantial and lasting change must flow 'downward' from the macro to the micro levels" (Chafetz, 1990, p. 108).

Unintentional Change

Like her understanding of gender stability, Chafetz divides her theory of how the structure of gender inequality changes into unintentional and intentional processes. Quite a bit of the change regarding the roles of women in society has been the result of unintended consequences. For example, the moves from hunter-gather to horticulture to agrarian economies were motivated by advances in knowledge and technology. But these moves also produced the first forms of gender inequality, as men came to protect the land (which gave them weapons and power) and to control economic surplus through inheritance and the control of sexual reproduction (women's bodies). The intent behind technological advancement wasn't gender inequality; the intent was first survival and then to make life less burdensome. However, in the end, economic developments produced gender inequality.

In explicating the unintended change processes, Chafetz is interested in what she terms the *demand side*. She argues that quite a bit of work on gender focuses on the *supply side*, which concentrates on the general attributes of women. For example, contemporary women tend to have fewer children, to be better educated, and to marry later than in previous generations. These supply-side attributes may influence what kinds of women become involved in the economy, but they "do not determine the rate of women's participation" (Chafetz, 1990, p. 122). For example, a woman may have a master's level education, but unless there is a structural demand for this quality, she will remain unemployed.

In addition, Chafetz assumes gender stability in theorizing about change. This means that males are the default to occupy a given position in the economy. The bottom line here is that "*as long as there are a sufficient number of working-age men available to meet the demand for the work they traditionally perform, no change in the gender division of labor will occur*" (Chafetz, 1990, p. 125, emphasis original). Thus, what we are looking for are structural demands that outrun the supply of male labor.

Chafetz lists four different kinds of processes that can unintentionally produce changes in the structure of gender inequality: population growth or decline, changes in the sex ratio of the population, and technological innovations and changes in the economic structure. The processes and their effects are listed below.

- *Population changes:* If the number of jobs that need to be filled remains constant, then the greater the growth in the working age population, the lower will be women's workforce participation. The inverse is true as well: as the size of the working population declines, women will gain greater access to traditionally male jobs, if the number of jobs remains constant.

- *Sex-ratio changes:* The sex ratio of a population, the number of males relative to the number of females, tends to change under conditions of war and migration. If in the long run there is a reduction in the sex ratio (more women than men) of a population, women will gain access to higher-paying and more prestigious work roles. Conversely, if there is an increase in the sex ratio (more men than women), the restrictions on women's workforce participation will tend to increase.

- *Economic and technological changes:* There are two general features of men's and women's bodies that can influence women's workforce participation: men on average tend to be stronger than women, and women carry, deliver, and nurse babies. When technological innovations alter strength, mobility, and length of employment requirements, then there are possibilities for changing the structure of gender inequality in those jobs. Women will tend to gain employment if new technologies reduce strength, mobility, and time requirements. In addition to work requirements, technology can also change the structure of the job market. As the economy expands due to technological innovations, women will tend to achieve greater workforce participation (holding population growth constant). On the other hand, if the economy contracts for whatever reason, women will tend to lose resource-generating work roles.

Intentional Change

In addition to unintentional change processes, Chafetz argues that there are specific ways in which people can act that help to address gender inequality. But before we get into those processes, we need to note that Chafetz maintains that gender change is particularly difficult for two reasons. First, women have more crosscutting influences than any other group. Think about it this way: Almost every social group is gendered. This means that women may be black, white, Chicano, Baptist, Buddhist, Jewish, pagan, homosexual, bisexual, heterosexual, homeless, professional, and so on. Women thus "differ extensively on all social variables except gender" (Chafetz, 1990, p. 170). These crosscutting group affiliations make it extremely difficult to form a woman's ideology that doesn't cut across or offend some of the women the ideology is trying to embrace. The second issue is that women, unlike many disenfranchised groups, do not live in segregated neighborhoods. Chafetz argues that this reduces women's political clout because they are "dispersed throughout all electoral districts" (p. 171). Because of these difficulties, change in the level of gender equality may occur more slowly or diffusely than other types of change.

Chafetz's model of gender change is depicted in Figure 8.3. As we talk through Chafetz's theory, be sure to trace through the effects on the model. There are two major independent variables in Chafetz's theory: macro-structural changes and elite support. Notice that the primary issues pushing intentional change exist outside of the direct control of women. Study Figure 8.3 and note where women's movements come into play. As you can see, the political activities of women are important, but they work more as a catalyst for change rather than the engine of it.

Chafetz's approach is similar to the way Karl Marx saw the production of class consciousness. According to Marxian theory, structure leads social change. Class consciousness is necessary for social change, but it is produced as dialectical elements of the economic structure, such as increasingly disruptive business cycles and overproduction, play themselves out. For Marx, it is the structure that pushes people together

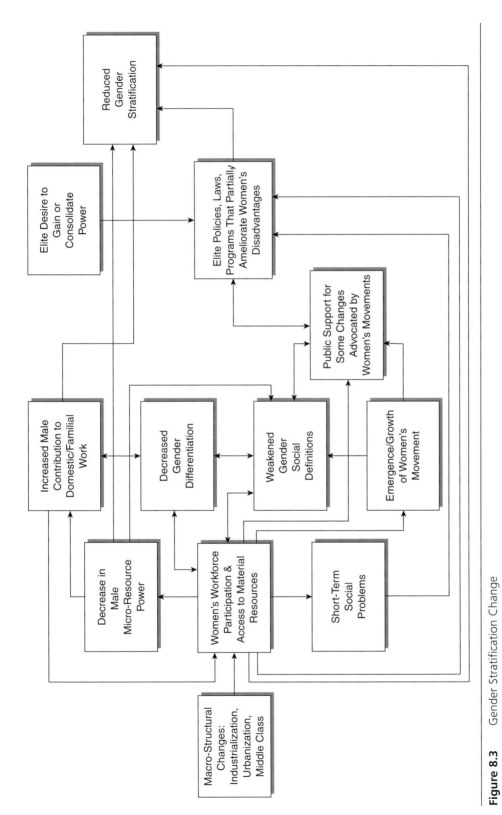

Figure 8.3   Gender Stratification Change

Source: From Chafetz, J. S., *Gender Equity: An Integrated Theory of Stability and Change*, copyright © 1990, Sage Publications, Inc. Reprinted with permission.

and gives them the ability to see their oppression, communicate with one another, and mount the resistance that will lead to the demise of capitalism.

The macro-structural changes that Chafetz focuses on are industrialization, urbanization, and the size of the middle class. Chafetz claims that, historically, almost all women's movements have been led by middle class women. These women are among the first to experience gender consciousness due to the effects of industrialization and urbanization. Industrialization initiates a large number of social changes such as increases in urbanization, commodification, the use of money and markets, worker education, transportation and communication technologies, and so forth. These all work together to expand the size of the middle class. The importance of this expansion isn't simply that there are more middle class people; it also means that there are more middle class jobs available. Many of these are nondomestic jobs that may be filled by women.

Industrialization thus structurally creates workforce demands that women can fill. Women are more likely to be called upon to fill these roles when the number of males is kept fairly constant, or when the number at least doesn't increase at the same rate as the demand for labor. The more rapid the growth of industrialization and urbanization, the more likely the demand for labor will outpace the supply of men.

As women increase their workforce participation, they increase their level of material and political resources as well, thus decreasing males' relative micro power and the level of gender differentiation, as well as weakening gender stereotypes, ideologies, and normative expectations (gender social definitions). In addition, as women's resources and thus their micro power increase, they are more able to influence the household division of labor and men's contribution to familial and domestic work. And, as men contribute more to domestic and familial work, women are more able to gain resources through workforce participation, as noted by the feedback arrow in Figure 8.3.

In addition to influencing these micro-level issues, industrialization, urbanization, and women's increased workforce participation do two other things: they increase women's experience of relative deprivation and the number of women's social contacts. While absolute deprivation implies uncertain survival, relative deprivation is a subjective, comparative sense of being disadvantaged.

Social movements are extremely unlikely with groups that experience absolute deprivation. They have neither the will nor the resources to organize political movements. Relative deprivation, on the other hand, implies a group that has resources and experiences rising expectations. Thus, women who are newly moving into the workforce will tend to experience relative deprivation. They will start to see that their salaries are not comparable to men's and they will begin to accumulate resources that can potentially be used to become politically active.

Urbanization and industrialization also increase a group's ability to organize. You'll remember that Ralf Dahrendorf (1957/1959) talked about this ability to organize as the principal difference between quasi-groups and interest groups. Quasi-groups are those collectives that have latent identical role interests; they are people that hold the same structural position and thus have similar interests but do not experience a sense of "belongingness." Interest groups, on the other hand, "have a structure, a form of organization, a program or goal, and a personnel of members" (p. 180).

The interest group's identity and sense of belonging are produced when people have the ability to communicate, recruit members, form leadership, and create a unifying ideology. Urbanization and industrialization structurally increase the probability that these conditions of interest group membership will be met. Women living and working in technologically advanced urban settings are more likely to come into contact with like others who are experiencing relative deprivation and the status and role dilemmas that come from women working a double workday.

As women begin to organize and as traditional gender definitions become weakened, public support for change is likely to arise. Chafetz argues that a significant portion of what women's movements have been able to achieve is related to articulated critical gender ideologies and radical feminist goals. While the public may not buy into a radical ideology or its set of goals, the ideologies and goals of women's movements will tend to justify changes in the structure of gender inequality more broadly. This support, along with pressure from short-term social problems and a direct effect from women's workforce participation, place pressure upon elites to create laws, policies, and programs to help alleviate the unequal distribution of scarce resources by gender.

Let's pause a moment and talk about short-term social problems. Chafetz argues that, as a result of women having and using greater levels of resources, short-term social problems are likely to arise. Chafetz uses the term "social problem" in a general sense to indicate the challenges that society at large has to overcome any time significant change occurs. In other words, change to the social system brings a kind of disequilibrium that has to be solved so that actions and interactions can once again be patterned. Such social problems tend to accompany any type of social change as a society adjusts culturally and socially. In the case of gender, the short-term problems are related to women having greater levels of resources and thus higher levels of independence and power. Examples of these kinds of issues include increases in the divorce rate and women's demands for control over their own bodies. Social disruptions such as these tend to motivate elite support of women's rights in order to restore social order.

As you can see from Figure 8.3, elite support shows up twice in Chafetz's theory. Sociologists have learned that every social movement eventually requires support from the elite. The elite not only pass laws and oversee their enforcement, they can also lend other material or political support, such as money and social capital (networks of people in powerful positions).

The first place elites show up is in the way we have been talking about it so far. Elites create new or support already established laws that help bring social order. In this case, the support comes mostly because elites perceive some form of social disorder. In other words, "they may perceive that basic problems faced by their society, which negatively affect large numbers of people and may possibly jeopardize their incumbency in elite roles, are exacerbated by a gender system that devalues and disadvantages women" (Chafetz, 1990, p. 152). This kind of change may be incremental and not specifically associated with women's movements.

The second place elite support shows up is as an exogenous variable (that is, one whose value is determined outside the model in which it is used) in the top right of the model. In this case, elite support may appear more directly tied to women's interests, though it may not necessarily be generated out of concern for women. Chafetz argues that we've made the mistake before of thinking of the elite as a single group with similar interests. In fact, the elite are divided into different factions, each struggling for power. In this kind of political environment, some elite groups are likely to see women as a potential resource and make promises designed to gather their support. At least some of these promises result in actual changes that reduce gender stratification.

I'm now going to restate Chafetz's theory of reducing gender inequality in brief propositions, since this is the discourse in which Chafetz works. Be sure to follow the elements of the propositions through the model in Figure 8.3. This way you'll be able to get a textual and visual rendering of the ideas.

> Proposition 1: Taken together, the level of reduction in gender stratification due to intentional efforts is a positive function of the level of male domestic labor, elite support, and women's control of material resources, and a negative effect of the level of male micro power.
>
> Proposition 2: Women's level of control over material resources is a positive function of industrialization, urbanization, and the size of the middle class. Elite support in general is a positive function of the level of women's control over material resources, the level of short-term social problems, the level of public support for changes advanced by women's movements, and elite competition.

Chafetz Summary

- In general, Chafetz argues that workforce participation and control over material resources both stabilize and change a system of gender inequality. Gender inequality is perpetuated when women's participation in the workforce is restricted, and reduced when women are allowed to work and control material resources.

- Chafetz argues that gender is stabilized more through voluntaristic actions than through the use of coercive power. When men control the division of labor in society, they are able to exercise authority at the meso level through assuring male incumbency in elite positions and at the micro level through women's exchange of deference and compliance for material resources. Men thus control gender social definitions that set up engenderment processes: psychodynamic structuring, gender socialization, and the idealized expression of gender through impression management and interaction. Engenderment and wifely compliance work in turn to solidify women's exclusion from the workforce, to impose double duty upon those women who do work, to stabilize the unequal distribution of opportunities and resources, and to define negative worker attributes for women but positive ones for men.

- Gender inequality is reduced as women are allowed greater participation in the workforce and increased control over material resources. These factors decrease women's reliance upon men and men's authority over women. Men then contribute more to domestic and familial work, which further frees women to participate in the workforce and weakens gender stereotypes, norms, and ideologies. In addition, women's access to resource-generating work roles increases the probability of women's political movements, which, along with weakened gender definitions, positively impacts public opinion and elite support for women. Short-term social issues that come about because of women's increased workforce participation also impact elite support, as well as the elite's desire to consolidate or gain political power. Gender stratification is reduced as women move into the workforce and control more material resources, as elites support women's rights and legislation, and as men's micro resource power is reduced and their domestic contribution is increased.

Looking Ahead

In Western European societies, inequalities of gender have been addressed through at least two waves of feminism. Inklings of the first wave began when ideas about equal rights for women emerged during the Enlightenment. The first significant expression of these concerns was Mary Wollstonecraft's book *A Vindication of the Rights of Woman* (1792). But the first wave of feminism didn't become organized until the 1848 Seneca Falls Convention, which called for equal rights to vote and own property, full access to educational opportunities, and equal compensation for work.

The second wave of feminism grew out of the civil rights movements of the 1960s. Publication of Simone de Beauvoir's *The Second Sex* (1949) and Betty Friedan's *The Feminine Mystique* (1963) were particularly important for this second wave of feminists, as was the founding in 1966 of the National Organization for Women (NOW). Central issues for this movement were pay equity; equal access to jobs and higher education; and women's control over their own bodies, including but not limited to sexuality, reproduction, and the eradication of physical abuse and rape. Janet Chafetz gives us a way of understanding gender that is in keeping with the structural issues raised by both the first and second waves of feminism and is closely aligned with mainstream sociology and positivistic theorizing.

However, in the early 1970s, two clear divisions began to appear among second-wave feminists. One camp emphasized the more traditional concerns of women's rights groups and generally focused on the similarities between men and women. Their primary concern was structural inequality. The other group moved to more radical issues. Rather than emphasizing similarities, they focused on fundamental differences between men and women. In some ways, the concerns of this group, described in the paragraph below, are more radical than equal opportunity. This more critical group is often described as a "third wave" of feminism.

The idea of third-wave feminism began to take hold around the intersection between race and gender—there are marked distinctions between the experiences of black and white women. But more recently, it has gained currency with reference to age. It appears that the experiences of young, contemporary feminists are different from those of second-wave feminists. Young feminists grew up in a social world where feminism was part of common culture. These young women are also playing out some of the postmodern ideas of fluid identities. The result is that many young feminists can best be described through contradiction and ambiguity. As Jennifer Drake (1997) says in a review essay, "[W]hat unites the Third Wave is our negotiation of contradiction, our rejection of dogma, our need to say 'both/and'" (p. 104). For example, third wavers might claim their right to dress sexy for fun while simultaneously criticizing patriarchy for objectifying women.

Our next theorist of gender, Dorothy E. Smith (Chapter 13), is among those who take the more radical view of feminism. For Smith, the life experiences and perspectives of men and women are different, and gender equality is more fundamental than structural equality. Smith focuses on the unique consciousness and lived experience of women. She argues that women have a bifurcated awareness of the world, split between the objective world of men and their own lived experiences. This type of control is deeply embedded in the mind and body, and it generally functions below the level of awareness. The key to sex or gender equality for Smith seems to lie in subversion—practices that critically expose the assumptions upon which the control of women is based.

Building Your Theory Toolbox

Knowing Theories of Race and Gender Inequality

After reading and understanding this chapter, you should be able to define the following terms theoretically and explain their theoretical importance to structural theories of race and gender inequality:

> Social structures, Marxian elite theory, race relations, exploitation, state–capitalist collusion, dark nations, split labor market, racial-caste system, Jim Crow segregation, urbanization, postindustrial, tight labor market: unpaid labor force, women's workforce participation, organizational personality, social exchange and micro power, gender definitions, intrapsychic structures, social learning theory, engenderment, unintentional and intentional changes in the structure of gender inequality

After reading and understanding this chapter, you should be able to

- Compare and contrast Marxian elite theory, split labor market theory, and Wilson's class–state theory

- Explain Wilson's three periods of American race relations and use Marxian elite theory, split labor market theory, and Wilson's class–state theory to analyze the different periods. In other words, you should be able to demonstrate, using Wilson's theory, that class is becoming more important in race relations than race itself.

- Explain why the workforce of women is so important to Chafetz's theory

- Explain how exchange processes work to produce gender inequalities. How do men use their micro power to their gendered advantage?

- Explain the differences between the intrapsychic structures of boys and girls. Explain how they are created and how they influence gender inequality.

- Explain how social learning theory and dramaturgy contribute to our understanding of how gender inequality is voluntaristically reproduced

- Explain how gender inequality unintentionally changes

- Tell which two characteristics of gender, according to Chafetz, make change difficult (Be sure to explain them fully.)

- Explain how women's movements form and how they influence changes in gender inequality. What does this tell us about social movements in general?

Learning More: Primary Sources

- Janet Chafetz:

  Chafetz, J. (1990). *Gender equity: An integrated theory of stability and change.* Newbury Park, CA: Sage, 1990.
  Chafetz is also the editor of the excellent *Handbook of the Sociology of Gender,* New York: Kluwer/Plenum, 1999.

- William Julius Wilson:

  Wilson, W. J. (1980). *The declining significance of race* (2nd ed.). Chicago: The University of Chicago Press.
  Wilson, W. J. (1987). *The truly disadvantaged.* Chicago: The University of Chicago Press.
  Wilson, W. J. (1997). *When work disappears.* New York: Vintage.

Learning More: Secondary Sources

  In 1993, Chafetz joined Rae Lesser Blumberg, Scott Coltrane, Randall Collins, and Jonathan Turner to produce a synthesized theory of gender stratification. This is their rare and powerful effort:
  Toward an Integrated Theory of Gender Stratification, *Sociological Perspectives, 36,* 185–216.

Theory You Can Use (Seeing Your World Differently)

- Using your favorite Internet search engine, type in "African American policy." Thoroughly review at least three sites. Would you say that the information and thrust of the sites are in keeping with Wilson's analysis and policy implications? In not, why do you think that is? If so, which policies do you think will have the greatest impact on structural inequality? How can you get involved in promoting this or other policy changes?

- Consult at least four reliable Internet sources to learn about the "separation of spheres." What is it and how does it figure into Chafetz's theory? List at least six ways that this historical, structural issue influences your life.

- Using your favorite search engine, type in "global gender inequality." Based on information from at least three different societies, prepare a report on the state of gender inequality in these diverse countries. Also, compare and contrast these societies with the one you live in. How applicable do you think Chafetz's theory would be in those other three societies?

- Volunteer at a local women's shelter or resource center.

Further Explorations—Web Links

- Check out the group, Sociologists for Women: http://newmedia.colorado .edu/~socwomen/. You can become involved in starting a local chapter or in specific student concerns.

- The following are just a few of the many excellent Web sites concerning women's rights and feminism: http://feminism.eserver.org/ or http://www .amnestyusa.org/women/index.do

- Wilson served as the eightieth president of the American Sociological Association. See http://www.asanet.org/galleries/default-file/Presidential Address1990.pdf for his presidential address concerning inner city social dislocations.

- As we've seen, Wilson argues that education and research are important factors in producing equality. In 2003, Wilson gave a brief talk at Amherst College addressing the place of social science: http://www.amherst.edu/ news/inauguration/panel/wilson.html

- Currently, Wilson is the director of the Joblessness and Urban Poverty Research Program at the John F. Kennedy School of Government. Check out the school, its mission, and what Wilson is doing there: http://www .ksg.harvard.edu/index.shtml

CHAPTER 9

Interactionist Theories

Herbert Blumer (1900–1987)

Erving Goffman (1922–1982)

Harold Garfinkel (1917–)

I'd like you to think about the last time you got together with a group of your friends. What happened? What did you do? If you could answer me, chances are you would say something like "we just hung out and talked" or "we watched the football game on TV." Now, how would you answer that question theoretically? To put it another way, how would you answer the question in sociological terms? How does watching a football game relate to society? How can you think about hanging out and talking in theoretical terms? What is going on when we are face-to-face with other people? On the surface, these may seem like silly questions, but they are far from it.

In a broad way, the three theories presented in this section all look at what happens in and around social situations, such as when you talk or watch TV with friends. The theories we'll be looking at don't present a complete understanding of what's going on in the situation, but they do give us amazing insights into the everyday world of social encounters. Blumer's work will help us see how meaning is negotiated and produced through symbolic interaction. Goffman will give us eyes to see how we manage the impressions that others have of us in face-to-face encounters. And, finally, Garfinkel will lift the curtain on the subtle ways through which we create a sense of social order and reality in every situation.

Herbert Blumer and Symbolic Interaction

Photo: © Reprinted with permission of the American Sociological Association.

The Essential Blumer

Biography

Herbert Blumer was born March 7, 1900. He completed his Ph.D. at the University of Chicago in 1928; his dissertation was titled *Methods in Social Psychology*. While at Chicago, Blumer studied under George Herbert Mead. When Mead became ill, he asked Blumer to take over his course, Advanced Social Psychology. Blumer was on the faculty at Chicago from 1927 to 1952, during which time he also played professional football, interviewed gang members, and mediated labor disputes. In 1952, Blumer moved to the University of California at Berkeley to chair its new department of sociology. Blumer edited one of sociology's premier journals, the *American Journal of Sociology*, from 1940 to 1952, and he was president of the American Sociological Society in 1955. Blumer died on April 15, 1987.

Passionate Curiosity

Blumer is most curious about meaning: What does it imply to say something is meaningful? How is meaning created? How is meaning utterly and completely social? Blumer also wants to give us a methodology, a way of seeing how the social world is continuously constructed out of negotiated interactions.

Keys to Knowing

Meaning, pragmatism, language, society, generalized other, role-taking, social objects, action, interaction, joint action, empirical science, reified concepts, definition of the situation, exploration, inspection

Seeing Further: Meaning and Interaction

Herbert Blumer is the key transition point between the work of George Herbert Mead and contemporary sociology. Blumer (1969) coined the term "symbolic interactionism" to refer to Mead's work, although he characterizes his term as a "somewhat barbaric neologism that I coined in an offhand way" (p. 1n). However, Blumer did more than simply give Mead's work a name: Blumer systematized Mead's work and gave it a coherent methodology. In a very real way, Blumer moved Mead's work from philosophy to sociology.

Having said that, I need to backpedal a bit. My statement about Blumer is true, if what we mean by symbolic interactionism (SI) is narrowly defined, as obviously I am doing. There are, however, other strong competing definitions within symbolic interactionism, when it is broadly defined. One such definition came from a man named Manford Kuhn. Kuhn's work is so different that eventually Kuhn's and Blumer's work became categorized respectively as the Iowa and Chicago schools of symbolic interactionism. The primary discrepancy between the two approaches lies in the notion of structure: Kuhn argues for structure, and Blumer sees the world more in terms of emergence.

I don't want us to get bogged down with academic considerations of how to classify theories; you can have fun with that later on in your academic career (and it really can be fun, if you don't take it too seriously!). My point here is that, however it is defined, the roots of symbolic interaction are found in the work of Mead and Blumer. The notion of emergence and Blumer's theoretical methodology are very important to know and be able to use in order to see social worlds.

Defining Symbolic Interactionism

Using Blumer's work to define SI, interactionists assume that human nature is based on meaning and its symbolic representation. In this sense, humans live in a world of their own creation—a meaningful world that is always something other than brute reality. Humans then act on the basis of meaning, rather than instincts or knee-jerk reaction. This type of meaningful action is what makes society possible. Meaning, however, is not an intrinsic feature of any social object or symbol; it thus *emerges* out of interactions. As such, face-to-face encounters are the only acting system: structures and institutions don't act; people act. The vehicle through which this action is carried out is the interaction.

Concepts and Theory: The Premises of Symbolic Interactionism

Humans Act on the Basis of Meaning

There are three premises or assumptions upon which symbolic interaction is built. The first assumption is that humans act on the basis of meaning. To understand its significance, let me ask you this: What's the meaning of meaning? It's an odd question, isn't it? At first glance, it looks like a play on words, kind of like "how much wood could a woodchuck chuck?" But it isn't a play on words; it's a serious question that has a significant answer. Asking it in the form that I did helps to point out something important about the answer. *Webster's* (1983) defines meaning as "1. that which exists in the mind . . . 2. that which is intended to be, or in fact is, conveyed, denoted, signified, or understood by acts or language" (p. 1115). We're going to put the first definition aside for the moment and concentrate on the second. Meaning is something that is conveyed, denoted, or signified by acts, words, or objects. Notice here what meaning *isn't*. Meaning isn't the act, word, experience, or object. Meaning, by definition, isn't the thing itself, whatever that thing might be.

Asking about the meaning of meaning puts in sharp relief the fact that meaning is something *other* than the object. Meaning always points to something else. If I asked you, what is the meaning of the Star Spangled Banner? you could take the song and explain to me the meaning of the flag, freedom, democracy, or any number of patriotic themes. You could point to what is being symbolized. Let's use another example. If I asked you, what does it mean that you play only black guitars? you might give me an explanation of Goth music and its commentary on contemporary life. But you would not tell me about the thing itself. Picture this kind of conversation: I ask, "Why do you play only black guitars?" You say, "Well, you see, the guitar is totally black. The knobs are black, the neck is black, the pick guard is black, and it even has black strings. The guitar is totally black. All my guitars are black." That kind of answer would be confusing, wouldn't it? My question isn't about the qualities of the thing itself, in this case the black guitar, but, rather, the meaning that the black guitars symbolize for you.

Human beings, by their very nature, can't accept things as they simply are. Humans must give meaning to things. In fact, we have to give meaning to everything—whatever it might be, it doesn't exist for us unless and until we give it meaning. And meaning is never the thing itself. Think about money—it isn't the paper that matters; it's the meaning. It's what the paper signifies. Think about killing a person—it isn't the actual act or the fact that a person's life has ended—it's the context wherein the killing takes place. Killing can be war, or terrorism, or murder, or accidental homicide, or suicide, or religious sacrifice, or first-degree murder, or execution in response to first-degree murder—killing can be legitimate or morally wrong depending on its meaning context. Human life doesn't matter in and of itself; it's the context that matters. Remember the movie or the history of Apollo 13? The eyes of the entire world were on the three men trapped in that spacecraft. America watched in horror the catastrophe unfolding and had heartfelt concern for the lives of those three Americans—while at the same time hundreds if not thousands of people were being killed or maimed in Vietnam by American servicemen. What matters isn't life; what matters is meaning.

But, why do we place such importance on meaning? Ah, that's a very good question. To put it another way, why can't human beings get away from meaning? The first part of the answer is because humans are utterly and completely social. Being social is how we as a species exist. Every species is defined by its method of survival or existence. Why are whales, lions, and hummingbirds all different? Because they have different ways of existing in the world. What makes human beings different from whales, lions, and hummingbirds? Humans have a different mode of existence.

But being social is only part of the answer. There are a number of species that exist socially, like ants and bees, so what makes humans different from them? What is different is the *magnitude* of our sociability, and, more importantly, the way in which we create our social bonds. Most other societal species instinctually create social bonds through a variety of things such as scent, physical spacing, and so on. Humans use symbols to create *meaningful* social bonds. Granted, there are some species that have a kind and degree of culture, but no other animal uses culture to the extent that humans do—and no other species uses symbols. We are primarily built for and oriented toward using signs and symbols. Culture and language are the reasons we have the brain structure that we do; culture is the reason we have the kind of vocal structures that we do; culture is why we have the kind of hands that we do—culture is the defining feature of humanity.

Meaning Arises Out of Interaction

Okay, if meaning isn't in the object, word, symbol, language, or event, then where is it? The quoted definition from *Webster's* says that meaning exists in the mind. However, as with most everything in a society infatuated with the individual, that definition is only partially correct. The individualized definition also lines up with an important philosophical school of thought: idealism. Idealism and realism are two sides in the philosophical debate about reality (ontology or metaphysics) and knowledge (epistemology). To even begin to scratch the surface of this debate would take much more time and text than we have (philosophers have been debating these issues for over 2,000 years), but we really don't have to. Symbolic interaction teaches us that in our day-to-day life we are not at all interested in ontology or epistemology. Meaning for humans is a pragmatic issue.

As we saw in Chapter 4, **pragmatism** argues that meaning and ideas are not meant to be representations of ultimate reality. In pragmatism, ideas and meanings are organizational instruments. That is, humans organize their behaviors based on ideas. Meaning, then, is a tool for action and it has value to us only insofar as it facilitates behavior. In other words, the "truth" of any idea or moral is not found in what people believe; truth can only be found in the actions of the people that hold the value and the practical benefits that they bring. Pragmatism argues that people hold onto what works, and what works is the only truth that endures for humans. Thus, the meaning of any idea, moral, word, symbol, or object is pragmatically determined, or determined by its practical use.

Let me give you an example: Who was Martin Luther King Jr.? You might answer, he was a good man, a civil rights leader who tried to bring equality to all humankind. You might also say, he is one of America's national heroes, worthy of a holiday, statues, and streets and parks being named after him. From a symbolic interactionist perspective, the truth or falsity of such statements is of no consequence. It doesn't matter who or what King "really was." From a pragmatic point of view, there is no ultimate truth about King. What matters is *how people use the idea* of Martin Luther King: How is the idea of Martin Luther King used in interactions? How do people organize their behaviors around the social object of MLK? From a pragmatic, emergent point of view, the meaning associated with King is not a function of who or what King might have been. The meaning of Martin Luther King emerges as people negotiate different practical (and political) issues through social interaction. (For an insightful study of this very topic, see Lilley & Platt, 1994.).

In true pragmatic form, symbolic interaction argues that meaning is found only in people's behaviors and that meaning emerges out of social interaction in response to adaptive concerns. Thus, while meaning comes to be in the mind, it is produced and exists within pragmatically emergent social interactions—that which is in the mind is only a residue of these social interactions.

This idea of emergent meaning is very important. Since meaning doesn't come from and isn't tied to the object itself, and because meaning is not made inflexible through structures and institutions, meaning is supple and can change. For example, the meanings of "female" and "black" have changed over time. That's obvious, but those are broad, historical examples. Can the meaning of "female" (or "black") change from one interaction to another? Sure. A woman may have an interaction wherein she is treated with respect as a professional colleague, but in the very next interaction she may be treated as a sexual plaything, an object of gaze. Meaning isn't simply a one-way street. All meanings are subject to negotiation within each and every interaction by all the participants. So the professional woman may see a man she finds attractive and try to get his attention so that she can be an object of gaze for him. The meaning of any social object thus depends on the interaction and how we use symbols. (And, according to symbolic interactionists, broad, historical changes in meaning occur through multiple, micro-level interactions.)

Individuals Interpret Meaning

The third assumption of SI is that meaning is individually interpreted. The inclusion of individual interpretation is important because it highlights the place Blumer gives to human agency. In the symbolic interactionist scheme, people are not "cultural dopes" or "social robots" that simply act out structure or the demands of the interaction (a critique that is leveled against structural functionalism). People make decisions about their action. They do it by interpreting the different meanings that emerge out of interaction. Well, of course, you say, people *interpret* things. That's what makes the individual so important. But let's notice what this idea of interpretation implies. An individual interprets things by indicating or pointing out the objects to his or her self. Another way to put it is that the individual communicates with his or her self about the meanings of objects using language.

Let's think a bit more about the significance of this internalized conversation and language. The obvious function of language is to communicate, but it does more than that. Language functions as an index for our subjective meanings. When we want to understand something that we experience, we search through our language looking for the words and symbols that seem to match the experience. We can use language to talk to ourselves about the experience, and we can talk to others about it as well.

Okay, that might sound obvious, but what most of us don't think about is that we use language to communicate with ourselves, and that language is the repository of *social experiences*. Language doesn't exist for us as individuals. We didn't create language so that we could think or understand subjective experiences. Humans use language to express social and cultural events, experiences, and pragmatic meanings. Language, therefore, exists outside of the individual; it is a social entity. Thus, when we use language to understand our own experiences, those experiences become social.

Blumer's point here is that individuals have agency: freedom and responsibility in action. This issue is a basic point of symbolic interactionism as it goes back to George Herbert Mead and especially pragmatism. The person in pragmatic, symbolic interactionism is the supreme democratic citizen, able to take in various ideas and ideologies and make reasoned decisions about her or his social actions. Yet, at the same time, Mead and Blumer want us to know that interaction—both internal and external—is in every sense socially based. Here is where we come back to the first definition that *Webster's* gave us for meaning: "that which exists in the mind." It does, but as we saw in Mead's theory, the mind itself is a social entity, an internalized conversation.

Concepts and Theory: The Root Images of Symbolic Interactionism

The Nature of Society

Based on the premises of symbolic interactionism, Blumer gives us six "root images" of SI. The first concerns the nature of society: What is society and how does it exist? These obviously ought to be central concerns for sociologists—that's what the word sociology means: the study of society. But too often we take the nature of society for granted. We talk about society as if everybody knows what it is, just as we use the words "institution" and "structure" without clear definitions. It's part of our taken-for-granted vocabulary. Blumer, however, takes this question seriously, because the nature of society is inexorably linked to symbolic interactionism.

Sociologists talk a lot about institutions like education and the economy, but are they objectively there like a rock? Can you touch, taste, hear, see, or smell social institutions or structures? Not really. Think about it this way: If I were to ask you to point to a tree, you would be able to do it without much trouble. But if I were to ask you to point to the institution of religion or institutional racism, you would have difficulty doing it. You could point to behaviors, but to actually point to the institution is impossible.

Your difficulty would indicate that these things don't really exist as separate, objective entities; they only exist in and as the result of interactions. We have to imagine that there is such a thing as the economic institution, and we have to point it out to ourselves in interaction. As we interact and behave as if a structure is real, we develop ways of acting toward it and it *becomes* real. Thus, society isn't a thing that exists objectively on its own, nor is it a causal structure that determines our behaviors. Society is a social object whose meaning emerges out of interaction. As Blumer (1969) says, "The essence of society lies in an ongoing process of action—not in a posited structure of relations" (p. 71).

What symbolic interactionists mean by society is best captured by the notion of the generalized other. You will recall from Chapter 4 that the **generalized other** is the set of thoughts, feelings, dispositions, values, perspectives, and behaviors associated with a group. The important thing to see is that the generalized other is not a structure that objectively organizes your behavior. Just as we do with individual people, we role-take with the generalized other to determine what is expected or common for the situation. We may or may not decide to adopt the behaviors, feelings, and so forth that the generalized other indicates.

The Nature of Social Interaction

Most social scientific perspectives take the interaction for granted: it is generally seen simply as a means of expression. For most sociologists, the interaction is the medium through which social structures and institutions are expressed. For most psychologists, the interaction is the field in which the individual's intrinsic psychological tendencies are expressed. But in the way Blumer conceptualizes meaning, the interaction is the central arena in which both social structures and individual personalities are created and sustained. In other words, social interaction is *the* acting unit—no other social entity acts.

The key to understanding what interaction is, is in the word itself: *inter-action.* The prefix "inter" means between, among, mutual, reciprocal, and intervening. In interactions, then, we weave our various individual actions together. As Blumer (1969) puts it, "One has to fit one's own line of activity in some manner to the actions of others" (p. 8). We accomplish this fitting through the use of symbolic gestures and role-taking. Symbolic gestures include both verbal and nonverbal cues that hold reciprocally understood meanings; reciprocally means that the meanings are shared by all concerned. So if I say "the mountains," you and I both know what I intend. But let me call your attention to something again: While we come into an encounter generally understanding the meanings associated with social objects, the pragmatic or practical meanings must emerge as the people in the interaction fit their words and actions together. Role-taking is the process through which we put ourselves in the position of another to view our own self and behaviors. "Such role-taking is the *sine qua non* [the one essential characteristic] of communication and effective symbolic interaction" (Blumer, 1969, p. 10). Role-taking allows us to *socially* interact; apart from role-taking, action would be more like reaction that is motivated by impulse.

It's also important to note that while all people can and must role-take, it is an acquired ability. Some people are quite good at it and can carry off and control interactions beautifully. Others are less skilled and tend to make blunders within interactions. As a result, a meaning emerges that many not be intended or "true." For example, a friend of mine is the director of human resources at a large corporation. She gets hundreds of inquiries about positions, many of them coming through email. Some of these emails come with addresses such as daddysgirl@ met.com (fictitious address; any likeness to a real email address is purely coincidental). My friend automatically discards those emails. The author could well be a college graduate, but she hasn't learned how to role-take very well.

The Nature of Objects

Blumer and the symbolic interactionists have a specific definition in mind for "objects." We usually think of an object as something material that we can perceive through our senses. However, for Blumer, objects are created through intention and indication—we must indicate the object to ourselves or others and we must intend to act toward the object. The first thing to notice about this definition is that objects don't exist on their own. Objects come to exist for humans as we indicate them and act toward them. This feature makes sense when you think about the importance that Blumer places on meaning and the quality of meaning. Remember, meaning isn't the thing-in-itself; meaning is attributed or placed over events, things, and entities. So it makes sense that objects have to be intended and indicated—it isn't the material thing that we respond to at all. The "object" is social and created through meaning attribution. Because the meaning, legitimated actions, and objective availability (they are objects because we can point them out as foci for interaction) of symbols are produced in social interactions, they are **social objects.**

The Nature of Human Action

The distinct trait of human action is that it is not *re*-action. We make decisions about our behaviors. We don't do it all the time, but a fully human act involves decision rather than simple reaction. In order to be truly human, we must first take our impulse as a social object. So, for example, we feel pain. Pain often leads to a knee-jerk reaction: we might have the urge to yell or get angry. But human action requires us to take that pain and consider the possible effects our reactions might have on those around us. The pain, then, becomes a social object—we internally call attention to it, name it, and attach legitimate lines of behavior to it.

We have all experienced this kind of internal conversation and the tension it creates. We have all felt the impulse of anger or irritability or, on a more positive note, spontaneous playful behavior. A part of us wants to react immediately to stimuli. We want to jump or shout or punch or kiss or run naked. We're hungry or we're tired or we're sexually frustrated; and sometimes we act on these impulses and drives. However, to the extent that we do act on them, we are not acting social. Thus, action rather than reaction is the basis of society.

The Nature of Joint Actions

As we've seen, humans interact; that is, we lace together various individual actions. Such interaction, Blumer argues, gives rise to joint action. **Joint action** has a distinctive character of its own and is greater than the mere adding together of the various individual actions. Let's use "a wedding" as an illustration to get at what Blumer is talking about. There are various islands of interaction that take place at a wedding. There is, of course, the interaction among the bride, groom, and official. There are also the interactions around things such as "picture taking," "cutting the cake," "breaking the glass," "throwing the rice," and so on. Some of these interactions extend in time, to the past or the anticipated future. Thus, weddings may involve "picking the florist and caterer," "choosing the dresses and tuxes," "getting the marriage license," and so on.

There are a few things to note about joint actions. First, joint actions are created as we link together various interactions. The nature of interactions lies in the way in which we intertwine the many actions that create the interaction (I hear my friend introduce you, you look at me and I look at you, I give my "nice to meet you" smile and extend my hand, you see my hand as it is extending and you extend your hand, our hands meet and we judge the firmness of the grip and carefully match what we are given, and so forth—or, we do something totally different; remember, interactions emerge, they aren't determined). The same is true of joint actions, except in that case we link together multiple interactions.

Second, joint actions may be spoken of as such. They stand as identifiable units to us, which is why I put parentheses around them in the above paragraph. We see them as distinct yet joined together—"choosing the dresses" is distinct from "getting the license," yet we link them together into the joint action of a wedding. Third, each of the interactions and joint actions must be created anew each time it is produced. Weddings are not simply expressions of a causal social structure called marriage. The structure of norms, values, and beliefs doesn't determine the wedding; you do, which is why something as simple as a beach ceremony and something as complex as the wedding of Prince Charles and Lady Diana can both qualify as weddings.

Concepts and Theory: Empirical Science and Symbolic Interactionism

The Premises of Empirical Science

According to Blumer, *empirical science* is based on the assumption that the universe is objective—it exists outside of the observer. Further, the universe stands over and against the observer; that is, the empirical world resists the onlooker. It contains its own elements, processes, and laws that the scientist must discover.

Blumer lists six characteristics of empirical, scientific inquiry:

1. Scientific inquiry uses theory. Scientific theory is a formal and logically sound argument explaining some empirical phenomenon in general or abstract terms.

2. Theory is used to decide the kinds of questions that are asked.

3. Theory shapes what data are relevant, and how the data will be collected and tested.

4. Propositions are born out of theory: theory informs the kinds of relationships among and between the variables that are to be tested.

5. The data are interpreted and brought back to change, modify, or confirm theory.

6. All theory and scientific research is based on concepts, which are the basic building blocks of theory. Concepts inform the way questions are asked; they are the source of data categories (sought and grouped); they form the relationships among and between the data; and they are the chief way in which the data are interpreted.

Blumer's main point is that each part of this procedure, particularly the concepts that are used, must be scrutinized to make certain that it conforms to the empirical world that is being studied. Blumer contends that this examination doesn't usually occur in the social and psychological sciences. More often than not, what sociologists and psychologists study are *reified concepts* of the world rather than the social world itself. Here the word "reified" means to convert an idea into something concrete or objective. It is a problem that is rampant in the social and behavioral sciences, according to Blumer. The two most notable examples are "attitudes" in psychology and "structures or institutions" in sociology. Neither of these concepts has a "clear and fixed empirical reference" (Blumer, 1969, p. 91), yet both are seen as having some causal force in determining human behavior.

Obviously, the issue of most concern for us is the critique that social structures are not empirical. As we've seen in previous chapters, many sociologists see social structures as connections among sets of positions that form a network. The interrelated sets of positions in society are most often defined in terms of status positions, roles, and norms. The social structure is generally seen as that which "accounts for much of the . . . patterns of human experience and behavior" (Johnson, 2000, p. 295).

Yet, Blumer says there's a problem in attributing causal influence to social structures, because social structures aren't empirical. As sociologists we talk a lot about structures and institutions and society. But, as we've seen, you can't point to the institution because it isn't empirically there to be seen. The only empirical, social thing we can point to are real people interacting one with another. According to Blumer (1969), anytime we appeal to psychological or social structures as the impetus behind human behavior, "the human being becomes a mere medium through which such initiating factors operate to produce given actions" (p. 73). In other words, in standard structural approaches, people are not seen as free-acting agents.

Methods of Symbolic Interactionism

For Blumer, the most macro-level social entity that is empirically discernable is the joint action: interactions weave together individual actions and joint actions can weave together various interactions. Let's take my theory class for an example. I'm writing this paragraph in the month of July. Within four weeks, I'll be standing in front of my students. Between now and then, I will have written out my syllabus, sent it to the enrolled students, and given them their first reading assignment. They will perform whatever interactions necessary to obtain the books (interact with Mom and Dad or financial aid, drive to the bookstore [the drive itself is a series of interactions], interact with the various clerks at the bookstore, and so on).

We'll have 15 class meetings, each of which is a separate joint action during which several unexpected ideas, conflicts, and relationships will emerge. We will link these class meetings together through the use of tests, papers, and discussions about theses, classes, ideas, and graduation. Those linkages will themselves be linked to other courses at the university (prerequisites and the like). All of these will be linked together through various kinds of interactions, such as theses defenses and graduation ceremonies. All of those joint actions will be linked to the history of interactions and joint actions that have occurred at the place where one of the students will apply for a job or for further graduate work—and on and on, ad infinitum. Most of what we mean by "society," then, exists in recurring patterns of joint action.

At every point of interaction or joint action, there is uncertainty. All human behavior has to be purposefully initiated: it may or may not occur. Once begun, inter- and joint action can be interrupted, changed, or abandoned. Even during the course of an interaction, participants may have different definitions of what is going on. If there is a common definition, there still may be differences in the way each interaction is carried out: an individual may perform his or her lines of action differently; changes can impose themselves on the interaction in such a way as to make people define and lean upon one another differently. In addition, new situations may arise that call into question the ways things have been done.

Symbolic interactionism shifts the question from the structure to the interaction. It entertains the possibility that each interaction and joint action could be different, and it grants freedom of choice to the participants. Yet, many interactions and joint actions do have a patterned quality about them. How that happens becomes a chief question for symbolic interactionists. There are, of course, some factors that tend to make interactions and joint actions stable.

First, most participants generally enter the majority of situations with a common definition of the situation. The **definition of the situation** gives meaning to the interaction and it implicitly contains identities and scripts for behavior. People use the definition of the situation to produce appropriate behaviors based on role-taking, and they use those roles to understand others' behaviors. Second, each interaction and joint action has a history, career, or line of action. A joint action's career may be quite long or fairly short. For example, after a few weeks in your theory class, you will have built up a history of how the class will be enacted. That history sets up expectations, or meaning attributions, that are formed in advance. But that history is relatively short when compared to the definition of the situation that we call "graduation." Generally speaking, when changes are expected in "graduation," it sets up a series of interactions (committee meetings, public hearings) around the possibility of change and the meaning that might be conveyed. All of these themselves form joint actions with which "graduation" will be laced.

Implications of Symbolic Interactionist Methods

There are at least four methodological ramifications of Blumer's argument. First, to state the obvious, an empirical science needs to investigate empirical phenomena. Blumer argues that most of what passes as social or behavioral science is not empirical because most "social scientists" have not critiqued the reified concepts they are using to create theory, propose relationships, gather and analyze data, and interpret findings. Second, we need to understand social interaction as a moving process. Many sociological perspectives understand society in terms of one central form, such as conflict theory, structural functionalism, exchange theory, and so on.

Blumer's point is that real people in real interactions will sometimes be conflictual, sometimes functional, or sometimes be engaged in exchange. We need to see that human behavior is meaningful and that humans can interpret meanings in multiple ways: society is a moving process, not a static object. Third, we need to understand social action in terms of the social actor. Here Blumer is emphasizing the agency of interactants. The person and the interaction are not simply modes through which social structure is expressed; rather, they are the true acting units of society. And fourth, we need to be careful of using reified concepts to understand social life. The concepts of "institution," "structure," and "organizations" all fall short of the direct examination of the empirical world. Macro-level issues need to be understood in terms of the career or history of joint actions. We must also keep in mind that the history itself is a meaning attribution, subject to negotiation and interpretation.

There are two methodological approaches with which Blumer leaves us: exploration and inspection:

> Exploration is by definition a flexible procedure in which the scholar shifts from one to another line of inquiry, adopts new points of observation as his study progresses, moves in new direction previously unthought of, and changes his recognition of what are relevant data as he acquires more information and better understanding. (Blumer, 1969, p. 40)

Exploration is grounded in the daily life of the real social group the investigator wants to study. Rarely does a researcher have firsthand knowledge of the social world she or he wants to study.

Thus, rather than entering another's world with preconceptions, as much as possible the researcher simply enters the other's world and searches for the social objects that the group regularly employs in producing meanings through interactions. The records of such social objects and interactions become comprehensive and intimate accounts of what takes place in the real world. These accounts are in turn inspected. During *inspection,* the researcher seeks to sharpen the concepts she or he is using to describe the social world, to discover generic relationships (those that appear to hold true in various settings), and to form theoretical propositions. Inspection, in other words, moves us up from the actual empirical instances to the formation of theoretical concepts. However, these concepts are kept firmly grounded in the empirical setting.

Blumer Summary

- As explained by Blumer, there are three premises of symbolic interactionism: humans act on meaning, meaning does not exist in language, and individuals interpret meaning through social interaction.

- Based on these premises, Blumer gives us six root images of social interaction. These are concerned with the nature of society, the nature of social interaction, the nature of objects, the nature of human action, humans as acting organisms, and the interlinking of action. Society, as a totality, exists symbolically and its meaning emerges out of interaction. Society thus does not determine our actions; the meaning of society and its influence on our behaviors are an accomplishment of interaction. Thus, humans act rather than react. As interactions move along, we make small decisions about meaning and action. These meanings and actions emerge as people negotiate in interactions and they become identifiable lines of behavior. Within interactions, social objects come to exist as they are indicated by the inter-actants and as particular kinds of actions are intended toward the thing. Social objects may be actual things, symbolic meanings, selves, or others. Social objects— whether the self, other, or society—exist only insofar as they are the subject of interaction. Therefore, interactions—the face-to-face blending of individual actions—are the only empirically acting unit in society. Society exists only as sets of generalized others with which we role-take.

- Blumer argues that the only empirical and acting part of society is the inter-action, and he cautions us against the danger of reifying concepts such as institu-tions or social structures. In analyzing society, then, we need to focus on the interaction, realizing that it is an ongoing and moving process wherein individual actors exercise agency. This analysis should be in two phases. The first is explo-ration: Because of the emergent nature of society and self, researchers must divest themselves as much as possible of preconceived notions of what might be happen-ing in any given situation. Theory ought to be grounded in the actual behaviors and negotiations in real interactions. Second, as theoretical concepts suggest themselves from the experience of the researcher in the field, these should be inspected to see if they might hold in other settings as well.

Erving Goffman and Dramaturgy

Photo: Collections of the University of Pennsylvania Archives.

The Essential Goffman

Biography

Erving Goffman was born on June 11, 1922, in Mannville, Alberta, Canada. He earned his Ph.D. from the University of Chicago. For his dissertation, he studied daily life on one of the Scottish islands (Unst). The dissertation from this study became his first book, *The Presentation of Self in Everyday Life,* which is now available in 10 different languages. In 1958, Herbert Blumer invited Goffman to teach at the University of California, Berkeley. He stayed there for 10 years, moving to the University of Pennsylvania in 1968, where he taught for the remainder of his career. Goffman also served as president of the American Sociological Association in 1981 and 1982. Goffman died on November 19, 1982.

Passionate Curiosity

Goffman was inquisitive about everything people did in face-to-face interactions. He watched them continually. The face-to-face encounter so enthralled Goffman that he was driven to probe the interaction itself: What are the requirements of an encounter? How do these requirements influence everything that people do when they meet?

Keys to Knowing

Dramaturgy, the interaction order, impression management, front, teams, setting, front and back stages, role distance, deference and demeanor, face-work, focused and unfocused encounters

Seeing Further: Managing Impressions

Scenario I

Dateline: Friday, May 2, 2003. ABOARD USS ABRAHAM LINCOLN (CNN)—President Bush made a landing aboard the USS *Abraham Lincoln* Thursday, arriving in the co-pilot's seat of a Navy S-3B Viking after making two fly-bys of the carrier.

It was the first time a sitting president had arrived on the deck of an aircraft carrier by plane. The jet made what is known as a "tailhook" landing, with the plane, traveling about 150 mph, hooking onto the last of four steel wires across the flight deck and coming to a complete stop in less than 400 feet.

Moments after the landing, the president, wearing a green flight suit and holding a white helmet, got off the plane, saluted those on the flight deck and shook hands with them. Above him, the tower was adorned with a big sign that read, "Mission Accomplished." (CNN, 2003)

Scenario II

One day while driving to a lunch appointment, I stopped at a stoplight. Like in most cities at that hour, the majority of cars were filled with businesspeople going to lunch. In front of me was a man in a black BMW with his windows rolled down. While I watched, he changed his shirt, coat, and hat, and he put on cologne (I could smell it after he sprayed). He also switched radio stations, from a news station to a hip-hop station.

What's going on in these scenarios may seem clear. In each case, the person is working at presenting a specific image that is meant to convey a certain kind of self. I chose the two examples in order to highlight the fact that this kind of impression management occurs in both formal and informal settings. However, Erving Goffman says there's actually more going on than might first meet the eye.

According to Goffman, the self isn't simply one of many possible social objects as symbolic interactionists would have it; *the self is the central organizing feature of all social encounters*. In other words, neither the president nor the man in the car had a choice about presenting a self; there can be no interaction apart from selves. Certainly there are choices about what kinds of selves to present to others, but in every social situation a self must be presented. Further, we'll see that this presentation of self is more complex than we might think. We, like the president or the man in the BMW, may want to give a certain impression, but how we do that and the ramifications of doing it are vast and generally unseen.

Defining Dramaturgy

Dramaturgy is a way of understanding social encounters using the analogy of the dramatic stage. In this perspective, people are seen as performers who are vitally concerned with the presentation of their character (the self) to an audience. There are three major premises to dramaturgy. First, all we can know about a person's self is what the person shows us. The self isn't something that we can literally take out and show people. The self is perceived indirectly through the cues we offer others. Because of this limitation, people are constantly and actively involved in the second premise of dramaturgy: impression management. *Impression management* refers to the manipulation of cues in order to organize and control the impression we give to others. We use staging, fronts, props, and so on to communicate to others our "self" in the situation, and they do the same for us.

Taken alone, impression management sounds at best strategic and at worst deceitful. If everybody around us is manipulating cues in order to present a specific self, then how can we believe that it is their "true self" that we see? The short answer is that we can never be sure we are seeing an authentic self; we always have to *assume* that the self we see is real. But notice something here: *for the interaction, it does not matter if the self we see is genuine or false*—whether authentic or fake, all selves are communicated in the exact same way, through signs that are specifically given or inadvertently given off.

The third premise of dramaturgy is that there are particular features of face-to-face encounters that tend to bring order to interactions. Specifically, the presentation of self places moral imperatives on interactions. Selves are delicate things and are easily discredited. If you have ever felt embarrassed, you know the painful reality of this truth. Selves depend upon not only our skill in presenting and maintaining cues, but also the willingness and support of others. Thus, the simple act of presenting a self creates a cooperative order.

Concepts and Theory: Impression Management

Different Kinds of Selves

One of the biggest issues concerning the self is where it lives: To whom or to what does the self belong? It is clearly linked to the individual, but the self is also just as obviously a social entity. This notion of a social versus an individual self is related to another central issue about the self, the idea of a *core self* versus a *situational self*. Many theorists argue that the self is like a structure that lives inside the individual. This structure is a quality of the individual that consists of a hierarchical scheme of roles developed through childhood socialization, is stable by the late teens/early twenties, and consistently informs everything the individual thinks and does (see Stryker, 1980, for this perspective). We can think of this as the core, transituational self—a self that remains the same across diverse situations. On the other side of the coin is the situated self. Here the self is seen as located more in the interaction than in the individual. Rather than an internalized, hierarchical scheme, the self is an idea that is used to organize the behaviors of individuals in an encounter. The self thus changes and emerges with the flow of interaction. We can think of this as the situated, emergent self.

This situated self is Goffman's (1959) main concern: "the self . . . is not an organic thing that has a specific location, whose fundamental fate is to be born, to mature, and to die: it is a dramatic effect arising diffusely from a scene that is presented" (pp. 252–253). However, it would be incorrect to conclude that he doesn't have a concept of the core self. In *Stigma,* Goffman (1963a) proposes a threefold typology of self identity: the social identity, the personal identity, and the ego identity. Each of these is like a story that is built up through social encounters. The *social identity* is the story that distant others can tell about us; for example, most of my students hold a social identity of me. The social identity is composed of social categories imputed to the individual by the self and others in defined situations. Each category has a complement of attributes felt to be ordinary and natural for members of a particular category. The category and the attributes form anticipations in given social settings. People in an encounter lean on these anticipations, transforming them into normative expectations and *righteously imputed demands.*

That last bit about demands is very important. One of the most interesting things about humans is that they are capable of anything. Our behaviors aren't predetermined by instincts and we have freedom of choice. However, the only way we can function around other humans is by having some way to predict their behaviors. According to Goffman, we use social categories and their accompanying attitudes to accomplish this. We use the category to presume something about how the person works inside. Further, our expectations come to have a moral or righteous feel to them. Again, the reason for this comes back to the unpredictability of humans. Since all we have are cultural expectations, we must make sure that people live up to them. So, when someone doesn't live up to those expectations—like a professor dating a student or not caring about education—we become morally offended.

A *personal identity* is held by people who are close to us. These are the people that have known us the longest, who have interacted with us in multiple situations, and to whom we have cued much of our idea of who we see ourselves to be. Personal identities have more or less abiding characteristics that are a combination of life history events that are unique to the person. The personal identity plays a "structured, routine, standardized role in social organization just because of its one-of-a-kind quality" (Goffman, 1963a, p. 57). What Goffman means is that the more you know about me (the more I present a self to you in different kinds of situations over long periods of time), the less free I am to organize my presentation of self in any old way. I am held accountable to the self image that I have presented.

Thus far the selves we've talked about are within the range of the situational self. It is in the ego identity that Goffman hints at something else. The *ego identity* is "first of all a subjective, reflexive matter that necessarily must be *felt by the individual* whose identity is at issue" (Goffman, 1963a, p. 106, emphasis added). This identity is of the individual's own construction; it is the story we tell ourselves about who we are to which we get emotionally attached. The ego identity is thus "the subjective sense of his own situation and his own continuity and character that an individual comes to obtain as a result of his various social experiences" (Goffman, 1963a, p. 105). Note that according to Goffman, the ego identity is made out of the same materials that others use to construct the personal and social identities that they hold. Thus, the ego identity isn't something that comes from inside us, from innate personality characteristics. Rather, we construct the story through which we see our self using the same cues and categorical expectations that others use.

Performing the Self

The basic concept Goffman uses to explain the presentation of self is that of a front. A **front** is the expression of a particular self or identity that is formed by the individual and read by others. The front is like a building façade. *Merriam-Webster's* (2002) first definition of façade is remarkably like Goffman's idea of a front: a façade is "a face (as a flank or rear facing on a street or court) of a building that is given emphasis by special architectural treatment." Like a façade, a front is constructed by emphasizing and deemphasizing certain sign vehicles. In every interaction, we hold things back, things that aren't appropriate for the situation or that we don't want those in the situation to attribute to our self; and we accentuate other aspects in order to present a particular kind of self with respect to the social role. A social front is constructed using three main elements: the setting, appearance, and manner.

The idea of the **setting** is taken directly from the theater: it consists of all the physical scenery and props that we use to create the stage and background within which we present our performance. The clearest example for us is probably the classroom. The chalkboard, the room layout with all the desks facing the front, the media equipment, and so on are all used by the professor to make claim to the role of teacher. All this is obvious, but notice that the way the setting is used cues different kinds of professorial performances. For instance, I may choose to "decenter" the role of teacher and instead claim the role of facilitator by simply rearranging the desks in a circle and using multiple, mobile chalkboards placed all around the room.

Settings tend to ground roles by making definitions of the situations consistent. For example, it would be more difficult (but not impossible) for me to use the classroom as a setting for the definition of a "sports bar." The "groundedness" of settings is part of what makes us think that roles, identities, and selves are consistent across time and space. There is a taken-for-grantedness about the definition of the situation when we are in a geographic location that becomes more pronounced the more "institutionalized" the location is. When I say that a location is institutionalized, I mean that the use of a specific place appears restricted: the front of the classroom looks as if students are restricted, yet it is available to many professors; the office of the CEO of Microsoft, on the other hand, is even more institutionalized and restricted. However, there is still a great deal of flexibility and creativity available to those that use the space.

There is one specific setting wherein there is little or no flexibility: total institutions. *Total institutions* are organizations that control all of an individual's behaviors, from the time the person gets up until he or she is fast asleep (and even then, behaviors are regulated). Clear examples of these institutional settings include psychiatric hospitals and military boot camps. Goffman (1961) uses the idea of the total institution to demonstrate the clear association between the setting and the self: "This special kind of institutional arrangement does not so much support the self as constitute it" (p. 168).

In addition to the physical setting, a front is produced by using appearance and manner. *Appearance* cues consist of clothing, hair style, makeup, jewelry, cologne, backpacks, attaché cases, piercings, tattoos, and so on—in short, anything that we can place upon our bodies. While appearance refers to those things that we do *to* our bodies, manner refers to what we do *with* our bodies. *Manner* consists of the way we walk, our posture, our voice inflection, how we use our eyes, what we do with our hands, what we do with our arms, our stride, the way we sit, how we physically respond to stimuli, and so on. Both appearance and manner function to signify the performer's social statuses and temporary ritual state. What we mean by social statuses should be fairly clear. Bankers and bikers have different social statuses and they dress differently. They don't dress differently because they have dissimilar tastes; they dress differently because different appearance cues are associated with different status positions.

Ritual states refers to at least two things. The most apparent is the ritual state associated with different life phases. We have fewer of these than do traditional societies, but we still mark some life transitions with rituals, like birthdays (particularly when they signify a change in social standing, like the twenty-first birthday in the United States), graduations, promotions, and retirement. The second idea that the notion of ritual state conveys is our readiness to perform a particular role. Our appearance tells others how serious we are about the role we claim. For example, if you see two people riding bicycles and they are dressed differently, one in normal street clothes and the other in matching nylon/lycra jersey and shorts along with cycling shoes and helmet, then you can surmise that one is really serious about riding and the other is less so.

As in the theater, fronts are prepared backstage and presented on the front stage. Most of what is implied in these concepts is fairly intuitive. For example, every day before you go to school you prepare your student-self in *backstage*. You pick clothes, shower, do your hair, put on makeup, or whatever it is that corresponds to the self that you want others to see and respond to. Your work backstage for school is different from your work backstage when preparing for a date (unless someone you're dating will see your performance at school). You then present the student-self that you've prepared in the *front stages* of class, the lunchroom, the hallway, and so forth. But the backstage of school extends further back than your morning preparations. All students will study, read, and write to a certain degree in preparation for class. This too is part of the backstage for class. Even if you don't read or study, you are preparing to perform as a certain kind of student. As I said, most of this is intuitive. However, we need to realize that there are multiple front- and backstages and that they can occur at almost any time and place.

Let me give you an example. As a professor, I occasionally have students come into my office crying. They may have lost a loved one or simply want to tell me about how this has been the week from hell and they can't get their paper in on time. The thing that strikes me about this experience is that I know the student didn't start out crying at his or her dorm room or classroom and walk across campus crying all the way. The student not only managed his or her impression, but the emotions as well, in a way that was suitable for public display. Then, somewhere between the elevator and my office, the student had a backstage moment in which he or she accessed those emotions and allowed crying to be part of the front. This story about crying implies something important about impression management: part of it involves emotion.

Often it is *performance teams* that move from front- to backstage. Just as in the theater, most performances are carried off by a troupe of actors. There are, of course, such things as one-person shows, but by and large, actors cooperate with one another to present a show to the audience. The same is true in social encounters. We can think of teamwork as being either tacit or contrived. Members of like social categories generally, though not always, assume that others within the category will cooperate in preserving the group face. For example, perhaps you and a friend went to the beach for the weekend. Monday you are both talking to a professor. Your friend, who feels she didn't do well on that morning's test, tells the professor, "I don't think I did well on the test this morning. I was sick all weekend and didn't have a chance to study." You say nothing. You're a team.

Performance teams can also be much more deliberate. Let's say you are married and are having another couple over for dinner. In the middle of dinner, your spouse asks you to help her in the kitchen. In the kitchen, you find out that what she wants your help with isn't the food, but, rather, the team performance. She tells you, in no uncertain terms, that you are not to talk about the fact that she, your wife, is looking for a new job. It turns out that the couple you are having over for dinner is friends with her current boss and she hasn't told her boss that she is leaving. You both go back to the table then, and smoothly change the topic.

Relating to Roles

Every definition of a social situation contains roles that are normal and regularly expected. Goffman considers *roles* as bundles of activities that are effectively laced together into a situated activity system. Some of these role-specific activities we will be pleased to perform and others we perhaps will not. We can thus distance ourselves from a role or we can fully embrace it. *Role distancing* is a way of enacting the role that simultaneously allows the actor to lay claim to the role and to say that she or he is so much more than the role. Let's take the role of student, for example. There are certain kinds of behaviors that are expected of students: They should read the material through several times before class; they should sit in the front row and diligently take notes; they should ask questions in class and consistently make eye contact with the professor; they should systematize and rewrite their notes at least every week; they should make it a point to introduce themselves to the professor and stop by during office hours to go over the material; and so forth. You already know the list, though you probably haven't taken the time to write it out.

But how many of you actually perform the role in its entirety? Why don't you? It isn't a matter of ignorance or ability; you know and can perform everything we've listed and more. So, why don't you? You don't perform the role to its fullest because you want to express to people, mostly your peers, that there is more to you than simply being a student. When presenting the role of student, it is difficult to simultaneously perform another role. Thus, in order to convey to others that you might be more than a student, you leave gaps in the presentation. These gaps leave possibilities and questions in the minds of others: Who else is this person? Sometimes we fill the gaps with hints of other selves (like "athletic female" or "sensitive male") that aren't necessarily part of the definition of the situation.

Contained within the idea of role distance is role embracement. In *role embracement,* we adhere to all that the role demands. We effectively become one with the role; the role becomes our self. We see and judge our self mainly through this role. We tend to embrace a role when we are new to a situation or when we feel ourselves to be institutional representatives (like a parent or teacher). When we do embrace a role like this, we *idealize* the situation and its roles. That is, we "incorporate and exemplify the officially accredited values of the society" (Goffman, 1959, p. 35). When we manage our front in such a way, we place claims upon the audience— first, to recognize the self that we are presenting as one that embodies society, and second, to present our self in such a way as to represent society as well.

Some of the decision to distance ourselves from or embrace a role is personal, but most of it is situational. For example, we expect university students to experiment and try out different things. It's a time between highly institutionalized spaces—the family home and the job world. This kind of time is sometimes referred to as "liminal space" (V. W. Turner, 1969). You are no longer a child fully under the demands of your parents, nor are you working at a job that fully demands your time, effort, and impression management. However, when you do become a full member of the economy, you won't have much time or occasion to experience different situations and the selves they entail. Your daily rounds will be more restricted and managed by others, and you will be expected to more fully embrace the self that work requires. Of course, role distancing is still possible, but we have to work harder at it and it is circumscribed by our situations.

Sacred and Stigmatized Selves

Charles Horton Cooley (1998) was probably the first to recognize that society rests on felt pride and shame. Both are emotions that are distinctly social in origin, and both help monitor human behaviors. Goffman also sees pride and shame at work in social interactions. In every interaction, we present a front, and every time we present a front, we put the self at risk. People read our cues, categorize us, attribute attitudes, and then expect us to live up to the normal traits and behaviors. We are held accountable to the role.

Often we are unaware that we are making moral demands of someone's role performance until they are violated. These demands constitute the individual's virtual self in the situation. Participants compare the virtual self to the actual self, the actual role-related behaviors. The differences between the virtual and actual selves—and there are always differences—create the possibility of *stigma*. The word "stigma" comes from the Greek and originally meant a brand or tattoo. Today, stigma is used to denote a mark of shame or discredit.

There are well-known and apparent stigmas, though if we are going to be politically correct, we don't talk usually about them as such today. Among the ones that Goffman mentions are disabilities and deformities. People with such apparent stigmas are *discredited* by those that Goffman calls "normals." Having someone with an obvious stigma creates tension in encounters. Normals practice careful "disattention" and the discredited use various devices to manage the tension, such as joking or downplaying.

There are also well-known but not so apparent stigmas. People in this category are *discreditable*: they live daily and in every situation with the potential of being stigmatized. One such category that is prominent today in the United States is that of homosexual. Being homosexual in this society is a stigmatized identity—the homosexual is viewed as having failed to live up to the expectations associated with being a sexual person. Many homosexuals manage the information and impression that others have in such a way as to pass as a "normal" (as heterosexual). *Passing* is a concerted and well-organized effort to appear "normal" based on the knowledge of possible discrediting; this impression management entails a level of directed work that isn't usual for normals.

The reason that stigmas can exist and are an issue is that identities are "collective representations" in the true Durkheimian sense of the word (see Durkheim, 1912/1995, pp. 436–440). Identities belong to and represent society's values and beliefs. The representational, symbolic character of identities and the way in which we interact around those identities indicate that identities and selves are sacred objects. Durkheim (1912/1995) defines sacred things as those objects "protected and isolated by prohibitions . . . things set apart and forbidden" (pp. 38, 44). Sacred things, then, represent society, are reserved for special use, and are protected from misuse by clear symbolic boundaries. The sacred self, like all sacred objects, has boundaries that are guarded against encroachment. The sacred quality of identities and selves is the source of shame: "As sacred objects, men are subject to slights and profanation" (Goffman, 1967, p. 31). The flip side of stigma and shame is equally related to this collective feature of identities: a sense of pride in the sacred self.

Goffman uses the idea of face to express the dynamics of the sacred self. *Face* refers to the positive social value that a person claims in an interaction. As we've seen, when we present cues and we lay claim to a social identity, others grant us the identity and attribute to us a host of internal characteristics. Through role distancing we can negotiate some of these attributions, but in order to make an effective claim on the situated self, we must keep our performance within a given set of parameters—otherwise we could not be identified. Every established identity has positive social values attached to it, and over time we can become emotionally involved with those values. As individuals, we experience these emotions as our ego identity, and every time we present an identity, we expose our face—our ego identity—to risk. We risk embarrassment, but the risk is necessary in order to feel pride and a strong sense of self.

Concepts and Theory: The Encounter

Interaction Ritual and Face-Work

This idea of putting our face at risk implies that the encounter or interaction is of a ritual order. Goffman (1967) argues that interactions are ritualized insofar as they represent "a way in which the individual must guard and design the symbolic implications of his acts while in the immediate presence of an object that has a special value to him" (p. 57). Encounters are highly ritualized social interactions, particularly around the issues of self and respect. Goffman refers to the behaviors oriented around respect as deference and demeanor rituals.

Deference is the amount of respect we give others. We may defer to or prefer another's wishes or opinions. Deference also refers to courteous, polite, or formal behavior. All of this deference is granted in different ways. For example, we can know how much or what kind of respect to give others based on known status positions. When introduced to our physician for the first time, most of us call her or him by the title that normally goes along with the position: Doctor. This seems like a function of the status structure—we're just reflecting what the structure says. However, according to Goffman's way of seeing the world, acts of deference are prompted by a person's demeanor.

Demeanor refers most directly to the way someone holds him- or herself, as in manner, but it also entails the entire spectrum of expressive equipment we use to present a front. As such, we can see it isn't so much the structure that tells us to call our physician "Doctor" as it is the way the doctor presents herself. If the physician you visit has her office in a professional building, has the usual waiting room with glass dividers between you and the staff, has examination rooms decorated with equipment and degrees and comes into the room wearing a white uniform and other appearance cues (like a stethoscope), then the entire atmosphere surrounding the situation screams to you to grant respect, not only in title, but in all your behaviors. On the other hand, if the office is in a remodeled home with no glass dividers, and if the exam rooms have pictures and comfortable chairs, and if the physician comes in wearing jeans and a T-shirt, and if she introduces herself as "Samantha Stevens" (rather than "Doctor Stevens"), then you will feel less concerned about exhibiting such ostentatious signs of respect.

Much of Goffman's concern with ritual has to do with *face-work:* actions oriented toward maintaining or modifying face (in this sense, "face" means our dignity or status/prestige). We tend to preserve our face not only because we are emotionally attached to it (we can experience pride or shame), but also because our face is bound up with the face of others in the situation. For example, if I see you once and only once, then I can present any self I desire. However, if we interact on a regular basis, then we tend to become attached to faces and we maintain consistency of behaviors. You tend to rely on my face in order to consistently present your face. Roles and faces do not come individually prepackaged; they come in sets: My being a teacher would be impossible without the role of student. And my face as a good teacher would be equally impossible without your face as a good student. Once we present an identity, we and others build later responses on it so that our faces are intrinsically bound up with one another.

The most basic kind of face-work is *avoidance:* we avoid contacts and requests where threats are likely to occur. For example, a male student might avoid encounters where the two women he is dating might meet, or where he might see the professor to whom he owes a paper. This kind of avoidance is rather obvious, but there are other less obvious avoidance procedures. I might, for example, change the topic of conversation because it is getting too near a place where my face might be threatened. Or, when I make a claim about my self, I can do it with belittling modesty. Let's say you see me walking from my truck carrying a guitar. The guitar cues the identity of musician and creates a certain level of expectations. First you ask to see the guitar and then you ask me to play something. I respond by saying, "Sure, but I haven't practiced for a long time." With such an account, I am simultaneously laying claim to the identity of musician and creating space where I might fail the normal expectations. Such a move not only allows me to avoid embarrassment, it also opens up the possibility of my meeting the normal expectations but being seen as exceptional for doing so. I set the performance expectations low so that when I play average, it will seem like I am playing well.

All of these avoidance procedures create a safe space within the interaction: we are able to make mistakes or not meet the expectations of the identity and still maintain our self, identity, and face. Sometimes, however, things happen that would disqualify us from making positive claims about self. When this occurs, the encounter suffers *ritual disequilibrium* and we normally engage in a corrective process that has four moves: *the challenge*—participants take on the responsibility of pointing out the action; *the offering*—the offender either renders the problem understandable by redefining the action or encounter ("I thought we were teasing.") or offers compensations or penance; *the acceptance*—the interactants believe the offering; *the appreciation*—the offender expresses gratitude and the interaction is repaired.

All of this necessary face-work implies the possibility of *aggressive face-work*. In aggressive face-work, we depend on others' reactions by intentionally introducing a threat to equilibrium: knowing the rules of ritualized interaction, we can manipulate the encounter to our own benefit. If we know that others will respond to self-effacing comments by praising us, then we can "fish for compliments." If we know that they will accept our account or apology for an offense, then we can safely offend them. And if we are particularly good at face-work, we can arrange for others to offend us so that they will be emotionally indebted to us.

Focused and Unfocused Encounters

All of what I've been describing is part of what Goffman calls focused encounters. *Focused encounters* occur when two or more individuals extend to one another "a special communication license and sustain a special type of mutual activity that can exclude others who are present in the situation" (Goffman, 1963b, p. 83). The last part of the definition is a clue to understanding what Goffman is talking about. Focused encounters can obviously happen when the group engaged in mutual activity is alone, but its particular qualities stand out when we consider it around others. All focused encounters have a membrane that includes some people and excludes others. It's an invisible line that marks the gathering as an encounter of social beings belonging to just those people.

The most defining feature of a focused encounter is *face engagement*. We engage one another's faces through a single visual and cognitive focus of attention, a sense of mutual relevance in our actions, and by granting preferential communication rights. This is easy to illustrate. Picture yourself walking down the hallway at school. You see someone you know and speak a ritualized opening—"Hi, Tom." Once Tom responds, you visually and cognitively focus on one another to the exclusion of all others in the hall. You see Tom's subsequent behaviors as mutually relevant to yours in a way those of the others in the hall are not. And you grant communication rights to Tom that you don't give to those surrounding you. Out of this comes an emergent "we" feeling (versus everybody else in the hall). The invisible wall surrounding you and Tom is apparent in the use of ritualized openings, closings, entrances, exits, transformations, and so on.

One of the main values of Goffman's conceptualization of focused encounters is that he can call our attention to how we manage *unfocused encounters*. We encounter people all the time, but often we are required to keep the encounter unfocused. Goffman shows us that when we are walking down the hall at school, or through the mall, or in any other public place, we are working hard to maintain the unfocused nature of our encounters. While it is our job in focused interactions to call attention to the self we are presenting, there are norms in unfocused encounters that prohibit bringing attention to our self. If for some reason we do, we must repair the encounter. For example, when a person is walking in public and stumbles, it is common practice for her or him to look back at the walkway where the trip occurred. Whatever this action does for us personally, socially it conveys to everyone around us that we are in control of our actions and we simply were tripped by some object in our path.

Goffman Summary

- Goffman understands interactions through the analogy of the stage. Dramaturgy assumes that all we can know about a person's self is what we can pick up by reading cues. Individuals manipulate cues through impression management, in order to claim a certain kind of self in the interaction. If people use dramaturgy, then the encounter itself may be seen as an activity system, like a particular kind of stage or background that places its own demands upon the performers. Goffman's entire analysis, then, is focused on the presentation of self that organizes the interaction order.

- One way of understanding impression management is to see it through the notion of social, personal, and ego identities. Social identities are biographies held by distant others, personal identities are stories that intimate others hold, and the ego identity is the biography that the individual holds about the self. Each of these stories, even the ego identity, is created out of the way the individual manages a dramatic front using the expressive equipment of appearance, manner, and setting. The self must be seen to be known, and it is only seen through impression management cues. Typically, this impression management is prepared backstage and presented to an audience on the front stage; and while there are undoubtedly soliloquies in life as on stage, most of the presentations of self are managed by teams or troupes of actors.

- In an interaction, participants depend upon cues to attribute an identity and its attendant attitudes to the individual. The identity and its attitudinal and behavioral expectations form righteously imputed expectations. Sensing this, most people are careful in the way they manage the impression they give others. This work can be seen to vary on a continuum from role distance to role embracement. In role distance, one manages impression in such a way as to simultaneously lay effective claim to the role, its virtual self, and a yet unseen self. The purpose of such work is to claim a self that is more than the role communicates. In role embracement, the individual disappears within the virtual self. Such work idealizes the situation and its roles.

- The longer we perform a particular role or the closer we come to role embracement, the greater is the possibility of embarrassment. Goffman refers to this emotional attachment to roles as face. Every interaction represents a risk to self: we can either lose or maintain face. As such, most interactions are ritualized around face-work. Most face-work is performed through avoidance procedures. For our self, we avoid settings and topics that represent threat, we initially present a front of diffidence and composure, and we make claims about self with belittling modesty. For others, we do such things as leaving unstated "facts" that may discredit them, deliberately turning a blind eye to behaviors that might discredit them, providing accounts for them when needed, and when making "belittling demands" we may use a joking manner.

- Face is closely linked to the ideas of ego identity and sacred self. The idea of the sacred self points out the facts that the self is a social entity and the sacredness of the self as well as our emotional attachment to the self are produced through ritualized interactions. We experience a sacred self with a sense of emotional attachment to an ego identity as a result of the nature and structure of our interactions. We experience a sense of a core, unchanging self because (a) identities and selves are grounded in geographic settings; (b) we regularly frequent a daily, weekly, or monthly round of settings; and (c) focused interactions within those settings are ritually oriented toward maintaining face.

Harold Garfinkel and Ethnomethodology

Photo: © Reprinted from www.workpractice.com. Photographer: Andrew Clement.

The Essential Garfinkel

Biography

Harold Garfinkel was born on October 29, 1917, in Newark, New Jersey. He grew up during the Depression and was discouraged from attending the university. While attending business classes at a local school, Garfinkel was exposed to the idea of "accounting practices," which he later saw as the primary feature of interaction. He also met and befriended a group of sociology students. Those experiences, along with a summer spent at a work camp building a dam, prompted Garfinkel to hitchhike to the University of North Carolina, Chapel Hill (UNCCH), where he was admitted to graduate school. He completed his master's at UNCCH, after which he was drafted into the army during WWII. After the war, Garfinkel went to Harvard to study for his Ph.D. under Talcott Parsons. In 1954, Garfinkel joined the faculty at the University of California, Los Angeles, where he stayed until his retirement in 1987.

Passionate Curiosity

Social order is an age-old question for sociologists, but Garfinkel's gaze penetrates beneath the usual sociological answer. For one, he isn't so much interested in whether or not social order is actually present; he's interested in how we achieve a sense that there is order and reality. He is also disturbed by how quickly sociologists explain order by referring to outside forces, such as social structures. Garfinkel wants to know how a sense of social order and meaning are produced in just this way and at just this time.

Keys to Knowing

Phenomenology, accounting/accountable, seen but unnoticed, reflexivity, indexical expressions, incorrigible assumptions, secondary elaborations of belief, documentary method, ad hocing

Seeing Further: Achieving Social Order

My initial exposure to Garfinkel's work came in my first semester of graduate school. I was a teaching assistant in an introduction to sociology class. The teacher was a visiting professor by the name of Eric Livingston. As is customary, on the first day of class the professor explained what the course was about. He told us that we were taking an ethnomethodological approach to sociology. At the time, nobody knew what that meant, but he did say one thing that struck us all as odd: We had a textbook and we were going to read it, but not in the usual fashion.

As you know, we usually read textbooks to understand the topic. We approach the text as an authoritative source: math books teach us math, history textbooks tell us the history of a people, and sociology texts teach us about society. It sounds pretty straightforward, right? Well, not in this case. Dr. Livingston told us that we were going to use the textbook as an example of how sociology organizes itself to be sociology. We were to read the book *not* to learn about society; rather, we were to study the text to see how members of the sociology field render situations knowable as sociology.

My second exposure to ethnomethodology came at a professional conference. Harold Garfinkel was scheduled to lead a paper session. Paper sessions at professional conferences usually go like this: Someone organizes the session and gathers a group of four to six researchers to present reports of their current studies. After the researchers present their material, a discussant comments on the papers and leads a question-and-answer session. For this panel, Garfinkel was both organizer and discussant—a fairly standard approach. Pretty straightforward, right? Not in this case. Everything went as usual until Garfinkel led the discussion. Rather than discussing the papers, Garfinkel analyzed what had happened in the session—the way the papers were presented and our responses—as an instance of how a paper session is organized *in just this way* so as to give the sense of being a professional paper session.

Defining Ethnomethodology

Ethnomethodology is a theoretical perspective in sociology that is based on phenomenology and began with the work of Harold Garfinkel. I briefly mentioned phenomenology in Chapter 2 while we were talking about Max Weber. I'm going to go into a bit more detail here because understanding phenomenology will go a long way in helping us grasp ethnomethodology. We will come across phenomenology again later in the book, so tuck this information away for later.

Phenomenology originated with Edmund Husserl, a German philosopher, who formed the phenomenological approach in response to both empiricism and rationalism. *Empiricism* is the philosophy that all knowledge comes from and is tested by sense data gathered from the physical world. *Rationalism,* on the other hand, posits that reason and logic are the tools through which true knowledge is attained. In this perspective, reason can lay hold of truths that lie beyond the grasp of sense perception.

Phenomenology cuts a middle road between the two by focusing on human consciousness. Consciousness is more than sense data. Husserl argues that the only things that can exist for humans exist in consciousness. That is, only those things of which we are intentionally aware can exist for us. This isn't quite as esoteric as it sounds. Think of it this way: The world of the frog is very different from the human world. Part of the difference is based on sense perceptions (like the fact that the frog can only see moving objects), but more importantly, much of the disparity concerns the human brain and its process of awareness—we can be intentionally aware in a way the frog cannot.

I don't want us to get caught up in philosophical discussions, but notice what Husserl is attempting to do: He is trying to reduce all phenomena to their most basic elements. For Husserl, those elements were to be found in human consciousness and intentionality. Social phenomenology, on the other hand, assumes that most if not all human experience of phenomena is essentially social and cultural. Social phenomenologists, then, attempt to describe the life-world as it presents itself to people in everyday life, apart from any formal theory or discipline.

Thus, phenomenology is concerned with discovering the contours of an incident or object in its purest state: How does a social phenomenon presents itself to us apart from any scientific or philosophical interpretation (Luckmann, 1973, p. 183)? This approach is distinct from most sociology in that phenomenologists reduce phenomena to the simplest terms possible. For example, a structural sociologist will study racial inequality and its effects. Race is simply a given in this kind of approach. A phenomenological approach will take race itself as the most basic problem to explain: How does "race" present itself to us in just such a way as to appear real, taken-for-granted, and intersubjective?

I want you to notice something very important here. Look at how that last question is phrased. It doesn't say, "How do individuals see race in social situations?" In many ways, that kind of question comes from a symbolic interactionist approach to this issue. The way I phrased the question asks about race itself: How does race present itself to us? Now, a phenomenologist wouldn't say that race is biological or genetic; that isn't the point here. Phenomenologists accept that race is a social construction. The issue isn't what race means or how the meaning emerges through symbolic interaction. The issue is race itself.

This is a really difficult distinction to make, but it's important. Let me use the black guitar example I gave during our discussion of symbolic interactionism. There the issue was the meaning of black guitars: What does it mean if someone only plays black guitars and how is that meaning brought to the attention of other people? Note that both the guitar player and others are consciously aware that meaning is at stake. The guitar player intentionally buys, displays, and plays only black guitars in order to convey certain meanings and images. But what if we aren't interested in the meaning of the black guitar? What if we are interested in the guitar itself?

This seems like a normal interest, doesn't it? We can easily ask, how was the guitar built? We could go even further and ask, what constitutes a guitar? In other words, what makes a guitar a guitar rather than a cello? And these kinds of questions are relatively easy to answer, too. You could consult books or you could go on a tour of the Gibson factory and see how guitars are built.

Phenomenology asks those kinds of questions, but they seem awkward to us because we don't usually think of how social situations are built or what makes one situation a wedding and another a university class. Yet these are the kinds of questions ethnomethodologists ask by using a phenomenological approach. Now, think about our example of race. An ethnomethodologist would ask about race itself: How is race constructed? What is it about race that makes it race rather than gender? Another reason why these questions seem awkward to us is that we can't go to the race factory to see how it is built. Well, that's not quite true. We can't *go to* the race factory because the race factory—the entity that constructs race—is right here in the situation. The sense we have that race is a real and ordered thing is an achievement of the people in the situation itself.

Ethnomethodology, then, is the study of the everyday methods that ordinary people use to make sense of, organize, and constitute the situations in which they find themselves. The central question of the perspective is that of social order. Some sociological perspectives explain social order by referring to cultural and social structures that impinge on the situation. Ethnomethodology, on the other hand, explains social order as a local achievement. As people in the situation act in a way that gives to themselves and observers the impression of such-and-such a social order, they in fact produce such an organization. In other words, the main idea of ethnomethodology is that the methods people use to organize a situation are identical to the methods used to render the situation accountable.

Concepts and Theory: Everyday Methods

Everyday Social Order as an Accomplishment

Garfinkel's theoretical concern is the same as that of Parsons—social order. But rather than looking at big issues like norms and structures, as Parsons would, Garfinkel calls our attention to such small things as saying "you know." The taken-for-grantedness of these behaviors is both the power and problem of the ethnomethodological account. It's powerful because it lets us see how we create society through ordinary actions; getting us to see these powerful insights is a problem because we do these things without thinking. Ethnomethodology is not like the sociology you learned in your intro textbook. It might help us, then, to approach ethnomethodology the way Garfinkel does, by comparing it to standard sociology. To do so, Garfinkel (1967) makes three points:

1. [E]very reference to the "real world," even where the reference is to physical or biological events, is a reference to the organized activities of everyday life . . .

2. in contrast to certain versions of Durkheim that teach that the objective reality of social facts is sociology's fundamental principle . . .

3. the objective reality of social facts as an ongoing accomplishment of the concerted activities of daily life, with the ordinary, artful ways of that accomplishment being by members known, used, and taken for granted, is, for members doing sociology, a fundamental phenomenon. (p. vii)

Let's start with the second point. Durkheim argues that the things we refer to as social structures and institutions are social facts—they have a facticity about them that is comparable to empirical facts. Durkheim (1895/1938) also argues that the "determining cause" of any social fact is other social facts, not individual people (p. 110). This kind of argument is typical of both a structuralist approach and sociology in general. Garfinkel's point is that when most sociologists come into a situation, they search for variables *outside* the immediate situation to explain what's happening *inside* it; they look for the social facts that strongly influence the observed phenomenon.

Consider the following quote from John J. Macionis's (2005) introduction to sociology textbook (one of the best-selling introductory texts ever):

Why do industrial societies keep castelike qualities (such as letting wealth pass from generation to generation) rather than become complete meritocracies? The reason is that a pure meritocracy diminishes the importance of families and other social groupings. (p. 251)

Turn a few pages in Macionis's book and you'll find the heading, "Who Are the Poor?" That heading is followed by a series of subheadings: Age, Race, and Ethnicity; Gender and Family Patterns; and Urban and Rural Poverty. Macionis is doing two things here: he is offering an explanation of poverty in industrialized societies, and he is listing the standard variables that describe poverty. Notice that in good Durkheimian sociological fashion, the cause of structural inequality is explained in terms of other social facts or structures: in Macionis's statement above, structural inequality persists because of the family structure. Also notice that the structure is differentiated by variables that are themselves seen as social facts: the structure of inequality varies by age, race, ethnicity, gender, family, and urban versus rural settings.

Garfinkel wouldn't challenge whether family influences the continuation of structured inequality, nor would he question whether or not inequality varies by age, race, gender, and so on. And he wouldn't provide a competing theory to explain structural inequality. Garfinkel leaves the question of the theoretical explanation untouched—because to ask such questions or to provide different theoretical explanations of the same thing requires an ethnomethodologist to take the perspective of a structural sociologist. In other words, *you can only give structural explanations in response to questions posed from a structuralist point of view.* In this sense, the institutional order is sociology's achievement, without question, and ethnomethodology can't claim to know better. (See Garfinkel, 1996, p. 6n.)

Thus, ethnomethodology can't say anything about Durkheimian social structures. The claim that ethnomethodology does want to make is found in Garfinkel's third statement: ethnomethodology sees social facts as an accomplishment of people within social situations. In this case, ethnomethodology is interested in discovering the methods through which sociologists make what they do appear as sociology. Let's use the Macionis example again. What does the textbook actually do? According to Garfinkel, it doesn't teach us about society; rather, it is an example of how sociologists make what they do appear as sociology.

Garfinkel's focus is more profound than it might first appear. In adhering to the "rules of sociological explanation," what do sociologists do? They simultaneously create an explanation that appears to be a sociological one and they create sociology itself. This issue is why Eric Livingston (whom I later found out is one of only a handful of people that have ever published with Harold Garfinkel) asked the intro class to take the textbook as an example of how members of the sociology field render situations knowable as sociology. But, you might say, doesn't sociology discover and explain what is really happening in society? Maybe it does, but how would you know? How can you tell if sociology is actually explaining the real world? What proof is given that sociology's explanation is correct? How is that proof produced and who produces it? The only people who ever try to substantiate sociology are sociologists using sociological methodologies. Sounds odd, doesn't it? It might, but it is a powerful insight that we will return to in a few moments.

Garfinkel's interest, then, is in the everyday procedures or methods people use to account for their behaviors. In fact, the term ethnomethodology means the study of folk methods. Garfinkel (1974) began using the term as a result of a study he did on jury deliberations. He noticed that there were distinct methods used to render the conversations, deliberations, decisions, and judgments "jury-like," rather than sounding like the mundane opinions of the person on the street. In the process of "becoming a juror," these people drew on multiple sources for information, but the people themselves did not change much. In fact, according to Garfinkel (1967), "a person is 95 percent juror before he comes near the court" (p. 110). The process of becoming a juror didn't change the people that Garfinkel observed. They basically acquired knowledge and made decisions in the same way they always did. What did change was the way people accounted for their decisions as jurors.

This kind of *accounting* takes place constantly. According to Merriam-Webster (2002), accounting means "to think of as; look upon as: rate, regard, or classify as." To account decisions (like the jurors) or other behaviors as something means to think of it or regard it as a specific kind of thing. As an example, let's go back to the first part of the Garfinkel quote in his three points I mentioned earlier: "every reference to the 'real world,' even where the reference is to physical or biological events, is a reference to the organized activities of everyday life." According to Garfinkel, when we refer to anything in "the real world," what we are actually referencing are the activates or methods that are used to make whatever it is appear real or ordered in just such a way. The emphasis in most ethnomethodology is on *every* reference. Ethnomethodologists will study anything—sometimes it's as formal and "important" as science; other times it's simply how a line or queue is formed.

Seen but Unnoticed

There is something unusual about the common or mundane situations that ethnomethodologists study: they are all moral affairs in a knowable and observable sense. One of the things that allow us to successfully relate to one another is an assumed *reciprocity of perspectives*. It seems true that humans can't actually experience the world from another's point of view. We all "know" this: just ask anybody what happens at a party when you start a secret story with one person and tell her or him to pass it on; or, as another example, "everybody knows" that if five people witness an accident, there will be five different retellings of it.

Yet, even though we all know this, we don't act like it. In fact, in every situation we act as if everybody knows exactly (or close enough) what we are talking about and doing. In order to successfully accomplish an interaction, we have to assume that our perspectives are reciprocal, despite whatever evidence there might be to the contrary. We assume that our standpoints are interchangeable with those of others. We assume that the world that has meaning for me also has the same meaning for you; the only difference is our particular relationship to the world. Thus, while there may be differences, I believe that you would interpret the world as I do if you stood in my place and had my experiences. We also assume that whatever differences may exist in our positions in the life-world are irrelevant for the purposes at hand.

In order to point out this assumption, Garfinkel (1967) had his students perform *breaching demonstrations* in order to discover "the socially standardized and standardizing, 'seen but unnoticed,' expected, background features of everyday scenes" (p. 36). In other words, Garfinkel wanted his students to notice the always seen but never noticed scaffolding around which daily life is attained. Let me quote a couple of the cases so we can get a sense of what Garfinkel (1967) is talking about (In the following dialogue, S = Subject; E = Experimenter.):

CASE 2

(S) Hi, Ray. How is your girlfriend feeling?

(E) What do you mean, "How is your girlfriend feeling?" Do you mean physical or mental?

(S) I mean how is she feeling? What's the matter with you? (He looked peeved.)

(E) Nothing. Just explain a little clearer what do you mean?

(S) Skip it. How are your Med School applications coming?

(E) What do you mean, "How are they?"

(S) You know what I mean.

(E) I really don't.

(S) What's the matter with you? Are you sick?

CASE 3

"On Friday night, my husband and I were watching television. My husband remarked that he was tired. I asked, 'How are you tired? Physically, mentally, or just bored?'"

(S) I don't know, I guess physically, mainly.

(E) You mean that your muscles ache or your bones?

(S) I guess so. Don't be so technical.

(After more watching)

(S) All these old movies have the same kind of old iron bedstead in them.

(E) What do you mean? Do you mean all old movies, or some of them, or just the ones you have seen?

(S) What's the matter with you? You know what I mean.

(E) I wish you would be more specific.

(S) You know what I mean! Drop dead!

CASE 4

During a conversation (with E's female fiancee [sic.]) the E questioned the meaning of various words used by the subject. . . .

> For the first minute and a half the subject responded to the questions as if they were legitimate inquiries. Then she responded with "Why are you asking me those questions?" and repeated this two or three times after each question. She became nervous and jittery, her face and hand movements . . . uncontrolled. She appeared bewildered and complained that I was making her nervous and demanded that I "Stop it." . . . The subject

picked up a magazine and covered her face. She put down the magazine and pretended to be engrossed. When asked why she was looking at the magazine she closed her mouth and refused any further remarks. (pp. 42–43)*

There are a few things that we can pick up from such tests. First, there is a great deal of "seen but unnoticed" work that goes on in organizing a setting. Second, many conversations are organized around the denial of strict rational discourse. In a rational discussion, asking for clarification would be permitted and expected. When I was buying my truck, for example, the salesperson and manager expected and cooperated with my questions that were designed to extract specific information. On the other hand, conversations organized around what Georg Simmel (1971) called "sociability," the kind of conversations that we mostly have, are specifically *not* organized around the sharing of exact information.

In sociability, we must not push for additional information, we must "wait for clarification" (that may never come), we must suspend any doubt that might come to mind as the conversation takes shape, and we must understand all statements as being of the indexical kind (referencing unseen worlds or understandings that may never materialize)—all done to give us the sense that what we are having is a "normal conversation." When we see the work that goes into making a dialogue appear to everybody as a simple conversation, we can appreciate why Garfinkel sees everything we do as an "achievement." Taken-for-granted, everyday conversations don't simply happen; they are achieved.

The third thing we can glean from these stories is that there are "sanctioned properties of common discourse" (Garfinkel, 1967, p. 41). As you probably know from your other sociology courses, sanctions are positive or negative behavioral reinforcements. So, we can see that there are some pretty strong norms at work in mundane conversations and surmise that there is a moral order. However, from the point of view of ethnomethodology, the moral basis isn't part of some cultural structure that exists outside the situation. No, in fact, we can *see* that there are strong norms at work. What do we see? We see sanctioning behavior. This is a very important point for understanding ethnomethodology. Such things as sanctions, norms, values, and beliefs are not preexisting social facts that tell us how to behave; "they are rather constitutive of the 'sense' of the circumstances, of 'what the circumstances are' in the first place" (Heritage, 1984, p. 98).

Concepts and Theory: Doing Society

Reflexivity and Indexicality

The notion of *reflexivity* is at the heart of Garfinkel's work and ethnomethodology in general. Something is reflexive if it can turn back on itself. For example, circles are created through reflexive movement: The beginning and end points are the same, and the line that connects them constitutes everything that is contained within the circle. The word "reflexive" is also sometimes used to describe the introspective action of the mind, as when we talk to ourselves.

*Reprinted with permission of Polity Press.

Reflexivity is important for Garfinkel because all social situations are reflexively ordered. In his words, ethnomethodology's

> central recommendation is that the activities whereby members produce and manage settings of organized everyday affairs are identical with members' procedures for making those settings "account-able." The "reflexive," or "incarnate" character of accounting practices and accounts makes up the crux of that recommendation. (Garfinkel, 1967, p. 1)

One of the primary ways in which scenes are reflexively organized is through indexical expressions. To index something is to make reference to it or to point to it. Think of your index finger: it's the finger you use to point with. This book has an index. In this case, the index points to all the important issues that may be found in the book. This last example is very important. An indexical expression is like an index entry in a book: it points to itself. The only way a book index makes any sense at all is within the context of the book. Sometime try using the index from one book to find important items in a different book; it won't work. **Indexical expressions,** then, are situated verbal utterances that point to and are understood within the situation.

This is a very different notion from what is commonly held. Most of us, including many social scientists, believe that language, including verbal language, is representative. If I say "tree," then I am using that word to point to the physical object. One of the difficulties associated with this idea is that language doesn't represent very well. This problem is exemplified by color (Heritage, 1984, pp. 144–145): The human eye can distinguish about 7,500,000 colors. Yet the language with the most color names (English) only has 4,500 words that denote color, and out of those, just 8 are commonly used. It's plain, then, that in our everyday language we are not very concerned with representation.

Part of what Garfinkel is talking about with indexical expressions is similar to social interactionism (SI), except what Garfinkel notes isn't the emergent quality of meaning; it is the *reflexive character* of meaning. For Garfinkel, the meanings of such phrases as "how's it going?" "that's a nice one," "he's a novice," and "what's up?" aren't negotiated through interaction. The meanings of indexical expressions don't emerge; they are found in the context itself. In fact, as we saw in the examples of breaching demonstrations, if you try to explicitly negotiate the meanings of indexical expressions, the chances are good that you'll be sanctioned or the setting will break up. The "one among many" in "that's a nice one" is assumed to be contextually given, as is the meaning of "nice." For example, if I am showing you the new guitar I just bought and you say, "that's a nice one," we both assume the meaning is given in the context or situation. Thus, indexical expressions are reflexive because they appear in and reference the unique context in which they occur.

However, indexical expressions are reflexive for another reason. They not only appear in and reference the situation; they also bring the situation into existence. Mehan and Wood (1975) give us the example of "hello." Let's say you see me in the hall at school. You say, "Hello." What have you done? You have initiated or created a social situation through the use of a greeting. When you said "hello," you immediately drew a circle around the two of us, identifying us as a social group distinct from the other people around us. That social situation, which we can call an encounter, interaction, or situated activity system, didn't exist until you said "hello." But notice something very important about "hello": it can only exist as a social greeting within social situations. Every time "hello" is used, a social situation is created. Yet "hello" is only found in social situations, either real ones or imaginary ones (like with our example). Thus, "hello" is utterly reflexive: it simultaneously creates, exists, and finds meaning within the social situation.

However, indexical expressions aren't limited to these sorts of catch phrases. At one point, Garfinkel asked his students to go home and record a conversation. They were to also report on the complete meaning of what was said. The following is a small snippet of one such report (Garfinkel, 1967, pp. 25–26):

	What was said:	*What was meant:*
HUSBAND:	Dana succeeded in putting a penny in a parking meter today without being picked up.	This afternoon as I was bringing Dana, our four-year-old son, home from the nursery school, he succeeded in reaching high enough to put a penny in a parking meter when we parked in a meter parking zone, whereas before he has always had to be picked up to reach that high.
WIFE:	Did you take him to the record store?	Since he put a penny in a meter that means that you stopped while he was with you. I know that you stopped at the record store either on the way to get him or on the way back. Was it on the way back, so that he was with you or did you stop there on the way to get him and somewhere else on the way back?

The first thing to point out, of course, is that what was actually said is incomprehensible apart from what each member could assume the other knew. There is an entire world of experience that the husband and wife share in the first statement about Dana that gives the statement a meaning that any observer would not be able to access. So, the first point is apparent: vocal utterances reference or index presumed shared worlds.

The second point may not be quite so obvious. Garfinkel's students had a difficult time filling out the right-hand column of their report. It was hard to put down in print what was actually being said and indexically understood. However, it became a whole lot tougher when Garfinkel asked them to *explain* what was said. Garfinkel wanted them to explain the explanation because the explanation itself assumed indexical worlds of meaning. Garfinkel (1967) reports that "they gave up with the complaint that the task was impossible" (p. 26). The task of explaining every explanation is impossible because all our talk is indexical. Many of us who are parents come up against this issue in the course of raising a two-year-old. All two-year-olds are infamous for asking the same insistent question: "Why?" And every parent knows that once started, that line of questioning never ends—every answer is just another reason to ask why. It never ends because our culture is indexical and reflexive.

Mehan and Wood (1975) further point out just how fundamentally reflexive our world really is. Every social world is founded upon incorrigible assumptions and secondary elaborations of belief. **Incorrigible assumptions** are things that we believe to be true but never question. They are incorrigible because they are incapable of being changed or amended. And these assumptions form the base of our social world. **Secondary elaborations of belief** are prescribed legitimating accounts that function to protect the incorrigible assumptions. In other words, they are ready-made stories that we use to explain why some empirical finding doesn't line up with our incorrigible assumptions.

In Figure 9.1, I've pictured two models of sociology. These models obviously are not theoretical or causal but are simply pictures of assumptions and activities. The first model is what most sociologists see themselves as doing. Most sociologists assume that there is an empirical, social world that sociological methods can be used to study. These methods produce a particular kind of inquiry known as sociology. This kind of inquiry leads to insights into and discoveries about the social world. However, the idea of reflexivity makes us look at things differently. First, notice that there isn't a social world preceding sociological methods or inquiry. The methods themselves produce sociological inquiry and the social world of sociology. In other words, the world that sociologists see is produced by their own methodologies. In adhering to those methods, sociologists produce sociology, which may or may not have anything to do with any other social world.

I want us to take one further step into this issue of reflexivity. Not only are cultures reflexively created and protected, evidence for any reality system is always

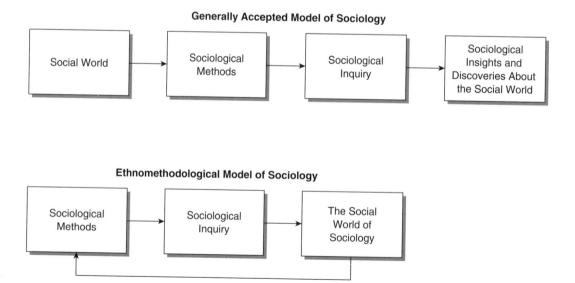

Figure 9.1   Garfinkel's Concept of Reflexivity

reflexively provided as well. Let us take the incident of two automobiles colliding. What would you call it? Most of us would call it an "accident." But what is implied in calling this collision an accident? Accidents can only exist if humans assume that there are no other, outside forces in back of events. But what if one of the drivers is a fundamentalist Christian? Then the episode, from the point of view of that driver, may well be defined as "God's will."

Let's ask the obvious question first: what is the incident really? According to ethnomethodology, it isn't anything *really*; it *becomes something meaningfully* as we make assumptions about the world and how it works. But once we make our assumptions, what can happen to the events around us? Think about the two cars colliding. In this example, the event becomes either an accident or God's will. Then, through a neat little trick, the event becomes proof of the system that defined it in the first place. The person who has had an "accident" is confirmed in her belief that shit happens. The person who has experienced "God's will" is confirmed in her belief in an omnipotent and merciful God. Either way, the collision is used to legit-imate an existing reality system—proof of the event's definition is provided by the self-same meaning system. The same is true for science (and sociology): what counts as "proof" for the validity of science is defined by science.

Accounting

The notion of accounts is central to Garfinkel's project and is explicitly tied to the issue of social order. Rather than seeing social order as something that is produced by a collective culture (Durkheim), institutionalized social systems (Parsons), or indi-vidually motivated exchanges (Chapter 10), Garfinkel sees social order as the result of members making settings "account-able." Garfinkel is using the term "account" in the sense of to regard or classify, such as "she was accounted to be a powerful senator." To make something **account-able,** then, is to make it capable of being regarded or clas-sified as a certain kind of object or event. As we render a situation accountable, we simultaneously produce social order and reality. Take a moment and let this idea soak in: social order and reality are in the accounting.

For example, a few years ago I was visiting my sister in San Diego and we went to Balboa Park, a gorgeous recreational area with museums, fountains, restaurants, street musicians, art exhibits, and so forth. While we were walking through some of the gar-dens, my sister said, "Oh, look, a wedding." How could my sister recognize what was happening as a wedding? That sounds like a simple and maybe silly question, but the implications are important for Garfinkel. We were all able to recognize the event before us as a wedding because the people who organized the setting did it in such a way that it would *appear* as a wedding, not only to others but specifically to themselves.

When we as a group organize ourselves to do something, whether it is forming a queue or waging war, there are "requirements of recognizability" (Rawls, 2003, p. 129) that must be met. In meeting those requirements, the situation is rendered accountable as a recognizable social achievement. This work of accounting is the primary job of the members. For example, if we had gone up to the people at the wedding at any time during the event (while preparing, setting up, performing, or celebrating) and asked them, "What are you doing?" their response would be something like, "We're having a wedding." This is Garfinkel's point: the members of any situation are cognitively aware of what they are doing—they are knowingly organizing their actions in just such a way as to create a sense of social order (a wedding). Because members are knowingly producing social order within a scene, and because answerability is the simplest explanation, it follows that *accounting is the primary force behind social organization.* (Please notice that we also used a wedding as an example of Blumer's idea of joint actions. In order to tease out some of the differences between SI and ethnomethodology, compare and contrast what Blumer's take is with Garfinkel's.)

Garfinkel (1967, pp. 18–24) tells a story of a research project he worked on at the UCLA Outpatient Clinic. The research was to determine the criteria by which applicants were selected for treatment. Two graduate students examined 1,582 clinic files. As is usually the case, the student coders were provided with a coding sheet and instructions for its use. And, as is usually the case, the findings were subjected to inter-coder reliability tests, which are used to determine the extent to which the coders agree with one another. It's generally thought that the higher the statistic, the greater the reliability. In other words, if I'm doing a study of television commercials and I have five different coders working from the same coding sheet and their inter-coder reliability is 85%, then I can be fairly certain that what they are coding actually exists in the commercials.

However, Garfinkel found that in order to code the contents of the folders, the coders actually assumed knowledge of the way in which the clinic was organized. This assumed knowledge base "was most deliberately consulted whenever, for whatever reasons, the coders needed to be satisfied that they had coded 'what really happened'" (Garfinkel, 1967, p. 20). Notice something important here: The coders were to find out how the clinic was organized, yet in order to fill out the coding sheet, the coders assumed knowledge about the way the clinic was organized. Thus, the coders' reliability rate wasn't due to their reliable use of the coding sheet to document what happened in the clinic; the reliability rate was due to something the coders themselves were doing, apart from the coding sheet or the folders.

Most researchers in Garfinkel's position would regard such issues as problems with the measurement instrument and as threats to the research. Garfinkel (1967) likens these responses to "complaining that if the walls of a building were only gotten out of the way one could see better what was keeping the roof on" (p. 22). Garfinkel is saying that most social scientists miss the boat: they don't see what's really going on because they are preoccupied in producing "sociology" or "psychology" rather than seeing the social world as it is. For Garfinkel, the graduate students' task as they saw it was to "follow the coding instructions." What the graduate students produced, then, was just that: a setting or scene that could be understood and accountable as "following the coding instructions." Garfinkel's (1967) ethnomethodological question in this case became, "What actual activities made up those coders' practices called 'following coding instruction'?" (p. 20).

This change in Garfinkel's question implies that we must attend to the practical, planned actions right here, right now, in just this way. Members continually demonstrate their accountability to the social scene. Their practical actions are intended to be seen and reported. When Garfinkel went back to the coders, he looked for precisely how, in just this way, at just this time, the coders' practices simultaneously (reflexively) produced and made accountable the action of "following the coding instructions."

Documentary Method

How do we make sense of or attribute meaning to an object or situation? Most sociological explanations explicitly or implicitly use the notion of common culture: being raised in the same society, we all share a common culture that we use to understand and create meaning. We've seen that SI argues it's the result of negotiated interactions. But Garfinkel would say that such approaches "gloss" over what is actually going on.

To understand the practices through which we claim to be competent interpreters, Garfinkel (1967) uses the idea of documentary method: "The method consists of treating an actual appearance as 'the document of,' as 'pointing to,' as 'standing on behalf of' a presupposed underlying pattern" (p. 78). Anytime we interpret something, we make a kind of identity statement: "this is that." We do this when we interpret conversations or when we recognize the person standing outside our door as the mail carrier, for example. Thus, the **documentary method** is the work we do when we take an object or event and set it in correspondence with a structure of meaning. We do this all the time, but *how* do we do it?

In order to put the documentary method in sharp relief, Garfinkel did an experiment. He brought in 10 undergraduates and told them that they were part of an experiment to explore a new, alternative method of psychotherapy. The students were given the opportunity to ask the "therapist" about anything they desired. The students needed to first provide the background to the problem and then phrase their questions in such a way that they could be answered yes or no. The therapist and students were in different rooms and communicated via an intercom.

The students were instructed to give the background, ask their question, listen to the therapist's answer (yes or no), and then turn the intercom off and give their reactions. The procedure was repeated for as many questions as the students wanted to ask. Of course, the hitch in the experiment was that there was no new therapy and the "therapist's" answers were given randomly. Thus, there was no real "sense" to the answers. The issue, then, was exactly how (using what methods) the students made sense out of the answers—how the students understood the answers as "standing on behalf of a presupposed underlying pattern."

Garfinkel (1967) gleaned several insights from this experiment; I'll list but a few:

- The students perceived the experimenter's responses as "answers-to-questions."
- After the first question, the questions the students asked were motivated by the experimenter's response—in other words, the students framed their questions by looking back at "answers" and anticipated future "helpful answers."

- When the meaning of the experimenter's response wasn't apparent, the student "waited for clarification" or engaged in an "active search" for the meaning.
- Incongruent answers were interpreted by imputing knowledge and motivation to the therapist.
- Contradictory answers prompted an "active search" for meaning in order to rid the answer of disagreement or meaninglessness.
- There was a constant search for patterns.
- The subjects made specific references to normatively valued social structures that were treated as if shared by both and as setting the conditions of meaningful decisions (for example, what "everyone knows" about family). (pp. 89–94)

Garfinkel's point is that the students rendered meaningful something that was not. The work of documenting—searching for and assigning a pattern—is performed by us all in every situation. A common culture or cognitive scheme isn't so much shared as the sense of commonality in documenting is achieved. The students gave us a clear case where there were no cultures or cognitive schemes shared. Nonetheless, in most cases a correspondence was achieved between the event and a meaningful structure. The students' descriptions of the events were given in such a way as to assure their "rights to manage and communicate decisions of meaning" (Garfinkel, 1967, p. 77). Further, notice that even though the individual students were doing all the work, it was perceived and reported by the students as group work—as work between the student and therapist.

Garfinkel refers to the kind of work the students were doing as "ad hocing." *Ad hoc* comes from New Latin and literally means "for this." We say something is ad hoc when it is made for just this occasion or with a particular end or purpose in mind. In the experiment, the students were deliberately kept in the dark about the meanings of what was being said (or lack thereof). In the face of such ignorance and with an assumed context (psychotherapy experiment), students "ad hoced": they used methods that allowed the conversation and social order to continue in the face of contrary or ambiguous dialogue. Read this carefully: They used their *retrospective–prospective* sense to place what was said in an ongoing context with a biographical past and future; they *waited for clarification* when they first heard something that seemed senseless; and they continually performed an *active search* for a meaning index. Notice that all this was done without calling anything into question and for the purpose of not interrupting the flow of events.

One final word about ethnomethodology and my experiences with Eric Livingston. About midway through the semester, I felt I was beginning to figure it out, and I was intrigued. One day while we were walking together to lunch, I asked him if we could get together for a couple of sessions. I wanted him to teach me ethnomethodology, and Dr. Livingston very much wanted me to learn it. But, he told me, it wasn't something that could be taught through reading or talking. We would need to spend many hours simply watching people—watching them at the malls, crossing streets, lining up at the movies, attending classes, and doing the thousand-and-one things that people do each and every day.

I was a little surprised. I thought I had asked a regular question, the same as if I'd asked Herbert Blumer to explain symbolic interactionism to me. And I was disappointed. I was carrying a heavy academic load in school and was a single parent; I just didn't have the time to invest in that kind of learning process. Plus, Dr. Livingston was only going to be at the university for one year, so it wasn't something that we could schedule for some other time. It all added up to our just having a few more conversations on the subject, but no people watching.

It didn't really strike me then, but I can see now why Dr. Livingston said we needed to spend much more time learning this theoretical perspective: Ethnomethodology is more caught than taught. Probably more than any other perspective in this book, ethnomethodology must be applied to be known. As I mentioned before, I think the chief difficulty with this perspective is that it deals with those things that are seen but unnoticed. We miss the power of ethnomethods because they are so commonplace—and they are powerful precisely because they are ordinary.

Ethnomethodology really is more caught than taught, so quit reading and go watch people. While you're doing so, ask yourself one simple question: what is this behavior or part of a conversation *doing*? Don't ask what it *means*; ask, rather, *what does this do*? Begin to think about interactional elements as mechanisms that achieve something in the social encounter. If you continually ask yourself this question, you'll soon begin to become aware of the seen but unnoticed foundations of social order.

Garfinkel Summary

- Garfinkel's perspective is unique among sociologists. He sees social order and meaning as achievements that are produced *in situ*. That is, Garfinkel sees social order and meaning as achieved *within its natural setting*—face-to-face interactions—and not through such things as institutions that exist outside the natural setting.

- The principal way this is done is through accounting. A basic requirement of every social setting is that it be recognizable or accountable as a specific kind of setting. Thus, the practical behaviors that create a setting just as it is are seen but not noticed for what they are; they are the very behaviors that achieve the setting in the first place.

- All settings and talk are therefore indexical; they index or reference themselves. The actions that create the situation of a wedding or a college class are simultaneously understood as meaningful, social activities within the situation. Human activity always references itself; it is thoroughly reflexive, based upon incorrigible assumptions, discovered through the documentary method, proven through indexical methods, and protected by secondary elaborations of belief.

Looking Ahead

Let's end with the same question with which we began: what did you do the last time you got together with your friends? Maybe you didn't start a social movement, but in that mundane situation you negotiated symbolic meanings; engaged in self-talk and evaluation though role-taking; managed and maintained your self and the self of everyone else present; achieved a sense of social order and reality; and linked up your one social situation with others occurring in the present, the past, and the future. It seems you were busier than you thought, but our theoretical journey into your interactions isn't finished.

We are going to expand what we know about the situation on several fronts. First, and most obvious, in the next chapter we are going to add the element of exchange. For social situations aren't simply places where we create meanings, produce and manage selves, and achieve social order; they are also places where we strategically seek to fulfill our goals in life. Thus, in Chapter 10 we will see that in many situations we are busy working out social exchanges. And, like most human action, social exchanges have unanticipated consequences—things happen as the result of exchange that dramatically impact our lives without us knowing it.

One thing that occurs is that, as exchange relations build up over time, we create certain kinds of macro-level structures. This additional element to our theoretical toolbox is important. So far in this book we've talked about social structures on the one hand and interactions on the other. But what links these two domains together? In the next chapter, we will get two answers to this problem. We'll see that one of the unanticipated consequences of exchange is power, and power is a key factor in the micro–macro link. In addition, we will find that we are generally unaware of what is really in back of all our exchanges—emotional energy and cultural capital—and that these residues of exchange actually form the glue that holds society together.

Later on in the book, the micro–macro link also gets played out in a different way. Anthony Giddens and Pierre Bourdieu (Chapter 11) both argue that the micro–macro "link" is a misnomer. There isn't a "link" that is missing and needs to be discovered; rather, social structures and situations, and human agency (free will), are created and maintained at the same moment and in the same behaviors.

Giddens also carries forward another issue we talked about in this chapter: the self. Giddens draws on Goffman, symbolic interaction, and phenomenology to argue that how we relate to and manage the self has changed in late modernity. To capture these differences, Giddens uses such terms as "the reflexive project of the self" and "plastic sexuality." Yet Giddens isn't the only one concerned with the person in more contemporary theorizing. Jean Baudrillard (Chapter 14), our postmodern representative, helps us see that the modern subject is dead, and what exists in its place is a kind of terminal where media images get projected in an endless field of play. In that same chapter, Michel Foucault will show us how our experiences of the self—our subjectivity—are historically contingent and are actually expressions of power and control.

Dorothy E. Smith (Chapter 14) further opens our eyes to see how these expressions of power are contained in the texts produced principally by the social sciences. She argues that the objectifying stance of sociology actually functions to alienate women from their life-world. Smith uses phenomenology to help us see that the day-to-day world of women is ignored and decidedly different from what social science can account for. She wants to bring the lived experiences of women into our sight so that we can see precisely how women are forced to negotiate male-dominated institutions and the books and articles that express them. So, let me ask again, what did you do the last time you got together with your friends?

Building Your Theory Toolbox

Knowing Interactionist Theory

After reading and understanding this chapter, you should be able to define the following terms theoretically and explain their theoretical importance to micro-level theories:

Meaning, pragmatism, language, society, generalized other, role-taking, social objects, action, interaction, joint action, empirical science, reified concepts, definition of the situation, exploration, inspection, dramaturgy, interaction order, impression management, front, teams, setting, front and back stages, role distance, deference and demeanor, face-work, focused and unfocused encounters, phenomenology, accounting/accountable, seen but unnoticed, reflexivity, indexical expressions, incorrigible assumptions, secondary elaborations of belief, documentary method, ad hocing

After reading and understanding this chapter, you should be able to

- Describe pragmatism and analyze its influence on symbolic interactionism

- Explain the importance of meaning to humanity and illustrate how it is produced in face-to-face interactions

- Explain how social objects are defined and used in interactions

- Analyze a social situation or event in terms of it being a part of a joint action

- Critique various methods used in the social sciences and recommend a more "empirical" way of approaching a research topic

- Explain how the interaction order is produced

- Illustrate how a front is organized and used

- Explain the different ways in which we can relate to social roles, and analyze how you relate to your role of student and why

- Differentiate between focused and unfocused interactions around campus

- Explain what Garfinkel means in analyzing events "in just this way and at just this time." What does his approach imply about social order and social structures?

- Describe how situations are reflexively organized. Specifically, how do accountability, indexicality, and the documentary method function to reflexively organize social events?

- Discuss how incorrigible assumptions and secondary elaborations of belief work to produce a sense of reality, and be able to analyze a conversation or news article using the terms

- Compare and contrast symbolic interactionism, dramaturgy, and ethnomethodology, and evaluate their abilities to analyze a social situation

Learning More: Primary Sources

- Herbert Blumer:

  Blumer, H. (1969). *Symbolic interactionism*. Berkeley: University of California Press.
  Blumer, H. (2000). *Selected works of Herbert Blumer: A public philosophy for mass society* (S. M. Lyman & A. J. Vidich, Eds.). Urbana: University of Illinois Press.

- Erving Goffman:

  Goffman, E. (1959). *The presentation of self in everyday life*. Garden City, NY: Anchor.
  Goffman, E. (1961). *Encounters*. Bobb-Merrill.
  Goffman, E. (1967). *Interaction ritual: Essays on face-to-face behavior*. New York: Pantheon.
  Goffman, E. (1983). The interaction order. *American Sociological Review, 48*, 1–17.

- Harold Garfinkel:

  Garfinkel, H. (1967). *Studies in ethnomethodology*. Cambridge, UK: Polity Press.
  Garfinkel, H. (1996). Ethnomethodology's program. *Social Psychology Quarterly, 59*, 5–21.

Learning More: Secondary Sources

- Herbert Blumer: There are a myriad of secondary sources for symbolic interactionism. Rather than listing those, I am going to encourage you to look at some of the new and exciting ways SI is being used:

  Affect control theory: Heise, D. (2002). Understanding social interaction with affect control theory. In J. Berger & M. Zelditch Jr. (Eds.), *New directions in contemporary sociological theory*. Boulder, CO: Rowman & Littlefield.

  Expectation states theory: Wagner, D. G., & Berger, J. (2002). Expectation states theory: An evolving research program. In J. Berger & M. Zelditch Jr. (Eds.), *New directions in contemporary sociological theory*. Boulder, CO: Rowman & Littlefield.
  Cultural studies: Denzin, N. K. (1992). *Symbolic interactionism and cultural studies: The politics of interpretation*. Oxford, UK: Blackwell.

- Erving Goffman:

  Fine, G. A., & Manning, P. (2000). Goffman. In G. Ritzer (Ed.), *The Blackwell companion to major contemporary social theorists*. Oxford, UK: Blackwell.

  Treviño, A. J. (Ed.). (2003). *Goffman's legacy*. Boulder, CO: Rowman & Littlefield.

- Harold Garfinkel:

  Heritage, J. (1984). *Garfinkel and ethnomethodology.* Cambridge, UK: Polity Press.
  Livingston, E. (1987). *Making sense of ethnomethodology.* London: Routledge & Kegan Paul.

Theory You Can Use (Seeing Your World Differently)

- Watch or record two different news programs, such as on CNN and Fox News, that cover the same story. You'll want to pick one with substance to it. Analyze how events or ideas are used pragmatically as social objects. Do you see differences in meanings? If so, what do these differences imply?

- Think about one of the more memorable interactions that you had last week. In what ways did you manage the impression others had of you? How did others respond through righteously imputed expectations? Were there any performance teams involved?

- From this point on in the semester, keep a daily journal about your role as a student. Overall, would you say you are closer to role distancing or embracement? List at least five ways you distanced yourself from or engulfed your self in your role as a student.

- Compare and contrast at least two of your professors using Goffman's idea of impression management. How are their offices different? How about the way they dress and talk? How do they use the setting differently? How do their demeanors communicate different levels of expected deference?

- Using Goffman's theory, explain Internet interactions. How are they both different from and similar to face-to-face interactions?

- Using ethnomethodology, describe how the textbook you have in your hands is an example of reflexively constructing sociology. Remember, get specific in your descriptions—"in just this way and at just this time."

- Write an ethnomethodological description of the social organization of grocery store checkout lines.

- How would a dramaturgical theory explaining class, race, gender, and sexual inequality be different from a symbolic interactionist theory that does the same? How would they compliment one another?

Further Explorations—Web Links

The Society for the Study of Symbolic Interaction (SSSI): http://sun.soci .niu.edu/~sssi/

CHAPTER 10

Exchange Theory

Peter M. Blau (1918–2002)

Karen S. Cook (1946–)

Randall Collins (1941–)

I magine, if you would, a world where everything you want is yours. It doesn't matter what it is; you have inalienable rights to everything in the world, and every need and desire is fulfilled because there are no restrictions on you. In this world you are alone, but loneliness isn't a problem; you're not even aware that other people are possible. (Keep this in mind as we go through our little thought experiment: in terms of emotional needs and ties, relationships aren't an issue.)

Now, imagine that same world but with two differences: there is one other person, and the "rights to everything in the world" are divided equally. At this point in your world, things shift dramatically. The most profound change is that there is the possibility of relationship. But, again, this relationship isn't emotional; it's rational. It is centered on goods and resources. In this case, the possibility of a relationship happens in either of two ways.

First, you may want something that the other person controls. If so, you have three options: do without it, steal it, or work out an exchange. With all three options, you will engage in a calculation of costs and benefits. There are costs associated with doing without, with stealing it, and with exchanging some good or service for what you want. The issue in each case is this: do the benefits outweigh the costs? But there's another issue that may not be as apparent: only one of the options maintains individuality, and that's doing without. The other two options establish some kind of relationship with the other, and here is where costs versus benefits can get a little tricky.

If you decide to steal it, you may get caught, and it then becomes a conflict. On the other hand, if you decide to exchange, there are two possible complications: equity in exchange and delayed gratification. Granted, these complications only become important if there is the possibility of other exchanges in the future, but it seems reasonable to assume that if you need one thing from this other person, you'll need more later. So, what happens if the exchange isn't equal? This could happen for any number of reasons, but for the moment let's just assume that there is inequity. What are the ramifications of such a situation?

The issue of delayed gratification concerns timeliness (or asynchronous exchanges). Let's suppose that you own the rights to all the cattle and the other person owns the rights to all the trees. In this scenario, she or he wants to eat and you want to build a house. The problem is that the trees aren't mature; they aren't big enough to cut for lumber, but the cattle are ready now. And the other person needs to eat now. In this situation, you either don't exchange or you give the person the cattle now with the promise of lumber later. What fundamental social quality is required for this situation to occur successfully? The answer is trust.

The second way a relationship could occur is if there are needs that can only be met through cooperation. Let's keep this simple and say that you both need a large boulder moved. You decide to work together to move it. What problem might arise? In order to see the potential problem more clearly, let's say that the world is divided up among six rather than two people, and you all decide to move the boulder. Now you have six people moving the boulder. What problem could exist?

It's an interesting thought experiment, and if you play around with it for a while, you might be able to tease out some more interesting implications about human beings. But the most important thing that we need to see right now is that *these very basic exchange relationships have unintended consequences.* In each of our scenarios, the aim is simply to acquire a good or meet a need, yet each of the possible relationships we explored implies additional features. The ramification of inequity in exchange is social power; the fundamental feature necessary for asynchronous exchange is trust; and the possible issue arising from the need being met through cooperation is the problem of free riders (people who reap the benefits without incurring the costs—one of the six could simply *pretend* to lift the boulder).

Peter Blau does one of the best jobs of clearly explaining how these unintended consequences come together to form society. Blau uses elements that come directly from exchange theory to explain how structures are built: in particular, general trust and the norm of reciprocity. Blau also gives us a theory of power based on exchange principles. Blau takes the exchange between two people (dyad) as the archetype of exchange. In contrast, Karen Cook focuses more on the *network* of exchange rather than the exchange itself. Both Cook and Blau give us theories about power in exchange, but as you'll see, Cook focuses on the structure of the exchange network.

Randall Collins, whom we first were introduced to in Chapter 7, gives us a unique perspective on exchange: he argues that the generalized values behind all exchanges are cultural capital and emotional energy. In other words, whatever else we may think we are exchanging—whether stocks and bonds or a shirt from Target—in back of it all are these more basic motivations for emotion and cultural capital. In addition, Collins gives us another way of seeing how exchanges are linked together. Rather than the exchange networks that Cook helps us see, Collins opens our eyes to become aware of chains of interaction rituals.

This chapter is exciting for another reason: it's our first introduction to the concept of a micro–macro link. This issue is one that is obvious in contemporary theory yet virtually unknown in classical theory. For some time, sociologists have thought about macro-level phenomena and micro-level interactions separately. In some ways, the two different domains seemed to discount one another. Micro-level theorists such as George Herbert Mead saw social institutions more in terms of symbols and ways of thinking and behaving, with their importance and influence emerging out of interactions. On the other hand, structuralists such as Émile Durkheim saw social facts as being created by other social facts or institutions. Eventually, sociologists began to see a theoretical issue here. If there are two separate fields—face-to-face interactions and social structures—then how are they related? In this chapter, we will get two answers to this problem, one from Blau and the other from Collins.

Peter M. Blau: Social Exchange Theory

Photo: Courtesy of Judith Blau.

The Essential Blau

Biography

Peter M. Blau was born on February 7, 1918, in Vienna, Austria, the year the Austro-Hungarian Empire fell. The son of secular Jews, he watched the rise of fascism in postwar Austria with growing concern. Blau became a U.S. citizen in 1943, and he served in the U.S. Army during WWII, earning the Bronze Star for valor. After the war, Blau attended school and was awarded his Ph.D. from Columbia University in 1952; Robert K. Merton was his dissertation chair. His dissertation was subsequently published as *The Dynamics of Bureaucracy,* and has since become a classic in organizational literature. Peter Blau held professorships at Chicago, Columbia, the State University of New York at Albany, and the University of North Carolina at Chapel Hill. He also taught at the Academy of Social Sciences in Tianjin, China, and he was president of the American Sociological Association in 1973. Blau published hundreds of articles and 11 books, and he received numerous awards for his contributions to sociology and to society at large. Peter Blau passed away March 12, 2002.

Passionate Curiosity

Blau was captured by Merton's infatuation with the idea of structural constraints, which Blau (1995) considers the "central subject matter of sociology" (p. 6). Thus, Blau wants to know how social structures impose themselves on the opportunities that people have and the choices they make. What kinds of things are beyond the control of people? How do these things influence how they live? Specifically, how do the forms of exchange influence how people act? How do factors that exist in the population, factors that came to exist before the individual was born, influence choice and opportunity?

Keys to Knowing

Utilitarianism, social exchange, exchange alternatives, power, norms of reciprocity and fair exchange, marginal utilities, social capital, value hierarchy, norm of compliance, secondary exchange relations

Seeing Further: Social Exchanges

Broadly speaking, Blau recognizes two main influences on human behavior: (1) the situational and personal factors that influence the preferences people have and the choices they make, and (2) those external conditions that restrict or enable those choices and preferences. Blau recognizes three factors that help determine the first set of influences. There are the psychological or personality aspects (individual likes and dislikes), the social-psychological factors (how social position/class and experience influence choice and preference), and the actual properties of exchange.

The idea of exchange as a social dynamic began with *utilitarianism,* an eighteenth-century philosophy that came out of the Age of Enlightenment's concern with human happiness and scientific calculation. The principles of utilitarianism were most clearly stated by Jeremy Bentham (1789/1996, pp. 11–16), who argued that happiness and unhappiness are based on the two sovereign masters of nature: pleasure and pain. The "utility" in utilitarianism refers to those things that are useful for bringing pleasure and thus happiness. Bentham developed the *felicific* or *utility calculus,* a way of calculating the amount of happiness that any specific action is likely to bring. The calculus had seven variables: intensity, duration, certainty/uncertainty, propinquity/remoteness, fecundity, purity, and extent. Knowing or memorizing Bentham's calculus isn't important; what is important is that Bentham introduced the idea of rational calculation being used to decide human behavior and the moral status of any act.

Georg Simmel was probably the first sociologist to consider exchange. He argued that most relationships for people are governed by exchange principles. The primary evidence that Simmel has for this claim is that the majority of our relationships are reciprocal. That is, most of our activities are at least a two-way give-and-take. It's important for us to see this definition of exchange because many exchange theorists insist that it is governed by rationality, the evaluation of costs and benefits. This issue has opened exchange theory up to criticism, for there are other theories that insist humans are usually not rational in their behaviors—humans minimize their cognitive efforts and often behave through routine or emotion. While Simmel sees pure economic exchange as governed by cost and benefit analysis, it isn't necessarily true for social exchange generally. Exchange, then, as understood through *reciprocation,* is the basic form of society: "value and exchange constitute the foundation of our practical life" (Simmel, 1971, p. 47).

Defining Exchange Theory

Social exchanges are distinct from economic exchanges in at least four ways. First, they lack specificity. All economic exchanges take place under the contract model. In other words, almost all of the elements of the exchange are laid out and understood in advance, even the simple exchanges that occur at the grocery store. Social exchanges, on the other hand, cannot be stipulated in advance; to do so would be a breach of etiquette. Imagine receiving an invitation for dinner that also stipulated exactly how you would repay the person for such a dinner ("I'll give you one dinner for two lunches."). Social exchanges, then, cannot be bargained and repayment must be left to the discretion of the indebted.

This first difference implies the second: Social exchanges necessarily build trust, while economic exchanges do not. Since social exchanges suffer from lack of specificity, we must of necessity trust the other to reciprocate. This implies that relationships that include social exchange—and almost all do—build up slowly over time. We begin with small exchanges, such as calling people on the phone, and see if they will reciprocate. If they do, then we perceive them as worthy of trust for exchanges that require longer periods of time for reciprocation, such as friendship.

The third difference between social and economic exchanges is that social exchanges are meaningful. The way we are using it here, meaning implies signification. In other words, an object or action has meaning if it signifies something beyond itself. What we are saying about exchange is that a purely economic exchange doesn't mean anything beyond itself; it is simply what it is: the exchange of money, or something else of equal value, for some good or service. Social exchanges, on the other hand, always have meaning. For an example, let's take what might appear as a simple economic exchange: prostitution. If a married man gives a woman who is not his wife money for sex, it is a social exchange because it has meaning beyond itself: in this case, the meaning is adultery.

Finally, the fourth difference between social and economic exchanges is that social benefits are less detached from the source. We use money in economic exchanges, but the value of money is completely detached from the person using it. I may use money every time I go to the music store, but the value of that money for exchange is a function of the United States government and has nothing to do with me—it is completely detached from me. However, the value in all social exchanges is dependent upon the participants in some way. For example, the social exchange between you and your professor requires you to fulfill the requirements of the course to get a grade. Someone else can't do the work for you and you can't legitimately buy the grade.

Taken together, these four unique features of social exchange create diffuse social obligations. For example, let's say you and your partner invite another couple over for dinner. You expect that the invitation will be reciprocated in some way, but exactly *how* the other couple is to reciprocate isn't clear—nor *can* it be clear; to make it clear would reduce it to an economic exchange. So you have a general, unspecified expectation that the other couple will reciprocate in some way. The reciprocation has to be in the indefinite future (the other couple can't initially respond to your invitation by scheduling their "repayment" dinner—then it would really look like a repayment in economic terms), yet it has to be repaid specifically by the couple (the other couple can't have a different couple invite you for dinner and have it count for them). The dinner is meaningful, but the meaning isn't clear as of yet (Are you all going to be friends? If so, what kind of friends?). Thus, you have to trust the other couple to provide the future unspecified meaning and reciprocation. The other couple is obligated to you, but in a very diffuse manner.

Generally speaking, then, social exchange theory focuses on acts of exchange that create the reciprocal relations among people. Exchange is based on some value hierarchy, where certain goods and services are more highly valued than others.

Value is a function of exchange relations and is generally determined by the sacrifice needed to gain a good or service and its availability. In exchange theory, people are usually viewed as exercising some degree of rationality, in terms of evaluating the costs and benefits of any exchange, and making decisions based on some level of profit. In addition, many of the things that sociologists are concerned with in exchange are unintended consequences.

Concepts and Theory: Exchange and Power

Basic Exchange Principles

There are three basic exchange principles that Blau gives us, and they concern motivations, alternatives, and marginal utilities. The first principle is that people are *rationally motivated* in exchanges to weigh out costs and benefits. In this respect, the ideal type of exchange is economic, where the calculations are specific and known. Yet, as we've seen, Blau argues that social exchanges are different from economic ones, and one implication of this is that our calculations will be different. They are not as specific or concrete as economic calculations. Thus, we have to view social exchange "rationality" in limited terms. It's a general motivation in back of our actions. We don't specifically think, "If I invite Bob to the barbeque, then I can borrow his truck next Thursday." It's more of a general and diffuse motivation—a broad desire for social profits and a sense of how we stand in our exchanges.

Alternatives are extremely important in exchange relations. We'll talk more about alternatives when we get to the section on power, but for now we should simply be aware that when alternatives are present, people will gravitate toward exchanges among equals. We tend to look for someone with whom we can balance out costs and benefits in the long run. These balanced exchange relations tend to reduce uncertainty and to lessen power differences. Blau also notes that balanced relationships tend to create unbalanced relations elsewhere. In this, Blau is positing that we have limited resources and to invest resources in one relationship is to deny them to another.

Let's use the example of dating and marriage. When we get married, one of the things we are doing is committing a large amount of our personal resources to one relationship. We do so because of the anticipation of rewards, both intrinsic and extrinsic, but we simultaneously take away the possibility of using those resources in an alternative relationship. If you are married, and one of your friends wants to exchange more than you are able due to the commitment of resources to your marriage partner, then that friendship will become an unbalanced exchange relation. Unbalanced or unequal exchange relations tend to be less strained if the differences are known, clear, and marked. We can see this easily in the marriage example: your unbalanced relationship with your friend who wants more will be very strained if you are unclear about what you are willing and unwilling to exchange. However, the strain is lessened as you are clear about what you will and will not do because you are married—the other person will be less likely to entertain unreal exchange expectations.

The principle of *marginal utilities* posits that people have satiation points regarding goods and services. In other words, too much of a good thing may not be a good thing. When we first enter into an exchange relationship, the profits that we glean have high value. However, repeated profits of the same kind have declining value. For example, the first time a man gives flowers to his partner, it has high value. And it probably has high value the second and third times it happens. But if the man brings flowers to his partner every Friday after work, the value of the gift declines as the partner becomes satiated. Thus, the value of any good or service is higher if there is some degree of uncertainty or a sporadic quality associated with it.

Exchange Norms and Social Power

There are two norms associated with social exchange: the norm of reciprocity and the norm of fair exchange. Blau sees exchange as the starting mechanism for social interaction and group structure. Before group identities and boundaries, and before status positions, roles, and norms are created, interaction is initiated in the hopes of gaining something from exchange. The fact that we are dependent upon others for reaction implies that the idea of reciprocation is central in exchange. The word "reciprocate" comes from Latin and literally means "to move back and forth." Exchange, then, always entails the give and take of some elements of value, such as money, emotion, favors, and so forth. As such, one of the first behaviors to receive normative power in a relationship is reciprocity. The *norm of reciprocity* also implies another basic feature of society, that of trust. As we've seen, social exchange requires that the reciprocated good or service be unspecified and that reciprocation is delayed to some undisclosed future. This lack of specificity obviously demands trust, which forms the basis of society and our initial social contact. Exchanges are also guided by *the norm of fair exchange*. Something is fair, of course, if it is characterized by honesty and free from fraud or favoritism. The expectation of fairness increases over the length of the exchange relation. So, the longer you interact with someone, the more honest and fair you expect him or her to be.

Because of the lack of specificity in social exchange and the norm of reciprocity, exchange creates bonds of friendship and establishes power relations. The basic difference between friendship and power relationships concerns the equity of exchange and is expressed in the amount of repayment discretion. Friendship is based on an equal exchange relationship: all parties feel that they give about as much as they take. This equal reciprocity among friends leads to a social bond built on trust and a high level of discretion in repayment. On the other hand, inequality in exchange leads to unfulfilled obligations that, in turn, grant power over repayment to the other.

According to Blau, there are four conditions that affect the level of social **power.** I've diagramed these conditions in Figure 10.1. In the diagram, we have a social exchange relation between A and B. In thinking through the way the model works, you can visualize yourself as either person and get a sense of the way the power flows in the relationship. Here *social capital* refers to the ability to participate in an exchange with goods or services that the other desires. As you can see, there is a negative relationship between social capital and power. In other words, the less ability Actor A has to control goods and services (exchange capital) that Actor B desires, the greater will be B's power over A. Alternatives are important for power as well. If I have a large number of alternatives through which I can obtain the social good or service that I desire, then others will have little power over me. However, the fewer the number of alternatives, the greater will be the power of others who control the social good. This inverse relationship is noted by the negative sign.

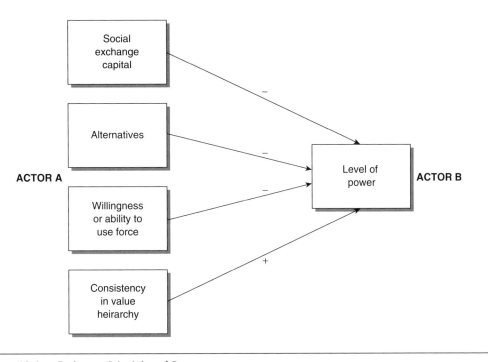

Figure 10.1   Exchange Principles of Power

There are two other important factors in establishing social power: the actor's willingness to use force and the consistency of the value hierarchy. If Actor A has the ability and willingness to use force, then by definition it is not an exchange relation and, thus, social power is impossible. This somewhat obvious condition tells us something important about social power: there is always a choice involved. With social power, "there is an element of voluntarism . . . the punishment could be chosen in preference to compliance" (Blau, 2003, p. 117). The relationship between Actor A's willingness to use force and Actor B's power is negative: the less likely Actor A is to use force to obtain social goods, the greater is the potential for social power for Actor B.

The last condition implies a consistent *value hierarchy.* If the value system of Actor A changes and she or he no longer values the social goods B controls, then there can be no power. If, on the other hand, Actor A consistently values the goods B has to offer, then social power will increase, if the other conditions hold as well. Try this thought exercise: Think about the relationships that you have with various professors as exchange relations. Use each of these four variables—social capital, alternatives, willingness to use force, and consistent value hierarchy—and ask the following: Who has more power and why? I believe you'll find that your professors have varying levels of power because of the different ways these factors align. At minimum, this implies that social power is not a simple function of bureaucratic position.

There are five possible responses to power. Four of the responses correspond to the four conditions of power and constitute attempts to change the balance of power. In other words, if you want to change the amount of power someone has over you, you could

- Increase your exchange capital by obtaining a good or service that the other person desires
- Find alternative sources to what you receive from the other person
- Get along without the good or service the other person controls
- Or, you could attempt to force the other person to give you what you need

The only other possible response is subordination and compliance. *Compliance*, unlike most features of social exchange, can be specified and functions like money: "Willingness to comply with another's demands is a generic social reward, since the power it gives him is a generalized means, parallel to money, which can be used to attain a variety of ends" (Blau, 2003, p. 22). The power to command is like a credit, an I.O.U. in social exchange. It is what we give to others when we can't participate in an equal exchange, yet we desire the goods they control.

Concepts and Theory: The Micro–Macro Link

Secondary exchange relations result from power and occur at the group level among those who are collectively indebted to someone or to another group. In order for secondary exchange relations to come into play, people who are individually indebted to a person or group must have physical proximity and be able to communicate with one another. For example, let's say you are tutoring a number of students in sociology without charging them money. Each of those individuals would be socially indebted to you, and each individual relationship would be subject to the dynamics of exchange (alternatives, marginal utilities, and norms).

As long as you only met with each person individually and they were unaware of each other, the exchange relations would remain individual. On the other hand, if you decide that your time would be better spent tutoring them as a group, then secondary exchange relations could come into play. You've provided them with the ability to become a group through physical proximity and communication. When a group like this is brought together physically and able to communicate, two possibilities exist with regards to power: the group may either legitimate or oppose your power as the tutor.

Blau gives us a very basic process through which power is either legitimated or de-legitimated. Both possibilities are in response to the group's perception of how those in power perform with regard to the norms of fair exchange and reciprocity. If the norms of reciprocity and fair exchange are adhered to, then social power will be legitimated. Blau posits that the path looks something like this: reciprocity and fair exchange with those in power → common feelings of loyalty → norm of compliance → legitimation and authority → organization → institutionalized system of exchange values.

As the group communicates with one another about the level of adherence of those in power to the norms of fair exchange and reciprocity, they collectively develop feelings of loyalty and indebtedness. Out of these feelings comes the *norm of compliance:* the group begins to sanction itself in terms of its relationship to those in power. This process varies in the sense that the more those in power are seen as consistently generous, that is, exceeding the norms of fairness and reciprocity, the more the group will feel loyal and the stronger will be the norm of compliance and sense of legitimation. Out of legitimation comes organization and an institutionalized system of values regarding authority (see Blau & Meyer, 1987, for Blau's treatment of bureaucracy).

In his theory of secondary exchange relations, Blau is giving us an explanation of the micro–macro link. We can see this move from individual exchanges to institutions in the path of secondary relations noted above. The "glue" that holds this path together consists of generalized trust and the norm of reciprocity. As we've already seen, all social exchanges are built on trust and the element of time, and lack of specificity in social exchanges demand trust. Organizations and institutions are, in Blau's scheme, long chains of indirect exchanges of rewards and costs. And, as we've seen, exchange intrinsically entails reciprocation. Every step, then, along the chain of exchanges is held together by the norm of reciprocity.

Secondary exchange relations can thus lead to legitimated authority, but they can also lead to opposition and conflict. If those in power do not meet the norms of fair exchange and reciprocity, and if those indebted are brought together physically and are able to communicate with one another, then feelings of resentment will tend to develop. These feelings of resentment lead to de-legitimation of authority. This path of secondary exchange relations looks like this: lack of reciprocity and fairness from those in power → feelings of resentment → communication (a function of physical proximity and communication technologies)→ de-legitimation of authority → ideology → solidarity → opposition → probability of change.

Groups experiencing a lack of fairness and reciprocity that are in close physical proximity and are able to communicate with one another will tend to develop a set of beliefs and ideas that justify both their resentment and their de-legitimation of authority. This ideology, in turn, enables group solidarity and overt opposition, thus increasing the probability of change. The last part of this path comes from the conflict theories of Marx and Weber. To this general theory of conflict and change, Blau adds the micro dimension of exchange: the beginning part of the path. Keep in mind that all of these factors function as variables and are therefore changeable.

Blau Summary

- Social exchanges are different from economic exchanges because they lack specificity, they require and build trust, they are meaningful, and social benefits are detached from the source. These differences imply that social exchanges create diffuse obligations, which in turn form the basis of society—social relations must be maintained in order to guarantee repayment of these obligations.

- There are three basic principles and two norms of exchange. The contours of all social exchanges are set by the principles of rational motivation, the presence of alternatives, and marginal utilities. Because of the peculiar properties of social exchanges, rationality within them is limited, as compared to economic exchanges. People are rational in social exchanges to the extent that they tend to repeat those actions they received rewards from in the past. People are also rational in the sense that they will gravitate toward exchanges that are equal, the equality of exchanges being determined by the presence of alternatives. And all social exchanges are subject to the principle of marginal utilities—a social good loses its value in exchange as people become satiated; in other words, value in exchange is determined to some extent by scarcity and uncertainty. All social exchanges are subject to the norms of reciprocity and fair exchange.

- Social actors achieve power through unequal exchanges, with inequality in exchange determined by four factors: the level of exchange capital, the number of potential source alternatives, the willingness and ability to use force, and consistency in value hierarchy. The first three are negatively related to power. That is, in an exchange relationship between Actor A and Actor B, as Actor A's capital, alternatives, and ability to use force go down, Actor B's power over Actor A increases. Consistency is a positive or at least a steady relationship—continuing to value the goods that Actor B controls places Actor A in a possible subordinate position.

- The presence of power sets up the possibility of secondary exchange relations. These occur at the group level between a supplier and consumer, where the consumer is a group of people who rely upon a single supplier. Under such conditions, the power of the supplier will either be opposed or legitimated. If the norms of reciprocity and fair exchange are adhered to, then the power of the supplier will be legitimated and feelings of loyalty and a norm of compliance will emerge. From this base, organizations and institutionalized systems of exchange are built.

Karen S. Cook: Power in Exchange Networks

Photo: Photo by Steve Castillo. Used with permission by the Institute for Research in the Social Sciences/Stanford University.

The Essential Cook

Biography

Karen S. Cook was born in Raton, New Mexico, July 25, 1946, and grew up in Austin, Texas. She did her undergraduate and graduate work at Stanford University, finishing her Ph.D. in 1973 with her dissertation, "The Activation of Equity Processes." She began teaching at the University of Washington in 1972 where she met and began her collaboration with Richard M. Emerson. Cook stayed at Washington for 23 years before moving to Duke University in 1995, where she was the James B. Duke Professor of Sociology, and back to Stanford in 1998 where she is currently the Ray Lyman Wilbur Professor of Sociology, director of the Institute for Research in the Social Sciences, and department chair. Along with Richard Emerson, Cook is one of the founding architects of what is sometimes called power-dependence theory or network exchange theory. Cook's most recent work looks at how people in a society where trust is waning can evoke cooperation.

Passionate Curiosity

Cook grew up in the southern United States during the time when racial segregation held sway and completed her university education in California during the civil rights and free speech movements. Cook explains that "coming of age professionally in the sixties as a sociologist marked my studies in subtle, but clear ways. I began with a focus on distributive justice and fairness and later moved to the topic of social power" (personal communication, May 10, 2006). As a scientific sociologist, then, Cook's passion is to understand the general social processes and structures that control the distribution of power and social justice.

Keys to Knowing

Social networks, exchange networks, positive and negative connections, decentralization principle, balance principle, exchange alliances

Seeing Further: Exchange Networks

Most of you have heard the phrase "six degrees of separation." It's the idea that the social distance between any two arbitrary people in the world is not that great; in fact, only six people separate you from any other person in the world. The phrase actually came from a quasi-experiment done by the psychologist Stanley Milgram in 1967 called "the small world experiment." Interestingly, this idea is continuing to be tested on the Internet. It appears that any two people on the Internet can be connected within five to seven steps.

We've also seen this notion played out in the Kevin Bacon game. The idea here is that any actor or actress can be linked to Kevin Bacon through the various movies they have been in. Every screen personality in this case has a Bacon number that corresponds to the number of films it takes to connect her or him to Kevin Bacon. It's an interesting game, and the University of Virginia Department of Computer Science maintains a Web site called "The Oracle of Bacon at Virginia." Check it out: http://www.cs.virginia.edu/oracle/

I'm using this idea to get you to see how Karen Cook sees the world. Cook blends exchange principles with network analysis. In exchange networks, it isn't the actual exchange or the person that is the focus. Rather, Cook sees the way people are connected as being paramount in determining how exchange relations function. But network theory isn't simply concerned with how many degrees of separation there are between you and Kevin Bacon; as you'll see, the kind of links, ties, and connections among the points (people) within a network is very important.

Social networks are defined by the patterns and positions involved in a group's interaction. For a quick example, think about an organizational chart. A person's position on that chart determines her or his general pattern of interaction. The same is true for any social group of which you are a member. Within a group, there are specific patterns of interaction: some members of the group interact more than others, and some members don't interact together at all, even though they are in the same group. Social networks, then, are a kind of structure that produces effects that don't originate with the individual, that don't come from social group membership or identity, and about which the individual may or may not be aware.

Concepts and Theory: Characteristics of Exchange Networks

Social *exchange networks* have five characteristics: (1) they are composed of a set of actors, either individual people or collective units; (2) among these actors, valued resources are distributed; 3) each actor has a set of exchange opportunities with others in the network; (4) there is some degree of commitment among the actors to use the exchange opportunities; and (5) the actors are connected and bonded in such a way as to form a single network.

Positive and Negative Relations

Beyond these five basic properties, exchange networks are established anytime two or more exchange relations are connected (notice that we are now talking about relations and not actors). That is, if A and B are in an exchange relationship and A and C are in an exchange relationship, B-A-C form an exchange network when exchanges between [A and B] and [A and C] become connected either positively or negatively. A *positive connection* exists if exchange in one direction is contingent upon exchange in the other direction. In other words, if the exchange between [A and B] is contingent upon the exchange between [A and C], then the connection is positive. A clear example of this kind of connection is the one among suppliers, manufacturers, and buyers. The exchange between the manufacturer and the buyer can only take place if the exchange between the manufacturer and supplier takes place. The supplier, manufacturer, and buyer thus form an exchange network with positive connections.

The other kind of relationship is a *negative connection,* which occurs when the exchange in one direction depends on non-exchange in the other. An example of this kind of connection exists among a buyer and two manufacturers, where the buyer can only purchase goods from one of the manufacturers at a time. These relations form a network in the sense that manufacturer B's sales are dependent on the level of exchange between buyer A and manufacturer C. I've depicted these two different kinds of connections in Figure 10.2.

You'll notice that I've represented the social actors as people. These kinds of relationships hold whether we are talking about companies or individuals—the important issue is the position within the network, not the actual actor. In the case

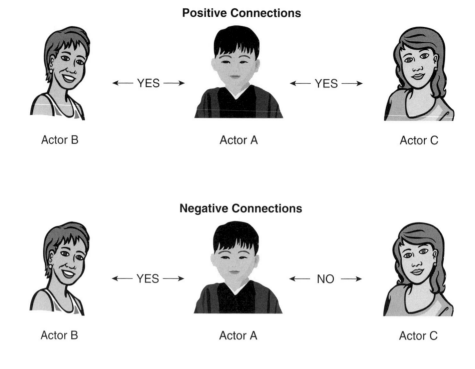

Figure 10.2   Exchange Network Connections

of people, we can think of a music band as an example of positive connections. In order to make music, the members all have to work cooperatively and exchange with one another, as noted by all connections being "yes." On the other hand, friendship or dating networks often have negative connections. For example, in order to watch horror movies with a friend, actor A has to get together with actor B rather than getting together with actor C to smoke cigars. An important thing to notice here is that actors B and C may not even be aware of one another, yet they would still experience the effects of the network. In addition, actor A may not have any idea of the other connections that actor B has.

Power and Balance in Exchange Networks

As we've seen, one of the main determinants of power in exchanges is the absence of *exchange alternatives,* or the level of dependency. Exchange networks with negative connections are obviously ones where alternatives exist, which is why "large networks completely positive in form are probably very rare because of the frequent existence of alternative sources" (Cook, Emerson, Gillmore, & Yamagishi, 1983, p. 278). One of Cook's central concerns is how exchange networks and alternatives influence one another. Cook is thus concerned with negatively related structures. In Figure 10.3, we have a diagram of such an exchange network. This diagram depicts a network wherein the same kind of goods or services is exchanged. Again, let me point out that the nature of the structure "transcends the occupants' knowledge" (Cook et al., p. 280). So, for example, A.3 is aware of its own connections (A.1, A.8, A.9, and A.10) but may be completely unaware of A.1's connections, though they have direct impacts on A.3's exchanges.

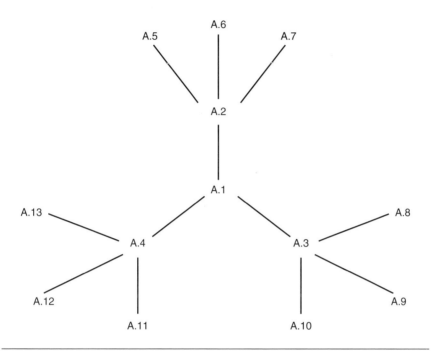

Figure 10.3   Exchange Network Positions

Source: Adapted from Cook, Emerson, & Gillmore, 1983, The distribution of power in exchange networks: Theory and experimental results. *American Journal of Sociology,* 89(2): 275–305. Reprinted with permission of The University of Chicago Press.

The network appears to be centralized around A.1, as all connections are directly or indirectly related to it. Yet, such networks are subject to a *decentralization principle.* Take a close look at Figure 10.3. Which positions structurally have the greatest number of alternatives? A.2, A.3, and A.4 have the greatest number and thus have more powerful positions than A.1. Let's use our band example again. Say that A.1, A.2, A.3, and A.4 are all members of the same music group. A.1 is the lead singer and brought the other members together. She thus appears to be the central character in the band. However, because the rest of the band members have alternatives and A.1 does not, A.1 is the least powerful of the members.

This situation changes if the network is positively related rather than negatively. Remember, in positive relationships the different positions in the network represent contingent or reciprocal exchanges. In such a scenario, A.1 becomes the most important and powerful position, because it serves as the only resource link for the entire network. One of the basic ideas that Cook works with is the *balance principle.* In the long run, exchange relations tend to balance themselves out. If you'll think back to our discussion of Blau, when we initially talked about power in exchange relations, we also talked about responses to inequality in exchanges. There are five of them, and the first four deal with balancing out the power. One of the theoretical benefits of seeing power related to dependence is that it allows for the "specification of ways in which dependencies can be altered to affect the balance of power in the exchange relation and in the network of connected exchange relations" (Cook & Rice, 2001, p. 705).

In terms of networks of exchange, power is primarily balanced through *exchange alliances*. Power differences in an exchange network prompt the formation of coalitions. Alliances that incorporate all weak actors balance out and add stability to the network. Unbalanced networks that do not form coalitions or do not include all weak members in the association tend to participate in fewer and fewer exchanges over time. The network, then, will tend to dissipate or reconfigure around different exchange relations. Cook (Cook & Gillmore, 1984) argues that changing the division of labor within a network can also balance out power, since it changes the distribution of resources. And, as between individuals, power differences can be mitigated by extending the network (thus increasing alternatives) or devaluing the good or service coming from the more powerful members. On the other hand, increasing competition among network members decreases the probability of coalition formation, because there will be fewer and fewer ways in which to align interests.

Cook Summary

- The work of Karen S. Cook moves us from considering individual exchanges and relationships to thinking about exchange networks. This move is theoretically powerful in that it attunes us to effects over which the individual has little control. Exchange networks contain two or more exchange relations that are positively or negatively linked. Negative connections between relations are ones wherein the exchange moves in one direction only; positive connections are those where the exchange in one direction is dependent upon exchange in the other direction.

- Power in negatively related networks structurally falls on the position with the greatest number of alternatives; power in positively related networks falls to the central position. In all exchange relations, power tends to balance out. Power in exchange networks may be balanced through coalitions of weak members, changes in the division of labor, extensions to the network, or reevaluation of resources. Increasing competition within the network tends to lessen the probability that coalitions will form.

Randall Collins: Ritual Exchange

Photo: Courtesy of Randall Collins.

The Essential Collins

Biography

Randall Collins was born in Knoxville, Tennessee, on July 29, 1941. His father was part of military intelligence during WWII and then a member of the state department. Collins thus spent a good deal of his early years in Europe. As a teenager, Collins was sent to a New England prep school, afterward studying at Harvard and the University of California, Berkeley, where he encountered the work of Herbert Blumer and Erving Goffman, both professors at Berkeley at the time. Collins completed his Ph.D. in 1969 at Berkeley. He has spent time teaching at a number of universities, such as the University of Virginia and the Universities of California at Riverside and San Diego, and has held a number of visiting professorships at Chicago, Harvard, Cambridge, and at various universities in Europe, Japan, and China. He is currently at the University of Pennsylvania.

Passionate Curiosity

In this area of his theorizing, Collins is focused on seeing how people feel their way to and through exchange-based interactions. Using this focal point of emotional energy, Collins wants to delineate the ways society is formed through ritualized interactions.

Keys to Knowing

Emotional energy; ritual; co-presence; mutual focus of attention; barrier to outsiders; shared emotional mood; rhythmic entrainment; group symbols; group solidarity; standards of morality; cultural capital; generalized, particularized, and reputational capitals; interaction rituals chains; market opportunities

Seeing Further: Exchange and Emotion

Randall Collins (1993b) is an interesting kind of exchange theorist: He sees emotion as the common denominator of rational action. To bring emotion in, he points to three long-standing critiques of exchange theory. First, exchange theory has a difficult time accounting for altruistic behavior. Merriam-Webster (2002) defines altruism as "uncalculated consideration of, regard for, or devotion to others' interests." If most or all of our interactions are exchange-based, and if all our exchanges are based on self-motivated actors making rational calculations for profit, how can altruism be possible? Collins claims that exchange theorists are left arguing that the actor is actually selfish in altruistic behavior—he or she gains some profit from being altruistic. However, just what that profit is has generally been left unspecified.

Second, evidence suggests that people in interactions are rarely rational or calculative. In support of this, Collins cites Goffman's and Garfinkel's work, the idea of bounded rationality in organizational analysis, as well as psychological experiments indicating that when people are faced with problems that should prompt them to be rational, they use non-optimizing heuristics instead. These heuristics function like approximate or sufficient answers to problems rather than the most rational or best answer. The third criticism of exchange theory is that there is no common metric or medium of exchange. Money, of course, is the metric and medium of trade for exchanges involving economically produced goods and services; however, money isn't general enough to embrace all exchanges, all goods, and all services.

Collins sees each of these problems as solved through the idea that emotional energy is the common denominator of rational action. This approach is rather adventuresome in that it combines two things that have usually been thought of as oil and water—emotion and rationality just don't mix. At least, they didn't before Collins came along. **Emotional energy** does not refer to any specific emotion; it is, rather, a very general feeling of emotion and motivation that an individual senses. It is the "amount of emotional power that flows through one's actions" (R. Collins, 1988, p. 362). Collins (2004) conceptualizes emotional energy as running on a continuum from high levels of confidence, enthusiasm, and good self-feelings to the low end of depression, lack of ambition, and negative self-feelings (p. 108). The idea of emotional energy is like that of psychological drive, but emotional energy is based in social activity.

Collins is arguing that emotional energy is general enough to embrace all exchanges. In fact, emotional energy is the underlying resource in back of every exchanged good and service, whether it's a guitar, a pet, a conversation, a car, a friend, your attendance at a show or sporting event, or anything else. More basic than money, emotional energy is the motivation behind all exchanges. Emotional energy can also be seen in back of social exchanges that might seem counterintuitive. Why would I exchange my free time to work at a soup kitchen on Sunday mornings? This, of course, is an example of altruistic behavior. Exchange theory, apart from the idea of emotional energy, is hard pressed to explain such behaviors in terms of exchange. Collins gives us a more general property of exchange in the form of emotional energy. People engage in altruistic behaviors because of the emotional energy they receive in exchange.

The idea of emotional energy also solves the problem of the lack of rational calculations. As Collins notes, people aren't generally observed making rational calculations during interactions. Rather than being rationally calculative, "human behavior may be characterized as emotional tropism" (R. Collins, 1993b, p. 223). A *tropism* is an involuntary movement by an organism that is a negative or positive response to a stimulus. An example of tropism is the response of a plant to sunlight. The stems and leaves react positively to the sun by reaching toward it, and the roots react negatively by moving away from it and deeper in the ground. Collins is telling us that people aren't cognitively calculative in normal encounters. Instead, people emotionally feel their way to and through most interactions, much like the way a plant reaches toward the sun.

Concepts and Theory: Interaction Ritual Chains (IRCs)

Rituals

For Randall Collins, **rituals** are patterned sequences of behavior that bring four elements together: bodily co-presence, barriers to outsiders, mutual focus of attention, and shared emotional mood. These elements are variables—as they increase, so also will the effects of ritualized behavior. There are five main effects of interaction rituals: group solidarity, group symbols, feelings of morality, individual emotional energy, and individual cultural capital. Collins's theory is diagrammed in Figure 10.4.

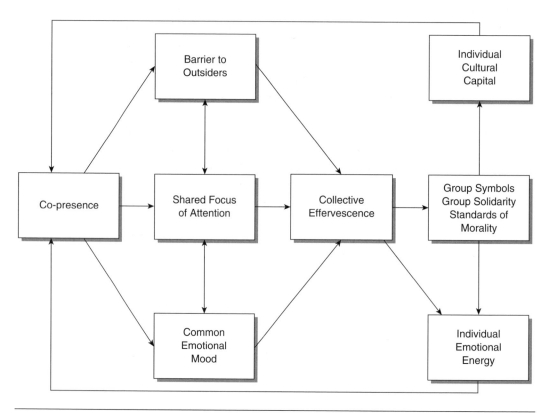

Figure 10.4   Randall Collins's Interaction Ritual

One of the first things that the model in Figure 10.4 calls our attention to is physical *co-presence*, which describes the degree of physical closeness. Even in the same room, we can be closer or further away from one another. The closer we get, the more we can sense the other person. As Durkheim (1912/1995) says, "The very act of congregating is an exceptionally powerful stimulant. Once the individuals are gathered together, a sort of electricity is generated from their closeness and quickly launches them to an extraordinary height of exaltation" (pp. 217–218). Bodily presence appears theoretically necessary because the closer people are, the more easily they can monitor one another's behaviors.

Part of what we monitor is the level of involvement or *shared focus of attention*, the degree to which participants are attending to the same behavior, event, object, symbol, or idea at the same time (a difficult task, as any teacher knows). We watch bodily cues and eye movements, and we monitor how emotions are expressed and how easily others are drawn away from an interaction. Members of similar groups have the ability to pace an interaction in terms of conversation, gestures, and cues in a like manner. Part of the success or intensity of an interaction is a function of this kind of rhythm or timing.

The key to successful rituals "is that human nervous systems become mutually attuned" (R. Collins, 2004, p. 64). Collins means that in intense interactions or ritual performance, we physically mimic one another's body rhythms; we become physically "entrained." *Rhythmic entrainment* refers to recurrent bodily patterns that become enmeshed during successful rituals. These bodily patterns may be large and noticeable as in hand or arm expressions, or they may be so quick and minute that they occur below the level of human consciousness.

It is estimated that human beings can perceive things down to about 0.2 seconds in duration. Much of this entrainment occurs below that threshold, or below the level of consciousness—which indicates that people literally feel their way through intense ritualized interactions. Randall Collins (2004, pp. 65–78) cites evidence from conversational analysis and audience–speaker behavior to show that humans become rhythmically coordinated with one another in interactions. Some research has shown that conversations not only become rhythmic in terms of turn-taking, but the acoustical voice frequencies become entrained as well. EEG (electroencephalogram) recordings have indicated that even the brain waves of interactants can become synchronized. In a study of body motion and speech using 16mm film, Condon and Ogston (1971) discovered that in interaction, "a hearer's body was found to 'dance' in precise harmony with the speaker" (p. 158).

A shared focus of attention and common emotional mood tend to reinforce one another though rhythmic entrainment. *Common emotional mood* refers to the degree to which participants are emotionally oriented toward the interaction in the same way. In ritual terms, it doesn't really matter what kind of emotion we're talking about. What is important is that the emotion be commonly held. Having said that, I want to point out that there is an upper and lower limit to ritual intensity.

One thing that tends to become entrained in an interaction is turn-taking. The rule for turn-taking is simple: one person speaks at a time. The speed at which turns are taken is vitally important for a ritual. The time between statements in a successful conversation will hover around 0.1 seconds. If the time between turns is too great, at say 1.0 seconds, the interaction will be experienced as dull and lifeless and no solidarity will be produced. If, on the other hand, conversational statements go in the other direction and overlap or interrupt one another, then the "conversation" breaks down and no feeling of solidarity results. These latter kinds of conversations are typically arguments, which can be brought about by a hostile common emotional mood. Randall Collins (2004) points out that generally at the micro level, "solidarity processes are easier to enact than conflict processes. . . . [T]he implication is that conflict is much easier to organize at a distance" (p. 74).

Being physically co-present tends to bring about the other variables, particularly, as we've seen, the shared focus of attention. Co-presence also aids in the production of ritual barriers. *Barriers to outsiders* refers to symbolic or physical obstacles we put up to other people attempting to join our interaction. The use of barriers increases the sense of belonging to the interaction that the participants experience. The more apparent and certain that boundary, the greater will be the level of ritualized inter-action and production of group emotional energy. Sporting events and rock con-certs are good illustrations of using physical boundaries to help create intense ritual performance.

Ritual Effects

Notice that there are a total of five effects coming out of the emotional efferves-cence that is produced in rituals. Let's first talk about the interrelated group effects: solidarity, group symbols, and standards of morality. *Group symbols* are those sym-bols we use to anchor social emotions. The greater the level of collective efferves-cence that's created in rituals, the greater will be the level of emotion that the symbol comes to represent. It's this investment of group emotion that makes it a collective symbol. The symbol comes to embody and represent the group. If the invested emotion is high enough, these symbols take on sacred qualities. We can think of the United States flag, gang insignia, and sport team emblems and colors as examples, in addition to the obvious religious ones. The symbols help to create group boundaries and identities. Group symbols have an important ritual function: They are used to facilitate ritual enactment by focusing attention and creating a common emotional mood.

Group solidarity is the sense of oneness a collective can experience. This concern originated with Durkheim (1893/1984, pp. 11–29) and meant the level of integra-tion in a society, measured by the subjective sense of "we-ness" individuals have, the constraint of individual behaviors for the group good, and the organization of social units. Collins appears to mean it in a more general way. Group solidarity is the feeling of membership with the group that an individual experiences. It's seeing oneself as part of a larger whole. One of the important things to see here is that the sense of membership is emotional. It is derived from creating high levels of collec-tive effervescence. Of course, the higher the level of effervescence, the higher will be the sense of belonging to the group that an individual can have.

Standards of morality refer to group-specific behaviors that are important to group membership and are morally enforced. Feelings of group solidarity lead people to want to control the behaviors that denote or create that solidarity. That is, many of the behaviors, speech patterns, styles of dress, and so on that are associ-ated with the group become issues of right and wrong. Groups with high moral boundaries have stringent entrance and exit rules (they are difficult to get in and out of). Street gangs and the Nazi party of WWII are good examples of groups with high moral boundaries.

One thing to notice about our example is the use of the word "moral." Most of us probably don't agree with the ethics of street gangs. In fact, we probably think their ethics are morally wrong and reprehensible. But when sociologists use the term moral, we are not referring to something that we think of as being good. A group is moral if its behaviors, beliefs, feelings, speech, styles, and so forth are controlled by strong group norms and are viewed by the members in terms of right and wrong. Because the level of standards of morality any group may have is a function of its level of interaction rituals, we could safely say that, by this definition, both gangs and WWII Nazis are probably more "moral" than we are, unless one of us is a member of a radical fringe group.

Concepts and Theory: The Micro–Macro Link

In Randall Collins's theory of interaction ritual chains, the individual is the carrier of the micro–macro link. There are two components to this linkage: emotional energy and cultural capital, the other two effects of ritual. Emotional energy is the emotional charge that people can take away with them from an interaction. As such, emotional energy predicts the likelihood of repeated interactions: if the individual comes away from an interaction with as high or higher emotional energy than she or he went in with, then the person will be more likely to seek out further rituals of the same kind. Emotional energy also sets the person's initial involvement within the interaction. People entering an interaction that are charged up with emotional energy will tend to be fully involved and more readily able to experience rhythmic entrainment and collective effervescence.

Cultural capital is a shorthand way of talking about the different resources we have to culturally engage with other people. The idea of cultural capital covers a full range of cultural items: it references the way we talk; what we have to talk about; how we dress, walk, and act—in short, anything that culturally references us to others. Collins lists three different kinds of cultural capital. *Generalized cultural capital* is the individual's stock of symbols that are associated with group identity. As Figure 10.4 illustrates, a great deal of this generalized cultural capital comes from interaction rituals. This kind of cultural capital is group specific and can be used with strangers, somewhat the way money can. For example, the other day I was in the airport standing next to a man wearing a handmade tie-dye T-shirt with a dancing bear on it. Another fellow who was coming off a different flight saw him and said, "Hey, man, where ya from?" These two strangers were able to strike up a conversation because the one man recognized the group symbols of Deadheads—fans of the band, The Grateful Dead. They were able to engage one another in an interaction ritual because of this generalized cultural capital.

Particularized cultural capital refers to cultural items we have in common with specific people. For example, my wife and I share a number of words, terms, songs, and so forth that are specifically meaningful to us. Hearing Louis Armstrong, for instance, instantly orients us toward one another, references shared experiences and meanings, and sets us up for an interaction ritual. But if I hear an Armstrong song around my friend Steve, it will have no social effect—there are no shared experiences (past ritual performances) that will prompt us to connect. From these two examples, you get a good sense of what cultural capital does: it orients people toward one another, gives them a shared focus of attention, and creates a common emotional mood—which are most of the ingredients of an interaction ritual.

The last kind of cultural capital Randall Collins talks about is *reputational capital.* If somebody knows something about you, he or she is more likely to engage you in conversation than if you are a complete stranger. That makes sense, of course, but remember that this is a variable. Mel Gibson, for example, has a great deal of cultural capital. If he were seen in a public space, many people would feel almost compelled to engage him in an interaction ritual.

Collins argues that interactions get "chained" together through cultural capital and emotional energy. Each person comes into an interaction with stocks of emotional energy (EE) and cultural capital (CC) that have been gleaned from previous interactions. The likelihood of an individual seeking out an interaction ritual is based on his or her levels of emotional energy and cultural capital; the likelihood of two people interacting with one another is based on both the similarity of their stocks and the perceived probability that they might gain either emotional energy or cultural capital from the encounter. The micro–macro link for Collins, then, is created as individual carriers who are charged up with emotional energy and cultural capital seek out other interaction rituals in which to revitalize or increase their stocks. If you think about your social encounters in this way, what you see is a trail of interactions that may form the beginning of an interaction ritual chain. As this chain builds up over time and space, a macro structure is formed.

One further point before we leave this idea: people have a good sense of their *market opportunities,* which are directly linked to cultural capital and indirectly to emotional energy. As we noted earlier, we exchange cultural capital in the hopes of receiving more of it back. Part of our opportunity in the cultural capital market is structured: our daily rounds keep us within our class, status, and power groups. However, our interpersonal markets are far more open in modern than in traditional societies. In these open markets, we are "rational" in the sense that we avoid those interactions where we will spend more cultural capital than we gain, and we will pursue those interactions where we have a good chance of increasing our level of cultural capital. We also tend to avoid those interactions where our lack of CC will be apparent. As a result, we tend to separate ourselves into symbolic or status groups.

Collins Summary

- Collins gives us a scientific theory that takes the situation where people are face-to-face as the primary social entity. He combines the idea of social exchange with Durkheim's theory of rituals to argue that emotional energy and cultural capital are primary elements that create and sustain society.

- Rituals are patterned sequences of behavior that entail four elements: bodily co-presence, barrier to outsiders, mutual focus of attention, and shared emotional mood. Each of these is a variable, with increasing levels of each leading to increasing ritual performance and effects. Rituals result in group solidarity, group symbols, feelings of morality, individual emotional energy, and individual cultural capital. The greater the level of ritual performance, the greater will be the levels of each of these outcomes. Emotional energy and cultural capital are particularly important because they create the links among chains of interaction rituals. As individuals move from one interaction to another, they carry differing levels of emotional energy and cultural capital. These differing levels strongly influence the likelihood and subsequent effects of further rituals.

Building Your Theory Toolbox

Knowing Exchange Theory

After reading and understanding this chapter, you should be able to define the following terms theoretically and explain their importance to exchange theorizing:

Utilitarianism; social exchange; exchange alternatives; power; norms of reciprocity and fair exchange; marginal utilities; social capital; value hierarchy; norm of compliance; secondary exchange relations; social networks; exchange networks; positive and negative connections; decentralization principle; balance principle; exchange alliances; emotional energy; ritual; co-presence; mutual focus of attention; barrier to outsiders; shared emotional mood; rhythmic entrainment; group symbols; group solidarity; standards of morality; cultural capital; generalized, particularized, and reputational capitals; interaction rituals chains; market opportunities

After reading and understanding this chapter, you should be able to

- Explicate the unique features of social exchange as compared to economic exchange

- Explain what power is and how it comes out of social exchanges. In your explanation, be certain to include both Blau's and Cook's ideas.

- Discuss the norms of reciprocity and fair exchange and illustrate how they are important

- Analyze secondary exchange relations and how they account for the micro–macro link

- Explain how positive and negative connections influence an exchange network

- Discuss how exchange relations tend to balance out in the long run

- Explicate the dynamics and effects of interaction rituals

- Explain how interaction rituals produce a micro–macro link

- Compare and contrast Blau, Cook, and Collins on the question of power

Learning More: Primary Sources

- Peter Blau

  Blau, P. (1968). Social exchange. In D. L. Sills (Ed.), *International encyclopedia of the social sciences.* New York: Macmillan.
  Blau, P. (2002). Macrostructural theory. In J. H. Turner (Ed.), *Handbook of sociological theory.* New York: Kluwer Academic/Plenum.
  Blau, P. (2003). *Exchange and power in social life.* Somerset, NJ: Transaction Publishers.

- Karen S. Cook

  Cook, K. S., & Emerson, R. M. (1978). Power, equity, and commitment in exchange networks. *American Sociological Review, 43,* 721–739.
  Cook, K. S., & Rice, E. R. W. (2001). Exchange and power: Issues of structure and agency. In J. H. Turner (Ed.), *Handbook of sociological theory.* New York: Kluwer.

- Randall Collins

  Collins, R. (1988). *Theoretical sociology* (pp. 188–203). *New York:* Harcourt Brace Jovanovich.
  Collins, R. (1990). Market dynamics as the engine of historical change. *Sociological Theory, 8,* 111–135.
  Collins, R. (1993). Emotional energy as the common denominator of rational action. *Rationality and Society, 5,* 203–230.
  Collins, R. (2004). *Interaction ritual chains.* Princeton, NJ: Princeton University Press.

Learning More: Secondary Sources

Calhoun, C. J., Meyer, M. W., & Scott, W. R. (Eds.). (1990). *Structures of power and constraint: Papers in honor of Peter M. Blau.* New York: Cambridge University Press.
Cook, K. S. (Ed.). (1987). *Social exchange theory.* Newbury Park, CA: Sage.
Willer, D. (Ed.). (1999). *Network exchange theory.* Westport, CT: Praeger.

Theory You Can Use (Seeing Your World Differently)

- One of the enlightening aspects of exchange theory is that it helps us understand our relationships in terms of exchange. For example, what kinds of exchange dynamics are at work in your family or significant relationships? How can you understand your relationship with your professor using exchange theory? Specifically, how is power achieved and how could you reduce the level of power?

- Garfinkel gives us a theory of how social order is achieved at the micro level. Blau gives us a theory of how power is achieved and maintained at the micro level. How can these two theories be joined? Is power a part of social order? If so, how? Illustrate your answer with examples from your own life.

- Using Collins's idea of emotional energy, explain why you "like" some people and not others.

- Analyze online interactions, such as blogs or chatrooms, using Collins's theory of interaction rituals. What is present and not present in online interactions? How do you think these differences might make face-to-face interactions different from online encounters? What might this imply for someone who tends to have more and more of his or her "social" encounters online or via cell phone?

- Go to a college sporting event. Analyze everything you see in terms of interaction rituals. Based on this analysis, can you make any suggestions as to how identities are formed around sports teams?

- We have now examined six different ways that we can conceptualize what happens when social actors get together: symbolic interaction, ethnomethodology, dramaturgy, social exchange, exchange networks, and interaction ritual chains. Come up with a one- or two-word description for what each perspective tells us is happening (for example, with Collins it could be "ritual"). Now, under each of the descriptors, list at least five points that make that perspective unique and five insights you would get using that theory (in other words, how do you see your world differently because of the theory?).

Section III
Introduction

Contemporary New Visions and Critiques

Throughout this book we have, in a very general way, considered the influence of historical changes on theory. Sociological theory as we know it today began as a result of the massive changes that Western society experienced during the eighteenth and nineteenth centuries. These were times of upheaval that were both transformative and energizing. Clearly there were problems in the move from traditional to modern society, and one of the chief motivations of our classical theorists was to identify these problems and speculate about how they might affect society. Marx focused on capitalism, Weber looked at rationality and bureaucracy, Durkheim was concerned with social diversity and decreasing cultural consensus, Simmel saw the problems associated with urbanization and increasing levels of objective culture, and Gilman and Du Bois pointed to the inconsistencies of inequality in a democratic society.

Yet, on the whole, these critiques were based on the hopes of modernity as well. Modernity is not simply a period of time; it is an attitude of mind—the belief that through reason human beings can control their world, and use that control to make progress to a better life, both socially and technically (for example, in medicine). Society thus was moving upward. As we moved into early twentieth-century theorizing we saw this confidence continue to be reflected in the way sociological theory was proceeding. Sociology busied itself with theory testing and cumulation. Knowledge of social things was valued on its own merit, much as pure research is valued in the hard sciences. Schools of thought were organized and research agendas launched.

But the twentieth century also brought doubt and despair and, with these, critical theory. The "war to end all wars" (WWI) didn't. World War II came and brought with it the Holocaust, a haunting testimony to the depths the human soul can sink. The twentieth century also initiated the atomic age and catastrophic risk—risk due not only to nuclear escalation but also to the ever-expanding potential of global ecological destruction. In addition, after WWII the last remnants of the British Empire fell, and with it came the end of colonialism: "for the peoples of most of the earth, much of the twentieth century involved the long struggle and eventual triumph against colonial rule" (Young, 2003, p. 3). Yet this one rule was replaced by another—the United States became hegemonic, and global dominion moved from colonialism to a triadic combination of capitalism, culture, and militarism. To keep and expand that rule, the United States has entered into a succession of never-ending and rarely "won" police actions.

In this century of doubt, the 1960s perhaps stands central. The contradictions of democracy and the failure of the projects of modernity, both social and technical, became clearer than ever before, and the decade brought worldwide upheaval. There were student movements not only in the United States but also around the globe, including in China, West Germany, Poland, Italy, Japan, Vietnam, Czechoslovakia, and Mexico. This period of time marks what has been called the first "world revolution." As we'll see when we get to Chapter 12, Immanuel Wallerstein identifies 1968 as the most significant turning point in the current world system.

Contemporary theory, then, is more and more characterized by criticism than hope. As we've seen, modernist theory is based on the twin beliefs in technical progress and secular salvation. The events of the twentieth cast deep doubts on these ideas of the Enlightenment. In the West, we became increasingly aware that modern democracy is built on inequality—it became painfully obvious that not all "men" are created equal. We also awakened to the destructive side of capitalism and science—the exploitation of people and land can have long-term and initially hidden effects. Contemporary theory is characterized by this same awareness. Just as criticism and doubt gradually dawned on people in the twentieth century, and just as these twin themes weave themselves in and out of the twenty-first-century consciousness, so it is with contemporary theory.

There are four chapters in this section, each with a specific theme. In Chapter 11, we'll be considering how societies are constructed. By and large, most of the theories we've looked at so far have taken society for granted. The intent of these theories is to explicate the social factors, processes, and experiences that are present in society. But neither Giddens nor Bourdieu takes society for granted. They want to first understand how society comes into existence. Their basic premise is that knowing how society exists will have a significant impact on what we pay attention to in our theorizing.

The theorists in Chapter 12 want us to change the way we look at society as well. But rather than looking at the essence of society, Wallerstein and Luhmann want to change the level of analysis. Let me ask you, at what level does society exist? Is it in the interaction, organizations, institutions, or interstructural relations? Wallerstein and Luhmann would say that we are casting our vision too low: society is a world system; it works at the global level.

In Chapter 13, we will consider the identity politics surrounding race and gender. We'll see that this approach to race and gender is decidedly different from the one we saw in Chapter 8. The focus in identity politics is on the construction of marginalized identities and the lived experiences that come with those identities, not on the structural inequalities that we studied in Chapter 8. Finally, in Chapter 14, we come to poststructuralism and postmodernity. If we know nothing else about these theories, we know that they come after structuralism and modernity simply by their names.

What isn't as apparent in the organization of these last few chapters is the critical edge they bring. Giddens, for example, while maintaining that society is still modern, argues that modernity by its nature is uncontrollable. For Giddens, the principal earmark of modernity is continual and accelerating change that is legitimated as progress. As such, modernity is like a runaway train: humankind might be able to steer it to some degree, but it also threatens to rush out of control and break itself apart. Bourdieu addresses the first concern of modernity that we were introduced to with Marx: class. Class in the modern age, as you might recall from our discussion of Habermas, was to be the social position par excellence that was achieved rather than inherited. That's one of the reasons that Marx was so critical. Rather than being a vehicle for equality, the capitalist class system creates and perpetuates new social divisions and inequality. Bourdieu takes Marx's critique one step further and shows us that class is structured far more deeply than we ever imagined.

Though they come from different perspectives—Marxism and functionalism, respectively—both Wallerstein and Luhmann argue that if we lift our eyes up out of our national parochialism and look to the global system, we will see that the world as we know it is teetering on the brink of disaster. Modernity may have intended to bring hope, but the processes of modernity have spelled its own demise. On a different level, identity politics point out a shift that has occurred in late or postmodernity. Modern identities almost by definition are emancipatory. We've seen that the social hope of modernity is freedom, equality, and happiness to all people, for all the social categories and identities that people can claim. There's a way in which identity politics asks us to look deeper than social structures and civil rights. Identity politics brings the effects of inequality home to the person. Oppression involves more than voting and employment rights; domination oppresses the very heart and soul of a people.

Our next theorist, Michel Foucault, is neither modern nor postmodern, per se. Foucault is a poststructuralist, yet he contributes significant ideas to the modern/postmodern debate. Poststructuralism argues that there are no structures, and that human behavior is neither determined nor caused by anything. It denies that there is any firm base for behaviors, society, or reality. Thinking about the building analogy from Chapter 8, structure is nothing but smoke and mirrors. All we as humans have are discourse and text. Further, while there is no reality or meaning behind them, texts and forms of knowledge exert tremendous power over every aspect of our lives, primarily through discourse. The insidious part is that we control, limit, and objectify ourselves through discourse. In response, poststructuralism deconstructs the text or produces a counter-history of knowledge, which reveals the underlying and subtle political power found within all histories and discourses.

Lastly, we turn to the postmodernist work of Jean Baudrillard. In some ways, the postmodern argument is like that of Giddens: modernity contains dynamics that continually push for change. The difference is that postmodernists say there has been a breach or rupture and we're no longer modern. Different theorists emphasize different social factors, but in my reading of the field there appear to be two effects with which postmodernists are most concerned: culture and the individual subject. Both are seen as simultaneously becoming more important and less real in some fashion.

CHAPTER 11

Constructing Society

Anthony Giddens (1938–)

Pierre Bourdieu (1930–2002)

While president of the United States, Ronald Reagan once said, "You can't help those who simply will not be helped. One problem that we've had, even in the best of times, is people who are sleeping on the grates, the homeless who are homeless, you might say, by choice." This quote certainly points out a hot topic in political discourse: are the poor in poverty because of social structures or because of their individual choices? The political parties of this country will generally fall on one side or the other of this kind of debate, but it is important for us to see that this issue is also a point of contention in the social sciences, though it is phrased somewhat differently.

In the social sciences, there is a central dichotomy that sets up some of the basic parameters of our discipline, such as the distinction between quantitative and qualitative methods and the divergence between structuralism and interactionism. This dichotomy also sets up one of the thorniest issues sociologists address: the link between the micro and macro levels of society. Like the issue in Reagan's quote, the dichotomy I am referring to is the dilemma of structure (objective) versus agency (subjective): how much agency or free will do individuals have in the face of social structures? We saw this issue in a different form in Chapter 10. There Blau and Collins attempted to bridge the gap by explaining how the two different domains are linked together through either exchange processes or elements of ritual.

Our two theorists in this chapter take different approaches. Pierre Bourdieu (1985) characterizes the dichotomy between structure and agency as one of the most harmful in the social sciences, and sees working to overcome this dichotomy as the most steadfast and important goal in the social disciplines (p. 15). As we'll see, Bourdieu's theory seeks to solve this dilemma by pointing out the creative tension that exists between structure and agency. Anthony Giddens agrees with Bourdieu's assessment, but solves the problem by doing away with both structure and agency, at least in the usual meanings of the words. In addition, both Giddens and Bourdieu have specific social issues to which they apply their theories. For Giddens, the issue is modernity; for Bourdieu, the topic is social class.

Anthony Giddens: Structuration and Modernity

Have you ever ridden a rollercoaster? One of the things that makes riding a rollercoaster fun is the way danger and security are mixed together. We wouldn't ride the rollercoaster if we didn't believe it was safe, but the rollercoaster wouldn't be fun if we didn't have a sense of danger. When we slam into the curves and plummet over a hundred feet down, we feel the possibility of death, but it is tempered by our sense of trust in the machine and the experts who built it. Anthony Giddens pictures modernity in much the same way, but with some important differences.

According to Giddens (1990), modernity is a juggernaut, "a runaway engine of enormous power which, collectively as human beings, we can drive to some extent but which also threatens to rush out of our control and which could rend itself asunder" (p. 139). The word "juggernaut" comes from the Hindi word *Jagannātha,* which refers to a representation of the god Vishnu or Krishna, the lord of the universe. Every year, in sections of ancient India the god's image was paraded down the streets amid crowds of the faithful, dancing and playing drums and clashing cymbals. It's thought that at times, believers would throw themselves under the wheels of the massive cart, to be crushed to death in a bid for early salvation. A juggernaut, then, is an irresistible force that demands blind devotion and sacrifice.

This image of an irresistible force conjures up the thrilling ride of the rollercoaster, with its twin sensations of trust and danger, but the juggernaut of modernity isn't as controllable or predictable as a rollercoaster. Here we can see a chief difference between Giddens and Jürgen Habermas (Chapter 7): for Habermas, rational control is central to modernity and imminently possible, but for Giddens, modernity is almost by definition out of control. The intent of modernity is progress, but the *effect* of modernity is the creation of mechanisms and processes that become a runaway engine of change. And we, like the devotees of Jagannātha, are drawn to modernity's power and promise.

> The ride is by no means wholly unpleasant or unrewarding; it can often be exhilarating and charged with hopeful anticipation. But, so long as the institutions of modernity endure, we shall never be able to control completely either the path or the pace of the journey. In turn, we shall never be able to feel entirely secure, because the terrain across which it runs is fraught with risks of high consequence. (Giddens, 1990, p. 139)

Photo: Courtesy of Anthony Giddens.

The Essential Giddens

Biography

Anthony Giddens was born January 18, 1938, in Edmonton, England. He received his undergraduate degree with honors from Hull University in 1959, studying sociology and psychology. Giddens did his master's work at the London School of Economics, finishing his thesis on the sociology of sport in 1961. From then until the early 1970s, Giddens lectured at various universities including the

University of Leicester, Simon Fraser University, the University of California at Los Angeles, and Cambridge. Giddens finished his doctoral work at Cambridge in 1976. He remained there through 1996, during which time he served as dean of social and political sciences. In 1997, Giddens was appointed director of the London School of Economics and Political Science. Giddens is the author of some 34 books, which have been translated into over 20 languages. Giddens is also a member of the Advisory Council of the Institute for Public Policy Research (London) and has served as advisor to British Prime Minister Tony Blair.

Passionate Curiosity

Giddens is a political sociologist, driven by both political questions and political involvement. While his early work certainly contained a typical Marxist interest in class, his later work is much more concerned with the political ramifications of globalization and what he characterizes as the juggernaut of modernity or the runaway world. Given the juggernaut of modernity, he asks, how are interactions and behaviors patterned over time? How can people become politically involved? In order to answer those questions, Giddens must first understand the essence of society. In this, Giddens seeks an ontology of the social world: What kinds of things go into the making of society? Precisely how does it exist?

Keys to Knowing

Structuration theory, duality of structure, social structures, normative rules, signification codes, authoritative resources, allocative resources, institutional orders, reflexive monitoring, discursive and practical consciousnesses, routinization, regionalization, time–space distanciation, modalities of structuration, ontological security, disembedding mechanisms, symbolic tokens, expert systems, globalization, reflexive project of the self, bodily regimens, plastic sexuality, emancipatory and lifestyle politics, pure relationships, mediated experiences

Seeing Further: Structuration Theory Part I: Recursive Structures

Duality of Structure

In structuration theory, Giddens unites the insights from theories that you are already familiar with: symbolic interaction, phenomenology, dramaturgy, and the classic sociological idea of structure. As we discuss Giddens's ideas, you will see him work back and forth with these theories. So keep what we have covered about interactions and structures in mind. Giddens is sometimes seen as difficult because he uses a lot of new words, but after all you've learned in this book so far, you are ready to think through these issues.

Giddens is arguing that the subject–object divide (or agent-structure) is a false dichotomy, created to explain away the complexity of human practice. Giddens

(1986) says, "Human social activities, like some self-reproducing items in nature, are recursive" (p. 2). It's like the chicken and egg question, which in some ways is a silly one. When you have the egg, you have the chicken. They are one and the same, just in different phases. In society it's something like this: social actors produce social reality, but the mere fact that you have "social actors" presumes an already existing social world.

The primary insight of Giddens's **structuration theory** is that social structures and agency are recursively and reflexively produced: they are continuously brought into existence at the same moment through the same behaviors. Rather than seeing structure and agency as a dualism, as two mutually exclusive elements, Giddens proposes a duality—two analytically distinguishable parts of the same thing. The *duality of structure* indicates that structure is both the medium and the outcome of the social activity or conduct that it reflexively organizes.

"This is a difficult concept to understand, so let me give you an example." And I just did. I put the first sentence of this paragraph in quotation marks because it is our example. Anytime we write or speak a sentence, we do a couple of things. First, and most obviously, we create the sentence. The second and less obvious thing we do is re-create the rules through which the sentence was made in the first place. This is a little tricky, so pay close attention. In order to put together a sentence, we have to follow the rules. If we don't, the sentence won't make any sense and it won't really be a sentence. Thus, in order to exist as a sentence, the line of words must be formed according to the rules. Yet, at the same moment we create the sentence, we also re-create the rules through which the sentence was made in the first place.

You might say, "Wait a minute, the rules existed before the sentence." Did they? You learned the rules in school and those rules are found in English grammar texts, right? In fact, we often refer to the grades between the primary levels and high school as "grammar school" because that's where you learn the rules of grammar. But is that really where you learned the rules? If it was, it would imply that you couldn't form a sentence before reaching that point in school. But the fact is, you could form sentences well before you "learned the rules"; further, studies have shown that five- and six-year-olds make use of very complex grammars. In truth, the rules you learned in school are the rules you already knew. The difference is that the rules found in grammar texts are formalized interpretations of the rules that already exist in the language itself; grammars and dictionaries are produced by academics based on the study of language. Note that grammarians study the language to *discover* the rules—the rules are already there in the language. The rules for making the sentence are in the sentence itself—the expression and the structure are created in the same moment. According to Giddens, the same is true about social agents and structures.

In order to talk about the duality of agency and structure, Giddens changes the definition of social structure. In structuration theory, *social structures* consist of rules and resources. There are two kinds of rules: normative rules and codes of signification. Keep in mind that in both cases these "rules" are fluidly embedded in social practices. They don't exist abstractly or independently. Giddens also notes that rules may be consistently or rarely invoked, tacit or discursive, informal or formal, and weakly or strongly sanctioned. You should be familiar with *normative rules*—they are those that govern behavior, such as the norm against littering. But signification codes require a bit of explanation.

Signification codes are rules through which meaning is produced. In the sentence illustration above, the signification code is lodged in the practices of speaking and writing. One example of the consequence of these codes or rules is the rhetoric of political spin-doctors. Spin-doctors want to guide us so that we interpret events in a specific manner, but in doing so they must abide by generally accepted rules of interpretation. If they don't, then chances are good that we won't buy their "spin." It's important to mention that these rules are historically and culturally specific. That's why interpretations can change over time.

There are also two kinds of resources: authoritative and allocative. *Authoritative resources* are made up of such things as techniques or technologies of management, organizational position, and expert knowledge. *Allocative resources* come from the control of material goods or the material world. Resources, then, involve the control of people and supplies.

I've pictured the duality of structure in Figure 11.1. As we've seen, Giddens argues that structure and agency are mutually formed in the same act. Just as in our sentence example, the rules and resources that are used in social encounters both create and are found in the interaction and structure. This is important: the act of social co-presence is possible only through the use of social rules and resources—and the rules and resources only exist in the act of social co-presence. Thus, structure and agency are mutually constructed through the use of the exact same rules and resources, as noted by all the two-headed arrows in the figure.

Modalities of Structuration

Notice the large arrow linking the two sets of interactions in Figure 11.1. The arrow indicates how behaviors and encounters are patterned over time. Giddens (1986) rephrases the problem of social order and patterning behaviors in terms of time–space distanciation: "the fundamental question of social theory . . . is to explicate how the limitations of individual 'presence' are transcended by the 'stretching' of social relations across time and space" (p. 35). The idea of **time–space distanciation** refers to the ways in which physical co-presence is stretched through time and space. This is a fairly unique and graphic way of thinking about patterning behaviors. We can think of Giddens's idea as an analogy: if you've ever played with Silly Putty or bubble gum by stretching it out, then you can see what he is talking about. What this analogy implies is that the interactions at Time-1 and Time-2 appear patterned because they are made out of the same materials, stretched out over time and space.

How this stretching out of time and space happens is Giddens's fundamental question. His answer is found in his idea of modalities of structuration. The word "modality" is related to the word "mode," which refers to a form or pattern of expression, as in someone's mode of dress or behavior. For example, in writing this book, I'm currently in my academic mode. **Modalities of structuration,** then, are ways in which rules and resources are knowingly used by people in interactions.

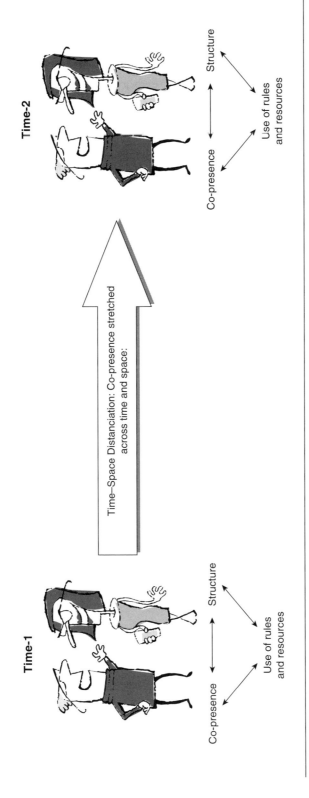

Figure 11.1 Duality of Structure

Graphic: Brian Barber/Istockphoto.com.

I've pictured a bit of what Giddens is getting at in Figure 11.2. Notice that there are three elements in the circle: social practices, modalities, and structures. Modalities of structuration are ways in which structure and practice (or agency) are expressed. I've indicated that relationship by the use of overlapping diamonds. In a loose way, we can think of structures as the music itself; the modalities as the mode of reproduction, as in analog or digital; and the social practices as the musician. As you can see, Giddens gives us three modalities or modes of expression (interpretive schemes, facilities, and norms), corresponding on the one hand to three social practices (communication, power, and sanctions), and on the other to structures (signification, domination, and legitimation).

This isn't as complicated as it might seem. Let's use the example of you talking to your professor in class. Let's say that in this conversation you refuse to take the test that she or he has scheduled. The professor reacts by telling you that you will fail the course if you don't take the test. What just happened? You can break it down using Giddens's modalities of structuration, following the model in Figure 11.2.

First, there were actual social practices that involved communication and sanctions. Your communication was interpreted using a scheme that both you and your

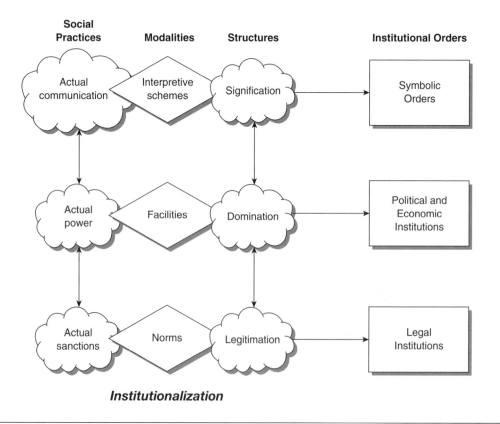

Figure 11.2   Modalities of Structuration

professor know. For convenience's sake, let's call this scheme "meanings in educational settings." You know this interpretive scheme because it is part of the general signification structure of this society at the beginning of the twenty-first century. Second, the professor invoked sanctions based on norms of classroom behavior. Again, you both know these norms because they are part of the legitimation structure of this society.

I'm sure you were able to follow this discussion through the model without any difficulty. I'm also sure you didn't have any trouble with any of the phrases I used in the above explanation. Things like "society at the beginning of the twenty-first century" sound reasonable and familiar. But remember, the first principle of structuration theory is *duality,* not dualism. So when we use the terms society or structure, we are not talking about something separate from social practices. Like our sentence illustration, the actual conversation between you and the professor and the interpretive schemes and the structure of signification all come into existence at the same moment. Apart from signification and interpretation, communication can't exist; likewise, without actual communication, interpretation and communication can't exist. Obviously, the same is true for the sanctions that the professor invoked.

In terms of Giddens's definition of structure as rules and resources, signification and legitimation are more closely tied to rules, and domination is more linked to resources (facilities). Domination is expressed as actual power through the facilities of authoritative and allocative resources. For example, part of the way the actual power of the university over me as a professor is expressed is through the facilities of classroom space, computer and Internet access, and so forth. By encircling all these elements together in Figure 11.2 and by using two-headed arrows, I'm indicating that all these processes—the social practices, modalities, and structures—are reflexive and recursive. That is, they mutually and continuously influence one another.

Part of what I want you to see in Figure 11.2 is the connectedness of all social practices, modalities, and structures. They are all tied up together and expressed and produced in the same moment. Further, the recursive and interpenetrative nature of these facets of social life are what Giddens means by *institutionalization,* or the stretching out of co-presence across time and space. Remember, human life is ongoing. The process that I've placed a circle around in Figure 11.2 works like a ball that just keeps rolling downhill. It is this continuity of recursive practices and structures that stretches interactions across time and space. Comparing Figures 11.1 and 11.2, we can think of the two men talking as moments in which we stopped the ball and looked inside. The arrow between the two sets of interactions depicts the movement of the ball between those two moments.

There's one other thing we need to notice from Figure 11.2: all of this action of institutionalization results in different *institutional orders.* While there are some terms in there that look familiar, like "economic institutions," they aren't the same as we usually think of them. Many sociologists think of institutions as substantive and distinct (or "differentiated," in functionalist terms). In other words, most sociologists treat institutions as if they are real, separate objects with independent effects. However, Giddens is saying that institutions don't exist as substantive things or objects and they aren't truly separate and distinct.

Notice that the different institutional orders are all made from the same fabric; they're just cut or put together differently in each case. They all draw from the same structures (rules and resources) of signification (S), domination (D), and legitimation (L), but emphasizing one of the elements over the others produces different kinds of institutional orders. The way Figure 11.2 is mapped allows us to work out the different orders. The structure that is most closely associated with the institutional order is the one that is most important.

Thus, in the case of symbolic institutional orders, the order is drawn as S-D-L. Think about a political ideology like democracy: it is primarily based on signification and meaning with domination and legitimation backing the meanings and symbols. In the case of legal institutional orders, the arrangement is reversed from that of symbolic orders. A legal order is drawn as L-D-S. Its primary structural source is legitimation, closely followed by domination and further back, signification. The other two institutional orders that Giddens gives us are drawn similarly to one another but with different kinds of domination. Thus, political institutions are D (authoritative)-S-L and economic institutions are drawn as D (allocative)-S-L.

Seeing Further: Structuration Theory Part II: Reflexive Actors

Levels of Awareness

If Giddens is giving us a theory that helps us see structures and agents as mutually constituted, then he must compliment his notion of structures with a theory of the social actor in interaction. Giddens argues that there are three important things going on in interactions: reflexive monitoring of action, rationalization of action, and motivation for action. Giddens thinks of these tasks as being "stratified," or as having different levels of awareness. The behavior that is most conscious is *reflexive monitoring*. In order to interact with one another, people must watch the behaviors of other people, monitor the flow of the conversation, and keep track of their own actions. As part of this routine accomplishment, we can also provide reasons for what we do; that is, we can provide a rationalization for our own actions.

In talking about this, Giddens makes a distinction between discursive and practical consciousness. The word "discursive" is related to discourse or conversation. But it has a deeper meaning as well: it's a discourse marked by analytical reasoning. So **discursive consciousness** refers to the ability to give a reasoned verbal account of our actions. It's what we know and can express about social practices and situations. This consciousness is clearly linked to reflexive monitoring of the encounter and the rationalization of action—discursive consciousness is our awareness of these two.

Practical consciousness refers to the knowledge that we have about how to exist and behave socially. However, people can't verbally express this knowledge. Social situations and practices are extremely complex, according to Giddens; they thus require a vast and nuanced base of knowledge, and we have to act more by intuition than by rational thought. As an illustration, we can think of the ability to perform an opening ritual ("Hey, how's it going?") as part of this practical consciousness. People know *how* to perform an opening ritual, but most people can't rationally explain *why* they do it.

There's something important for us to see here: discursive and practical consciousnesses aren't necessarily linked. At first glance, it might appear that discursive consciousness is our ability to explain what practical consciousness tells us to do. But notice what I said above about practical consciousness: "people can't verbally express this knowledge." So discursive consciousness (the explanation) isn't necessarily associated in any real way with practical consciousness (the actions). We know how to act and we know how to explain our action, but both of these issues are part of the social interaction, not part of the unconscious motivations of the actor.

Unconscious Motivation

Practical consciousness is bound up with the production of routine. It's like driving a car or riding a bicycle; most of what is involved is done out of habit, or practical consciousness. In the same way, most of what we do socially on a daily basis is routine. *Routinization* "is a fundamental concept in structuration theory" (Giddens 1986, p. xxii) and refers to the process through which the activities of day-to-day life become habitual and taken-for-granted. Routinization, then, is a primary way in which face-to-face interactions are stretched across time and space (Figure 11.1). Put another way, routinization is one of the main ways through which the modalities of structuration are institutionalized (Figure 11.2). Part of the way we routinize activities is through *regionalization,* which is the zoning of time and space in relation to routinized social practices. In other words, because we divide up physical space, we can more easily routinize our behaviors. Thus, certain kinds of social practices occur in specific places and times. Regionalization varies by form, character, duration, and span.

The *form of the region* is given in terms of the kinds of barriers or boundaries that are used to section it off from other regions, and it allows greater or lesser possible levels of co-presence. When you stop and talk with someone in the hallway, there is a symbolic boundary around the two of you that is fairly permeable; it is very possible that others could join in. However, when you go into the men's or women's restroom, there is a physical and symbolic barrier that explicitly limits the possibility of co-presence.

The *character of the region* references the kind of social practices that can typically take place within a region. For example, people have lived in houses for centuries, but the character of the house has changed over time. In agrarian societies, the home was the center of the economy, government, and family, but in modern capitalist societies, the home is the exclusive domain of family and is thus private rather than public.

The *duration and span of the region* refers to the amount of geographic space and to the length or kind of time. Certain regions are usually available for social practices only during certain parts of the day or for specific lengths of time; the bedroom is an example in the sense that it is usually associated with "sleep time." Regions also span across space in varying degrees. For example, a coliseum gives unique opportunities for co-presence and social activities when compared to an airplane.

We come now to the principal force behind Giddens's structuration and time–space distanciation. The word "ontology" refers to the study of existence. The point in using it here is that the way in which humans and their world exist is unique; as we've seen, it's *meaningful*. The reality of the human world is existentially moored in meaning, which is fallible, mutable, and uncertain (see the earlier discussion of meaning in Chapter 9). According to Giddens, if people ever notice this about their reality, they will suffer deep psychological angst. We are motivated, then, as a result of this unconscious psychological insecurity about the socially created world, to make the world routine and thus taken-for-granted. Note that this anxiety is unconscious—it isn't usually experienced, but when it is, it is felt as a diffuse, general sense of unease.

Many students have a tendency to discount this idea of Giddens's. Most of us don't feel anxious about reality, so it seems that ontological insecurity doesn't exist. But the problem in thinking about ontological angst isn't that it doesn't exist; it is that we're very good at meeting this anxiety through regionalization and routinization. However, imagine, if you would, a person who wholeheartedly believes in God, does everything in the name of God, and sees the entire universe as given meaning and order because of God. Now imagine that this person's faith is utterly shattered in a moment. Such a person would have no basis for understanding life, no reason to organize his or her behaviors in any specific manner, and would be left questioning the reality of any belief system. Such a person would experience ontological insecurity. Giddens (1986), like Garfinkel, argues that these "critical situations" reveal the underlying basis of everyday life; he points to prisoners' experiences in WWII concentration camps at Dachau and Buchenwald as examples (pp. 60–64).

According to Giddens (1990), **ontological security** refers to the feelings of "confidence that most humans [*sic*] beings have in the continuity of their self-identity and in the constancy of the surrounding social and material environments of action" (p. 92). Giddens argues that the fundamental trust of ontological security is generally produced in early childhood and maintained through adult routines. Because most of the social practices in our lives are carried out by routine, we experience trust in the world, due to its routine character, and we can take for granted the ontological status of the world. In premodern societies, trust and routine in traditional institutions covered up the conditional character of the social world. Kinship and community created bonds that reliably structured actions through time and space; religion provided a cosmology that reliably ordered experience; and tradition itself structured social and natural events, because tradition by definition is routine.

In modern societies, however, none of these institutional settings produces a strong sense of trust and ontological security. According to Giddens, those needs are met differently: routine is integrated into abstract systems, pure relationships substitute for the connectedness of community and kin, and reflexively constructed knowledge systems replace religious cosmologies—but not with the certainty or the psychological rewards of premodern institutions. The result is that ontological insecurity—anxiety regarding the "existential anchoring of reality" (Giddens, 1991, p. 38)—is a greater possibility in modern rather than traditional societies.

Defining Structuration Theory

In brief, then, structuration theory is the theoretical perspective developed by Anthony Giddens. The perspective is more an analytical framework—or ontological scheme—than a complete theory. Structuration tells the theorist-researcher what kinds of things exist socially and what to pay attention to. It denies the existence of structure and free agency (seen as a false dualism) and argues that these generally reified concepts form an active duality: two parts of the same thing. In order to act, social actors must use known rules and resources. Giddens conceptualizes rules (normative rules and codes of signification) and resources (authoritative resources and allocative resources) as structural elements, which means that actors constitute or enact structure through their agency (actions). Giddens applies this concept to the issue of time–space distanciation. Thus, in structuration, local interactions are linked with distant others through the use of known rules and resources. This way of seeing things avoids the reification of agency or structure, places the structuring (patterning) elements within the observable interaction, and encourages an historical sociology. The latter is important for Giddens as he explains the ways through which modern society stretches out time and space, or links local interactions with distant ones, and allows him to explicate some of the unique features of modernity.

Concepts and Theory: The Contours of Modernity

We have now laid the groundwork for Giddens's understanding of how society works in general: Actors are motivated to routinize social actions and interactions by the psychological need for ontological security. These routines serve to stretch out face-to-face encounters through time and space as actors use different modalities to express the social structures of signification, domination, and legitimation through their social practices. This constant structuration produces different institutional orders that, along with regionalization, work to stabilize routine. Routinization and the institutional orders that it generates stabilize time–space distanciation and thus give the individual a continual basis of trust in her or his social environment, which, in turn, provides the individual with ontological security.

Thus, in Giddens's scheme, society isn't structured—it doesn't exist as an obdurate object with an independent existence. The important point here is that society by its nature is continually susceptible to disruption or change. This constant possibility is, of course, what creates the diffuse and unconscious sense of insecurity that people have about the reality of society. However, this possibility is also what makes modernity an important issue, for both the process of structuration and for the person. In the next section, I will ask you to think about how living in modernity influences your experience of yourself and others. But for now, simply think about how the dynamic quality of modernity radically changes structuration and time–space distanciation. There are four analytically distinct factors that produce the dynamism of modernity: radical reflexivity, the separation of time and space, disembedding mechanisms, and globalization. As we'll see, though we can separate these areas analytically, they empirically reinforce one another.

Radical Reflexivity

Giddens sees reflexivity as a variable, rather than a static condition. He argues that modernity dramatically increases the level of reflexivity. Previous to this time, people didn't think much about society. In fact, the entire idea of society as an entity unto itself wasn't really conceived of until the work of people such as Montesquieu and Durkheim. Today, we are quite aware of society and we think deliberatively about our nation and the organizations and institutions in which we participate.

Progress and reflexivity are thus intrinsically related. It only takes a moment's reflection to see that progress demands reflexivity. It is endemic in modernity because every social unit must constantly evaluate itself in terms of its mission, goals, and practices. However, the hope of progress never materializes—the ideal of progress means that we never truly arrive. Every step in our progressive march forward is examined in the hopes of improving what we have achieved. Progress becomes a motivating value and a discursive feature of modernity, rather than a goal that is never reached.

Here's a real-life example: Chances are good that you are attending an accredited college or university. Schools of higher education are certified by regional accrediting organizations. Being accredited allows you as a student to qualify for federal financial aid and to transfer credits from one college to another, and it allows professors to apply for federal grants for research. At my university, we just finished our reaccreditation self-study, which took two years to complete. Even though this seems like a long time, we actually began preparing for the self-study the two years previous by evaluating our mission statement in light of the new criteria for accreditation.

Out of the earlier study came a new mission statement that was then used during the following study to reevaluate every aspect of the university (notice the reflexive element). The self-study produced recommendations for the next 10 years, and the study and its recommendations were scrutinized by a committee of academics and administrators sent by our regional affiliation. Changes were and will be implemented as a result of the study. The interesting thing to me is that 80 to 90% of the study deals with things that are only tangentially related to actual learning, which is what we think the university is about. Most of the study addresses symbolic or political issues that have little to do with what happens in the classroom or in your learning experience. The greater proportion of the changes, then, would not have come about except for their symbolic or political values and reflexive organization.

As Weber pointed out, modern organizations are bureaucratic in nature and are thus bound up with rational goal setting, recursive practices, and continual reflexivity. For example, the reaccreditation study I just mentioned will be repeated in 10 years and every 10 years thereafter. This year, my department is doing its self-study, and it gets repeated every five years. When I worked for Denny's restaurants as a manager, we had corporate plans that helped form the regional plans that helped create the unit plans, which strongly influenced my personal plans as a manager. Depending on the level, those plans were systematically evaluated every one to five years. Modern organizations, institutions, and society at large are thus defined through the continued use of reflexive evaluation.

One further point about radical reflexivity: it forms part of our basic understanding of knowledge and rational life. Modern knowledge is equivalent to scientific knowledge, and part of what makes knowledge scientific is continual scrutiny and systematic doubt. This understanding of knowledge is woven into the fabric of our culture. Every child in the United States receives training in what is called "scientific literacy." According to the National Academy of Sciences (1995), "This nation has established as a goal that all students should achieve scientific literacy. The *National Science Education Standards* are designed to enable the nation to achieve that goal. They spell out a vision of science education that will make scientific literacy for all a reality in the 21st century." Thus, children in the United States are systematically trained to be reflexive about knowledge in general.

Emptying Time and Space

In this section, it is very important for you to keep in mind what Giddens means by time–space distanciation—it's his way of talking about how our behaviors and actions are patterned and are thus somewhat predictable. Therefore, whatever happens to time and space in modernity influences the patterns of interaction that make up society. With that in mind, Giddens argues that the *separation of time and space* is crucial to the dynamic quality of modernity.

In order to understand how time and space can be emptied and separated, we have to begin by thinking about how humans have related to time and space for most of our existence. Up until the beginnings of modernity, time and space were closely linked to natural settings and cycles. People have always marked time, but it was originally associated with natural places and cycles. The cycle of the sun set the boundaries of the day, the cycle of the moon marked the month, and the year was noted by the cycles of the seasons. But the week, which is the primary tool we use to organize ourselves today, exists nowhere in nature—it's utterly abstract in terms of the natural world. Something similar may be said about the mechanical clock. Previous to the invention and widespread use of the mechanical clock, people regulated their behaviors around the moving of the sun (see McCready, 2001; Roy, 2001, pp. 40–45.).

Thus, in modern societies, time and space have become abstract entities that have been emptied of any natural connections. Further, the concept of place itself has become stretched out and more symbolic than physical. As I mentioned, modernity is distinguished by the belief in progress. Progress implies change, and the emptying of time and space "serves to open up manifold possibilities of change by breaking free from the restraints of local habits and practices" (Giddens, 1990, p. 20). Making time and place abstract has also aided in another distinctive feature of modernity, the bureaucratic organization. Our lives are subject to rational organization precisely because time and place are emptied of natural and social relations. My students and I can all meet at 9:45 A.M. in the Graham building, Room 308, for class because time and place have been emptied. Similarly, Boeing airplane manufacturing in California can order parts from a steel plant in China to be ready for assembly beginning in January because time and place are abstract.

The emptying of time and place means that time–space distanciation can be increased almost without limit, which is one of the defining characteristics of modernity. Traditional societies are defined by close-knit social networks that create high levels of morality, and an emphasis on long-established social practices and relationships. Any social form that could break with the importance of tradition would have to be built upon something other than close-knit social groups. Modernity, then, is defined as the time during which greater and greater distances are placed between people and their social relations. As we'll see, increasing time–space distanciation and escalating reflexivity mutually reinforce one another, and together they create the dynamism of modernity—the tendency for continual change.

Institutions and Disembedding Mechanisms

In discussing the transition from traditional to modern society, many sociologists talk about structural differentiation, especially functionalists. The problem that Giddens sees in institutional differentiation is that it can't give a reasoned account of a central feature of modernity: radical time–space distanciation. However, thinking about institutions in terms of disembedding mechanisms can do so. Thus, Giddens claims that the distinction between traditional and modern institutions isn't differentiation so much as it is embedding versus disembedding. **Disembedding mechanisms** are those practices that lift out social relations and interactions from local contexts. Again, let's picture a kind of ideal type of traditional society where most social relationships and interactions take place in encounters that are firmly entrenched in local situations. People would live in places where they knew everybody and would depend upon people they knew for help. Distant situations, along with distant others, were kept truly distant. There are two principal mechanisms that lifted life out of its local context: symbolic tokens and expert systems.

Symbolic tokens are understood in terms of media of exchange that can be passed around without any regard for a specific person or group. There are a few of these kinds of tokens around, but the example par excellence is money. Money creates a universal value system wherein every commodity can be understood according to the same value system. Of necessity, this value system is abstract; that is, it has no intrinsic worth. In order for it to stand for everything, it must have no value in itself. The universal and abstract nature of money frees it from constraint and facilitates exchanges over long distances and time periods. Thus, by its very nature, money increases time–space distanciation. The greater the level of abstraction of money, such as through credit and soft currencies, the greater will be this effect.

The other disembedding mechanism that Giddens talks about is *expert systems.* Let's again think about a traditional community. If you were a woman who lived in a traditional community and were going to have a baby, to whom would you go? If in the same group you experienced marital problems, where would you go for advice? If you wanted to know how to grow better crops or appease the gods or construct a building or do anything that required some form of social cooperation, where would you go? The answer to all these questions, and all the rest of the details of living life, would be your social network. If you wanted to grow better crops, you might go to your friend Paul whose fields always seem to thrive and produce abundantly. For marital advice, you would probably go to your grandparents; for childbirth help, you'd go to the neighbor's wife who had been practicing midwifery for as long as you can remember.

Where do we go for these things today? We go to experts—people that we don't personally know, who have been trained academically in abstract knowledge. But we don't really have to "go to" an expert to be dependent upon expert systems of knowledge. For instance, I have no idea how to construct a building that has many levels and can house a myriad of classrooms and offices, yet I'm dependent upon that expert knowledge every time I go to my office or teach in a classroom. Every time we turn on a computer or flick a light switch or start our car or go to buy food at the grocery store—in short, every time we do anything that is associated with living in modernity—we are dependent upon abstract, expert systems of knowledge. Systems of expert knowledge are disembedding because they shift the center of our life away from local contexts and toward dependence on abstract knowledge and distant others, who sometimes never even appear.

Globalization

Giddens argues that four institutions in particular form the dynamic and the time period of modernity: capitalism, industrialism, monopoly of violence, and surveillance. In terms of the dynamic of modernity, capitalism stands out. Capitalism is intrinsically expansive. It is driven by the perceived need for profit, which in turn drives the expansion of markets, technologies, and commodification.

Industrialization is of course linked to capitalism, but it has its own dynamics and relationships with the other institutional spheres. Industrialism, the monopoly of coercive power, and surveillance feed one another and create what is generally referred to as the industrial-military complex. A military complex is formed by a standing army and the parts of the economy that are oriented toward military production. Once a coercive force begins to use technology, it not only becomes dependent upon industrialism, but it also provides a constant impetus for more and better technologies of force and surveillance. A military complex by its very existence is not only available for protection, it is also in its best interests to instigate aggression whenever possible in order to expand its own base and the interests of its institutional partners.

Thus, the institutional dimensions of modernity are explicitly tied up with globalization. *Globalization* is a term that was coined in the early 1990s to describe

> the closer integration of the countries and peoples of the world which has been brought about by the enormous reduction of costs of transportation and communication, and the breaking down of the ratification barriers to the flows of goods, services, capital, knowledge, and (to a lesser extent) people across borders. (Stiglitz, 2003, p. 9)

While the boundaries of what is to be included in the definition of globalization are unclear, it is most commonly seen as an economic phenomenon, one that is focused on free trade. In this economy, as in most economists' models, market forces and invisible hands operate like devices of natural selection to control prices and the behaviors of firms.

That last part is important for understanding what sociologists do with the term. Rather than assuming market forces, sociologists generally define globalization more precisely around explicit social factors. Three theorists in this book talk about globalization: Giddens, Niklas Luhmann, and Immanuel Wallerstein. As you'll see, each of these theorists defines the term somewhat differently. Because globalization is possibly one of the most important social processes that will influence your life through the twenty-first century, I encourage you to keep track of the different elements of this idea as we consider Giddens, Luhmann, and Wallerstein. Synthesizing these theories will give you a fuller understanding of how globalization works.

Giddens (1990) defines globalization as "the intensification of worldwide social relations which link distant localities in such a way that local happenings are shaped by events occurring many miles away and vice versa" (p. 64). Globalization is thus defined in terms of a dialectic relation between the local and the distant that *further stretch out co-presence through time and space.* The four dimensions of globalization, according to Giddens, are the world capitalist economy, the world military order, the international division of labor, and the nation-state system.

In order to help us get a handle on what Giddens is arguing, I've drawn out the chief processes that we've been talking about in Figure 11.3. Most all of the factors on the far left of the model are interrelated in some way. For example, the use of bureaucratic, rational management increases in the presence of world capitalism and expert systems. But to draw out all the relationships at that level would defeat the purpose of the model as a heuristic device. I have indicated the mutual effects at the next level. All of these dynamics—radical reflexivity, separation of time and place, disembedding, globalization—mutually imply and affect one another. For example, as time and place are separated from the actual, institutions can further remove the social from the local, which in turn allows more abstract connections at the global level. These all mutually reinforce one another and build the dynamism of modernity. Collectively, this figure and all that it implies answers the question, "Why are change and discontinuity endemic in modernity?" Use Figure 11.3 to think through that question and Giddens's theory of modernity.

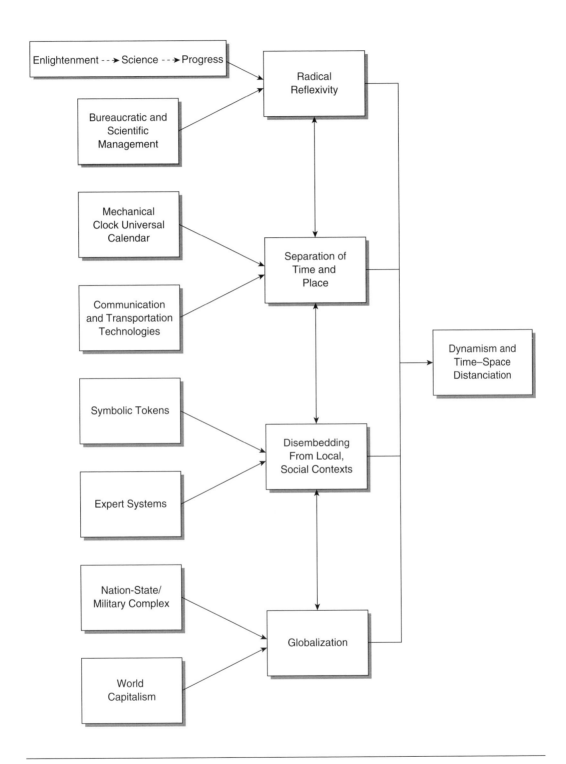

Figure 11.3   Dynamism of Modernity

Concepts and Theory:
The Experience of Modernity

It is extremely difficult to see the effects of modernity in our own lives. We live our lives as if they are essential, as if there were nothing more to us than our inner personality and experiences. Yet sociology teaches us that we are social beings created for and out of social relations, and the sociological imagination encourages us to see the intersections of biography, history, and society. Giddens paints a portrait of the modern individual and asks us to look behind (or in front of) our own subjective experiences and understand them as finding their roots in a particular social organization called modernity. Modernity is characterized by endemic reflexivity and time–space distanciation. What this implies for the person is that the individual and her or his subjective experiences have been lifted out of densely packed social networks and required to do increasing amounts of work on their own. Quite a bit of this personal work centers on the reflexive project of the self, life politics, and intimate relations.

The Reflexive Project of the Self

Recall that in Chapter 4, Mead argued that reflexivity is a necessary state of existence for the self. In his understanding of reflexivity, Giddens defines it as a variable: while some level of reflexivity is essential, humans can be more or less reflexive.

In times previous to modernity, the self was deeply embedded in the social. People were caught up in and saw themselves only in terms of the group. The self was an extension of the group just as certainly as your arm is an extension of your body. The individual life was not only seen as part of the group, but its trajectory was also plotted and marked socially. So, for example, a boy knew for certain when he had changed from a boy to a man—he went through a rite of passage. Such is not the case today.

In modernity, the individual stands alone. The individual is "free." You, for example, are free to express yourself in any number of ways. You have a plethora of potential identities and experiences open to you. But at the same time, you no longer have any institutional markers to guide you or to define your "progress," and there are no institutions that are directly responsible for you. In the *reflexive project of the self,* you have to reflexively define your own options and opportunities with regard to social and personal change. The self is no longer an entity embedded in known and firm social and institutional relationships and expectations. This shift from the traditional, social self with clear institutional guidelines to the individual reflexive project was brought about because of the dynamics of modernity that we reviewed in the previous section.

The body is drawn into this reflexive project as well. Before radical modernity, the body was, for the most part, seen as either the medium through which work was performed or a vehicle for the soul. In either case, it was of little consequence and received little attention unless it became an obstacle to work or salvation. In radical modernity, on the other hand, the body becomes part of self-expression and helps to sustain "a coherent sense of self-identity" (Giddens, 1991, p. 99). The body becomes wrapped up with the reflexive project of the self in four possible ways: appearance, demeanor, sensuality, and through bodily regimes. We covered the first two ways in the chapter on Goffman, so here we will just review the last two.

The body is involved in the self-project through *bodily regimes*. In radical modernity, "we become responsible for the design of our own bodies" (Giddens, 1991, p. 102). The body is no longer a simple reflection of one's work but can become a canvas for a self-portrait. Capitalism, mass media, advertising, fashion, and medical expert knowledge have produced an overabundance of information about how the body works and what kinds of behaviors result in what kinds of body images. We are called upon to constantly review the look and condition of our body and to make adjustments as necessary. The adjustments are carried out through various body regimens of diet, exercise, stress-reducing activities (yoga, meditation), vitamin therapies, skin cleansing and repair, hair treatments, and so forth.

With the *organization of sensuality*, Giddens has in mind the entire spectrum of sensual feeling of the body, but the idea is particularly salient for sexuality. Together, mass education, contraceptive technologies, decreasing family size, and women's political and workforce participation created the situation where "today, for the first time in human history, women claim equality with men" (Giddens, 1992, p. 1). Giddens links women's freedom with the creation of an "emotional order" that contains "an exploration of the potentialities of the 'pure relationship'" and "plastic sexuality" (pp. 1–2). The idea of *plastic sexuality* captures a kind of sexuality that came into existence as sex was separated from the demands of reproduction. Plastic sexuality is an explicit characteristic of modernity. For the first time in history, sexuality could become part of self-identity. We should also note that since sexuality is part of the reflexive project of the self, it is subject to reflexive scrutiny and intentional exploration.

Pure Relationships

To begin our discussion of pure relationships, let's think about friendship. Giddens points out that early Greeks didn't even have a word for friend in the way we use it today. The Greeks used the word *philos* to talk about those who were the most near and dear, but this term was used for people who were in or near to family. And the Greek *philos* network was pretty well set by the person's status position; there was little in the way of friends as we think of them, as personal choices.

In traditional societies in which languages that did have a word for friend, these friends were seen within the context of group survival. Friends were the in-group and others were the out-group. The distinction was between friend and enemy, or, at best, stranger. Keep in mind that groups were far more important then than they are now because individual survival was closely tied to group affiliations and resources. A friend was someone you turned to in time of need; thus, the values associated with friendship were honor and sincerity. Today, however, because of disembedding mechanisms and increased time–space distanciation, not all friends are understood in terms of in-group membership and actual assistance. The individual can have distant friends and is enabled and expected to take care of her- or himself (the reflexive project).

A fundamental change has thus occurred in friendships: from friendship with honor based on group identity and survival, to friendship with authenticity based on a mutual process of self-disclosure. Rather than trust being embedded in social networks and rituals, trust in modernity has to be won, and the means through which this is done is self-evident warmth and openness. By implication, this authenticity and self-regulation provide the personal, emotional component missing in trust in the abstract systems of modernity.

Intimate relations in modernity are thus characterized by *pure relationships*— friendships and intimate ties that are entered into simply for what the relationship can bring to each person. Remember that traditional relationships were first set by existing networks and institutions and the motivation behind them was usually social, not personal. For example, most marriages were motivated by politics or economics (not by love) and were arranged for the couple by those most responsible for the social issues in question (not by the couple themselves). This is the way in which modern relationships are pure: they occur purely for the sake of the relationship. Most of our relationships today are not anchored in external conditions, like the politically or economically motivated marriage. Rather, they are "free-floating." The only structural condition for a friendship or marriage is proximity: we have to be near enough to make contact. But with modern transportation and communication technologies, our physical space is almost constantly in motion and can be quite far-ranging, and we have "virtual" space at our fingertips.

In addition to the free-floating and pure nature of these relations, they are also reflexively organized, based on commitment and mutual trust, and focus on intimacy and "self"-growth. Like the reflexive project of the self, relationships are reflexively organized; that is, they are continually worked at by the individuals, who tend to consult an array of sources of information. The number of possible sources for telling us how to act and be in our friendships and sexual relations is almost endless. Daytime television is filled with programming that explores every facet of relationships; the magazine rack at the local supermarket is a cornucopia of surveys and advice on how to have the best communication/sex life/romance, or any other aspect of an intimate relationship; it's estimated that over 2,000 new self-help book titles are published every year in the United States; and the Internet resources available for improving relationships are innumerable. Most of us have taken a relationship quiz with our partner at some point (if you haven't, just wait—it's coming), and all of us have asked of someone the essential question for relationships that are reflexively organized: "Is everything all right?" This kind of communication is a moral obligation in pure relationships; the gamut of communication covers everything from the mundane ("How was your day at work?") to the serious ("Do you want to break up with me?").

Choice and Life Politics

Along with the accelerating changes in modernity, there has been a shift from emancipatory politics to life politics. *Emancipatory politics* is concerned with liberating individuals and groups from the constraints that adversely affect their lives. In some ways, this type of political activity has been the theme of modernity—it was the hope that democratic nation-states could bring equality and justice for all. And, in some respects, this theme of modernity has failed. We are more than ever painfully aware of how many groups are disenfranchised.

Life politics, by way of contrast, is the politics of choice and lifestyle. It is not based on group membership and characteristics, as is emancipatory politics; rather, it is based on personal lifestyle choices. We have come to think of choice as a freedom we have in the United States. But it is more than that—it has become an obligation, a fundamental element in contemporary living. This principality of choice is based on disembedding mechanisms and time–space distanciation, and results in, as we've seen, the reflexive project of the self. Part of that project comes to be centered on the politics of choice.

Mass media also play a role in creating choice by facilitating mediated experiences. *Mediated experiences* are a contrast to social experiences that take place in face-to-face encounters and are created as people are exposed to multiple accounts of situations and others with whom they have no direct association through time and space. Every time you watch television or read a newspaper, you have experiences that are "mediated." You are exposed to lives to which you have absolutely no real connection, and like so many other features of modernity, this stretches out co-presence but it also creates a collage effect. The pictures and stories that we receive via the media do not reflect any essential or social elements. Instead, stories and images are juxtaposed that have nothing to do with one another. The picture we get of the world, then, is a collage of diverse lifestyles and cultures, not a direct representation.

As a result of being faced with this collage, what happens to us as individuals? One implication is that the plurality of lifestyles presented to us not only *allows* for choice, it *necessitates* it. In other words, what becomes important is not the issue of group equality, but rather, the insistence on *personal* choice. What is at issue in this milieu is not so much political equality (as with emancipatory politics) as inner authenticity. In a world that is perceived as constantly changing and uprooted, it becomes important to be grounded in one's self. Life politics creates such grounding. It creates "a framework of basic trust by means of which the lifespan can be understood as a unity against the backdrop of shifting social events" (Giddens, 1991, p. 215). Life politics, then, helps to diminish the possibility and effects of ontological insecurity.

A good example of life politics is veganism—the practice of not eating any meat or meat byproducts. Not only is eating flesh avoided, but also any products with dairy, eggs, fur, leather, feathers, or any goods involving animal testing. One vegan I know summed it up nicely when she said, veganism "is an integral component of a cruelty-free lifestyle." It is a political statement against the exploitation of animals, and for some it is clearly a condemnation of capitalism—capitalism is particularly responsible for the unnatural mass production of animal flesh as well as commercial animal testing. Yet, for most vegans, it is a lifestyle, one that brings harmony between the outside world and inner beliefs, and not necessarily part of a collective movement.

However, it would be wrong to conclude that life politics are powerless because they do not result in a social movement. Quite the opposite is true. Life politics springs from and focuses attention on some of the very issues that modernity represses. What life politics does is to "place a question mark against the internally referential systems of modernity" (Giddens, 1991, p. 223). Life politics asks, "Seeing that these things are so, what manner of men and women ought we to be?" In traditional society, morality was provided by the institutions, especially religion. Modernity has wiped away the social ground upon which this kind of morality was based. Life politics "remoralizes" social life and demands "renewed sensitivity to questions that the institutions of modernity systematically dissolve" (p. 224). Rather than asking for group participation, as does emancipatory politics, life politics asks for self-realization, a moral commitment to a specific way of living. Rather than being impotent in comparison to emancipatory politics, life politics "presage[s] future changes of a far-reaching sort: essentially, the development of forms of social order 'on the other side' of modernity itself" (p. 214).

Giddens Summary

- According to Giddens, the central issue for social theory is to explain how actions and interactions are patterned over time and space; or, to use Giddens's terms, social theory needs to explain how the limitations inherent within physical presence are transcended through time–space distanciation. There are two primary ways through which this occurs: the dynamics of structuration and of routinization.

- Structuration occurs when people use specific modalities to produce both structure (rules and resources of signification, domination, and legitimation) and practice (physical co-presence). Thus, the very method of structuration reflexively and repeatedly links structure and person and facilitates time–space distanciation.

- Routinization is psychologically motivated by a diffuse need for ontological security. The reality of society is precarious because it depends on structuration, which is reflexive and recursive. In other words, the process of structuration doesn't reference anything other than itself and it depends on ceaseless interactional work. This precariousness is unconsciously sensed by people, which, in turn, motivates them to routinize their actions and interactions and to link their routines to physical regions and institutional orders, which further adds stability.

- Routinization was unproblematically achieved in traditional societies. People rarely left their regions and the institutional orders were slow to change. Modernity, however, is characterized by dynamism and increasing time–space distanciation. Dynamism and time–space distanciation are both directly related to radical reflexivity, extreme separation of time and place, the disembedding work of modern institutions, and globalization. These factors are related to the proliferation of science and progress, bureaucratic management, the mechanical clock and universal calendar, communication and transportation technologies, symbolic tokens and expert systems of knowledge, the military complex, and world capitalism.

- As a result of radical modernity, the individual is lifted out of the social networks and institutions that socially situated the self by acquiring certain identities, knowledge, and life-course markers. The modern individual is given the reflexive project of the self that is only internally referential. As part of the reflexive project of the self, the individual involves him- or herself in strategic life planning using expert systems of knowledge and mediated experiences, all of which are permeated with doubt. The reflexive project of the self involves constant evaluation and reevaluation based on possible new information (ever revised by experts and available through mass media) and self-reflection (How am I doing? Should I be feeling this way?). The reflexive project includes lifestyle politics in which the individual must reflexively work her or his way through continuously presented and expanding arenas of social existence. Individuals, then, become hubs for social change as they reflexively order their life in response to a constantly changing political landscape.

Pierre Bourdieu: Constructivist Structuralism and Class

One of the wonderful things about working in academia is that it is part of my job to learn things. In preparing to write this chapter, I read Craig Calhoun's (2003) chapter on Bourdieu. The first section is entitled "Taking Games Seriously." In it, Professor Calhoun talks about Bourdieu's life as a former rugby player and how it influenced his theory. Throughout Bourdieu's writing, he uses such terms as "field," "game," and "practice," and he talks about the bodily inculcation of culture. I had read Bourdieu and approached such terms and ideas as theoretical concepts. For some reason, it never occurred to me to understand their use as a kind of analogy— the analogy of the game. But as I read Calhoun's three pages about Bourdieu's fascination with rugby, his ideas and terms all came alive for me in a new way.

So, thanks to Craig Calhoun, the first thing we will talk about in introducing Bourdieu is his analogy of the game. It's important to keep in mind that Bourdieu's use of the analogy doesn't come from a background in playing cards. Bourdieu was a rugby player. Rugby is a European game somewhat like American football, but it is considered by most to be much more grueling. In rugby, the play is continuous with no substitutions or time-outs (even for injury). The game can take anywhere from 60 to 90 minutes, with two halves separated by a 5-minute halftime. An important part of the game is the scrum. In a *scrum,* eight players from each side form a kind of inverted triangle by wrapping their arms around each other. The ball is placed in the middle and the two bound groups of players struggle head to head against each other until the ball is freed. To see the struggle of the scrum gives a whole new perspective on Bourdieu's idea of social struggle.

Rugby matches take place on a field, involve strategic plays and intense struggles, and are played by individuals who have a clear physical sense of the game. Matches are of course structured by the rules of the game and the field. The field not only delineates the parameters of the play, but each field is also different and thus knowledge of each field of play is important for success. The rules are there and, like in all games, come into play when they are broken, but a good player embodies the rules and the methods of the game. The best plays are those that come when the player is in the "zone," or playing without thinking.

Trained musicians can also experience this zone by jamming with other musicians. Often when in such a state, the musician can play things that he or she normally would not be able to, and might have a difficult time explaining after the fact. The same is true for athletes. There is more to a good game than the rules and the field; the game is embodied in the performer. And, finally, there is the struggle, not only against the other team, but also the limitations of the field, rules, and one's own abilities.

You may not know it, but I just gave you a brief overview of Bourdieu's theory through the use of analogy. I will occasionally mention the game analogy as we work our way through the material, but I think if you keep it consistently in mind, you'll find it much easier to grasp the intent of Bourdieu's thinking.

Photo: © Corbis.

The Essential Bourdieu
Seeing Further: Constructivist Structuralism
    Overcoming Dichotomies
    Defining Constructivist Structuralism
Concepts and Theory: Structuring Class
    Capitals
    Habitus
    Fields
Concepts and Theory: Replicating Class
    Linguistic Markets
    Symbolic Struggle
Bourdieu Summary

The Essential Bourdieu

Biography

Pierre Bourdieu was born August 1, 1930, in Denquin, France. Bourdieu studied philosophy under Louis Althusser at the École Normale Supérieure in Paris. After his studies, he taught for three years, 1955–1958, at Moulins. From 1958 to 1960, Bourdieu did empirical research in Algeria (*The Algerians*, 1962) that laid the groundwork for his sociology. In his career, he published over 25 books, one of which, *Distinction: A Social Critique of the Judgment of Taste*, was named one of the twentieth century's 10 most important works of sociology by the International Sociological Association. He was the founder and director of the Centre for European Sociology, and he held the French senior chair in sociology at Collège de France (the same chair that had been held by sociologist and anthropologist, Marcel Mauss, Durkheim's nephew). Craig Calhoun (2003) writes that Bourdieu was "the most influential and original French sociologist since Durkheim" (p. 274).

Passionate Curiosity

Bourdieu's passion is intellectual honesty and rigor. He of course is concerned with class, and particularly the way class is created and re-created in subtle, non-conscious ways. But above and beyond these empirical concerns is a driving intellect bent on refining critical thinking and never settling on an answer: "an invitation to think with Bourdieu is of necessity an invitation to think beyond Bourdieu, and against him whenever required" (Wacquant, 1992, p. xiv).

Keys to Knowing

Constructivist structuralism; misrecognition; economic, social, symbolic, and cultural capitals; objectified, institutionalized, and embodied cultural capital; symbolic and empirical fields; habitus; taste; distance from necessity; pure gaze; linguistic markets; symbolic violence; self-sanctioning mechanism

Seeing Further: Constructivist Structuralism

While Bourdieu's work covers a diverse landscape, I think it is fair to say that his focus is on the replication of class. In this, his work is Marxist, and there is a sense in which Bourdieu's theory may be seen as the mirror image of Marx. According to Marx, the economy and class are two of the most important structures in society. Bourdieu's theory begins with material class, but he clearly moves the reproduction of class structures into the symbolic realm. In the reproduction of class, it is the symbolic field and the relations expressed by and through what he calls habitus that have the greater causal force. Both the symbolic field and habitus are unique kinds of structures that are in tension one with another, and this tension is generative— it not only creates and reproduces class, it also allows for new and unexpected behaviors. But I'm getting ahead of myself. What I want to point out here is that

Bourdieu uses a new theoretical approach to understand how class positions are reproduced: constructivist structuralism.

Overcoming Dichotomies

Bourdieu brings the two sides of the structure–agent dichotomy together in *constructivist structuralism* (or structuralist constructivism—Bourdieu uses the term both ways). In this scheme, both structure and agency are given equal weight. Concerning social structures, Bourdieu (1989) says that within the social world there are "objective structures independent of the consciousness and will of agents, which are capable of guiding and constraining their practices or their representations" (p. 14). He thus sees social structures in much the same way that we talked about them in Chapter 8—they both facilitate and restrain human behavior.

Yet, at the same time, Bourdieu emphasizes the agent and subjective side. In Bourdieu's (1989) scheme, the subjective side is also structured in terms of "schemes of perception, thought, and action" (p. 14), which he calls habitus. Part of what Bourdieu does is to detail the ways through which both kinds of structures are constructed; thus, there is a kind of double structuring in Bourdieu's theory and research. But Bourdieu doesn't simply give us an historical account of how structures are produced. His theory also offers an explanation of how these two structures are dialectically related and how the individual uses them strategically in linguistic markets.

In preserving both sides of the dichotomy, Bourdieu has created a unique theoretical problem. He doesn't want to conflate the two sides as Giddens does, nor does he want to link them up in the way that Collins and Blau do. He wants to preserve the integrity of both domains and yet he characterizes the dichotomy as harmful. He is thus left with a sticky problem: how can Bourdieu keep and yet change the dichotomy between the objective and subjective moments without linking them or blending them together?

Let's take this issue out of the realm of theory and state it in terms that are a bit more approachable. The problem that Bourdieu is left with is the relationship between the individual and society. Do we have free choice? Bourdieu would say yes. Does society determine what we do? Again, Bourdieu would say yes. How can something be determined and yet the product of free choice? I've stated the issue a bit too simplistically for Bourdieu's theory, but I want you to see the problem clearly. Structure and agency, or the objective and subjective moments, create tension because they stand in opposition to one another. And that tension is exactly how Bourdieu solves his theoretical problem.

Bourdieu argues that the objective and constructive moments stand in a dialectical relationship. Understanding dialectical processes is extremely important here. Therefore, if you're at all hazy about how it works, please go back and review the sections where we've talked about it previously (see Chapters 1 and 7). Bourdieu's dialectic occurs between what he calls the field and the habitus. Both are structures; *habitus* is "incorporated history" and the field is "objectified history" (Bourdieu,

1980/1990, p. 66). We will go into more depth later on, but for now think of habitus as that part of society that lives in the individual as a result of socialization, and think of the field as social structures. They are both much more, but what I want us to see now is Bourdieu's dialectic. The tension of the dialectic is between the subjective and objective structures. The dialectic itself is found in the individual and collective struggles or practices that transform or preserve these structures through specific practices and linguistic markets.

In other words, Bourdieu is arguing that a number of different elements in our lives are structured, and among them are the habitus of the individual (schemes of thought, feeling, and action) and the social field (structured social positions and the distribution of resources). These different structures dialectically exert force upon one another through the strategic actions and practices of people in interaction. And, as with most dialectics, the tension can produce something new and different out of the struggle; these differences can then influence the structures of habitus and field.

Defining Constructivist Structuralism

Briefly then, constructivist structuralism is Pierre Bourdieu's theoretical perspective. It's a way of seeing the social world that does away with dualisms such as object/subject and structure/agent. Bourdieu sees a dialectical tension between constructive and structuring social elements. The dialectic indicates that the elements are in tension (structuring and constructing) with one another and that the outcome includes both but is different from either.

Concepts and Theory: Structuring Class

Capitals

The basic fact of capitalism is capital. Capital is different from either wealth or income. Income is generally measured by annual salary and wealth by the relationship between one's assets and debt. Both income and wealth are in a sense static; they are measurable facts about a person or group. Capital, on the other hand, is active: it's defined as accumulated goods devoted to the production of other goods. The entire purpose of capital is to produce more capital.

Bourdieu actually talks about four forms of capital—economic, social, symbolic, and cultural—all of which are invested and used in the production of class. Bourdieu uses *economic capital* in its usual sense. It is generally determined by one's wealth and income. As with Marx, Bourdieu sees economic capital as fundamental. However, unlike Marx, Bourdieu argues that the importance of economic capital is that it strongly influences an individual's level of the other capitals, which, in turn, have their own independent effects. In other words, economic capital starts the ball rolling; but once things are in motion, other issues may have stronger influences on the perpetuation of class inequalities.

Social capital refers to the kind of social network an individual is set within. It refers to the people you know and how they are situated in society. The idea of social capital can be captured in the saying, "it isn't what you know but who you know that counts." The distribution of social capital is clearly associated with class. For example, if you are a member of an elite class, you will attend elite schools such as Phillips Andover Academy, Yale, and Harvard. At those schools, you would be afforded the opportunity to make social connections with powerful people—for example, in elections over the past 30 years, there has been at least one Yale graduate running for the office of president of the United States. But economic capital doesn't exclusively determine social capital. We can build our social networks intentionally, or sometimes through happenstance. For example, if you attended Hot Springs High School in Arkansas during the early 1960s, you would have had a chance to become friends with Bill Clinton.

Symbolic capital is the capacity to use symbols to create or solidify physical and social realities. With this idea, Bourdieu begins to open our eyes to the symbolic nature of class divisions. Social groups don't exist simply because people decide to gather together. Max Weber recognized that there are technical conditions that must be met for a loose collection of people to form a social group: people must be able to communicate and meet with one another; there must be recognized leadership; and a group needs clearly articulated goals to organize. Yet, even meeting those conditions doesn't alone create a social group. Groups must be symbolically recognized as well.

With the idea of symbolic capital, Bourdieu pushes us past analyzing the use of symbols in interaction. Symbolic interactionism argues that human beings are oriented toward meaning, and meaning is the emergent result of ongoing symbolic interactions. We're symbolic creatures, but meaning doesn't reside within the symbol itself; it must be pragmatically negotiated in face-to-face situations. We've learned a great deal about how people create meaning in different situations because of symbolic interactionism's insights. But Bourdieu's use of symbolic capital is quite different.

Bourdieu recognizes that all human relationships are created symbolically and not all people have equal symbolic power. For example, I write a good number of letters of recommendation for students each year. Every form I fill out asks the same question: "Relationship to applicant?" And I always put "professor." Now, the *meaning* of the professor–student relationship emerges out of my interactions with my students, and my professor–student relationships are probably somewhat different from some of my colleagues as a result. However, neither my students nor I *created* the professor–student relationship.

Recall our earlier discussion about Bourdieu's critique of sociology's dichotomy. Here we can see both Bourdieu's critique and his answer: social phenomenology can't account for the creation of the categories it uses, and social physics reifies the categories—Bourdieu (1991) tells us that objective categories and structures, such as class, race, and gender, are generated through the use of symbolic capital: "Symbolic power is a power of constructing reality" (p. 166).

Bourdieu (1989) characterizes the use of symbolic capital as both the power of constitution and the power of revelation—it is the power of "world-making . . . the power to make groups. . . . The power to impose and to inculcate a vision of divisions, that is, the power to make visible and explicit social divisions that are implicit, is political power par excellence" (p. 23). This power of world-making is based on two elements. First, there must be sufficient recognition in order to impose recognition. The group must be recognized and symbolically labeled by a person or group that is officially recognized as having the ability to symbolically impart identity, such as scientists, legislators, or sociologists in our society. Institutional accreditation, particularly in the form of an educational credential (school in this sense operates as a representative of the state), "frees its holder from the symbolic struggle of all against all by imposing the universally approved perspective" (p. 22).

The second element needed to world-make is some relation to a reality—"symbolic efficacy depends on the degree to which the vision proposed is founded in reality" (Bourdieu, 1989, p. 23). I think it's best to see this as a variable. The more social or physical reality is already present, the greater will be the effectiveness of symbolic capital. This is the sense in which symbolic capital is the power to consecrate or reveal. Symbolic power is the power to reveal the substance of an already occupied social space. But note that granting a group symbolic life "brings into existence in an instituted, constituted form . . . what existed up until then only as . . . a collection of varied persons, a purely additive series of merely juxtaposed individuals" (p. 23). Thus, because legitimated existence is dependent upon symbolic capacity, an extremely important conflict in society is the struggle over symbols and classifications. The heated debate over race classification in the U.S. 2000 census is a good example.

There is a clear relationship between symbolic and cultural capital. The use of symbolic capital creates the symbolic field wherein cultural capital exists. In general, **cultural capital** refers to the informal social skills, habits, linguistic styles, and tastes that a person garners as a result of his or her economic resources. It is the different ways we talk, act, and make distinctions that are the result of our class. Bourdieu identifies three different kinds of cultural capital: objectified, institutionalized, and embodied. *Objectified cultural capital* refers to the material goods (such as books, computers, and paintings) that are associated with cultural capital. *Institutionalized cultural capital* alludes to the certifications (like degrees and diplomas) that give official acknowledgment to the possession of knowledge and abilities. *Embodied cultural capital* is the most important in Bourdieu's scheme. It is part of what makes up an individual's habitus, and it refers to the cultural capital that lives in and is expressed through the body. This function of cultural capital manifests itself as taste.

Taste refers to an individual preference or fondness for something, such as "she has developed a taste for expensive wine." What Bourdieu is telling us is that our tastes aren't really individual; they are strongly influenced by our social class—our tastes are embodied cultural capital. Here a particular taste is legitimated, exhibited, and recognized by only those who have the proper cultural code, which is class specific. To hear a piece of music and classify it as baroque rather than elevator music implies an entire world of understandings and classification. Thus, when individuals express a preference for something or classify an object in a particular way, they are simultaneously classifying themselves. Taste may appear as an innocent and natural phenomenon, but it is an insidious revealer of position. As Bourdieu (1979/1984) says, "Taste classifies, and it classifies the classifier" (p. 6). The issue of taste is "one of the most vital stakes in the struggles fought in the field of the dominant class and the field of cultural production" (p. 11).

Habitus

Taste is part of habitus, and habitus is embodied cultural capital. Class isn't simply an economic classification (one that exists because of symbolic capital), nor is it merely a set of life circumstances of which people may become aware (class consciousness)—class is inscribed in our bodies. **Habitus** is the durable organization of one's body and its deployment in the world. It is found in our posture, and our way of walking, speaking, eating, and laughing; it is found in every way we use our body. Habitus is both a system whereby people organize their own behavior and a system through which people perceive and appreciate the behavior of others.

Pay close attention: this system of organization and appreciation is felt in our bodies. We physically feel how we should act; we physically sense what the actions of others mean, and we approve of or censure them physically (we are comfortable or uncomfortable); we physically respond to different foods (we can become voracious or disgusted); we physically respond to certain sexual prompts and not others—the list can go on almost indefinitely. Our humanity, including our class position, is not just found in our cognitions and mental capacity; it is in our very bodies.

One way to see what Bourdieu is talking about is to recall the rugby analogy. I love to play sand volleyball, and I only get to play it about once every five years, which means I'm not very good at it. I have to constantly think about where the ball and other players are situated. I have to watch to see if the player next to me is going for the ball or if I can do so. All this watching and mental activity means that my timing is way off. I typically dive for the ball 1.5 seconds too late, and I end up with a mouthful of sand (but the other bunglers on my team are usually impressed with my effort). Professional volleyball players compete in a different world. They rarely have to think. They sense the ball and their teammates, and they make their moves faster than they could cognitively work through all the particulars. Volleyball is inscribed in their bodies.

Explicating what he calls the Dreyfus model, Bent Flyvbjerg (2001) gives us a detailed way of seeing what is going on here. The Dreyfus model indicates that there are five levels to learning: novice, advanced beginner, competent performer, proficient performer, and expert. Novices know the rules and the objective facts of a situation; advanced beginners have concrete knowledge but see it contextually; and the competent performer employs hierarchical decision-making skills and feels responsible for outcomes. With proficient performers and experts, we enter another level of knowledge. The first three levels are all based on cognitions, but in the final two levels, knowledge becomes embodied. Here situations and problems are understood "intuitively" and require skills that go beyond analytical rationality. With experts, "their skills have become so much a part of themselves that they are not more aware of them than they are of their own bodies" (p. 19).

Habitus thus works below the level of conscious thought and outside the control of the will. It is the embodied, non-conscious enactment of cultural capital that gives habitus its specific power,

> beyond the reach of introspective scrutiny or control by the will . . . in the most automatic gestures or the apparently most insignificant techniques of the body . . . [it engages] the most fundamental principles of construction and evaluation of the social world, those which most directly express the division of labour . . . or the division of the work of domination. (Bourdieu, 1979/1984, p. 466)

Bourdieu's point is that we are all, each one, experts in our own class position. Our mannerisms, speech, tastes, and so on are written on our bodies beginning the day we are born.

There are two factors important in the production of habitus: education and distance from necessity. In *distance from necessity,* necessity speaks of sustenance—the things necessary for biological existence. Distance from the necessities of life enables the upper classes to experience a world that is free from urgency. In contrast, the poor must always worry about their daily existence. As humans move away from that essential existence, they are freed from that constant worry, and they are free to practice activities that constitute an end in themselves. For example, you probably have hobbies. Perhaps you like to paint, act, or play guitar as I do. There is a sense of intrinsic enjoyment that comes with those kinds of activities; they are ends in themselves. The poorer classes don't have that luxury. Daily life for them is a grind, a struggle just to make ends meet. This struggle for survival and the emotional toll it brings are paramount in their lives, leaving no time or resources for pursuing hobbies and "getting the most out of life."

Think of distance from necessity as a continuum, with you and I probably falling somewhere in the middle. We have to be somewhat concerned about our livelihood, but we also have time and energy to enjoy leisure activities. The elite are on the uppermost part of the continuum, and it shows in their every activity. For example, why do homeless people eat? They eat to survive. And if they are hungry enough, they might eat anything, as long as it isn't poisonous. Why do members of the working classes or nearly poor people eat? For the same basic reason: the working classes are much better off than the homeless, but they still by and large live hand to mouth. However, because they are further removed from necessity, they can be more particular about what they eat, though the focus will still be on the basics of life, a "meat and potatoes" menu. Now, why do the elite eat? You could say they eat to survive, but they are never aware of that motivation. Food doesn't translate into the basics of survival. Eating for the elite classes is an aesthetic experience. For them, plate presentation is more important than getting enough calories.

Thus, the further removed we are from necessity, the more we can be concerned with abstract rather than essential issues. This ability to conceive of form rather than function—aesthetics—is dependent upon "a generalized capacity to neutralize ordinary urgencies and to bracket off practical ends, a durable inclination and aptitude for practice without a practical function" (Bourdieu, 1979/1984, p. 54). This aesthetic works itself out in every area. In art, for example, the upper class aesthetic of luxury prefers art that is abstract while the popular taste wants art to represent reality. In addition, distance from economic necessity implies that all natural and physical desires and responses are to be sublimated and dematerialized. People in the working class, because it is immersed in physical reality and economic necessity, interact in more physical ways through touching, yelling, embracing, and so forth than do the distanced elite. A lifetime of exposure to worlds so constructed confers cultural pedigrees, manners of applying aesthetic competences that differ by class position.

This embodied tendency to see the world in abstract or concrete terms is reinforced and elaborated through *education*. One obvious difference between the education of the elite and the working classes is the kind of social position in which education places us. The education system channels individuals toward prestigious or devalued positions. In doing so, education manipulates subjective aspirations (self-image) and demands (self-esteem). Another essential difference in educational experience has to do with the amount of rudimentary scholastics required—the simple knowing and recognizing of facts versus more sophisticated knowledge. This factor varies by number of years of education, which in turn varies by class position. At the lower levels, the simple recitation of facts is required. At the higher levels of education, emphasis is placed on critical and creative thought. At the highest levels, even the idea of "fact" is understood critically and held in doubt.

Education also influences the kind of language we use to think and through which we see the world. We can conceive of language as varying from complex to simple. More complex language forms have more extensive and intricate syntactical elements. Language is made up of more than words; it also has structure. Think about the sentences that you read in a romance novel and then compare them to those in an advanced textbook. In the textbook, they are longer and more complex, and that complexity increases as you move into more scholarly books. These more complex syntactical elements allow us to construct sentences that correspond to multileveled thinking—this is true because both writing and thinking are functions of language. The more formal education we receive, the more complex are the words and syntactical elements of our language. Because we don't just think *with* language, we think *in* language, the complexity of our language affects the complexity of our thinking. And our thinking influences the way in which we see the world.

Here's a simple example: Let's say you go to the zoo, first with my dog and then with three different people. You'd have to blindfold and muzzle my dog, but if you could get her to one of the cages and then remove the blinders, she would start barking hysterically. She would be responding to the content of the beasts in front of her. All she would know is that those things in front of her smell funny, look dangerous, and are undoubtedly capable of killing her, but she's going to go down fighting. Now picture yourself going with three different people, each from a different social class and thus education level. The first person has a high school education. As you stand in front of the same cage that you showed to my dog, he says, "Man, look at all those apes." The second person you go with has had some college education. She stands in front of the cage and says, "Gorillas are so amazing." The third person has a master's level education and says, "Wow, I've never seen *gorilla gorilla, gorilla graueri,* and *gorilla berengei* all in the same cage."

Part of our class habitus, then, is determined by education and its relationship to language. Individuals with a complex language system will tend to see objects in terms of multiple levels of meaning and to classify them abstractly. This type of linguistic system brings sensitivity to the structure of an object; it is the learned ability to respond to an object in terms of its matrix of relationships. Conversely, the less complex an individual's classification system, the more likely are the organizing syntactical elements to be of limited range. The simple classification system is characterized by a low order of abstractedness and creates more sensitivity to the *content* of an object, rather than its structure.

Bourdieu uses the idea of habitus to talk about the replication of class. Class, as I mentioned earlier, isn't simply a part of the social structure; it is part of our body. We are not only categorized as middle class (or working class or elite), we also *act* middle class. Differing experiences in distance from necessity and education determine one's tastes, ways of seeing and experiencing the world, and "the most automatic gestures or apparently most insignificant techniques of the body—ways of walking or blowing one's nose, ways of eating or talking" (Bourdieu, 1979/1984, p. 466). We don't choose to act or not act according to class; it's the result of lifelong socialization. And, as we act in accordance with our class, we replicate our class. Thus, Bourdieu's notion of how class is replicated is much more fundamental and insidious than Marx and more complex than Weber.

However, we would fall short of the mark if we simply saw habitus as a structuring agent. Bourdieu intentionally uses the concept (the idea originated with Aristotle) in order to talk about the creative, active, and inventive powers of the agent. He uses the concept to get out of the structuralist paradigm without falling back into issues of consciousness and unconsciousness. In habitus, class is structured but it isn't completely objective—it doesn't merely exist outside of the individual because it's a significant part of her or his subjective experience. In habitus, class is *structured but not structuring*—because, as with high-caliber athletes and experts, habitus is intuitive. The idea of habitus, then, shows us how class is replicated subjectively and in daily life, and it introduces the potential for inspired behaviors above and beyond one's class position. Indeed, the potential for exceeding one's class is much more powerful with Bourdieu's habitus than with conscious decisions—most athletes, musicians, and other experts will tell you that their highest achievements come under the inspiration of visceral intuition rather than rational processes. It is through habitus that the practices of the dialectic are performed.

Fields

As we talk about Bourdieu's notion of the *field*, keep the rugby analogy in mind. Just like in rugby, fields are delineated spaces wherein "the game" is played. Obviously, in Bourdieu's theoretical use of field, the parameters are not laid out using fences or lines on the ground. The parameters of the theoretical field are delineated by networks or sets of connections among objective positions. The positions within a field may be filled by individuals, groups, or organizations. However, Bourdieu is adamant that we focus on the relationships among the actors and not on the agents themselves. It's not the people, groups, or even interactions that are important; it's the relationships among and between the positions that set the parameters of a field. For example, while the different culture groups (such as theater groups, reading clubs, and choirs) within a region may have a lot in common, they probably do not form a field because there are no explicit objective relationships among them. On the other hand, most all the universities in the United States do form a field. They are objectively linked through accreditation, professional associations, federal guidelines, and so forth. These relationships are sites of active practices; thus, the parameters of a field are always at stake within the field itself. In other words, because fields are defined mostly through relationships and relationships are active, which positions and relationships go into making up the field is constantly changing. Therefore, what constitutes a field is always an empirical question.

Fields are directly related to capitals. The people, groups, and organizations that fill the different objective positions are hierarchically distributed in the field, initially through the overall volume of all the capitals they possess and secondly by the relative weight of the two particular kinds of capital, symbolic and cultural. More than that, each field is different because the various cultures can have dissimilar weights. For example, cultural capital is much more important in academic rather than economic fields; conversely, economic capital is more important in economic fields than in academic ones. All four capitals or powers are present in each, but they aren't all given the same weight. It is the different weightings of the cultures that define the field, and it is the field that gives validity and function to the capitals.

While the parameters of any field cannot be determined prior to empirical investigation, the important consideration for Bourdieu is the correspondence between the empirical field and its symbolic representation. The objective field corresponds to a symbolic field, which is given legitimation and reality by those with symbolic capital. Here symbolic capital works to both construct and recognize empirical, social positions; it creates and legitimates the relations between and among positions within the field. In this sense, the empirical and symbolic fields are both constitutive of class and of social affairs in general. It is the symbolic field that people use to view, understand, and reproduce the objective.

I've pictured Bourdieu's basic theory of class structuring in Figure 11.4. When reading the model, keep in mind Bourdieu's intent with open concepts. This model is simply a heuristic device—something we can use to help us see the world. The figure starts on the far left with the objective field and the distribution of capitals. But for Bourdieu, the objective field isn't enough to account for class reality and replication, and that is where many sociologists stop. Bourdieu takes it further in that the objective field becomes real and potentially replicable through the use of symbolic capital. The use of symbolic capital creates the symbolic field, which in turn orders and makes real the objective field. The exercise of symbolic capital, along with the initial distribution of capitals, creates cultural capital that varies by distance from necessity and by education. Cultural capital produces the internal structuring of class: habitus. But notice that the model indicates that the potential of habitus to replicate is held in question—it is habitus exercised in linguistic markets and symbolic struggles that decides the question.

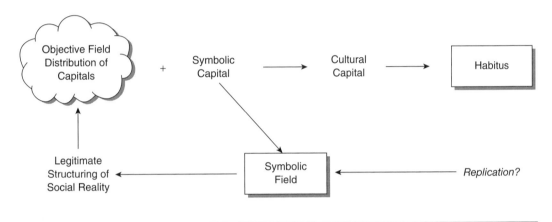

Figure 11.4   Habitus and the Replication of Class

Concepts and Theory: Replicating Class

Linguistic Markets

Bourdieu (1991) says that "every speech act and, more generally, every action" is an encounter between two independent forces (p. 37). One of those forces is habitus, particularly in our tendency to speak and say things that reveal our level of cultural capital. The other force comes from the structures of the linguistic market. A **linguistic market** is "a system of relations of force which impose themselves as a system of specific sanctions and specific censorship, and thereby help fashion linguistic production by determining the 'price' of linguistic products" (Bourdieu & Wacquant, 1992, p. 145).

The linguistic market is like any other market: it's a place of exchange and a place to seek profit. Here exchange and profit are sought through linguistic elements such as symbols and discourses. The notion of a free market is like an ideal type: it's an idea against which empirical instances can be measured. All markets are structured to one degree or another; and linguistic markets have a fairly high degree of structuring. One of the principal ways they are structured is through formal language.

Every society has formalized its language. Even in the case where the nation might be bilingual, such as Canada, the languages are still formalized. Standard language comes as a result of the unification of the state, economy, and culture. The education system is used to impose restrictions on popular modes of speech and to propagate the standard language. We all remember times in grammar school when teachers would correct our speech ("There is no such word as *ain't*."). In the university, this still happens, but mostly through the application of stringent criteria for writing.

Linguistic markets are also structured through various configurations of the capitals and the empirical field. As we've seen, empirical fields are defined by the relative weights of the capitals—so, for example, religious fields give more weight to symbolic capital and artistic fields more import to cultural capital, but they both need and use economic capital. The same is true with linguistic markets. Linguistic markets are defined through the relative weights of the capitals and by the different discourses that are valued. For example, the linguistic market of sociology is heavily based on cultural capital. In order to do well in that market, you would have to know a fair amount about Karl Marx, Émile Durkheim, Michel Foucault, Pierre Bourdieu, Dorothy Smith, and so forth. Linguistic markets are also structured by the empirical field, in particular by the gaps and asymmetries that exist between positions in the field (by their placement and position of capitals, some positions in a field are more powerful than others). These empirical inequalities help structure the exchanges that take place within a linguistic market.

When people interact with one another, they perform speech acts—meaningful kinds of behaviors that are related to language. In a speech act, habitus and linguistic markets come together. In other words, the person's embodied class position and cultural capital are given a certain standing or evaluation within the linguistic market. The linguistic market contains the requirements of formal language; the salient contour of capitals; and the objective, unequal distribution of power within the empirical field.

Let me give you three examples from my own life. When I go to a professional conference, I present papers to and meet with other academics. My habitus has a number of different sources, among them are training in etiquette by a British mother and many years spent studying scholarly texts and engaging in academic discourse. The linguistic market in academia is formed by the emphasis on cultural and symbolic capital, and by the positions in the empirical field held by academics such as those at the conference; some people have more powerful positions and others less so. Each encounter, each speech act, is informed by these issues. In such circumstances, I tend to "feel at home" (habitus), and I interact freely, bantering and arguing with other academics in a kind of "one-upmanship" tournament.

This weekend, I will be going to the annual Christmas party at my wife's work. Here the linguistic market is different. Economic capital and the cultural capital that goes along with it are much more highly prized, and the empirical field is made up of differing positions and relationships achieved in the struggle of American business. Because of these differences, my market position is quite different here from what it was at the professional conference. In fact, I have no market position.

Worse, my habitus remains the same. The way I talk—the words I use and the way I phrase my sentences—is very different from the other people at this event. The tempo of my speech is different (it's much slower) as is the way I walk and hold myself. In this kind of situation, I try and avoid speech acts. When encounters are unavoidable, I say as little as possible because I know that what I have to say, the way I say it, and even the tempo of my speech won't fit in.

These two different examples illustrate an extremely important point in Bourdieu's theory: individuals in a given market recognize their institutional position, have a sense of how their habitus relates to the present market, and anticipate differing profits of distinction. In my professional conference example, I anticipate high rewards and distinction, but in the office party example, I anticipate low distinction and few rewards. In situations like the office example, anticipation acts as a *self-sanctioning mechanism* through which individuals participate in their own domination. Perhaps "domination" sounds silly with reference to an office party, but it isn't silly when it comes to job interviews, promotions, legal confrontations, encounters with government officials, and so forth. I gave you an example that we can relate to so that we can more clearly understand what happens in other, more important speech acts.

These kinds of speech acts are the arena of symbolic violence. **Symbolic violence** is the exercise of violence and oppression that is not recognized as such. More specifically, "symbolic power is that invisible power which can be exercised only with the complicity of those who do not want to know that they are subject to it or even that they themselves exercise it" (Bourdieu, 1991, p. 164). For example, for quite some time, patriarchy had been seen as part of the natural order of things. Yet, in believing in her husband's right to rule, a woman participated in and blinded herself to her own oppression. Here's another example: In believing that schools should be locally controlled and funded and that education is the legitimate path to upward social mobility, we actively participate in the perpetuation of the class system in the United States.

More insidious for Bourdieu is the way language is used to inflict symbolic violence. Have you ever been around someone of higher social status that you wanted to talk to but didn't? Why didn't you? If you're like me, you didn't because you were afraid of making a fool out of yourself. I had a professor in graduate school that I so admired, but I never talked to him unless it was absolutely necessary. I just knew that I would misspeak and say something foolish. Every social group has specific languages. It is easily seen with such pop culture groups as hip-hop, skaters, and graffiti taggers. But it is also true with experts and people in high-status positions, including the elite class. They typically have specialized languages. And, while we don't know the language, we *know* that we don't know the language, "which condemns [us] to a more or less desperate attempt to be correct, or to *silence*" (Bourdieu, 1991, p. 97, emphasis original).

In society, power is seldom used as coercive force (see Chapter 2 for a discussion of power and legitimacy), but is translated into symbolic form and thereby endowed with a type of legitimacy. Symbolic power is an invisible power and is generally misrecognized: we don't see it as power; we see it as legitimate. In recognizing as legitimate the hierarchical relations of power in which they are embedded, the oppressed are participating in their own domination:

> All symbolic domination presupposes, on the part of those who submit to it, a form of complicity which is neither passive submission to external constraint nor a free adherence to values. . . . It is inscribed, in a practical state, in dispositions which are impalpably inculcated, through a long and slow process of acquisition, by the sanctions of the linguistic market. (Bourdieu, 1991, pp. 50–51)

My third example is from a conversation with a friend. In most conversations among equals, formal linguistic markets have little if any power. We talk and joke around, paying no attention to the demands of proper speech. I'm certain that you can think of multitudes of such speech acts: talking with friends at a café or at the gym or in your apartment. Those kinds of speech acts will always stay that way, unless one of you has a higher education or a greater distance from necessity—that is, unless your habitus is different from that of your friends. Even in such cases, however, linguistic markets usually won't come into play. But they can. "Every linguistic exchange contains the *potentiality* of an act of power, and all the more so when it involves agents who occupy asymmetric positions in the distribution of the relevant capital" (Bourdieu & Wacquant, 1992, p. 145, emphasis original). Bourdieu tells us that in such situations, where the market position is different or the habitus is different, the potential for power and symbolic violence is only set aside for the moment.

Symbolic Struggle

Social change for Bourdieu is rooted in symbolic struggle, which makes sense given Bourdieu's emphasis on symbolic capital and power. Part of that struggle occurs within the speech act or encounter. As we've seen, encounters are structured by markets of differing distinction, and habitus expresses itself naturally within those markets. We will feel at home or foreign in an encounter; we will speak up or silence ourselves, all without thought. However, we also have to keep in mind that habitus is embodied and expresses itself through intuitive feelings. And sometimes our intuitions can lead us to brilliant moves, whether on the sports field, the game board, the music stage, or in the speech act. Just so, our habitus at times can lead us to speech acts that defy our cultural, symbolic, economic, or social standings.

This kind of symbolic struggle can bring some incremental change. Bourdieu gives us hints about how more dynamic change can occur, but keep in mind that his isn't a theory of social change or revolution. Bourdieu allows that there are two methods by which a symbolic struggle may be carried out, one objectively and the other subjectively. In both cases, symbolic disruption is the key. Objectively, individuals or groups may act in such a way as to display certain counter-realities. His example of this method is group demonstrations held to manifest the size, strength, and cohesiveness of the disenfranchised. This type of symbolic action disrupts the taken-for-grantedness that all systems of oppression must work within—it offers an objective case that things are not what they seem.

Subjectively, individuals or groups may try and transform the categories constructed by symbolic capital through which the social world is perceived. On the individual level, this may be accomplished through insults, rumors, questions, and the like. A good example of this approach is found in bell hooks's (1989) book *Talking Back:* "It is that act of speech, of 'talking back,' that is no mere gesture of empty words, that is the expression of moving from object to subject—the liberated voice" (p. 9).

Groups may also operate in this way by employing more political strategies. The most typical of these strategies is the redefinition of history—that is, "retrospectively reconstructing a past fitted to the needs of the present" (Bourdieu, 1989, p. 21). But notice with each of these kinds of struggle, a response from those with symbolic capital would be required. These disruptions could bring attention to the cause, but symbolic power would be necessary to give it life and substance within the symbolic field first and then the objective field.

Bourdieu Summary

- Bourdieu's basic approach is constructivist structuralism. With this idea, Bourdieu is attempting to give us a point of view that gives full weight to structure and agency. There is tension between constructivism and structuralism, between agency and structure, and it is that tension that Bourdieu uses to understand how both can coexist. The tension is a dialectic and is played out in symbolic markets and social practices.

- Bourdieu is specifically concerned with the reproduction of class. In contrast to Marx, Bourdieu sees class replicated through symbolic violence rather than overt oppression. Bourdieu argues that there are four types of capital: economic, social, cultural, and symbolic. The latter two are his greatest concern. Symbolic capital has the power to create positions within the symbolic and objective fields. The objective field refers to social positions that are determined though the distributions of the four capitals. But these positions don't become real or meaningful for us unless and until someone with symbolic capital names them. This naming gives the position, and the individuals and groups that occupy it, social viability. The symbolic field has independent effects in that it can be manipulated by those with symbolic capital. Also, people use the symbolic field to view, understand, and reproduce the objective field.

- Cultural capital refers to the social skills, habits, linguistic abilities, and tastes that individuals have as a result of their position in the symbolic and objective fields. Cultural capital is particularly important because it becomes embodied. This embodiment of cultural capital becomes the individual's habitus: the way the body exists and is used in society. Distance from necessity and level of education are two of the most important ways in which habitus is structured, both of which are related to economic capital. Class position, then, is replicated through the embodied, non-conscious behaviors and speech acts of individuals.

- Habitus is expressed in linguistic markets. Linguistic markets are structured by different weightings of the various capitals. One's position within the market is determined by different rankings on the capitals and the embodied ability to perform within the market. Linguistic markets are played out in speech acts where individuals sense how their habitus relates to the market and thus anticipate differing profits of distinction. This non-conscious sense provides the basis for symbolic violence: anticipating few rewards in acts where they are "outclassed," individuals simultaneously sanction themselves and legitimate the hierarchical relations of class and power.

- There is, however, the possibility of symbolic struggle. The struggle involves symbolic disruption. First, individuals or groups can act in such a way as to objectively picture alternative possibilities. This is what we normally think of as social movements or demonstrations. But because Bourdieu sees the importance of symbolic power in the replication of class, he understands these demonstrations as pictures—they are objective images of symbolic issues that disrupt the taken-for-grantedness in which oppression must operate. Second, individuals and groups can challenge the subjective meanings intrinsic within the symbolic field. In daily speech acts, the individual can disrupt the normality of the symbolic field through insults, jokes, questions, rumors, and so on. Groups can also challenge "the way things are" by redefining history.

Building Your Theory Toolbox

Knowing Constructivist Theorizing

After reading and understanding this chapter, you should be able to define the following terms theoretically and explain their theoretical importance to Giddens and Bourdieu:

Structuration theory; duality of structure; social structures; normative rules; signification codes; authoritative resources; allocative resources; institutional orders; reflexive monitoring; discursive and practical consciousnesses; routinization; regionalization; time–space distanciation; modalities of structuration; ontological security; disembedding mechanisms; symbolic tokens; expert systems; globalization; reflexive project of the self; bodily regimens; plastic sexuality; emancipatory and lifestyle politics; pure relationships; mediated experiences; constructivist structuralism; misrecognition; economic, social, symbolic, and cultural capitals; objectified, institutionalized, and embodied cultural capital; symbolic and empirical fields; habitus; taste; distance from necessity; pure gaze; linguistic markets; symbolic violence; self-sanctioning mechanism

After reading and understanding this chapter, you should be able to

- Explain time–space distanciation and why it is the central question for Giddens

- Explicate how modernity affects time–space distanciation

- Discuss Giddens's idea of social structures and the modalities of structuration

- Discuss the three institutional orders and explain how they are created

- Identify incidences of practical and discursive consciousnesses and analyze how they constitute reflexive monitoring

- Identify the primary unconscious motivation in human interaction and explicate the specific processes that come about due to this motivation

- Define each social process of modernity and explain how it contributes to the dynamic character of modernity

- Define the reflexive project of the self, pure relationships, plastic sexuality, and life politics and explain why these issues are exclusive to modernity

- Explain Bourdieu's constructivist structuralism approach. In your answer, be certain to define and explain the problems associated with the social physics (structural) and social phenomenology (subjective) approaches and how Bourdieu's approach counters both.

- Discuss how symbolic fields are produced

- Define habitus and explain how it is produced

- Explain how class inequalities are replicated, and how class is contingent. In your answer, be certain to explain linguistic markets, symbolic violence, and the role that habitus plays.

Learning More: Primary Sources

- Anthony Giddens

  Giddens, A. (1986). *The constitution of society.* Berkeley: University of California Press.

  Giddens, A. (1991). *Modernity and self-identity: Self and society in the late modern age.* Stanford, CA: Stanford University Press.

  Giddens, A. (1992). *The transformation of intimacy: Sexuality, love and eroticism in modern societies.* Stanford, CA: Stanford University Press.

- Pierre Bourdieu

  Bourdieu, P. (1984). *Distinction: A social critique of the judgment of taste.* Cambridge, MA: Harvard University Press.

  Bourdieu, P. (1989). Social space and symbolic power. *Sociological Theory, 7,* 14–25.

  Bourdieu, P. (1991). *Language and symbolic power.* Cambridge, MA: Harvard University Press.

  Bourdieu, P. (1992). *An invitation to reflexive sociology.* Chicago: University of Chicago Press.

  Bourdieu, P. (1999). Acts of resistance: Against the tyranny of the market. New York: New Press.

Learning More: Secondary Sources

Jenkins, R. (2002). *Pierre Bourdieu* (Key sociologists). New York: Routledge.

Mestrovic, S. G. (1998). *Anthony Giddens: The last modernist.* New York: Routledge.

Swartz, D. (1998). *Culture and power: The sociology of Pierre Bourdieu.* Chicago: University of Chicago Press.

Theory You Can Use (Seeing Your World Differently)

- Giddens is one of the architects and proponents of what is known as the "third way" in politics. Using your favorite Internet search engine, look up "third way." What is the third way and how is Giddens involved? How can you see it related to his theory?

- Keep a journal for a week, month, or year. In this journal, analyze your life in terms of the reflexive project of the self, pure relationships, plastic sexuality, and life politics.

- Compare and contrast Giddens's theory of structuration with Bourdieu's constructivist structuralism approach. Specifically, how are patterns of behavior replicated in the long run? How does each one overcome the object–subject dichotomy? Do you find one approach to be more persuasive? Why or why not?

- Compare and contrast Habermas's and Giddens's views of modernity. I recommend you start with their defining characteristics of modernity and review the issues that are implied in the definitions. After you've worked your way through these comparisons, define modernity and its chief problems.

- Check the index in this book and look up the different definitions and explanations of "social structures." Evaluate each of these approaches and create what you feel to be a clear, robust, and correct definition/explanation of social structures. Justify your theory.

- Use Bourdieu's theory to describe and explain the differences between the way you talk with your best friend and the way you talk with your theory professor.

- Using Bourdieu's take on misrecognition, analyze the following ideas: race, gender, and sexuality. What kinds of things must we misrecognize in order for these to work as part of the symbolic violence of this society? Look up Bourdieu's notion of *doxa* (1972/1993, p. 3; 1980/1990, p. 68). How do doxa and symbolic violence work together in the oppression of race, gender, and sexuality?

- Choose the structure of inequality that you know best (race, gender, sexuality, religion). Using what you already know, analyze that structure using Figure 11.4. How would approaching the study of inequality change using Bourdieu compared to more conventional sociological approaches, such as those found in Chapter 8?

Further Explorations—Web Links

- Anthony Giddens

  http://www.theory.org.uk/giddens.htm (An interesting potpourri of information about Giddens)
  http://news.bbc.co.uk/hi/english/static/events/reith_99/ (Participate in debates that Giddens has sparked)

- Pierre Bourdieu

  http://pubs.socialistreviewindex.org.uk/isj87/wolfreys.htm (An interesting portrait painted by the International Socialism Journal
  http://www.theglobalsite.ac.uk/times/202calhoun.htm (*In memoriam* by two prominent social theorists: Craig Calhoun and Loïc Wacquant)

World-Systems

Immanuel Wallerstein (1930–)

Niklas Luhmann (1927–1998)

I want you to think about the last time you were in class. Where exactly were you? To answer my question you might say, "I was in the third row from the back." Okay, but where's the third row from the back (and don't say in class, because that's the question we started with)? "Well," you might say, "it's in the Graham Building." Good; now, where's the Graham Building? "On the campus of UNCG," you respond. Excellent. Now—you got it—where's UNCG? With a big sigh and slight glare in your eye, you say, "UNCG is part of the University of North Carolina university system, which is (obviously) in North Carolina, which is in the United States, which is in North America, which is part of the earth, and the earth is in the solar system . . ."

In a very rough way, we just ran down the basic insight of sociology: everything takes place in a context and context matters. You could do this with anything. You could think about the conversation you had down at the Oldtown Bar last night or the apple you ate for lunch, and you could extrapolate out the basic social contexts that made each one possible. But let me ask you another question: in moving outward, do you ever reach a point where what you are talking about is substantially different from what you began with? In our first example we started with you in class, which is a form of interaction. At a later point we reached UNCG. Is your classroom interaction different from UNCG? Probably so, but in what ways? How do we know when we reach a different kind of social entity? And how are these social entities related to one another?

You've probably guessed that I'm leading us up to thinking about society in system terms. Both Immanuel Wallerstein and Niklas Luhmann bring the idea of social systems to the forefront and ask us to think about some massive processes and relations. Not only do they ask us to consider a single society as a system, as do functionalism and Marxism, but they move us up to think about the world-system. We hear a lot about globalization today, but how is it that separate nations can exist and function together as a system? Remember, systems aren't simply a collection of independent things; rather, a system is a group of interdependent and mutually constituting parts. In this sense we could ask, how are the United States and Brazil part of each other? How do they depend upon each other? More importantly, how do they influence and create one another?

But that's not all. Wallerstein and Luhmann also ask us to think about history. There are multiple ways of talking about history. We can talk about the stages of civilization from the advent of ideographic writing to the dawn of the computer age. Or, we can categorize time periods in terms of economic technologies, with our epochs then running from hunting and gathering societies to agrarian to industrial and postindustrial. Or, we can divide history up in terms of culture, such as the Medieval, Renaissance, Classical, and Romantic periods.

Wallerstein and Luhmann talk about our historical period as modernity. We discussed modernity in the chapter on Giddens, and you know that the concept of modernity was a leading concern for classical, social theorists. So central was this issue that in order to understand the transition from traditional societies to modern societies, some theorists constructed social typologies as well, such as Tönnies's *gemeinschaft* and *gesellschaft,* Durkheim's mechanical and organic solidarity, and Simmel's organic and rational group memberships. In addition, most of the classical theorists had an idea of modernity's central dynamic: for Marx it was capitalism, for Weber it was bureaucracy and rationalization, and for Durkheim it was structural and social diversity.

But, no matter how we divide it up, one thing is certain: historical periods influence life, and in a fundamental way our historical epoch matters. Think about it this way: You as an individual exist within social situations. Those situations exist within social structures, which in turn exist within society. Societies exist within systems of societies, and those systems are historically specific. Unfortunately, these contexts and influences are the ones that most people are least aware of but they have the greatest influence in our lives.

Immanuel Wallerstein: Global Capitalism

Wallerstein is going to ask us to stretch our minds. He asks big questions, but he uses a perspective and language that we are familiar with. Wallerstein is a Marxist and he wants to know how capitalism influences society. But he moves us up from the societal level to thinking about global capitalism: how does the globalization of capitalism influence not only the United States but also the entire world? Wallerstein also wants to know why the revolution that Marx predicted hasn't yet occurred. And—here's the interesting part—he is particularly concerned with the end of the world as we know it.

Photo: Courtesy of Immanuel Wallerstein.

The Essential Wallerstein

Biography

Immanuel Wallerstein was born in New York City on September 30, 1930. He attended Columbia University where he received his bachelor's (1951), master's (1954), and Ph.D. (1959). Wallerstein has also formally studied at various universities around the globe, including the Université Paris 7–Denis-Diderot, Université Libre de Bruxelles, and Universidad Nacional Autónoma de México. His primary teaching post was in New York at Binghamton University (SUNY), where he taught from 1976 to his retirement in 1999. However, he has also held visiting professor posts in Amsterdam, British Columbia, and the University of Hong Kong, as well as several other locations. In addition to many professional posts, he has served as president of the International Sociological Association and director of the Fernand Braudel Center for the Study of Economies, Historical Systems, and Civilizations.

Passionate Curiosity

Wallerstein is driven to first critically understand (through a Marxist perspective) how the nations of the world are joined together in a global system of capitalism, and second to find ways to politically act to change that system.

Keys to Knowing

World-systems; division of labor; exporting exploitation; overproduction; capitalist states and quasi-monopolies; core, semi-periphery, and periphery states; external areas; world-empires and world-economies; Kondratieff waves

Seeing Further: World-Systems Theory

Think about history for a moment. If you're like me, you'll think about the classes you've taken, such as United States History. In such a class, you're told a longitudinal story of the major historical events and people of the United States. In this story, you were undoubtedly told about other nations as well, such as those encountered during the U.S. involvement in WWII. And this involvement is followed by other involvements, such as in the Korean War. But there may be an entirely different way of thinking about history—a history of global systems rather than events.

Globality and Historicity

If social actors such as nations, institutions, and groups are related to each other through a specific system, then the history of that system is extremely important for understanding how the system is working presently. Wallerstein picked up the notions of structural time and cyclical process from French historian and educator

Fernand Braudel. Braudel criticized event-dominated history, the kind with which we are most familiar, as being too idiographic and political.

The prefix "ideo" implies ideas and individuality. Ideographic knowledge, then, is concerned with ideas, but more importantly with understanding unique phenomena. An example of this event history approach is the kind of history I talked about earlier—U.S. history in terms of things like Abraham Lincoln and the Civil War and Martin Luther King Jr. and the civil rights movement. Such an understanding doesn't see changes through history as the result of systematic social facts; rather, it perceives historical change as occurring through unique events and political figures.

Braudel felt that this kind of history is dust and tells us nothing about the true historical processes. Yet Braudel also criticized the opposite approach, nomothetic knowledge. The word "nomothetic" is related to the Greek word *nomos*, which means law. The goal in seeking *nomothetic* knowledge, like that of science, is to discover the abstract and universal laws that underpin the physical universe. According to Braudel, when nomothetic knowledge is sought in the social sciences, more often than not it creates mythical, grand stories that legitimate the search for universal laws instead of explaining social history.

Wallerstein's idea of *historicity* lies between the ideographic focus on events and the law-like knowledge of science. Rather than focusing on events, Wallerstein's approach concentrates on the history of structures within a world-system. He intentionally uses the hyphen in world-system to emphasize that he is talking about systems that constitute a world or a distinct way of existing. Wallerstein captures this systemic approach with the idea of *globality*, an analysis that looks at the relationships among nations and other political entities that form a system. Thus, the systemic factors cut across cultural and political boundaries and create an "integrated zone of activity and institutions which obey certain systemic rules" (Wallerstein, 2004, p. 17).

For example, capitalism is a world-system that has its own particular history. There have always been people who have produced products to make a profit, but the capitalism of modernity, the kind that Weber (1904–1905/2002) termed "rational capitalism," is unique to a particular time period. An account of rational capitalism from its beginnings, from around the sixteenth century, that would include all the principal players (such as nations, firms, households, and so forth) and their systemic relations is what Wallerstein has in mind.

Historicity thus includes the unique variable of time. In taking account of world-systems rather than event history, historicity is centered upon *structural time* and the *cyclical time* within the structures. Wallerstein is telling us that structures have histories, and it is the history of structures with which we should be concerned, rather than events, because structures set the frames within which human events take place. Structures have life spans; they are born and they die, and within that span there are cyclical processes. Here we begin to see Wallerstein's Marxist roots clearly. The idea of structural change occurring through cycles comes from Marx's notion of the dialectic. As we move through Wallerstein's theory, keep in mind the idea of structural change through dialectical oppositions—pay attention to the contradictions that are intrinsic to capitalism.

Unidisciplinarity

With unidisciplinarity, the world-systems' critique is aimed at the political underpinning of knowledge. Wallerstein argues that the configuration of the modern university system corresponds to the political systems of the age. Modernity brought a new world, one driven by markets, political states, and societal change. New academic disciplines came into existence to understand and control these arenas through the new knowledge of science. Thus, societies and economies were seen to be subject to law-like principles because science became the knowledge of most worth. At the core of the social sciences was the idea that modernity is the touchstone against which all other social forms would be tested. Thus, anthropology and Orientalism (the study of the Orient as distinct from the West) came into existence to measure, understand, and make distinct those other worlds from modernity. When the world politically divided again in 1945, the university changed as well. The world divided into first-, second-, and third-world countries. The world was seen developmentally, with two forms of development politically competing—capitalism and communism. The universities developed area studies so that the knowledge of the university would be politically useful and relevant.

Against this political division of knowledge, world-systems proposes *unidisciplinarity.* The prefix "uni" means singular, as in unicycle. Rather than implying the bringing together of the disciplines, the term instead denotes the denial of the disciplines altogether. As Wallerstein (1999) notes,

> if there were historically emergent and historically evolving processes in the world-system, what would lead us to assume that these processes could be separated into distinguishable and segregated streams with particular (even opposed) logics? The burden of proof was surely on those who argued the distinctiveness of the economic, political, and sociocultural arenas. (p. 195)

To reiterate, unidisciplinarity is not multidisciplinarity. The multidiscipline approach, common in American universities today, still maintains the distinctions among the disciplines. World-systems advocates a holistic approach, one that not only sees the world as a totality, but that also sees knowledge as a whole. World-systems analysis, then, transcends sociology, political science, and economics. It takes a longer view than does conventional history and divides the world not in response to political needs and configurations but according to structural time and cyclical processes.

Defining World-Systems Theory

World-systems theory is a critical theoretical perspective that is most closely associated with Immanuel Wallerstein. The perspective employs three unique concepts: globality (conceptualizing the world as linked through a single system), historicity (history defined in terms of structural cycles rather than events or people), and unidisciplinarity (denial of multiple disciplines with knowledge and research conceptualized holistically). Wallerstein conceptualizes two world-systems: world-empires and world-economies. In world-empires, connections among countries are created through military dominance and tribute; in world-economies, nations are linked through the capitalistic division of labor, accumulation, and exploitation. Each world-system is characterized by a core country with peripheral dependent countries and by dialectical cycles that create the conditions for the system's demise.

Wallerstein (1999) summarizes the world-systems approach with a series of questions, which must keep in mind in order to keep the perspective's critical edge:

- What is the distinctive nature of social science knowledge? How can its parameters and social role be defined?
- What are the theoretical relationships between social science and social movements, and social science and social power?
- Is it possible that multiple systems are at work at any one time? If so, what are their defining features?
- Do such historical, social systems have a "natural" history to them? If so, how do they work; that is, are they evolutionary or some other form?
- How are time and space socially constructed? What ramifications does this construction have for social science?
- What kinds of processes are at work in the transition from one historical system to another? What kinds of metaphors can we use to understand these processes?
- What is the theoretical relationship between the quest for knowledge and the quest for a just society?
- How can we conceive of our existing world-system? What can be said about how it started, its current structure, and how it will end, in light of our answers to the other questions? (pp. 199–200)

I've deliberately given these questions in the order that Wallerstein presents them. Notice the predominance of critical questions, questions that interrogate the way knowledge is constructed and used. Also notice that we don't get to the question we might think world-systems analysis would privilege until the very last one.

Concepts and Theory: The Dialectics of Capitalism

Wallerstein's critique is essentially a Marxist one. Marx did what Wallerstein says needs to be done: he focused on structures moving through cyclical time. He was particularly interested in capitalism—and, according to Wallerstein, the elements of capitalism are in fact the only features that can truly create a world-system today. Certain of Marx's concepts, then, have special importance in explaining and critiquing the world-system. Among them are the division of labor, exploitation, accumulation, and overproduction.

The Division of Labor and Exploitation

For Wallerstein, the importance of the *division of labor* is that it is the defining characteristic of an economic world-system. Labor, of course, is an essential form of human behavior; without it we would cease to exist. By extension, the division of labor creates some of the most basic kinds of social relationships, and these relationships are, by definition, relations of dependency. In our division of labor, we depend upon each other to perform the work that we do not. I depend upon the farmer for food production, and the farmer depends upon teachers to educate her or his children. These relations of dependency connect different people and other social units into a structured whole or system. Wallerstein argues that the world-system is connected by the current capitalist division of labor: world-systems are defined "quite simply as a unit with a single division of labor and multiple cultural systems" (Wallerstein, 2000, p. 75). Multiple cultural systems are included because world-systems connect different societies and cultures.

The important feature of this division of labor is that it is based on exploitation. I've already gone into some detail defining exploitation, so I won't do it again now. But as we go through the next section, keep in mind two things: First, exploitation is a measurable entity: it is the difference between what a worker gets paid and what he or she produces. Different societies can have different levels of exploitation. For example, if we compare the situation of automobile workers in the United States with those in Mexico, we will see that the level of exploitation is higher in Mexico. The second thing to keep in mind is that exploitation is fundamental to capitalism. Surplus labor and exploitation are the places from which profit comes and are thus necessary for capitalism.

What is important to see here is that profit is based on exploitation and there are limitations to exploitation. Yet the drive for exploitation doesn't let up; capitalists by definition are driven to increase profits. The search for new means of exploitation, then, eventually transcends national boundaries: capitalists *export exploitation*. Because of the limitations on the exploitation of workers in advanced capitalist countries—due primarily to the effects of worker movements, state legislation, and the natural limitations of technological innovation—firms seek other labor markets where the level of exploitation is higher. Marx had a vague notion of this, but Wallerstein's theory is based upon it. It is the exportation of exploitation that structures the division of labor upon which the world-economy is based.

Accumulation and Overproduction

We all know what modern capitalism is: it is the investment of money in order to make more money (profit). As Wallerstein (2004) says, "we are in a capitalist system only when the system gives priority to the *endless* accumulation of capital" (p. 24, emphasis original). We see the drive to make money in order to make more money all around us, but most people only think about the personal effects this kind of capitalism has (like the fact that Bill Gates is worth $46.6 billion). But what are the effects on the economy? Most Americans would probably say that the effect on the economy is a good one: continually expanding profits and higher standards of living. Perhaps, but Wallerstein wants us to see that something else is going on as well. In order to fully understand what he has in mind, we need to think about the role of government in the endless pursuit of the accumulation of capital.

It's obvious that for capitalism to work, it needs a strong state system. The state provides the centralized production and control of money; creates and enforces laws that grant private property rights; supplies the regulation of markets, national borders, inter-organizational relations; and so forth. But there is something else that the state does in a capitalist system. We generally assume that the firm that pays the cost enjoys the benefits, as in the capitalist invests the money so she or he can enjoy the profit. However, the state actually decides what proportion of the costs of production will be paid by the firm. In this sense, capitalists are subsidized by the state.

There are three kinds of costs that the state subsidizes: the costs associated with transportation, toxicity, and the exhaustion of raw materials. Firms rarely if ever pay the full cost of transporting their goods; the bulk of the cost for this infrastructure is borne by the state, for such things as road systems. Almost all production produces toxicity, whether noxious gases, waste, or some kind of change to the environment. How and when these costs are incurred and who pays for them is always an issue. The least expensive methods are short-term and evasive (dumping the waste, pretending there isn't a problem), but the costs are eventually paid and usually by the state. Capitalist production also uses up raw materials, but again firms rarely pay these costs. When resources are depleted, the state steps in to restore or re-create the materials. Economists refer to the expenses of capitalist production that are paid by the state as *externalized costs,* and we will see that in this matter not all states are created equal.

However helpful these externalized costs are to the pursuit of accumulation, states that contain the most successful capitalist enterprises do more: they provide a structure for *quasi-monopolies.* A monopoly is defined as the exclusive control of a market or the means of production. Quasi-monopolies don't have exclusive control but they do have considerable control.

Wallerstein argues that totally free markets would make the endless accumulation of capital impossible. Totally free markets imply that all factors influencing the means of production are free and available to all firms, that goods and services flow without restriction, that there is a very large number of sellers and a very large number of buyers, and that all participants have complete and full knowledge. "In such a perfect market, it would always be possible for the buyers to bargain down the sellers to an absolutely minuscule level of profit," which would destroy the basic underpinnings of capitalism (Wallerstein, 2004, pp. 25–26). The converse of a totally free market is a monopoly; and monopolized processes are far more lucrative than those of the free market. Thus, the perfect situation for a capitalist firm is to have monopolistic control; it would then be able to pursue the endless accumulation of capital with the greatest efficiency and success.

The most important way in which states facilitate quasi-monopolies is through patent laws that grant exclusive production rights for an invention for a certain number of years. This state guarantee allows companies to gain high levels of profit in a monopolistic market for long enough to obtain considerable accumulation of capital. The practice of granting patents also results in a cycle of leading products. The largest and most successful firms actively market a patented product as long as the profit margin is high. As soon as the product becomes less profitable through more open competition, the product is given over to less profitable companies, with the original firm creating new leading products. Producers of the unpatented product engage in freer competition but with less profit.

You'll recall from Chapter 1 that capitalism is subject to *overproduction*. Because capitalists are driven to accumulate ever-increasing levels of capital, and because, unlike other animals, human beings can create new needs, capitalists will continue to create new and produce existing commodities until the market will no longer bear it. The cycles of overproduction and exploitation work in tandem, both of them driven by accumulation. Accumulation increases the demand for labor and product innovation. State protection through patent rights, tax incentives, and the like creates a state-sanctioned quasi-monopoly that in and of itself increases accumulation, better enabling the firm to engage in product innovation and increasing the demand for labor.

Over time, the demand for labor decreases the size of the labor pool, which drives wages up and profits down, which, in turn, precipitates an economic slow-down, the collapse of small businesses, and the search for new methods of exploitation through technological innovation in the work process or exporting exploitation. In the medium run, exporting exploitation is the more efficient of the two because technologies become diffused throughout the business sector. Exporting exploitation implies the movement of specific goods outside the national boundaries, and product movement from most profitable to less profitable firms explicitly entails such a shift. Both processes, then, move goods and labor from advanced capitalist countries to rising capitalist countries. And both processes lead to the collapse of small businesses and the centralization of accumulation—that is, capital held in fewer and fewer hands.

Concepts and Theory: The End of the World as We Know It

World-Empires and World-Economies

Very few people living in any specific world think about their world ending, but they all do. The great Mesopotamian, Greek, and Roman empires are gone; the sun has set on the British Empire; and even more recently, the USSR crumbled and is no more. Of course, just like the Phoenix, new worlds arise out of the ashes and history moves on. But what of our world? History tells us all worlds fail—when will our world fail?

Wallerstein asks us to consider the possibility that our world is failing and that we are in a chaotic period between historical moments. Perhaps shockingly for some of us, Wallerstein argues that this shift in historical epochs will lead to the demise of the United States as we know it. So, let's take these questions seriously: In what historical epoch do we live and how is it affecting our world?

Wallerstein argues that there have been two types of world-systems throughout history, one with a common political system and one without. Systems with a common political entity are called world-empires. *World-empires* exist through military dominance and economic tribute (money paid from one country to another as acknowledgment of submission). The political influence of one government is spread and held in place through a strong military, but this sets up a cycle that eventually leads to the demise of the empire. Maintaining a standing army that is geographically extended costs quite a bit of money. This money is raised through tribute and taxation. Heavy taxes make the system less efficient, in terms of economic production, and this increases the resistance of the populace as well. Increasing resistance means that the military presence must be increased, which, in turn, increases the cost, taxation, and resistance, and it further lowers economic efficiency. These cycles continue to worsen through structural time (see historicity above) until the empire falls. Examples of such world-empires include Rome, China, and India.

These world-empire cycles continued until about 1450, when a world-economy began to develop. Rather than a common political system, *world-economies* are defined through a common division of labor and through the endless accumulation of capital. As we saw earlier, in the absence of a political structure or common culture, the world-system is created through the structures intrinsic to capitalism. The worldwide division of labor created through the movement of products and labor from advanced capitalist nations to rising capitalist nations creates relationships of economic dependency and exploitation. These capitalist relationships are expressed through three basic types of economic states: core, semi-periphery, and periphery.

Briefly, *core states* are those that export exploitation; enjoy relatively light taxation; have a free, well-paid labor force; and constitute a large consumer market. The state systems within core states are the most powerful and are thus able to provide the strongest protection (such as trade restrictions) and capitalist inducements, such as externalizing costs, patent protection, tax incentives, and so on. *Periphery states* are those whose labor is forced (very little occupational choice or worker protections) and underpaid. In terms of a capitalist economy and the world-system, these states are also the weakest—they are able to provide little in the way of tax and cost incentives and they are the weakest players in the world-system. The periphery states are those to which capitalists in core states shift worker exploitation and more competitive, less profitable products. These shifts result in "a constant flow of surplus-value from the producers of peripheral products to the producers of core-like products" (Wallerstein, 2004, p. 28).

The relationship, then, between the core and the periphery is one of production processes and profitability. There is a continual shift of products and exploitation from core to periphery countries. Furthermore, there are cycles in both directions: periphery countries are continually developing their own capitalist-state base. As we've seen, profitability is highest in quasi-monopolies and these, in turn, are dependent upon powerful states. Thus, changing positions in the capitalist world-economy is dependent upon the power of the state.

Over time, periphery economies become more robust and periphery states more powerful: worker protection laws are passed, wages increase, and product innovation begins to occur. The states can then begin to perform much like the states in core countries—they create tax incentives and externalize costs for firms, they grant product protection, and they become a more powerful player in the world-system economy. These nations move into the semi-periphery. *Semi-periphery* states are those that are in transition from being a land of exploitation to being a core player, and they both export exploitation and continue to exploit within their own country.

A good illustration of this process is the textile industry. In the 1800s, textiles were produced in very few countries and it was one of the most important core industries; by the beginning of the twenty-first century, textiles had all but moved out of the core nations. A clear and recent example of this process is Nike. Nike is the world's largest manufacturer of athletic shoes, with about $10 billion in annual revenue. In 1976, Nike began moving its manufacturing concerns from the United States to Korea and Taiwan, which at the time were considered periphery states. Within four years, 90% of Nike's production was located in Korea and Taiwan.

However, both Korea and Taiwan were on the cusp, and within a relatively short period of time they had moved into the semi-periphery. Other periphery states had opened up, most notably Bangladesh, China, Indonesia, and Vietnam. So, beginning in the early 1990s, Nike began moving its operations once again. Currently, Indonesia contains Nike's largest production centers, with 17 factories and 90,000 employees. But that status could change. Just a few years ago, in 1997, the Indonesian government announced a change in the minimum wage, from $2.26 per day to $2.47 per day. Nike refused to pay the increase and in response, 10,000 workers went on strike. In answer to the strike, a company spokesperson, Jim Small, said, "Indonesia could be reaching a point where it is pricing itself out of the market" (Global Exchange, 1998).

Yet the existence of the semi-periphery doesn't simply serve as a conversion point—it has a structural role in the world-system. Because the core, periphery, and semi-periphery share similar economic, political, and ideological interests, the semi-periphery acts as a buffer that lessens tension and conflict between the core and periphery nations. "The existence of the third category means precisely that the upper stratum is not faced with the *unified* opposition of all the others because the *middle* stratum is both exploited and exploiter" (Wallerstein, 2000, p. 91, emphasis original).

Kondratieff Waves

Since 1450, world-economies have moved through four distinct phases. These phases occur in what are called Kondratieff waves (K-waves), named after Nikolai Kondratieff, a Russian economist writing during the early twentieth century. Kondratieff noticed patterns of regular, structural change in the world-economy. These waves last 50 to 60 years and consist of two phases: a growth phase (the A-cycle) and a stagnation phase (the B-cycle).

Much of what drives these phases in modern economic world-systems comes from the cycles of exploitation and accumulation that we've already talked about. During the A-cycle, new products are created, markets are expanded, labor is employed, and the political and economic influence of core states moves into previously external areas—new geographic areas are brought into the periphery for labor and materials (imperialism). At 25 to 30 years into the A-cycle, profits begin to fall due to overproduction, decreasing commodity prices and increasing labor costs. In this B-cycle, the economy enters a deep recession. Eventually, the recession bottoms out and small businesses collapse, which leaves fewer firms and greater centralization of capital accumulation (quasi-monopolistic conditions), which, in turn, sets the stage for the next upswing in the cycle (A_2-cycle) and the next recession (B_2-cycle). Historically, these waves reach a crisis point approximately every 150 years. Each wave has its own configuration of core and periphery states, with generally one dominant state, at least initially.

Wallerstein sees these waves as phases in the development of the world-system. Within each phase, three things occur: the dominant form of capitalism changes (agricultural → mercantilism → industrial → consolidation); there is a geographic expansion as the division of labor expands into external areas; and a particular configuration of core and periphery states emerges. There have been four such phases thus far in the world-system. In Figure 12.1, I've outlined the different phases and their movement through time. I've also noted some of the major issues and the hegemonic core nations for easy comparison. Wallerstein (2004) uses the term *hegemonic* to denote nations that for a certain period of time

> were able to establish the rules of the game in the interstate system, to dominate the world-economy (in production, commerce, and finance), to get their way politically with a minimal use of military force (which however they had in goodly strength), and to formulate the cultural language with which one discussed the world. (p. 58)

I'm not going to go into much historic detail here. You can read Wallerstein's (1974, 1980, 1989) three-volume work for the specifics. But briefly, Phase 1 occurred roughly between 1450 and 1640, which marks the transition from feudalism and world-empires to the nation-state. Both the Ottoman Empire and the Hapsburg dynasty began their decline in the sixteenth century. As the world-empires weakened, Western Europe and the nation-state emerged as the core, Spain

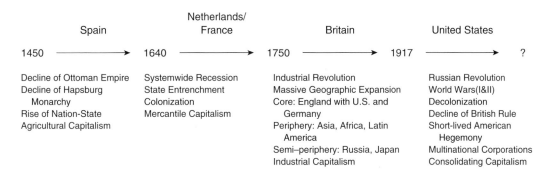

Figure 12.1   Wallerstein's World-Systems Phases

and the Mediterranean declined into the semi-periphery, and northeastern Europe and the Americas became the periphery. During this time, the major form of capitalism was agricultural, which came about as an effect of technological development and ecological conditions in Europe.

The second phase lasted from 1640–1750 and was precipitated by a systemwide recession that lasted approximately 80 years. During this time, nations drew in, centralized, and attempted to control all facets of the market through mercantilism, the dominant form of capitalism in this phase. Mercantilism was designed to increase the power and wealth of the emerging nations through the accumulation of gold, favorable trade balances, and foreign trading monopolies. These goals were achieved primarily through colonization (geographic expansion). As with the previous period, there was a great deal of struggle among the core nations, with a three-way conflict among the Netherlands, France, and England.

The third phase began with the Industrial Revolution. England quickly took the lead in this area. The last attempt by France to stop the spread of English power was Napoleon's continental blockade, which failed. Here capitalism was driven by industry and it expanded geographically to cover the entire globe. Wallerstein places the end of the third phase at the beginning of WWI and the beginning of the fourth phase at 1917 with the Russian Revolution.

The Russian Revolution was driven by the lack of indigenous capital, continued resistance to industrializing from the agricultural sector, and the decay of military power and national status. Together these meant that "the Russian Revolution was essentially that of a semi-peripheral country whose internal balance of forces had been such that as of the late nineteenth century it began on a decline towards a peripheral status" (Wallerstein, 2000, p. 97). During this time, the British Empire receded, due to a number of factors including decolonization, and two states in particular vied for the core position: Germany and the United States. After the Second World War, the United States became the leading core nation, a position it enjoyed for two decades.

Hegemonic or leading states always have a limited life span. Becoming a core nation requires a state to focus on improving the conditions of production for capitalists, but staying hegemonic requires a state to invest in political and military might. Over time, other states become economically competitive and the leading state's economic power diminishes. In attempts to maintain its powerful position in the world-system, the hegemonic state will resort first to military threats and then to exercising its military power (note the increasing U.S. military intervention over the past 25 years). The "use of military power is not only the first sign of weakness but the source of further decline," as the capricious use of force creates resentment first in the world community and then in the state's home population as the cost of war increases taxation (Wallerstein, 2004, pp. 58–59).

Thus, the cost of hegemony is always high and it inevitably leads to the end of a state's position of power within the world-system. For the United States, the costs came from the Cold War with the USSR; competition with rising core nations, such as Japan, China, and an economically united and resurgent Western Europe; and such displays of military might as the Korean, Vietnam, Gulf, and Iraqi Wars. The decline of U.S. hegemony since the late 1960s has meant that capitalist freedom has actually increased, due to the relative size and power of global corporations. There are many multinational corporations now that are larger and more powerful than many nations. These new types of corporation "are able to maneuver against state bureaucracies whenever the national politicians become too responsive to internal worker pressures" (Wallerstein, 2000, p. 99). The overall health of world capitalism has also meant that the semi-periphery has increased in strength, facilitating growth into the core.

The Modern Crisis

There are several key points in time for the world-system, such as the Ottoman defeat in 1571, the Industrial Revolution around 1750, and the Russian Revolution in 1917. Each of these events signaled a transition from one capitalist regime to another. Wallerstein argues that one such event occurred in 1968, when revolutionary movements raged across the globe, involving China, West Germany, Poland, Italy, Japan, Vietnam, Czechoslovakia, Mexico, and the United States. So many nations were caught up in the mostly student-driven social movements that, collectively, they have been called the "first world revolution."

As you'll recall from the introduction to this section, part of what defines the period of time in which we live is a critique of the projects of modernity, both social and technical. Wallerstein tells us that the upheavals of 1968 were directed at the contradictions and failures of society to fulfill the hope of modernity: liberation for all. Students by and large rejected much of the benefits of technological development and proclaimed society had failed at the one thing that truly mattered: human freedom. The material benefits of technology and capitalism were seen as traps, things that had blinded people to the oppression of blacks, women, and all minorities. And this critique wasn't limited to technologically advanced societies. "In country after country of the so-called Third World, the populaces turned against the movements of the Old Left and charged fraud. . . . [The people of the world] had lost faith in their states as the agents of a modernity of liberation" (Wallerstein, 1995, p. 484). The 1968 movements in particular rejected American hegemony because of its emphasis on material wealth and hypocrisy in liberation.

In Wallerstein's (1995) scheme, the collapse of Communism was simply an extension of this revolt, one that most clearly pointed out the failure of state government to produce equality for all: "even the most radical rhetoric was no guarantor of the modernity of liberation, and probably a poor guarantor of the modernity of technology" (p. 484). Interestingly, Wallerstein sees the collapse of Leninism as a disaster for world capitalism. Leninism had constrained the "dangerous classes," those groups oppressed through capitalist ideology and practice. Communism represented an alternative hope to the contradictions found in capitalist states. With the alternative hope gone, "the dangerous classes may now become truly dangerous once again. Politically, the world-system has become unstable" (p. 484).

Structurally, the upheavals of 1968 occurred at the beginning of a K-wave B-cycle. In other words, the world was standing at the brink of an economic downturn or stagnation, which lasted through the 1970s and 1980s. As we've seen, such B-cycles occur throughout the Kondratieff wave, but this one was particularly severe. The 20-year economic stagnation became an important political issue because of the prosperity of the preceding A-cycle. From 1945 to 1970, the world experienced more economic growth and prosperity than ever. Thus, the economic downturn gave continued credence and extra political clout to worldwide social movements. Economically, the world-system responded to the downturn by attempting to roll back production costs by reducing pay scales, lowering taxes associated with the welfare state (education, medical benefits, retirement payments), and re-externalizing input costs (infrastructure, toxicity, raw materials). There was also a shift from the idea of developmentalism to globalization, which calls for the free flow of goods and capital through all nations.

However, while the world-system is putting effort into regaining the A-cycle, there are at least three structural problems hindering economic rebound. First, as we've noted, there are limits to exporting exploitation. Four hundred years of capitalism have depleted the world's supply of cheap labor. Every K-wave has brought continued geographic expansion, and it appears that we have reached the limit of that expansion. More and more of the world's workforce is using its political power to increase the share of surplus labor or profit it receives (see Chapter 1 for discussion of surplus labor). Inevitably, this will lead to a sharp increase in the costs of labor and production and a corresponding decrease in profit margins. Remember, capitalism is defined by continual accumulation. This worldwide shift, then, represents a critical point in the continuation of the current capitalist system.

Second, there is a squeeze on the middle classes. Typically, the middle classes are seen as the market base of a capitalist economy. And, as we've seen, a standard method of pulling out of a downturn is to increase the available spending money for the middle classes, either through tax breaks or through salary increases. This additional money spurs an increase in commodity purchases and subsequently in production and capital accumulation. However, this continual expanding of middle class wages eventually becomes too much for firms and states to bear. One of two things must happen: either these costs will be rolled back, or they will not. If they are not reduced, "both states and enterprises will be in grave trouble and frequent bankruptcy" (Wallerstein, 1995, p. 485). If they are rolled back, "there will be significant political disaffection among precisely the strata that have provided the strongest support for the present world-system" (p. 485).

In the United States, indications are that the costs are being rolled back. Between 1967 and 2001, the income of the middle 20% of the population dropped from 17.3% to 14.6% of the total, while the upper 20% increased from 43.8% to 50.0%. Further, between 1981 and 1999, there was a 340% increase in middle class bankruptcies. And, in 2001, 1.4 million Americans lost their health insurance—over half of those had an annual household income above $75,000, clearly indicating that the majority were middle to upper-middle class. Granted, these are only isolated examples, but they give an indication of what might be happening in the United States.

Third, as we've noted, accumulation is based on externalizing costs. Two of those costs—raw material depletion and toxicity—have natural limits, and it appears that we might be reaching them. Global warming, ozone rupture, destruction of the rain forests, and land degradation from waste are themes with which we are all familiar. Nowhere does the idea of natural limits come out more clearly than in the work of Peter Vitousek, professor of biosciences at Stanford University. Vitousek, Ehrlich, Ehrlich, and Matson (1986) argue that directly (through consumption) and indirectly (through toxic waste), human beings presently use up about 40% of the world's net primary production (NPP), which represents the rate of production of biomass that is available for consumption by all plants and animals. In other words, of the total amount of energy available for life on this planet, human beings use 40% of it. Predicting the Earth's long-term ability to support human life is difficult to calculate, because it depends on the wealth of the population and the kinds of technologies supporting it, but we can see that humans use up a hugely disproportionate amount of the Earth's resources (we are but one of some 5–30 million animal species on the planet), and we can see that the resources of the Earth are finite.

But limits aren't the only concern; toxic waste has been going on for years. Typically, firms take the cheapest way of handling waste—dump it on someone else's or public property—until public outcry motivates governments to pass laws restricting dumping. But the laws are not retroactive and it appears difficult to assign responsibility. The result is that government, not industry, tends to pay for the bulk of cleanup. According to a recent article in *The Washington Post*, "the number of toxic or hazardous sites requiring federal attention continues to grow, and Congress will have to spend at least $14 billion to $16.4 billion over the coming decade just to keep pace with the problem" (Pianin, 2001, p. A19).

Structural and Cultural Signs of the End

Wallerstein argues that world-systems enter a time of chaos during transition periods. How things change or into what form is not predictable. A world-system runs its cyclical courses through the Kondratieff wave, with periods of growth and stagnation, finally ending in collapse. A new configuration emerges out of this rubble, but, unlike Marx, Wallerstein offers no clear predictions. However, Wallerstein does argue that the uprising of 1968 marked the beginning of the end of the current world-system. We can see not only the clear marks of the dialectical cycles near the end of a 150-year Kondratieff wave, we can also see that the structural supports upon which capitalism has been built are limited and nearing exhaustion.

There are also cultural and structural signs that indicate the system is in the uncertainty of transition. Wallerstein points to two cultural signs: the introduction of complexity theory in science and postmodern theory in the social sciences. In the past 15 years or so, a significant number of physical scientists and mathematicians have turned against the causal predictability of Newtonian physics, which postulated a universe run according to universal laws—laws that in all time and in every place could explain, predict, and control the physical features of the cosmos.

Currently, many scientists are saying that Newtonian physics is a special case of reality; it fits only in circumstances that are clearly circumscribed or limited. The tools of science must therefore incorporate more flexible schemes with wider scopes of application. Thus today we hear of complexity theory, chaos theory, strange attractors, fuzzy logic, and so on. Wallerstein's (1995) point is this: "The natural world and all its phenomena have become historicized" (p. 486). That is, the scientific view of the universe has historically changed: the old science was built on a mechanistic, linear view of the universe; the new science is not linear or mechanistic.

The idea of a historicized science is an oxymoron, at least from the initial perspective of science. Science assumed that the universe is empirical and operates according to law-like principles. These principles could be discovered and used by humans to understand, predict, and control their world. Science was in the business of producing *abstract* and *universal* truths, not truths that only hold under certain conditions. The historicity of society has always been an argument against the possibility of social science, precisely because the factors that influence human behavior and society change according to the context. The hard, laboratory sciences have now become susceptible to the same critique: according to complexity theory, all knowledge is contextual and contingent, and nothing is universal and certain. "Hence the new science raises the most fundamental questions about the modernity of technology" (Wallerstein, 1995, p. 486).

The scene in the social sciences has followed suit and become even less certain than it was before. In the past 25 years, the most vocal and influential voice in the social sciences has been postmodernism (see Chapter 14 of this volume). Postmodernism in its most radical form, as it was brought into the social sciences, argues that the social world in technologically advanced societies is a virtual or hyper-real world. The cultural signs, symbols, and images that we use aren't connected to any social reality. Most of them come not from real social groups in face-to-face interaction, but are, in fact, produced by media and advertising concerns.

As a result of this cultural fragmentation and the new doubts in science, all grand narratives are held in distrust. Grand- or *metanarratives* are stories that attempt to embrace large populations of people (see Chapters 3 and 5). Typically, grand narratives are generated by political groups (as in nationalism and national identities). In their place, postmodernism advocates *polyvocality*, or many voices. Postmodernism argues that all voices are equal and should be given equal weight. These voices are of course linked to specific groups, such as men, women, blacks, Chicanos, and all the subdivisions within the groups, such as bisexual-Chicano-Catholic-males. There is thus an ethical dimension to postmodernism: "it is a mode of rejecting the modernity of technology on behalf of the modernity of liberation" (Wallerstein, 1995, p. 487).

The two structural signs that indicate we are in a time of chaotic transition are financial speculation and worldwide organization of social movements. There has been limited success in rolling back costs and reducing the press on profits, but not nearly what was needed or hoped for. As a result, capitalists have sought profit in the area of financial speculation rather than production. Many have taken great profits from this kind of speculation, but it also "renders the world-economy very volatile and subject to swings of currencies and of employment. It is in fact one of the signs of increasing chaos" (Wallerstein, 2004, p. 86).

On the political scene, since 1968 there has been a shift from movements for electoral changes to the "organization of a movement of movements" (Wallerstein, 2004, p. 86). Rather than national movements seeking change through voting within the system, radical groups are binding together internationally to seek change within the world-system. Wallerstein offers the World Social Forum (WSF) as an example. It is not itself an organization, but rather a virtual space for meetings among various militant groups seeking social change.

Another indicator of this political decentralization is the increase in terrorist attacks worldwide, such as the strike on the World Trade Center, September 11, 2001, and the bomb attacks on London, July 7, 2005. The terrorist groups themselves are decentralized, non-state entities, which makes conflict between a state like the United States and these entities difficult. Nation-states are particular kinds of entities defined by a number of factors, most importantly by territory, rational law, and a standing military. These factors and the political orientation they bring mean that nation-states are most efficient at confronting other nation-states, ones with specified territories, which legitimate rational law, and have modern militaries. Almost everything about the terrorist groups that the United States is facing is antithetical to these qualities of the nation-state. The United States is a centralized state and the terrorists are decentralized groups. These differences in social structure and relation to physical place make it extremely difficult for the United States to engage the terrorists—there is no interface between the two—let alone defeat or make peace with them.

But more than that, the attacks of September 11 have energized politically right-wing groups in the United States. It has allowed them to cut ties with the political center and "to pursue a program centered around unilateral assertions by the United States of military strength combined with an attempt to undo the cultural evolution of the world-system that occurred after the world revolution of 1968 (particularly in the fields of race and sexuality)" (Wallerstein, 2004, p. 87). This, along with attempts to do away with many of the geopolitical structures set in place after 1945 (like the United Nations), has "threatened to worsen the already-increasing instability of the world-system" (p. 87).

What will follow the 400-year reign of capitalism is uncertain. World-systems theory, as Wallerstein sees it, is meant to call our attention to thinking in structural time and cyclical processes; it is meant to lift our eyes from our mundane problems so that we can perceive the world-system in all its historical power to set the stage of our lives; it is intended to give us the critical perspective to have eyes to see and ears to hear the Marxist dynamics still at work within the capitalist system; and, finally, it is intended to spur us to action.

Unlike C. Wright Mills (1956), Wallerstein is not saying that "great changes are beyond [our] control" (p. 3). Rather, he means that "fundamental change is possible . . . and this fact makes claims on our moral responsibility to act rationally, in good faith, and with strength to seek a better historical system" (Wallerstein, 1999, p. 3). According to Wallerstein (1999), because the system is in a period of transition where "small inputs have large outputs" (p. 1) and "every small action during this period is likely to have significant consequences" (Wallerstein, 2004, p. 77), we must make diligent efforts to understand what is going on; we must make choices about the direction in which we want the world to move; and we must bring our convictions into action, because it is our behaviors that will affect the system.

> We can think of these three tasks as the intellectual, the moral, and the political tasks. They are different, but they are closely interlinked. None of us can opt out of any of these tasks. If we claim we do, we are merely making a hidden choice. (Wallerstein, 2004, p. 90)

Wallerstein Summary

- Wallerstein sees his work more in terms of a type of analysis than a specific theory. His point is that it is the principles of analysis that drive the theorizing rather than the other way around. There are two main features of Wallerstein's perspective: globality and historicity. Globality conceptualizes the world in system terms, which cut across cultural and political boundaries. Historicity sees history in terms of structural time and cyclical time within the structures, rather than focusing on events, people, and linearity.

- In terms of theory, Wallerstein takes a Marxist approach. He focuses on the division of labor, exploitation, and the processes of accumulation and overproduction. In Marxist theory, exploitation is the chief source of profit. Thus, capitalists are intrinsically motivated to increase the level of exploitation. Since wages tend to go up as capitalist economies mature, reducing the level of exploitation and profit, there is a constant tendency to export exploitation to nations that have a less developed capitalist economy, thus increasing the worldwide division of labor.

- Capitalist accumulation implies that capital is invested for the purpose of creating more capital, which in turn is invested in order to create more capital. In modern capitalism, this process of accumulation is augmented by the state. The state specifically bears the costs associated with transportation, toxicity, and the exhaustion of materials. More powerful states additionally provide conditions that facilitate quasi-monopolies, thus increasing capitalists' profits and the rate of accumulation.

- Overproduction is endemic to capitalism as well. Because they are driven by the capitalist need for accumulation, commodification (the process through which material and nonmaterial goods are turned into products for sale) and production are intrinsically expansive. Capitalists will continue to create new and produce existing commodities until the market will no longer bear it, thus creating more supply than demand.

- Taken together, the processes of exploitation and the division of labor and the dynamics of accumulation and overproduction create a scenario in which there is a continual movement of products and labor from more powerful to less powerful nations.

- In the world-economy, there are four types of nations: the core, semi-periphery, the periphery, and external areas. In general, exploitation and mass production of least-profitable goods move from the core to the external areas. However, because this is a system, there is also a move of nations as they transition from external to peripheral to core. Eventually, there will be no more areas to exploit with low-profit mass production, which will lead to system breakdown.

- The world-economy thus tends to go through cycles of expansion, depression, and breakdown. These cycles reach a crisis about every 150 years. According to Wallerstein, the world is now in its fourth phase of world-economies. The last phase began in 1917, with the United States as the world-economy's core nation. Wallerstein marks the beginning of the end of this phase at the social upheavals of 1968. In addition to the social movements, the world-economy entered a cycle of depression that was particularly harsh and lasted for about 20 years. While the world-economy is actively trying to come back from this economic depression, there are three factors that are inhibiting this attempt: the system limits to exploitation, the middle class squeeze, and the limited ability of states to pick up externalized costs. Thus, Wallerstein argues that the world-system is on the brink of collapse and is currently experiencing the chaotic period that always precedes such an end.

Niklas Luhmann:
Social Systems and Their Environments

If you have been trained in sociology, Luhmann presents you with an opportunity to remake your mind. As I said, Wallerstein asks us to stretch our minds. We're somewhat used to thinking about interactions and society, and Wallerstein lifts us up to the world-system level. Yet he does so with terms and ideas that we are generally familiar with. Luhmann, on the other hand, asks us to think differently. He gives us a new perspective—systems theory—and he uses terms that are going to be somewhat new for us (thought not as many as Giddens did). But Luhmann isn't simply being academic and making things harder than they need to be. He wants to give us a fundamentally different way of thinking about society. I want you to pay close attention to what he says. Grapple with his ideas and work through his theory. If Luhmann is right, it's quite possible that sociology has been looking at society all wrong.

Photo: Courtesy of University of Bielefeld.

The Essential Luhmann

Biography

Niklas Luhmann was born on December 8, 1927, in Lüneburg, Germany. He initially studied law and worked as a public administrator for over 10 years. On a work sabbatical in 1961, Luhmann studied Talcott Parsons's theories at Harvard. He began lecturing in 1962 at the University for Administrative Sciences in Speyer, Germany, and published his first of over 30 books in 1964, a study in formal organizations. In 1968, Luhmann took his first position as professor of sociology at the University of Bielefeld in Germany where he stayed until his retirement in 1993.

Passionate Curiosity

Luhmann's concern is sociology's big questions: What is society, and how does it work as a whole? His first exposure to the issue of society and how it functions came through Talcott Parsons. He thus is keenly concerned with how society functions as a system. Luhmann is also specifically concerned with modern society and argued vigorously with Jürgen Habermas about the differences between "rational modernity" and modern society as a complex system.

Keys to Knowing

System environment; smart and dumb systems; open and closed systems; risk and complexity; meaning; autopoiesis; self-thematization; segmentary, stratified, and functional differentiation; modern society; positive law

---

Seeing Further: Society as a System

We should note at the beginning that Luhmann is generally considered a "neo-functionalist." The preface "neo" simply means new and it indicates that the theory is based on important previous work but with some modifications from other intellectual traditions. In Chapter 7, we saw some neo-Marxian theories, especially with Habermas. Immanuel Wallerstein in this chapter and Jean Baudrillard in Chapter 14 are two other clear examples of neo-Marxian theorists. Luhmann is our best example of neo-functionalism in this book. So, we'll see some similarities between Parsons and Luhmann, but many differences as well. The differences come from Luhmann's emphasis on systems theory.

Defining Systems Theory

While Parsons uses the idea of the social system to understand how social structures and institutions relate to one another, Luhmann bases his theory on the most elementary feature of a system: its relationship to the environment. This seems simple enough, especially because Parsons talked about a system's relationship to the environment, but, as we'll see, this small shift produces some very different ways of thinking about society.

We've already covered the basic elements of functional theorizing in our introduction to Parsons, so keep those in mind as we look at the general features of systems theory. In brief, there are at least four qualities of a system. First, a system is made up of interrelated parts. Your car and your body are both examples of systems. In this way, functionalism is a systems theory. It's fundamentally concerned with the relationships among the parts and of the parts with the whole. But, as you'll see, functionalism is a limited kind of systems theory.

The second characteristic of a system is that it exists in an *environment*. We are all somewhat familiar with this idea; we talk about computer environments and "the environment" (meaning the ecological system). In terms of their environments, systems can be more or less open. An important word of caution, however: a system cannot be completely open or closed. One of the major points of Luhmann's theory is that a system is formed by boundaries between it and the environment. A completely open system would have no boundaries and would thus be part of the environment and not a system. Totally closed systems are impossible as well. A completely closed system would have to be a perpetual motion machine, with no loss of energy. So we need to think about the relationship between systems and their environments as running on a continuum.

Basically, *open systems* take in information or energy from their environment and *closed systems* do not. A good example of an open system is your body. It takes in energy and information (food, air, sense data) and is thus directly influenced by the environment. If the environment changes too rapidly, your body will die. Relatively closed systems take in less information and energy. As a system, your car is more closed than your body. At this moment, the chances are good that your car is just sitting somewhere. It's inactive yet remains a system because its parts are related to one another. Your body, on the other hand, is never inactive. It is always taking in and processing information and energy from the environment.

Third, systems are dynamic. Of course this is a variable, but all systems involve processes. As we've already noted, systems take in or process energy and information. The energy can be in the form of food for organic systems or electricity for mechanical systems. Parsons actually touches on this issue a little in his cybernetic hierarchy of control. Energy moves up the hierarchy from the organic system and information moves down. However, there is more to this idea of process than the presence of energy and information.

Dynamic systems have *feed-forward and feedback dynamics*. The basic distinction has to do with how the system processes information. Can the system adjust to changes in the environment? If so, there is some kind of feedback process in place. Your car is a good example of a system that lacks feedback mechanisms. It's primarily a feed-forward process. The system doesn't feed back information from the environment so that the car can make adjustments. Considering the advances in computer technologies, this is probably a limited example, but you get the point. Your body, on the other hand, has a number of feedback systems in place.

For example, it self-adjusts to changes in the external temperature. "Cold-blooded" organisms do not have this feature.

The fourth defining characteristic of a system is that systems can be smart or dumb. Generally speaking, feedback systems are smart, but not always. In addition to feedback, a system must have a goal and explicit mechanisms in place to make adjustments based on incoming information and the system's goal. Obviously, your body is a smart system and your car a dumb system. But mechanical systems can be smart.

The heating and cooling system in your home is a good example of a smart mechanical system (Collins, 1988, pp. 49–50). It's smart because it has a thermostat. The thermostat has three important elements: a goal state (the temperature you set it at), an information mechanism (its ability to read the temperature in your home), and a control mechanism through which it turns the air conditioner or heater on and off. Smart systems such as the thermostat tend toward equilibrium. It balances out the forces of hot and cold through its control mechanism to keep your house at a comfortable 73 degrees, or whatever temperature it was set to.

As I mentioned, there are a number of differences between functionalism and systems theory. Below is a short list. As we'll see, Luhmann's concern is primarily focused on the first few issues.

- Systems theory pays attention to the relationship between the system and its environment; structural functionalism generally does not.
- There are no requisite needs in systems theory; functionalism is defined by the delineation of such needs.
- Systems do not necessarily tend toward equilibrium; functionalism generally posits an equilibrated state. For systems theory, a state of equilibrium is a consequence of a system being smart.
- Systems theory is focused on processes; functional theory is focused on structures. Functional theory thus tends to reify its concepts and systems theory does not.

Concepts and Theory: Defining Social Systems

System Boundaries

For Luhmann, the concepts of function and functional analysis no longer belong to the system itself, as with functionalism, but, rather, "*to the relationship between system and environment. The final reference of all functional analyses lies in the difference between system and environment*" (Luhmann, 1984/1995, p. 177, emphasis added). It is important to note that system environments are made up of other systems. Luhmann sees systems as interdependent and thus mutually constitutive. This is easy to illustrate: Did the collapse of the Soviet Union influence the United States? It obviously did. It changed the entire global environment for the United States and all other nations as well—each nation is a system within a global system. For example, the idea of first-, second-, and third-world countries is no longer viable. In other words, with the demise of the USSR, the concept of "third-world countries" ceased to exist, and the United States has become something different from a first-world country (the other first-world country was Soviet Russia) because such terms have become obsolete.

Thus, for Luhmann, the important beginning point for functional analysis isn't the system itself, but, rather, the boundary between the system and its environment. This idea has two implications. First, it means that systems are defined in terms of boundaries. A system exists only if it is different from its environment, that is, if there's some kind of boundary between it and the environment. This means that neither the system nor the environment is more important than the other—it's the relationship and boundary that are important. The second implication is that "this leads to a radical de-ontologizing of objects as such" (Luhmann, 1984/1995, p. 177). Luhmann is saying that shifting analysis to the boundary means that treating social structures as real objects becomes extremely difficult, because we cannot treat difference as an object or thing. Thus, in Luhmann's neo-functionalism there are no objective, social structures.

That being said, Luhmann argues that the boundary between system and environment is created by *reducing complexity and risk*. Again, this is by definition. Systems must be less complex than their environments because environments are composed of other systems. We should also note that reducing complexity and risk are related. Systems reduce the risk of being overwhelmed by their environment by reducing complexity. Moreover, the reduction of risk and complexity are active issues, both of which are tied up with survivability. Systems survive in their environments because they reduce risks; they do this by reducing the complexity of the environment so that certain elements can be controlled.

As an example, let's again use the relationship between an individual nation and the global system of nations. The global system is much more complex than any single nation. In order to survive within that worldwide system, a nation must reduce the complexity of its environment. For example, a nation can't trade all goods with all nations—the complexity of such a task would render it impossible. Nations come up with rules that regulate trade and those rules reduce the complexity of the system. The same is true for any organization or system—all the possibilities of the environment must be reduced to manageable limits. How different systems do this is what makes them distinct. Here's an obvious example: you and your car are different kinds of systems. Why? Because the automobile system and your biological system reduce the complexity of the environment in different ways.

Meaning and Social Systems

As we've seen, systems are defined in terms of boundaries that reduce risk and complexity in the environment. The social system evolved a very specific way of doing this. In fact, this evolution actually involved two systems: the social system and the psychic system. In other words, people and society need one another; society is impossible without people and people are impossible without society.

At any point, the one is the necessary environment of the other. Luhmann (1984/1995) argues that this co-evolution came about because of a common achievement: "We call this evolutionary achievement 'meaning'" (p. 59). The social system is created through *meaning*, which is the elemental nature of human beings—meaning is what makes the human psyche and social system unique.

In reducing risk and complexity, humans have to address three central issues: time, space, and symbols. Now here's an important point: Because the social system is created through meaning, time, space, and symbols have endless horizons. In other words, for humans, time, space, and signification are all potentially infinite. Let's take time, for example. Most animals live for today and have very little sense of past or future. Humans, on the other hand, can think about the past and plan for the future. Not only that, but we can communicate with one another about the beginning of the universe through the language of physics. We can even extend our reach further: through religion we can talk about before the beginning of time and after time ceases. We humans can divide up time almost any way we want to. Just think of all the different calendars that humanity has come up with and discarded. Of course, the infinite possibilities of time and space are based on the ability of humans to use meaning, and meaning itself (symbols) must be held back from its endless possibilities. Thus, social systems are defined and produced when meanings are created that orient actors to a specific past, present, and future; that delineate certain spatial relationships; and that restrict the endless possibilities of symbolic worlds.

The Reflexive Nature of Social Systems

One of the issues that meaning introduces is reflexivity or self-reference. Take a moment to think about our past and all the crazy things we humans have believed: the earth is flat; the earth is the center of the universe; gods live on Mount Olympus; the universe was created by the water god Nu and the sun god Atum; hekura live inside shaman and devour the souls of their enemies; it's the manifest destiny of the white man to dominate the world. The list is endless and that's the point—meaning refers only to itself. This, according to Luhmann (1984/1995), is "the fundamental law of self-reference" (p. 37).

Luhmann uses the term *autopoiesis* to talk about the issues surrounding self-reference. The word is made up of two Greek words: *auto* meaning self, and *poiesis* meaning creation. Autopoietic systems, then, are self-producing: they produce the basic elements and organization that make up the system. The term originated in the work of two Chilean biologists, Humberto Maturana and Francisco Varela (1991). A clear example of an autopoietic system in biology is the cell.

Biological cells are made up of biochemical components and bounded structures. In life, the cells use these components and structures to convert an external flow of energy and molecules into their own components and structures. Thus, the elements of the biological cell reproduce themselves. Think of it this way: biological cells are made up of certain elements. Those elements have a rather short life span, which implies that biological cells should die off rather quickly. But they don't. The reason they don't is that they reproduce the necessary elements. Thus, the components of a cell at time-1 will be different than at time-2 (because the components at time-1 have died off) yet be exactly the same (because the cells have reproduced). According to Luhmann, society does the same thing.

How would this look in society? Well, look at your school. After teaching at my university for about five years, something struck me as I returned from summer break: "Different faces, same students." There are obviously some personality differences and so on, but by and large the university produces the same students year after year. They are freshmen, sophomores, juniors, and seniors; they are sociology, biology, and business majors; they are struggling through the same curriculum and face the same kinds of decisions. Thus, the students are basically the same, year after year.

Social systems, then, are self-referencing and this reflexive nature implies three things. First, "only meaning can change meaning" (Luhmann, 1984/1995, p. 37). People exist in systems of meaning and they make decisions that influence the social system in which they are working. Actors in this sense are free agents: Their decisions are not constrained. However, people can only use meaning to make decisions about meaning. This sounds circular because it is. If you make a decision about what courses to take or what major to have, you do it within the education system. Even if people change the meaning of something—as gays and lesbians are working to do with the meaning of marriage—it is done within an already existing meaning system.

The second thing that self-referencing systems imply is that they must be continually remade. Specifically, the boundary between the system and the environment must be maintained. Think about an obviously human-made artifact dug up from a recently found ancient archeology site. In order to figure out what it is, the researchers must attempt to reconstruct the culture and society that it came from. Obviously, the meaning of the object isn't in the object itself. As Luhmann (1984/1995) puts it, "system events disappear from moment to moment and subsequent events can be produced only via the difference between system and environment" (p. 177).

The third implication is that societies self-thematize. *Self-thematization* is based on the idea that social systems can be reflective. That is, society can think about itself. Societal thematization is much like individual thematization. You undoubtedly would be able to give me a clear and coherent answer if I asked, who are you? There's a theme about you; it's your self-identity. It's the story line around which you organize ideas and experiences about who you are. This theme does a few things for you. First, it makes you different from everybody around you (your environment); second, it gives you meaning (by reducing the complexity of your real experiences through continuous time and space); and third, it forms the basis upon which you make decisions (for example, you decided to come to school because you saw the decision as part of your understanding of who you are). Social themes function in the same way. Self-thematization implies that social systems are self-organizing: using meaning, they organize their environmental boundary as well as the boundaries within the system (such as between religion and education).

Concepts and Theory: Social Evolution

Three Societal Systems

Luhmann gives us three different societal systems: interactional, social (society), and organizational. The foundation of all social systems is communication. As such, the basic unit of the societal system is the interaction, where two or more people meaningfully interrelate their actions: "As soon as any communication whatsoever takes place among individuals, social systems emerge" (Luhmann, 1982, p. 70). For Luhmann, face-to-face encounters provide the opportunity for an inter-locking relationship of action through symbolic communication; speaking to one another automatically sets up a boundary and this boundary reduces complexity from all possible communications. In addition, face-to-face communication is self-limiting in the sense that only one person generally talks at a time and only one topic can be dealt with at a time.

The limitations of the *interactional system* force movement to a system of another type. In other words, there has to be a communication system that can con-nect your face-to-face interaction with other interactions. Society, then, *"is the com-prehensive system of all reciprocally accessible communicative actions"* (Luhmann, 1982, p. 73, emphasis original). Society coordinates communication with and among all possible actors missing from a single case (your interaction), and society regulates or systematizes through the principle of possible communication.

Society, or the *social system,* then, is the meaning system that is capable of embracing a number of interactions. This is accomplished symbolically, through such things as language and self-thematization. To see how this works, let's set up three interactions at two different times. We'll say that the first set of interactions happened 600 years ago, with one interaction taking place in Tenochtitlan (the Aztec capital), one in York (England), and the third in Luoyang City (China). Each of these interactions would take place using different languages and different societal themes. However, all the interactions taking place within those cities would be meaningfully linked and thus constitute a society—so, all the interactions in Tenochtitlan would be Aztec, all in York would be English, and all the ones in Luoyang City would be Chinese.

Now let's move those interactions into our time. Since Tenochtitlan no longer exists, we'll use the Mexican city of Tecate, but both York and Luoyang still stand. Now, what are the differences? Are the interactions still as separate, thus constituting different social systems? Chances are good that the three interactions will still be in three different languages. But are there themes that link the interactions? Chances are better today that there are such themes, and the chances increase as we move to greater population centers, such as London, Mexico City, and Hong Kong. There are two strong themes that cut across large numbers of interactions today: capitalism and democracy. Such themes prompt Luhmann (1982) to conclude, "Today, there is only one societal system—society is a world society" (p. 73). We will come back to this notion of a world society when we consider differentiation and modernity.

But for now I want you to notice how unique Luhmann's idea of society is. Using the ideas of system boundaries and meaning, Luhmann is able to give us a much more flexible and robust definition of society. This definition escapes the limitations of seeing society in terms of a territory, language, and state (the Weberian approach); the drawbacks of defining society in structural-functional terms (Durkheim and Parsons); the limitations of defining society on the basis of economic relations (the Marxian approach); and the restrictions of conceiving of society as a set of structures. In Luhmann's theory, society is almost organic, moving and changing as people redefine their meanings and change their interactions.

The third social system is organization. This system is "inserted" between the societal and interactional systems. The purpose of *organizational systems* is to sustain artificial behaviors for long periods of time in order to accomplish specific goals. The behaviors are "artificial" in the sense that they aren't directly motivated by either aesthetic values or moral demands. For example, most people working in a McDonald's kitchen do not do so for the work's intrinsic value.

In addition to motivating people to work by providing rewards—generally money in capitalist societies—there are two other methods of obtaining compliance: *role specification* and entrance and exit rules. All organizations have explicit behavioral expectations of their members. For example, the university of which you are a part expects you to behave as a student; that's your role in the organization. Further, the university counts on you to internalize a good portion of these role expectations, such as the values of academic honesty and scholastic truth. Of course, the specificity and commitment demanded vary by type of organization and position. If you are planning to go on to medical school, for example, the role of doctor is more highly specified, and the attitudinal and motivational expectations are greater, than your current role as student.

Organizations also have explicit *entrance and exit rules* that help manage members' commitment to work. Let's continue to use the university example to see how this works. The university that you are attending has demanding entrance rules. In order to become a student, you had to meet several criteria, such as minimum GPA and SAT scores. Entrance rules such as these create investment in organizational roles. Many organizations also have exit rules. The university is again a good example: You must fulfill specific requirements concerning courses and quality of work to successfully exit the university. As you can see, role specification and entrance and exit rules make you personally invested in the organization. Note also that the exit rules from the university form a good part of the entrance rules for your next organizational position: most of the jobs you'll be applying for carry a minimum requirement of a bachelor's degree. Your entrance position is predicated upon how well you completed the exit requirements for your degree. Notice also that your investment in roles and rules of the university carries right over as well.

Evolutionary Processes

As we saw with classic functionalism, evolution means differentiation. Luhmann agrees. However, because Luhmann defines functional issues in terms of the relationship between a system and its environment, his concerns are different. We'll see the differences as we talk our way through his theory.

Generally speaking, evolution occurs through three processes: variation, selection, and stabilization. Systems increase the possibility of their survival by having the ability to create different options. There are two important qualities that define the social system's ability to create *variation*. First, social systems, because they are based on meaning and communication, aren't limited to organic constraints. Humans now evolve symbolically, by dreaming imaginary worlds and bringing them into existence. Second, the "*capacity for evolutionary variation* is guaranteed because language always offers the option of saying 'no'" (Luhmann, 1982, p. 266, emphasis original). Luhmann argues that communication systems are based upon codes, which have a binary quality to them. Binaries always contain opposites, such as good/bad, male/female, and so on. Since every idea by definition contains its opposite, "we can communicate new, surprising, and unsettling messages, and will be understood" (p. 266).

The next evolutionary principle is *selection*. Luhmann talks about this in terms of the differences between language and media of communication. Language itself offers almost endless possibilities. Thus, I can say that "I'm a six-foot tall rabbit with the ears of an elephant." But one of the things that selection does is to "de-realize" some of what is possible. I can *say* almost anything, but not everything I say will be understood, which is the basic requirement for communicative success and selection.

Communicative success, and thus selection, is governed by recognized media. A medium is a means of bringing about or conveying something. When we are talking about a generalized medium of communication or exchange, we are referencing such things as the ideas and beliefs surrounding truth, love, money, political power, art, and so on. These are legitimate codes or discourses that we reference, which in turn give a statement intelligible space. Thus, the things I can intelligibly say through a medium of "faith" might be very different from what will have communicative success through a medium of "political power."

The third process in evolutionary change is *stabilization,* and this requires the formation of a system—every social change must be systemically stabilized. The media that I mentioned above, faith and power, are understandable to you because they are already a part of a system of communication. Faith exists within the communicative system of religion and political power within the system of government.

Patterns of Differentiation

Sociocultural evolution occurs initially through separating the three systems we noted above—interaction systems, organizational systems, and societal systems. Thus, the greater the level of differentiation, the greater will be the independence of these systems. As differentiation between levels is achieved, social reality becomes more complex and the systems can assume separate functions and set themselves off from one another.

Differentiation also occurs within each system. Luhmann argues that this takes place through replication. Systems differentiate internally along the same path that they used to differentiate externally. In other words, systems replicate themselves internally. For instance, organizational systems will differentiate internally by proposing, selecting, and systematizing different organizational forms. "Differentiation is thus understood as a reflexive and recursive form of system building. It repeats the same mechanism, using it to amplify its own results" (Luhmann, 1982, pp. 230–231).

As I mentioned earlier, Luhmann asks us to look at systems and their environments. Part of what this means is that every differentiated subsystem has three references: (1) the external environment common to all subsystems; (2) its relation to other subsystems within the larger system; and (3) its relationship to itself. For example, each state within the United States has a common external environment (the federal government), is differently related to each of the other states (subsystems), and each state has its own unique configuration of state and local governments (relation to self).

The implication is that Luhmann's theory of evolutionary change is much more dynamic than the previous ones. Classic functionalism did not give sufficient weight to issues of environment. Luhmann, however, recognizes the movements between systems and their environments, both internally and externally. These multiple relationships in the long run tend to increase the level of differentiation and system building exponentially. In other words, complexity breeds complexity. Initial differentiation will be small, but because of the environmental relationships of systems to systems, differentiation in one system makes for a more complex environment for other systems, which then have to differentiate, which further adds to the complexity of the environment, which again prompts system differentiation.

Luhmann argues that there are three primary patterns of differentiation:

1. *Segmentary differentiation,* which differentiates society into equal and alike subsystems. In this case, a primitive society using kinship as its principal organizational form will tend to duplicate or extend kinship when differentiation is needed. This results in a system that is large but not very complex, which in the long run reduces the number and kind of variations that the system can produce (evolution is thus hampered).

2. *Stratification differentiation,* which differentiates society into unequal subsystems. The organization of society becomes hierarchical, with some subsystems having greater power or status than others. While segmentation only duplicates its systems, stratification creates diverse systems. This kind of differentiation does two things: it increases the number and diversity of possible variations and adaptive systems, and it creates pressures for increased communication and generalized media of exchange. A more abstract medium is needed to facilitate communication among different kinds of groups, and communication increases because of the diversity, both within and between strata.

3. *Functional differentiation,* which organizes communication around special functions to be fulfilled at the level of society. This is the type of differentiation with which classic functionalism is concerned. In functional differentiation, there are institutional fields that link up different organizations or subsystems. You are

undoubtedly familiar with the names of these institutions, such as education, government, family, and so on. Let's use education for an illustration. "Education" is really a group of different organizations linked by a particular culture and communication system. In education there are school districts, university systems, textbook and journal publishers, organizations that produce chalk and blackboards, and so on. These organizations are functionally related to one another and to other institutional fields (through more abstract means of communication). Notice that creating these institutional domains produces entirely new environments, each with its own set of issues.

I've outlined Luhmann's ideas about system evolution and differentiation in Figure 12.2. In comparison to the actual world, the diagram is fairly simple. But it should work to give us an idea of how Luhmann thinks about societal evolution and provide us with at least a sense of the complexity involved. As you can see, the basis of society is the interaction. It forms a system by creating a boundary between it and the environment, which I am noting with a thick line. In this "before differentiation" phase, all tasks are fulfilled through this one interactive grouping.

Due to shifts in the environment, the interaction throws out different possibilities of variation, selection, and stabilization. Society then differentiates, but segmentally by duplicating itself (the interactions are all of the same type). The line between two interactions (Org.1) represents an organization. These organizations could take place among almost any of the interactions. This is still a simple social system, but notice that a new environment has been created. There is still the general, external environment, but now there is an internal one as well, formed by the communicative relations among the interactions and organizations.

Again, there is variation, selection, and stabilization, which move society to *stratified differentiation*. Notice that now there are different types of interactions (Interactions 1, 2, and 3), with any number of organizations occurring among them. These lines of like interactions constitute different social strata, as noted by the box surrounding each type. This is where society develops groups with different levels of resources to address different system concerns (in this case there might be religious and political elites and lower-level economic workers). As a result of stratified differentiation, there is a corresponding increase in environmental complexity. There is still the external environment, but the internal social environment has expanded to include all interactions. In addition, there is also a social environment of like interactions (noted by the horizontal rectangles surrounding the different interactions).

With functional differentiation, things change dramatically. Notice that the social environment among like interactions still exists, as do the larger social system and the external boundary. Added to these are the institutional environments. I've noted two such environments, A and B, and connected all the organizations that constitute a functional domain. As you can see, some of the organizations are parts of two domains. An example of such an organization is Boeing, which is part of the airline industry as well as the military. The number and diversity of these crossover organizations increases as organizations include multinational, diversified corporations.

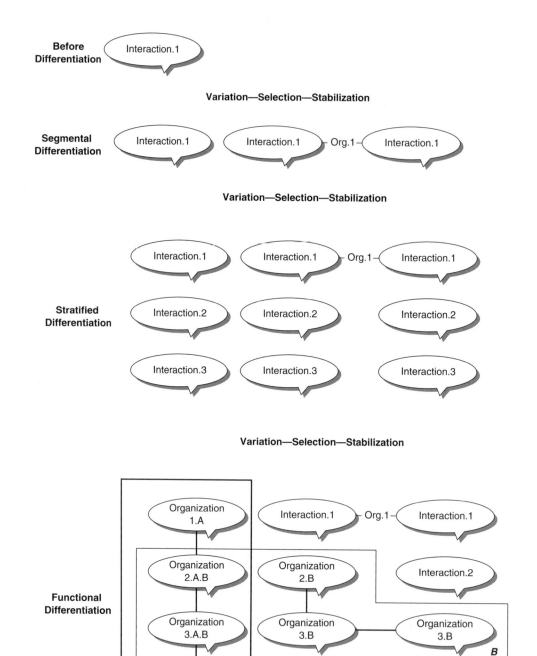

Figure 12.2   Luhmann's System Differentiation

These patterns of differentiation roughly correspond to social evolution from simple to complex societies: archaic societies are differentiated primarily through segmentation, high-culture societies through social strata or class, and modern societies by functional differentiation. What is unique about Luhmann's approach is that the complexity of society isn't gauged simply by the degree of differentiation, as classic functionalists would have it, but also by the incorporation of previous forms. Modern society is most complex because it incorporates the other two forms in addition to functional differentiation.

Concepts and Theory: Modernity

To begin our discussion of modernity, we need to remind ourselves that Luhmann argues that modern society is global. Again, one of the theoretically powerful gains from using Luhmann's theory is the way it conceptualizes society. It clearly frees us from being constrained by territories or political systems. Political systems are certainly a type of social system, but in modernity they are a subsystem within the larger system of society. Keep this in mind as we review the general contours of modern social systems. These issues can be applied to any level, including the global social system.

As we noted earlier, initial differentiation occurs among the three different social systems: interactional, organizational, and societal. Differentiation among the three system levels also means that society can intervene in interactional or organizational systems without threatening its own survival. However, while they are differentiated, they are not disassociated. A total disjunction of the three levels is impossible since "all social action obviously takes place in society and is ultimately possible only in the form of interaction" (Luhmann, 1982, p. 79).

Social Integration

The above implies that the three subsystems are *nested* within and functionally dependent upon each other, and this interdependency keeps conflict to a minimum. Conflict is also minimized because of the tendency toward *piecemeal involvement*. As societies differentiate, the individual becomes a nexus of diverse expectations that she or he cannot completely fulfill. You've undoubtedly experienced this already, with demands from parents, significant others, peers, work, school, religious organizations, and so on. The end result is that overt conflict between groups becomes increasingly difficult because people and their emotional involvements are spread too thin.

In addition, differentiation allows members to be indifferent to the roles of others. Because our individual roles mean less to us in an environment of complex and irresolvable expectations, we don't have the strength of commitment to an identity to take exception with others. In other words, Luhmann sees modernity as a time where there is greater equality and acceptance. However, notice why this comes about. It isn't due to utopian beliefs about egalitarianism; it is simply due to the effects of the complex differentiation of society.

In terms of integration, one other important effect of differentiation remains—the *autonomy of law*: "the emergence of a functionally specific legal system appears to possess special significance as an achievement of social evolution: it is a condition for all further social evolution" (Luhmann, 1986/1989, pp. 129–130). Law represents specific communicative codes that persons and organizations can use to understand and regulate relationships. In this sense, the law is a special kind of medium of exchange within society. Society is based on the constraint of acts of communication and law itself is a way of constraining communication. The separation of law as a function serves as a kind of evolutionary catalyst, accelerating differentiation by facilitating communication across organizations and societies.

For social evolution and integration, it is important that law become able to reflexively produce its own themes, apart from social ideologies. This results in what Luhmann calls *positive law*. Before modernity, and even through the Enlightenment, law was seen as founded on a natural order or theocracy (rule by God). The well-known line from the Declaration of Independence—"We hold these truths to be self-evident, that all men are created equal, that they are endowed by their Creator with certain unalienable Rights"—is an expression of deistic natural law.

Natural law and divine law are absolute in every way: absolute truth and absolutely enforced. However, in the long run, this basis of legitimation hampers social evolution. America provides a case in point: women and people of color were not included in the above quotation. If the United States had continued to believe in the natural or divine truth of the above statement, women, people of color, and so on would still not be considered citizens. Positive law rests solely upon the legislative and judicial decisions that make it law, and thus it is more readily adaptable to a rapidly changing society. In contrast to natural law, the validity of positive law rests on the principle of variation.

One of the things to notice about modern social integration is that it is based on a generalized and flexible medium of exchange and that systems are integrated as a result of *unintended consequences of differentiation:* modern society has no center and thus orderliness does not hinge upon structural (state, religion) or cultural (identity, nationalism) centrality. There's a way in which the mechanism for integration in highly differentiated societies is avoidance. Contrary to what Durkheim and Parsons argued, society doesn't need a single set of shared value commitments—it has become too complex. Modern society is a highly abstract communicative network that simply defines vague and lax conditions for social compatibility. Moreover, modern society involves abstraction of and indifference to multiple aspects of the lives of individuals.

Problems of Integration

Of course, modern society isn't without its problems. All social forms are ultimately dependent upon the interaction system. This dependence restricts the abilities of organizations and society. Since coordination of all the levels and units is achieved via direct interaction, a bottleneck or logjam can occur in the flow of information, because interactions are self-limiting (topics, turn-taking, etc.).

Another important issue of modern differentiation is that problems and issues can become displaced from the level of society to a subsystem. The result is that the societal subsystem responsible may not have the communicative tools (codes and themes) to deal with the issue. As an illustration, let's think about modern capitalism, the ecological environment, and peripheral (or underdeveloped) nations. The communicative scheme of capitalism is at odds with such humanitarian concerns as the poor or the environment. The profit motivation is intrinsic to capitalism—it is its defining theme. Without it, capitalism would not be capitalism.

Profit is accrued through expanding markets and increasing commodification, both of which are objectifying and amoral. Thus, we should expect capitalist organizations to be oriented toward maximizing profit through expanding markets and commodification. Capitalist themes and codes are oriented toward bottom-line considerations, even in such global capitalist entities as the World Trade Organization, which is assumed to have oversight over ecological issues. This thematic value of profit ought not to be seen as bad in and of itself. It is the result of capitalism's self-thematization. But it does tell us why ecological destruction is receiving so little real attention: the organizations that have been given the responsibility for it do not have the themes or communication systems to address it in any real way.

Luhmann Summary

- Luhmann uses systems theory to argue that the primary issue for a social system is the boundary between it and its environment. Functional analysis should take place at this boundary point, not internal structural relations (as functionalism argues). This means that every system is different according to its environment and that at least part of a system's environment is made up of other systems. Systems theory pays attention to processes, not structures; systems vary by their degree of complexity.

- Environmental risk and complexity are reduced as systems differentiate and become more complex themselves. The risk and complexity that social systems must deal with revolve around time, space, and symbols. System differentiation is an evolutionary process that entails variation, selection, and stabilization. Human systems are created through communication; thus, social evolution involves variation in communication (provided by linguistic opposition), selection of new communicative forms (recognized media), and stabilization through creating new systems of communication.

- Because social systems create their environments through meaning, they are inherently self-referencing. The reflexivity of the social system implies that systems are self-organized (principally through self-thematization), self-produced, and are continuously remade.

- Societal systems have three distinct subsystems: the interaction system, the social system, and the organization system. Interaction systems are made up of face-to-face communication. Social systems (society) are comprehensive communication systems that link all reciprocally accessible communicative actions. Organizations are formal collectives with specific entrance and exit rules, roles, and goals that sustain artificial behaviors for long periods of time.

- Societal evolution differentiates among and within these subsystems through three different patterns: segmentation, stratification, and function. These patterns roughly correspond to increasing complexity in societal types: archaic, high-culture, and modern. Through each phase of differentiation, societies become more complex, first because each pattern increases the amount and diversity of communication and second because each evolutionary type contains the previous differentiation pattern. However, each evolutionary type has a dominant pattern: archaic—segmentation, high-culture—stratification, and modern—function.

- Differentiated societal systems integrate because society can influence interactional and organizational systems without endangering itself, because the different subsystems are nested, and because of positive law. Integration problems include interactional bottlenecks and problem dispersal to subsystems without the necessary communicative tools.

Building Your Theory Toolbox

Knowing Systems

After reading and understanding this chapter, you should be able to define the following terms theoretically and explain their theoretical importance to analyzing the world as a system:

World-systems; globality; historicity; structural and cyclical time; division of labor; exporting exploitation; overproduction; capitalist states and quasi-monopolies; hegemony; core, semi-periphery, and periphery states; external areas; world-empires and world-economies; Kondratieff waves; system environment; smart and dumb systems; open and closed systems; risk and complexity; meaning; autopoiesis; self-thematization; interactional, organizational, and societal systems; role specification; entrance and exit rules; variation; selection; stabilization; replication; segmentary, stratified, and functional differentiation; modern society; nesting; piecemeal involvement; positive law

After reading and understanding this chapter, you should be able to

- Describe what makes world-systems analysis unique

- Explain the central features that link national economies into a global system

- Illustrate how states externalize costs and help create quasi-monopolies

- Explicate the Marxian economic dynamics in back of the relationships among the core, semi-periphery, periphery, and external areas and explain how they are related to the demise of capitalism

- Explain Wallerstein's crisis of modernity, beginning with the events in 1968, and delineate the structural and cultural signs that the current world-system is failing

- Explain in Luhmann's terms the defining qualities of a system

- Describe what three things human society must address, in terms of reducing risk and complexity. In your description, be certain to explain why these are specific problems for social systems.

- Explain Luhmann's three processes of evolution and how they work specifically in society

- Review the processes through which modern societies are integrated and the specific problems that are associated with modern societies

- Compare and contrast Wallerstein's and Luhmann's understanding of the global system and modernity

Learning More: Primary Sources

- Immanuel Wallerstein: Wallerstein built his theory through three volumes of historical data (*The Modern World-System,* I, II, and III). The historical breadth is impressive and convincing, but I would suggest you begin your reading of Wallerstein with his later works:

  Wallerstein, E. (1999). *The end of the world as we know it: Social science for the twenty-first century.* Minneapolis: University of Minnesota Press.
  Wallerstein, E. (2004). *World-systems analysis: An introduction.* Durham, NC: Duke University Press.

- Niklas Luhmann

  Luhmann, N. (1982). *The differentiation of society.* New York: Columbia University Press.
  Luhmann, N. (1995). *Social systems.* Stanford, CA: Stanford University Press.
  Luhmann, N. (2000). *The reality of mass media.* Stanford, CA: Stanford University Press.

Learning More: Secondary Sources

  Chase-Dunn, C. (2002). World-systems theorizing. In J. H. Turner (Ed.), *Handbook of sociological theory.* New York: Kluwer.
  Rasch, W. (2000). *Niklas Luhmann's modernity: The paradoxes of differentiation.* Stanford, CA: Stanford University Press.

Theory You Can Use (Seeing Your World Differently)

- In the face of the crisis of modernity, what recommendations does Wallerstein have for political involvement? Be certain to explain his rationale for saying these recommendations will influence the system. After reading Wallerstein, how will you engage your world?

- Consider the approaches to inequality given us by Wilson, Chafetz, Bourdieu, and now Wallerstein, and then answer the following questions: What kinds of inequalities or scarce resources are the most important in each theory? What kinds of structures perpetuate these inequalities? How is social change possible for each of these theorists? Can these three theories complement one another?

- How do improvements in communication technologies (such as the Internet and cell phones) affect the limitations imposed by face-to-face communication? Are there different boundary issues? Do such things as Internet chat rooms reduce or remove the limitations of turn-taking and topic?

- Here's a brainstorming activity: Take Luhmann's idea of communicative success and pick two issues you care about (such as race or gender). What are the recognized media of these issues? Have the media changed over time? How do the different media set the parameters of what can and cannot be done politically? For example, how does the discourse of gender change if the medium is essentialist versus constructivist?

- Given Luhmann's theory, how do you think worldwide ecological concerns can be addressed?

- Many of the theorists we've covered so far have unique definitions of society. Prepare a list of the different definitions. What do you make of the fact that there are so many different understandings of society? What do you think society is? Why?

CHAPTER 13

Identity Politics

Dorothy E. Smith (1926–)

Cornel West (1953–)

Patricia Hill Collins (1948–)

dentity politics encompasses "a way of knowing that sees lived experiences as important to creating knowledge and crafting group-based political strategies. Also, [it is] a form of political resistance where an oppressed group rejects its devalued status" (P. H. Collins, 2000, p. 299). Notice that identity politics focuses on lived experiences. Previously, under what Giddens calls emancipatory politics, the general focus in studying inequality was on the structural effects of disenfranchisement, such as we saw in Chapters 7 and 8. For example, in Chapter 8 Chafetz doesn't explain what it is like to live as a woman; she is concerned with emancipation and the social structures that prevent full political and economic equality. In contrast, what you'll see as we explore the work of the three theorists in this chapter is that the lived experiences of women are put front and center. But this isn't a reduction to personal feelings—the lived experiences of women demand certain kinds of political strategies that may be different from those demanded when structural inequality is central.

Also notice in the quote from Patricia Hill Collins that identity politics creates its own forms of political resistance. One specific form of resistance is expressing and insisting upon authentic representations of a group's lived experiences. This sense of identity is intentionally cultivated, as an important part of identity politics is members' focused efforts to understand, explore, express, and claim the distinctive qualities of their group. What is it like to be black in America? What is it like to be gay in America? What is it like to be a black *woman* in America? Identity politics insists on the cultural space to exist and practice a culture without interference or judgment from others. The practice of identity politics is increasing, probably because it is particularly fitting in what we could call late or post-modernity. Identities are more flexible and permeable today than they were at the beginning of the twentieth century. Yet at the same time they are also more important.

In this chapter, we will be looking at three different theorists. Dorothy E. Smith is going to help us see that the lived experiences of women are distinctly different from those of men. That in and of itself isn't a problem. The trouble comes in because, generally speaking, men hold the majority of ruling positions and create a specific kind of knowledge and way of knowing. This knowledge form is imposed upon women in such a way that it actually "colonizes" their consciousness.

Cornel West is going to open our eyes to the distinctive position that African Americans have come to occupy as a result of becoming the target of massive marketing efforts by capitalists. The success of these efforts has resulted in the disruption of traditional black community structures and has produced black nihilism, a growing sense of meaninglessness and uselessness.

Patricia Hill Collins is also going to lift the curtain on what she calls intersectionality and matrices of domination. Like Weber and Simmel, Collins tells us that no one holds just one social position. Every African American is also a gendered person, so studying just one system of inequality doesn't give us the whole picture. Race and gender collectively create social status positions that are also informed by national, class, sexual, and religious identities. Power and oppression, then, are exercised through various matrices of domination.

As we work our way through these theories, keep in mind everything that we've learned so far about power, class, race, and gender inequalities. Think about how Marx, Weber, Du Bois, Gilman, Dahrendorf, Wilson, Chafetz, Blau, R. Collins, Cook, Bourdieu, and Wallerstein have all talked about these issues. You will, of course, find the discourse of inequality changing over time. But if you keep these ideas in your mind as we think through what Smith, West, and P. H. Collins have to say, then you will have the tools to create a very powerful and robust understanding of how inequality works in society.

Dorothy E. Smith: Gendered Consciousness

Photo: Courtesy of Dorothy E. Smith.

The Essential Smith

Biography

Dorothy E. Smith was born in Northallerton in North Yorkshire, Great Britain, where she earned her undergraduate degree in 1955 from the London School of Economics. In 1963, Smith received her Ph.D. from the University of California at Berkeley. She has taught at Berkeley, the University of Essex, and the University of British Columbia, and is currently a professor emeritus at the University of Toronto, where she has been since 1977. In recognition of her contributions to sociology, the American Sociology Association honored Smith with the Jessie

Bernard Award in 1993 and the Career of Distinguished Scholarship Award in 1999. Her book *The Everyday World as Problematic* has received two awards from the Canadian Sociology and Anthropology Association: the Outstanding Contribution Award and the John Porter Award, both given in 1990.

Passionate Curiosity

Smith's passion is found at the intersection of text and life. Smith argues that the social and behavioral sciences have systematically developed an objective body of knowledge about the individual, social relations, and society in general. This body of knowledge claims objectivity and thus authority, "not on the basis of its capacity to speak truthfully, but in terms of its specific capacity to exclude the presence and experience of particular subjectivities" (Smith, 1987, p. 2). Smith wants to begin with and center social and behavioral research on the actual lived experiences of people and their encounter with texts, rather than on the texts that deny the very voices they claim to express.

Keys to Understanding

Standpoint, relations of ruling, new materialism, institutional ethnography, lived experience, text, fault line, bifurcated consciousness

Seeing Further: Standpoint Theory

Not Theory—Method!

In 1992, *Sociological Theory*, the premier theoretical journal of the American Sociological Association, presented a symposium on the work of Dorothy E. Smith. Though Smith had been publishing for quite some time, her dramatic impact on sociology came with the publications in 1987 of *The Everyday World as Problematic* and in 1990, *The Conceptual Practices of Power*. Being the subject of a special issue in *Sociological Theory* so soon after the publication of two major works attests to the impact that Smith's perspective was having on sociology. Among the commentators in that special issue were Patricia Hill Collins, Robert Connell, and Charles Lemert, each a significant theorist in her or his own right. However, Smith (1992) critiqued each of these theorists as having misinterpreted her work, saying "each constructs her or his own straw Smith" (p. 88).

Of course, Collins, Connell, and Lemert had their own individual issues, but Smith (1992) argues that they universally misconstrued her work as theory rather than method. "It is not . . . a totalizing theory. Rather it is a *method of inquiry*, always ongoing, opening things up, discovering" (p. 88, emphasis original). This is obviously an important point for us to note at the beginning of our discussion of Smith's work. She doesn't give us a general theory, not even a general theory of gender oppression. Smith gives us a method, but it isn't a method in the same sense as data analysis—Smith's is a *theoretical* method. It's a method grounded in a theoretical understanding of the world that results in theoretical insights. Further, for Smith these theoretical insights are themselves continually held up to evaluation and revision.

In general, Smith's work is considered "standpoint theory." As we'll see below, that's a fairly accurate description of what she does. But Smith argues that thinking about standpoint theory theoretically makes the idea too abstract and it defeats the original intent. Like Pierre Bourdieu, Smith is very interested in the practices of power. She is interested in what happens on the ground, "where the rubber meets the road," in the lived experiences of women, more than the abstract words of sociological theory.

Let me give you an example that might help us see the distinction that Smith is making. Not long ago I was talking to a friend of mine who plays and builds drums. We were talking about the special feeling that comes from building the instrument you play. There's a kind of connection that develops between the builder and the wood, a connection that is grounded in the physical experience of the wood. I agreed with what he said and told him that kind of knowledge is called "kinesthetic." But I was painfully aware that there was a real difference between what we were each talking about. He has actually worked with the wood out of which he builds his drum kits; though I play guitar, I have never experienced that kind of connection with my instrument. I had the word for what he was talking about, but he had the actual experience.

Smith is arguing that something happens when we formalize and generalize our concepts. We can quickly move out of the realm of real experience. As such, it is possible for concepts to play a purely discursive role. Just like in my example of kinesthetic knowledge, we can talk about things of which we only have discursive or linguistic knowledge. Thus, I can talk about the intuitive connection that exists between a musician and an instrument that he or she has built, but I have no actual knowledge of it. It's purely theoretical for me.

Obviously, there are no significant consequences of my woodworking example. But in the social world, there can be important ramifications, and that's the point that Smith wants us to see. Standpoint theory isn't a theory per se; it's a method of observation that privileges the point of view of actual people over theoretical, abstract knowledge. That may sound commonsensical and you may agree with it, but Smith would contend that most of what you and I know about the social world is like my knowledge of building a musical instrument.

In thinking about Smith's approach, it is important to note that she doesn't see herself as arguing against abstractions. To one degree or another, theory is usually abstracted. When we talk about theory being abstract, we mean that it is not simply a statement or restatement of the particulars. In a fundamental way, then, most theories and theoretical terms exist outside of the actual situation as generalizations. For example, there is a significant difference between saying "LaToya went to Food Lion to do the food shopping" and "women generally do the grocery shopping." The first statement is particular; it refers to the behaviors of a specific person at a definite location and time. In that sense, the statement is limited and not theoretically powerful. The second statement, because it is abstract, is more theoretically powerful. Most theory is at least somewhat abstract; it's the best way for us to say something significant about what is going on. Because she focuses on the actualities of lived experience, Smith's standpoint theory can be read to mean that abstractions are themselves bad. But that isn't her intent.

Nor is she interested in simply discrediting or deconstructing the knowledge or relations of ruling. Quite a bit of critical theory is aimed at these issues. For example, chances are good that much of what you've learned in other classes about gender or race is an historical account of how patriarchy or racism came about and how it functions to oppress people. The intent in these courses is to discredit sexism or racism by deconstructing its ideological and historical basis. But discrediting isn't Smith's specific intent either.

Smith argues that in both these cases, abstractions and ideological deconstruction, the critique by itself isn't enough; it doesn't tell the actual story. Theory in both forms plays itself out in the everyday, actual world of people, and that is Smith's concern. Insofar as theory and ideology mean anything, they mean something in everyday life, whether that life is the researcher's or that of ordinary women. Like I said, Smith is interested in where the rubber meets the road. In this case, the "rubber" is made up of theoretical abstractions and ideological knowledge that governs, and the "road" is the actual experiences of women. Thus, Smith isn't interested in doing away with abstractions per se, nor is she simply interested in exposing the relations of ruling; doing so is not enough and it runs the risk of replicating the problem, as we will soon see.

Facts and Texts: The New Materialism

As we've seen in previous chapters, Marx's materialism argues that there is a relationship between one's material class interests and the knowledge one has. Smith proposes a *new materialism*, one where facts and texts rather than commodification produce alienation and objectification. With Marx, commodities and money mediate the relationships people have with themselves and others. That is, we relate and come to understand our self and others through money and products. Marx's theory was specific to industrialized capitalism—the economies of more technologically advanced societies may be different. Some of the important changes include shifts from manufacturing to "service" economies, increases in the use of credentials and in the amount and use of expert knowledge, advances in communication and transportation technologies, exponential increases in the use of advertising images and texts, and so on. In such economies, relationships and power are mediated more through texts and "facts" than commodities and money. Further, just as people misrecognized the reality in back of money and commodities, so today most people misrecognize the relations of power in back of texts and facts. Texts and the facticity that text produces are the primary medium through which power is exercised in a society such as the United States.

Text

Though the idea of text is gaining usage and popularity, it, like culture, is one of the more difficult words to define. Winfried Nöth (1985/1995), in her *Handbook of*

Semiotics, says that given that textuality is defined by the researcher, "it is not surprising that semioticians of the text have been unable to agree on a definition and on criteria of their object of research" (p. 331). Smith, however, gives us a broad, clear, and useful definition of "text" that includes three elements: the actual written words or symbols, the physical medium through which words and symbols are expressed, and the materiality of the text—the actual practices of writing and reading.

Smith is specifically concerned with texts that are officially or organizationally written and read. She gives us the example of two different texts that came out of an incident in 1968 involving police and street people in Berkeley, California. One text came in the form of a letter to an underground newspaper and was written by someone who was marginally involved in the altercation. His text was "written from the standpoint of an actual experience" (Smith, 1990, p. 63) and contained specific references to people, places, times, and events. It was embedded in and expressed actual life experiences as they happened. This was a personal account of a personal experience that reflexively situated the writer in the event.

The other text was the official incident report that came from the mayor's office. The standpoint of this second text is organizational. Rather than being an account of a personal experience, it is written from the point of view of anonymous police officers who are portrayed as trained professionals and organizational representatives. In addition, the official report embedded the text within "sequences of organizational action extending before and after them" (Smith, 1990, p. 64) using reports from police, courts, and probation officers. In other words, the official text brought in many elements that exist outside of the actual situation and experience. In the end, every element of the actual experience was given meaning through these extra-local concepts rather than the experience itself.

Facts

The obliteration of the historical and specific sources is part of the process of creating facts (Smith, 1990, p. 66). The facticity of a statement is thus not a property of the statement itself. A statement simply proposes a state of affairs such as "the earth is flat." For a statement to become fact, there must be a corresponding set of practices that provide its plausibility base—a group of people, beliefs, and practices that give substance to the statement. **Facticity,** then, "is essentially a property of an institutional order mediated by texts" (p. 79). Facts and texts are organizational achievements, not independent truths of the world. These are the texts and facts in which Smith is interested: the ones that are written and read as part of organizational method and relations of power. They create an objective reality whose existence is dependent upon specific institutionalized practices.

Defining Standpoint

As you'll see, we actually have two standpoint theorists in this chapter: Dorothy Smith and Patricia Hill Collins. In general, standpoint theory addresses the issue of which kind of knowledge carries the greatest value. Most people have a tendency to accept knowledge in a taken-for-granted manner: knowledge may be incorrect, but those mistakes can be fixed and knowledge progresses onward. Standpoint theory, like Marxian and other critical theories before it, points out that knowledge is specific to social structure and position. In other words, there are many forms of knowledge, but some of them are privileged over others. Obviously, the privileged forms of knowledge benefit the power elite and serve to suppress others.

Standpoint theory argues that groups standing outside the place of privilege actually have a more authentic knowledge of the social system. This is true first because they are in a better position to see the whole system at work. For example, most white people aren't aware that they are not the subject of police scrutiny. Most blacks, on the other hand, have firsthand knowledge of how surveillance works in the shopping malls and streets of America. The second reason it is more authentic is that it intrinsically recognizes the political nature of all knowledge and ways of knowing. Rather than seeing information as pure and free from ethical considerations, standpoint recognizes that all knowledge exists because of a specific kind of sociopolitical configuration of social structures and interests. Further, the use to which information is put is always tainted by values and politics.

As a form of critical knowledge, then, standpoint theory seeks to

- privilege the lived experiences of those who are outside the relations of ruling
- represent the social world from the standpoint of the oppressed
- make the studies and accounts of disenfranchised groups accessible to those who are the subjects of the studies
- create knowledge that can be used by the oppressed to subvert and change their social world

Concepts and Theory: The Standpoint of Women

Smith argues that the distinction between abstract knowledge (or text) and lived experience holds for all people, whether male, female, black, white, Chicano, or anyone else. However, women's experience and knowledge is specifically important. Generally speaking, there are a few reasons why this would be accurate. First, as we've already seen, knowledge of oppressed peoples is in some ways truer than that of the ruling groups. Because it crosscuts all other social categories, the oppressive system par excellence is gender. Thus, women's knowledge is uniquely suited to help us see an oppressive structure for what it is.

Another reason to favor women's knowledge is that women are particularly embodied. For example, the beginnings of gender stratification are undoubtedly linked to the control of women's sexuality and bodies. Obviously, these beginnings are clouded by time and are fairly complex. Yet, we can get a sense of how women's inequality and their bodies became linked by looking at a few of these issues. One of the most important factors in establishing this link is the control of wealth. In order to dominate wealth, men had to control inheritance. Until DNA paternity testing became a reality, a man's ability to legitimize his lineage was largely dependent upon exclusive access to the woman; thus, men had to regulate women's sexual behaviors in order to control wealth.

Women's bodies also became important politically in at least two other ways. First, women were used to form political alliances through marriage. Obviously, a man was involved in this marriage, but in Western civilizations it was generally the woman who left her home and became part of her husband's realm. In exchange terms, she was the "good" that was traded, and the quality of that good, in terms of sexual purity, was of utmost importance. Second, women's bodies were used in warfare as a way of demoralizing the enemy, through such things as systematic rape, a practice termed "rape warfare" by Beverly Allen (1996) and "mass rape" by Alexandra Stiglmayer (1994) in their analyses of incidents in Bosnia-Herzegovina.

There is no doubt that women's bodies continue to be a primary site of gender inscription, as countless studies and films (such as Jean Kilbourne's *Killing Us Softly* series) document. For Smith (1987), a woman's body is significant because it "is also the place of her sensory organization of immediate experience; the place where her coordinates of here and now, before and after, are organized around herself as center" (p. 82). Thus, women are likely more aware of and more centered in their bodies than are men.

A third reason for privileging women's experiences is the position they play relative to men and men's relationship to objective text. We'll consider this again in the section on the fault line, but it bears mentioning here. While what Smith is saying about objective knowledge on one hand and subjective experience on the other is true about men, it is also true that women by and large take care of most of the details of life (such as cooking, cleaning, childrearing, and so on). These "details" are what allow "men's life" to be lived. Because women take care of the actualities, men are allowed to think that life is really about the abstract, general knowledge they construct and believe. Women thus typically provide a buffer between men and the actual demands of life, and thus, women's knowledge is more materially real.

Because of her emphasis on standpoint, Smith argues that her project is not an ideological representation or movement. Often when we think of feminism, we think of a social movement with a specific agenda and ideology. While liberation from oppression is certainly part of what Smith (1987) wants to attain, she doesn't offer us "an ideological position that represents women's oppression as having a determinate character and takes up the analysis of social forms with a view to discovering in them the lineaments of what the ideologist already supposes that she knows" (pp. 106–107). Whether it comes from social science or feminism, Smith is concerned about knowledge that objectifies, that starts from a position outside the everyday world of lived experience, as generally sociology does.

Smith gives us an example of walking her dog. When walking her dog, she needs to be careful that he doesn't "do his business" in places that are inappropriate. Smith points out that her behavior in this situation would generally be understood in terms of norms. From the normative perspective, she would simply be seen as conforming to the social norms of walking a dog. However, Smith (1987) contends that the idea of norm "provides for the surface properties of my behavior, what I can be seen to be doing" (p. 155). In other words, the normative approach can only give us a surface or simplistic understanding of what is going on. What is ignored in seeing the norm is "an account of the constitutive work that is going on" (p. 155).

In this case, "constitutive work" refers to the efforts Smith must put forth in conforming to the norm. And in the process of conforming, there are any number of contingencies, including the kind of neighborhood, the type of neighbors, the kind of leash, the breed of dog, the weather, her subjective states, and so on. All of the contingencies require practical reasoning that in turn produces a specific kind of reaction to the norm. The issue for Smith is that the normative account ignores the actual experiences of the person: how, when, and why the individual conforms to, negotiates, or ignores the demands made by the norm.

Disregarding the site of constitutive work is how "the very intellectual successes of the women's movement have created their own contradictions" (Smith, 1992, p. 88). The contradictions arise, according to Smith, as feminism becomes its own theory—a theory that is seen to exist apart from the lived experiences of the women it attempts to describe. For Smith, resistance and revolution do not—indeed, cannot—begin in theory or even sociology. Such a beginning would simply replace the ruling ideas with another set of ruling ideas. In order to create a sociology of women, or to bring about any real social change, it is imperative to begin and continue in the situated perspectives of the people in whom we are interested.

Thus, Smith's intent is to open up the space of actual experience as the site of research. This is exactly what Smith means by the title of her 1987 book, *The Everyday World as Problematic*. Most social research takes on problems that are guided by the literature, the researcher's career, or by the availability of funds. According to Smith, this practice results in a body of knowledge that more often than not only references itself or the relations of ruling that fund it. In Smith's work, it is the everyday world of women that is problematized. It's the actual experience of women that sets the problems and questions of research and provides the answers and theory. "Inquiry does not begin within the conceptual organization or relevances of the sociological discourse, but in actual experience as embedded in the particular historical forms of social relations that determine that experience" (Smith, 1987, p. 49).

Another way to put this issue is that most social research assumes a reciprocity of perspectives. One of the things that ethnomethodology (see Chapter 9) has taught us about the organization of social order at the micro level is that we all assume that our way of seeing things corresponds fairly closely to the way other people see things. More specifically, we assume that if another person were to walk in our shoes, they would experience the world just like we do. This is an assumption that allows us to carry on with our daily lives. It lets us act as if we share a common world, even though we may not and we can never know for sure if we do. According to Smith, social science usually works in this way, too, but she wants us to problematize that assumption in sociology. She wants us to ask, "What is it like to be *that* person in *that* body in *those* circumstances?"

Sociology and the Relations of Ruling

Smith (1990) talks about the practices, knowledge, and social relations that are associated with power as relations of ruling. Specifically, *relations of ruling* include "what the business world calls *management,* it includes the professions, it includes government and the activities of those who are selecting, training, and indoctrinating those who will be its governors" (p. 14, emphasis original). In technologically advanced societies that are bureaucratically organized, ruling and governing take place specifically through abstract concepts and symbols, or text. As Michel Foucault explains, knowledge is power; it is the currency that dominates our age. Authority and control are exercised in contemporary society through different forms of knowledge—specifically, knowledge that objectifies its subjects.

The social sciences in particular are quite good at this. They turn people into populations that can be reduced to numbers, measured, and thus controlled. Through abstract concepts and generalized theories, the social sciences empty the person of individual thoughts and feelings and reduce her or him to concepts and ideas that can be applied to all people grouped together within a specific social type. The social sciences thus create a textual reality, one that exists in "the literature" outside of the lived experience of people.

Much of this literature is related to data generated by the state, through such instruments as the U.S. Census or the FBI's Uniform Crime Reporting (UCR) Program. These data are accepted without question as the authoritative representation of reality because they are seen as *hard data*—data that correspond to the assumptions of science. These data are then used to "test" theories and hypotheses that are generated, more often than not, either from previous work or by academics seeking to establish their names in the literature. Even case histories that purport to represent the life of a specific individual are rendered as documents that substantiate established theoretical understandings.

Thus, most of the data, theory, and findings of social science are generated by a state driven by political concerns, by academics circumscribed by the discipline of their fields, by professors motivated to create a vita (resume) of distinction, or by professionals seeking to establish their practice. All of this creates "textual surfaces of objective knowledge in public contexts" that are "to be read factually . . . as evidences of a reality 'in back of' the text" (Smith, 1990, pp. 191, 107). Therefore, a sociology that is oriented toward abstract theory and data analysis results in a discipline that "is a systematically developed consciousness of society and social relations . . . [that] claims objectivity not on the basis of its capacity to speak truthfully, but in terms of its specific capacity to exclude the presence and experience of particular subjectivities" (Smith, 1987, p. 2).

These concepts, theories, numbers, practices, and professions become relations of ruling as they are used by the individual to understand and control her own subjectivity, as she understands herself to be a subject of the discourses of sociology, psychology, economics, and so on. We do this when we see ourselves in the sociological articles or self-help books we read, in the written histories or newspapers of society, or in business journals or reports. With or without awareness of it, we mold ourselves to the picture of reality presented in the "textual surfaces of objective knowledge."

Smith points out that this process of molding becomes explicit for those people wanting to become sociologists, psychologists, or business leaders. Disciplines socialize students into accepted theories and methods. In the end, these are specific guidelines that determine exactly what constitutes sociological knowledge. For example, most of the professors you've had are either tenured or on a tenure track. Whether an instructor has tenure or not is generally the chief distinction between assistant and associate professors. And when a sociology professor comes up for tenure and promotion, one of the most important questions asked about his or her work is whether or not it qualifies as sociology. Not everything we do is necessarily sociology—it has to conform to specific methodologies, assumptions, concepts, and so on to qualify as sociology.

There is something reasonable about this work of exclusion. If I wrote an article with nothing but math concepts in it, it probably shouldn't be considered sociology. Otherwise there wouldn't be any differences among any of the academic disciplines. However, Smith's point is that there is more going on than simple definitions. Definitions of methods and theory are used by the powerful to exclude the powerless. What counts as sociology and the criteria used to make the distinctions are therefore reflections of the relations of ruling. Sociology and all the social sciences have historically been masculinist enterprises, which means that what constitutes sociology is defined from the perspective of ruling men. The questions that are deemed important and the methods and theories that are used have all been established by men: "how sociology is thought—its methods, conceptual schemes, and theories—has been based on and built up within the male social universe" (Smith, 1990, p. 13).

Let me give you an example to bring this home, one that has to do with race, but the illustration still holds. In the latter part of the 1990s, two colleagues and I were untenured in our department. One of those colleagues is black. All three of us were worried about tenure and promotion—there was quite a bit of contradictory information circulating about how we could get tenure. So we had a meeting with the man who was department head at the time. Each of us had specific concerns. My black colleague's concern was about race. As a result of some of the things the department head said, I asked him point blank: "Will the articles that [my black colleague] has published count for tenure and promotion or not?" The head answered that he wasn't sure because the articles were published in black journals and may not therefore "count as sociology." As you can see, what counts as "sociology" is defined by those in power.

The Fault Line

Smith argues that since the motivations, questions, and data come out of the concerns of those that govern and not the actual experiences of those living under the relations of ruling, masculinist knowledge is by default objective and objectifying—from beginning to end, it stands outside of the actual experience of those other than the ruling. Smith's sociology is thus not specifically concerned with what usually passes as prejudice or sexism, that expressed through negative stereotypes and discrimination. Rather, "we are talking about the consequences of women's exclusion from a full share in the making of what becomes treated as our culture" (Smith, 1987, p. 20).

One of those consequences is the experience of a fault line for those women training as social scientists. The idea of a fault line comes from geology where it refers to the intersection between a geologic fault (a fracture in the earth's crust) and the earth's surface. Many fault lines are dramatically visible. (If you've not seen one, use an Internet search engine to find an image of a fault line.) Smith's analogy is quite striking. She is arguing that the **fault line** for women is conceptual; it occurs between the kind of knowledge that is generally produced in society, specifically through the social sciences, and the knowledge that women produce as a result of their daily experiences. There is a decisive break between the two.

We generally think there are some differences between objective culture or knowledge and the lives that people live. But because the current relations of ruling produce masculinist knowledge, men do not sense a disjuncture between what they live and what they know of the world. Part of the reason for this is that many of the activities of men match up with or correspond to abstract, objectifying knowledge. A male sociologist "works in the medium he studies" (Smith, 1990, p. 17). But even for men, there is still a clear distinction between objective knowledge, "the governing mode of our kind of society" (p. 17), and daily life. Thus, while there may be a correspondence for men, there is also a place "where things smell, where the irrelevant birds fly away in front of the window, where he has indigestion, where he dies" (p. 17). In other words, Smith is arguing that even for men there is a break between objective forms of knowledge and daily life as it is subjectively experienced. The difference is that generally men don't sense the disjunction. But *why* don't men sense or experience it?

The reason, Smith informs us, is that women have traditionally negotiated that break for men. Let's think about the usual distinction between boss and secretary. Generally speaking, the secretary is there to do the menial labor, to take care of the mundane details through which an organization functions, and to keep the boss free from intrusions from the outside world by screening all calls and letters. Think also about the traditional division of labor in the home. Men go to work while women take care of the "small details" of running a household: grocery shopping, cooking, cleaning, and taking care of the kids. Both of these examples picture the mediation role that Smith tells us women play—women intervene between men and the actual lived world and they take care of the actualities that make real life possible. In doing so, they shelter men from the "bifurcation of consciousness" that women experience (Smith, 1987, p. 82).

Standpoint and Text-Mediated Power

Bringing all this together, we end up with a rather new way of doing sociology, one that focuses on the experience of women as it is mediated through various texts, particularly those produced through the relations of ruling. I've diagrammed my take on Smith's ideas in Figure 13.1. As with any such model, especially one constructed to reflect a critical perspective, it is a simplification. But in some ways I think that a simplification is exactly what Smith is after. Her argument entails elements from existentialism, phenomenology, symbolic interactionism, ethnomethodology, and Marxist theory. The argument is thus not simplistic. It is quite complex and nuanced and it can and will inspire intricate and subtle thought and research. But her point is rather straightforward—social research and theory need to be grounded in the actual lived experiences of people, particularly women.

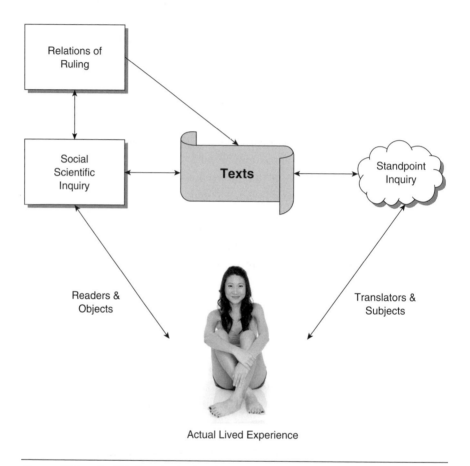

Figure 13.1   Smith's Standpoint Inquiry

Photo: Phil Date/Istockphoto.com

The first thing I'd like for you to notice about Figure 13.1 is the central position of both actual lived experience and text. Smith (1992) argues that text forms "the bridge between the actual and discursive. It is a material object that brings into actual contexts of reading a fixed form of meaning" (p. 92). Notice the distinction that Smith is making between the "fixed form" of the text and the "actual context of reading." The actual context is our place of lived experiences. It's a place where life is unfixed, spontaneous, meaningful, and subjective. The text, however, is fixed.

When we become aware of the texts that surround our lived reality, they form the bridge that Smith is talking about. There are two ways through which these texts can influence us. First, we may become directly aware of them, generally through higher education but also through the media. At this point, the discursive text directly enters the everyday life of people. This kind of text is generally authoritative; it claims to be the voice of true knowledge gained through scientific or organizational inquiry. However, as Smith points out, social scientific research is based outside of actual lived experience. Its position outside is in fact what makes this knowledge appear legitimate, at least in a culture dominated by scientific discourse. It is this appearance that prompts us to privilege the objective voice above our own. But there is more to these texts, as you can see from the left side of Figure 13.1.

The relations of ruling have a reciprocal relationship with social scientific inquiry, as noted by the double-headed arrow. We believe that legitimate research produces the only real knowledge, and government finances, directs, and thus defines the kinds of research that are seen as legitimate. Social scientific inquiry then produces the kinds of data and knowledge that reinforce and legitimate the ruling. The single-headed arrow from relations of ruling to text implies the top-down control of knowledge that Marx spoke of: the ruling ideas come from the ruling people, in this case men. The arrow between scientific inquiry and text, however, is two-headed. This means that the questions and theories that social scientific research uses come from the literature rather than the real lives of people. It also implies that social science is in a dialogue with itself, between its texts and its inquiry.

The second way we can become aware of these texts is through social scientific inquiry itself. Have you ever answered the phone and found that someone wanted you to respond to a survey? Or have you ever been stopped in a mall and "asked a few questions" by someone with a clipboard? Have you ever filled out a census survey? Through all these ways and many others, we are exposed to objectifying texts by social scientific inquiry.

Notice that the arrow coming from social scientific inquiry has only one head, going toward the actual world of women, and notice that the arrow has two nouns: readers and objects. This one-way arrow implies that social scientific research produces both readers and objects. The readers are the researchers; they are trained to read or impose their text onto the actual world. They see the lived experience of women through the texts and methods of scientific research. They come to real, actual, embodied life with a preexisting script, one that has the potential to blind them to the actualities of women. Further, when the questions and methods of science are used to understand women, women are made into objects, passive recipients of social science's categories and facts.

The right side of the model depicts Dorothy Smith's approach. There are two important things to notice. First, there are no relations of ruling controlling standpoint inquiry. Part of this is obvious. As I've mentioned, Smith says that this way of seeing things is applicable to all types of people, but it is particularly salient for women. The reason for its importance for women is that the relations of ruling are masculine in a society such as ours. Men control most of the power and wealth and thus control most of the knowledge that is produced. And while there is a difference between objective knowledge on the one hand and the lived experience of men on the other, women mitigate that discrepancy.

But this issue of ruling isn't quite that clear-cut for Smith. Relations of ruling are obviously associated with men. However, there is a not-so-obvious part as well. The work of women or feminists (who can be either men or women) can fall prey to the same problem that produces social scientific inquiry. This can happen when women reify the ideas, ideology, or findings of feminist research. Anytime research begins outside of the lived experience of embodied people, it assumes an objective perspective and in the end creates abstract knowledge. This is how women's movements "have created their own contradictions." It's possible, then, for women's knowledge to take on the same guise as men's. In Smith's approach, there are no relations of ruling, whether coming from men or women. Standpoint inquiry must continually begin and end in the lived experiences of people.

The other thing I'd like to call your attention to is that all the arrows associated with standpoint inquiry are double-headed. Rather than producing readers and objects, standpoint inquiry creates space for translators and subjects. In standpoint, the lives of women aren't simply read; that is, they aren't textually determined. A researcher using standpoint inquiry is situated in a never-ending dialogue with the actual and the textual. There is a constant moving back and forth among the voice of the subject, the voice of authoritative text, and the interpretations of the researcher. Smith (1992) sees this back-and-forth interplay as a dialectic: "The project locates itself in a dialectic between actual people located just as we are and social relations, in which we participate and to which we contribute, that have come to take on an existence and a power over against us" (pp. 94–95).

Notice that the dialectic is between actual experience and social relations. Smith is arguing that in advanced bureaucratic societies, our relationships with other people are by and large produced and understood through text. For example, you have a social relationship with the person teaching this class. What is that relationship? To state the obvious, the relationship is that of professor–student. Where is that relationship produced? You might be tempted to say that it is produced between you and your professor, but you would be wrong, at least from Smith's point of view. The relationship is *practiced* between you and your professor, but it is *produced* in the university documents that spell out exactly what qualifies as a professor and a student (remember, you had to apply for admittance) and how professors and students are supposed to act.

This textuality of relationships is a fact of almost every single relationship you have. Of course, the relations become individualized, but even your relationship with your parents (How many books on parenting do you think are available?) and with the person you're dating (How many articles and books have been written about dating? How many dating-related surveys have you seen in popular magazines?) are all controlled and defined through text. However, as we've already seen, Smith argues that even in the midst of all this text, there is a reality of actual, lived experience. Smith is explicitly interested in the dialectic that occurs between abstract, objectifying texts on the one hand, and the lived actualities of women on the other.

We thus come to the core of Smith's project. Recently (2005), Smith has termed this project "institutional ethnography." The "ethnography" portion of the term is meant to convey its dependence upon lived experience. Smith's project, then, is one that emphasizes inquiry rather than abstract theory. But, again, remember that Smith isn't necessarily arguing against abstractions and generalizations. She herself uses abstractions. Notice this quote from Smith (1987) concerning the fault line: "This inquiry into the implications of a sociology for women begins from the discovery of a point of rupture in my/our experience as woman/women within the social forms of consciousness" (p. 49). In it she uses both abstractions and particulars: my/our, woman/women. To say anything about women—which is a universal term—is to already assume and use a theoretical abstraction. Thus, Smith uses abstract concepts, so she isn't saying that in and of themselves they are problematic—the issue is what we do with them. Her concern is for when abstractions are reduced to "a purely discursive function" (Smith, 1992, p. 89). This happens when concepts are reified or when inquiry begins in text: "To begin with the categories is to begin in discourse" (p. 90).

There are, I think, two ways that Smith uses and approaches abstractions. First, in standpoint inquiry, concepts are never taken as if they represented a static reality. Lived experience is an ongoing, interactive process in which feelings, ideas, and behaviors emerge and constantly change. Thus, the concepts that come out of standpoint inquiry are held lightly and are allowed to transform through the never-ending quest to find out "how it works."

The second and perhaps more important way that Smith approaches theoretical concepts is as part of the discursive text that constitutes the mode through which relations of ruling are established and managed. As we've seen, "the objectification of knowledge is a general feature of contemporary relations of ruling" (Smith, 1990, p. 67). A significant principle of standpoint inquiry is to reveal how texts are put together with practices at the level of lived experience. "Making these processes visible also makes visible how we participate in and incorporate them into our own practices" (Smith, 1992, p. 90) and how we involve ourselves in creating forms of consciousness "that are properties of organization or discourse rather than of individual subjects" (Smith, 1987, p. 3).

It's at this point that Smith's use of the word "institutional" is relevant. It signals that this approach is vitally concerned with exploring the influences of institutionalized power relations on the lived experiences of their subjects. Institutional ethnography is like ethnomethodology and symbolic interactionism in that it focuses on how the practical actions of people in actual situations produce a meaningful social order. But neither of these approaches gives theoretical place to society's ruling institutions, as Smith's method does. In that, it is more like a contemporary Marxist account of power and text. Thus, **institutional ethnography** examines the dialectical interplay between the relations of ruling as expressed in and mediated through texts, and the actual experiences of people as they negotiate and implement those texts.

Smith uses the analogy of a map to help us see what she is getting at. Maps assist us to negotiate space. If I'm in a strange city, I can consult a map and have a fair idea of how to proceed. Maps, however, aren't the city and they aren't our experience. Smith (1992) wants sociology to function like a map—a map that gives an account of the person walking and finding her or his way (lived experience) through the objective structures of the city (text). This kind of sociology "would tie people's sites of experience and action into accounts of social organization and relations which have that ordinarily reliable kind of faithfulness to 'how it works'" (p. 94).

Specifically, Smith is interested in finding out just how the relations of ruling pervade the lives of women. These relations, as we've seen, come through texts and researchers. But in most cases, the relations of ruling are misrecognized by women. They are rendered invisible by the normalcy of their legitimacy. Part of what these maps can do, then, is make visible the relations of ruling and how they impact the lived experiences of women.

Smith is also interested in how actual women incorporate, respond to, see, and understand the texts that are written from a feminist or standpoint perspective. This is an important issue. Looking at Figure 13.1, we might get the impression that standpoint inquiry automatically and always produces translators and subjects. That is, it appears as if standpoint inquiry is a static thing, as if once done, the inquiry stands as the standpoint forever. This is certainly not what Smith is arguing. Notice again that double-headed arrow between "Standpoint Inquiry" and "Texts." Once standpoint inquiry is expressed in text, there is the danger that it will be taken as reality and become discursive. Smith's is thus an ongoing and ever-changing project that takes seriously the objectifying influence of text.

> For me, then, the standpoint of women locates a place to begin inquiry before things have shifted upwards into the transcendent subject. Once you've gone up there, settled into text-mediated discourse, irremediably stuck on the reading side of the textual surface, you can't peek around it to find the other side where you're actually *doing* your reading. You can reflect back, but you're already committed to a standpoint other than that of actual people's experience. (Smith, 1992, p. 60, emphasis original)

Smith Summary

- Smith argues that in contemporary society, power is exercised through text. Smith defines text using three factors: the actual words or symbols, the physical medium, and the materiality of the text. It is the last of the three with which Smith is most concerned—the actual practices of writing and reading. Most, if not all, of the texts produced by science, social science, and organizations achieve their facticity by eliminating any reference to specific subjectivities, individuals, or experiences.

- These texts are gendered in the sense that men by and large constitute the ruling group in society. Men work and live in these texts and thus accept them as taken-for-granted expressions of the way things are. Women's experience and consciousness, on the other hand, are bifurcated: they experience themselves within the text, as the ruling discourse of the age, but they also experience a significant part of their lives outside of the text. And it is in this part of women's lives where the contingencies of actual life are met, thus giving these experiences a firmer reality base than the abstract, ruling texts of men. Further, men are enabled to take objective, ruling texts as true because women provide the majority of the labor that undergirds the entire order.

- The bifurcated consciousness becomes particularly problematic for those women trained in such disciplines as business, sociology, psychiatry, psychology, and political science. In these professions, women are trained to write and read ruling texts, ignoring the lived experiences of women at the fault line.

- Smith proposes a theoretical method of investigation (standpoint inquiry, or institutional ethnography) that gives priority to the lived experiences of women. In this scheme, texts are not discounted or done away with; rather, they are put into the context of the embodied, actual experiences of women. Smith thus opens up a site of research that exists in the dialectic interplay between text and women's experience.

Cornel West: Race Matters

Cornel West turns our attention to matters of race. As we saw in Chapter 8 and Wilson's work, the way in which race influences those of African heritage in the United States has changed over time. Specifically, Wilson argued that since the civil rights movement in the 1960s, the black population in the United States has been split as never before by class. Most of the changes that have occurred with reference to race have thus benefited rising middle class blacks and have left those African Americans at the poverty level and below as the "truly disadvantaged."

West picks up Wilson's theme but moves it more into the realm of culture. Since the 1960s, the upwardly mobile black population in the United States has increasingly become the target of capitalist markets. Not only did capitalists discover a new market when African Americans entered the middle class, they also in a sense tried to make up for lost time. Capitalists have had over two centuries of marketing to whites, but blacks have constituted a strong and viable market for only the past 40 years or so. Given the historical background of the black community in the United States, this concentrated market force has had unique effects on African Americans of all classes. In order to see into this distinctive process, West creates a unique theoretical perspective by blending critical Marxism, pragmatism, existentialism, and Christianity.

Photo: © Richard Howard/Time Life Pictures/Getty Images.

The Essential West

Biography

Cornel West was born in Tulsa, Oklahoma, on June 2, 1953. He began attending Harvard University at 17 and graduated three years later, magna cum laude. His degree was in Near Eastern languages and literature. West obtained his Ph.D. at Princeton, where he studied with Richard Rorty, a well-known pragmatist. West has taught at Union Theological Seminary, Yale Divinity School, the University of Paris, and at Princeton. He is currently a professor at Harvard.

Passionate Curiosity

Cornel West's life is committed to not only the race question in America but also to the democratic ideals of open and critical dialogue, the freedom of ideas and information, and compassion for and acceptance of diverse others.

Keys to Knowing

Pragmatism, existentialism, Christianity, culture, market saturation, black nihilism, market moralities, ontological wounds, black leadership crisis, politics of conversion, racial reasoning, moral reasoning, existential angst, black cultural armor, coalition strategy, market dynamics, market saturation

Seeing Further: Prophetic Democracy

Critical Culture

For West (1999), Marxist thought is an "indispensable tradition for freedom fighters" (p. 212). West is particularly interested in Georg Lukács's expansion of the Marxist relationship between commodities and false consciousness. Lukács (1923/ 1971) developed Marx's ideas and argued that the process of commodification affects every sphere of human existence and is the "central, structural problem of capitalist society in all its aspects" (p. 83). Under these circumstances, value is determined not by any intrinsic feature of human activity or relationships, but rather by the impersonal forces of markets, over which individuals have no control. The objects and relations that will truly gratify human needs are hidden, and the commodified object is internalized and accepted as reality. Thus, commodification results in a consciousness based on reified, false objects. This Lukácsian "reified mind" does not attempt to transcend its false foundation, but rather, through rationalization and calculation, "progressively sinks more deeply, more fatefully and more definitively into the consciousness of man" (Lukács, 1923/1971, p. 93)—in other words, people caught in consumerism tend to justify unending buying by any means possible.

One of the classic Marxian problems has to do with overcoming false consciousness with class consciousness. For both Marx and Lukács, the only group of people who can overcome false consciousness is the working class. The only way to truly see the whole is by standing outside of it—thus, only the proletariat can conceive of the social system in its entirety: "the proletariat represents the true reality, namely the tendencies of history awakening into consciousness" (Lukács, 1923/1971, p. 199). The working class by its very position of alienation is capable of seeing the true whole, the knowledge of class relations from the standpoint of the entire society and its system of production and social relations. Bourgeois thought is simply an ahistorical acceptance of its own ideology and status quo.

Obviously, there is a tension here. On the one hand, commodification results in false consciousness, and on the other hand, the workers—those who suffer the effects of commodification the most—are the only ones who can see the problem for what it is. There is also another problem: the "overwhelming resources of knowledge, culture and routine which the bourgeoisie undoubtedly possesses" (Lukács, 1923/1971, p. 197). Thus, though the working and middle classes are capable of grasping the whole, they are susceptible both to psychological false consciousness and the cultural resources of the elite. However, these cultural resources are not controlled by the elite exclusively. West draws from another mid–twentieth-century Marxist, Antonio Gramsci (1928/1971), to argue that the ability to rule does not depend on material relations alone, as in classic Marxist thought. Social change will involve a war of cultural positions—a battle for people's minds, not overt conflict.

Culture, then, takes on a critical value for West. Generally, this cultural position is important in race theory in two ways. First, it is the place of revolutionary work. As we've seen, according to this tradition of Marxist thinking, there is first a battle for the mind in any social change. There must be a critical intelligence. West sees this type of consciousness as part of the project of modernity itself, specifically the American democratic experiment. "My conception of what it means to be modern is shot through with a sense of the dialogical—the free encounter of mind, soul and body that relates to others in order to be unsettled, unnerved and unhoused" (West, 1999, pp. xvii–xviii). As we will see, this kind of culture doesn't just create itself, especially in capitalist countries. Critical awareness comes out of specific kinds of culture-producing practices, most notably community, religion, and public discourse.

The second way culture is important for race theory is because of its long heritage. Many of the early publications by people of color and women were in the form of literature—fiction, poetry, and song. Harriet Beecher Stowe's *Uncle Tom's Cabin* and Charlotte Perkins Gilman's *The Yellow Wallpaper* are good examples. While the work of W. E. B. Du Bois is obviously scholastic, it is permeated with poems, songs, and literary references. For example, Du Bois' *The Souls of Black Folk* begins with the lyrics and music notes of a song. Du Bois' other major work, *Darkwater*, begins with a credo, a confession of faith.

The choice on the part of these authors to produce literary work may have been due to the structural constraints placed upon political minorities, but it is also true that the plight of the disenfranchised often may be best communicated through means that can impact emotion and not simply cognition. Oppression isn't simply a fact; it is a profound experience. West (1999) points this out when he says that "our great truth tellers [are] mainly artists" (p. xix).

Further, the troubles themselves cry out for creative release. Note the way West (1999) describes music, the "highest expression" of human history:

> Music at its best achieves this summit because it is the grand archaeology into and transfiguration of our guttural cry, the great human effort to grasp in time . . . our deepest passions and yearnings as prisoners of time. Profound music leads us—beyond language—to the dark roots of our scream and the celestial heights of our silence. (p. xvii)

Now notice how Marx (1844/1978c) describes religion: "*Religious* suffering is at the same time an *expression* of real suffering and a *protest* against real suffering. Religion is the sigh of the oppressed creature, the sentiment of a heartless world, and the soul of soulless conditions. It is the *opium* of the people" (p. 54, emphasis original). The last part of that quote is what most people are familiar with, but taken in context it achieves a much fuller meaning. These intense forms of culture—music and religion—both articulate something about the human condition that is difficult to express elsewhere. This link between profound experience and deep cultural expression may be one reason why music and religion have occupied such prominent places in African American culture.

Philosophy: Pragmatism and Existentialism

Pragmatism

We've already discussed pragmatism, but let's review it briefly here. *Pragmatism* is the American philosophy that developed out of the cultural devastation of the Civil War as a way of understanding truth. Truth in pragmatism is specific to a community. It arises out of the collective's physical work and communication. Thus, pragmatism is based on common sense and the belief that "truth and knowledge shifts to the social and communal circumstances under which persons can communicate and cooperate in the process of acquiring knowledge" (West, 1999, p.151).

In pragmatism, human action and decisions aren't determined or forced by society, ideology, or preexisting truths. Rather, decisions and ethics emerge out of a consensus that develops through interaction. This is the pragmatism that West (1999) latches onto. It is a

> culture of creative democracy...where politically adjudicated forms of knowledge are produced in which human participation is encouraged and for which human personalities are enhanced. Social experimentation is the basic form, yet it is operative only when those who must suffer the consequences have effective control over the institutions that yield the consequences. (p. 151)

Existentialism

I want to start our consideration of existentialism by asking you a question: have you ever wondered why you exist, or what the meaning of life is? If so (and most of us have), you've experienced the essence of existentialism. Existentialism starts with the problem of being or existence and argues that the very question or problem itself creates human existence. As far as we know, human beings are the only animal that questions its existence; all other animals simply exist. But human beings ask, and in asking we create human existence as a unique experience. Thus, the answer to the question can only be found in the being asking it.

Here's a simple analogy: Let's say I come to you with a cup of coffee and declare that it tastes like mud. I then ask, "Why does this coffee taste like mud?" You taste the coffee and tell me that it tastes fine. "Okay," I say, "but why does it taste like mud?" Startled, you respond by telling me once again that the coffee doesn't taste like mud. But I insist, "Why does the coffee taste like mud?" Eventually, you would probably get frustrated and tell me that the muddy coffee is my problem, not yours. Why is it my problem? Because I'm the only one that thinks the coffee tastes like mud. Both the problem and its solution can only exist inside of me. The same is true for the question of existence.

Two ideas are prominent in *existentialism*: struggles or suffering, and authenticity or being (existence). These ideas find space between such tensions as the individual versus the social, subjective versus objective, liminal experiences (such as death) and nihilism, and situated thought versus reason. Whatever the tensions may be, the resolution is found within the person her- or himself. As we've seen, pragmatism argues that there is no absolute truth, only truths that are arrived at socially, truths that provide practical answers to real problems. Truth for pragmatists, then, is socially practical. In contrast to pragmatism, existentialism argues that there are no social truths and that truth exists emotionally, not objectively. Truths such as history and science—social, practical truths—belong to society or to the crowd. As such, they are objective and indifferent. Truth in existentialism is determined by individual authenticity—by a passionate, and generally absurd, leap of faith.

This issue of subjective authenticity is exemplified by the biblical story of Abraham. The story goes that one day God came to Abraham and commanded him to kill his son. For Abraham, there appeared to be little to justify such a command. In fact, Abraham had every reason to doubt that the command even came from God, since even Satan himself can appear as an angel of light. Even so, Abraham prepared to kill his son and offer his life on an altar of burning wood. With little or no objective evidence, what drove Abraham to make the sacrifice? Biblical believers would say faith, and existentialism agrees. Truth isn't something that exists objectively; truth is that which is embraced passionately in its uncertainty, in all its absurdity.

To be human—*to exist as a human*—then, is to embrace existence passionately in spite of the affront brought by irresolvable struggles. To be human is to hold tight to faith when there is no reason. Truth is discovered and exists only in the passionate embrace of authenticity, existing in the truth of one's own self. This existential faith is not found in anything outside the individual; it is found within the limits of the human soul. West (1999) puts it this way: "To be human is to suffer, shudder and struggle courageously in the face of inevitable death" (p. xvi).

West combines both pragmatism and existentialism in a call to democratic faith. According to West, the issues we struggle with are social issues—the oppression of humanity by humanity, found in such inequalities as racism and sexism. And these problems must be struggled with and resisted collectively, through community and religion and dialogue. But the faith that West (1999) is interested in comes from the individual soul, because the struggle will always be there. It's part of being human and modern: "the very meaning of being modern may be the lack of any meaning, that our quest for such meaning may be the very meaning itself—without ever arriving at any fixed meaning" (p. xviii). West isn't necessarily discounting the possibility of solving the problem of racism. But he recognizes these struggles as an essential part of being human. Thus, we cannot pin our hopes and actions on the solution, on the democratic victory over racism. Rather, hope comes from within, as a result of maintaining faith when there is no objective reason for faith.

Christianity

Cornel West is a Christian whose faith informs everything he is, thinks, and does—but he is a particular kind of Christian. Because Christianity is a broad, multifaceted religion, there are any number of ways that we can understand the faith. One way is to see where it focuses its energies. If we look at Christianity in this way, we can create a continuum. On one end is the kind of Christianity that is centered on the afterlife. This kind of Christianity takes its lead from Paul's injunction to the Colossians: "Since, then, you have been raised with Christ, set your hearts on things above, where Christ is seated at the right hand of God." At the other end of this continuum is the Christianity that focuses on this present life. This opposite pole takes as its model the Christ in the temple who in rage overturned the tables of the moneychangers and drove them out with a whip.

The first perspective is an example of the kind of religion that Marx criticized. Its model is the "meek and mild Jesus" that taught that the proper response to oppression is to turn the other cheek. Marx argued that this form of religion serves to maintain the status quo. It defers social action to the hope of a better life. Marx didn't see the possibility of the second perspective. On this end of the continuum, religion is an agent of social change, and righteousness is a social quality, not simply an individual one. Of course, these two poles of my continuum and the examples of Christ can overlap in any individual, as they do in Cornel West. West is clearly oriented toward the social change pole, yet he finds inspiration in both models of Jesus.

West finds deep insight in the example of Christ as the long-suffering Son of God on earth. Remember that West's perspective is informed by existential authenticity: passionate faith in the presence of continuous struggle. For West (1999), the life of Jesus is the ideal model for such a faith: "the concrete example of the love and compassion of Jesus rendered in the biblical Gospel narratives constitutes the most absurd and alluring mode of being in the tragicomic world" (p. xvi). One of the principal forms of expression for this absurd existence is love. West's (1993/2001) dedication of his book *Race Matters* to his son is a moving example of this tenet of his faith: "To my wonderful son Clifton Louis West who combats daily the hidden injuries of race with the most potent of weapons—love of self and others" (p. vii).

The second way West's Christianity is expressed is through prophetic thought. This is where we find the Christ in the temple. We would be wrong to see the love, compassion, and patience that West espouses as what we usually think of as meek and mild. In West's Christianity, love and anger coexist in equal amounts. The most potent weapon may be love, but West (1993/2001) also chastens the black leaders of today for their "relative lack of authentic anger" (p. 58). He also doesn't pull punches with the state of religion: West (1999) tells us that he wrote his second book, *Prophetic Fragments,* in order to "convey my moral outrage at the relative indifference of American religion to the challenging of social justice beyond charity" (p. 357). West is angry, but it is a slow anger, a settled rage based on a sense of social righteousness.

West's Christianity, then, is a double-edged sword. On the one side, there is love of self and of others, and on the other side, there is prophetic outcry over humankind's inhumanity toward humankind. In both cases, West's spirituality is oriented to this side of the veil. For West, a heavenly religion is a useless religion.

His orientation toward this present life gives West's Christianity one further characteristic: skepticism. His prophetic utterances are not born out of any sense of absolute truth; rather, their genesis is found in the pragmatic truth and suffering of this life. "For me," he says, "there is always a dialectic of doubt and faith, of skepticism and leap of faith" (West, 1999, p. 215). In the end, this "inescapable demon of doubt" enables West to love his way "through the absurdity of life and the darkness of history" (p. xvi) and to see the American experiment as "fueled with a religion of vast possibility" (p. xviii) rather than a closed dogmatism. West thus sees faith and his prophetic anger as intertwined with skepticism, resulting from human interaction within communities and religions, founded on uncertainty and struggle, rather than based on unshakable truth. Here we clearly see West's philosophy and Christianity come together.

Concepts and Theory: Race Matters

Markets, Mobility, and Community

West gives us two basic structural influences on blacks in the United States: the economic boom and expansion of civil rights for blacks in the 1960s, and the saturation of market forces. In terms of the economic and political well-being of blacks in the United States, West is simply saying that they both increased, particularly during the boom of the 1970s. For example, the U.S. Census Bureau (n.d.) reports that black, male, median income increased from $9,519 in 1950 to $17,055 in 1970, as measured in 2003 dollars. These changes helped define blacks as a viable market group, one with disposable income and market-specific products. In some obvious ways, these changes have benefited the black experience in America. However, the development of black economic markets has also had significant negative effects.

In the broadest sense of the word, markets are defined by the exchange of goods and services, and in this sense are fundamental to most if not all economic activity. However, there are different kinds of markets. A good illustration of these differences is the distinction made between "market economies" and "command economies." The difference between command and market economies is one of control. Command economies are top heavy: the government has a strong hand in determining what kinds of things are bought and sold and the method of their exchange. The former Soviet Union had such an economy. Free market economies are marked by less state involvement and, more importantly, an ideology of free markets.

Capitalist *market dynamics* with relatively little control have certain features. They are amoral, expansive, and tend toward greater levels of abstraction. There's a sense in which markets are without any morals whatsoever: they aren't restricted by any kind of ethic—they can be used to sell bibles or guns to terrorists. This quality makes them applicable to any situation or product. However, as we will see, the absence of any ethical restrictions implies and creates a morality of its own. Capitalist markets are also expansive. They expand vertically (accessories for an existing product), horizontally (new products within a market), and geographically (extending existing markets to new social groups). Market expansion drives commodification, the process through which more and more of the human life-world becomes something that can be bought and sold. A chief result of capitalist markets, then, is the growing number of "needs" that are met through commodities. Markets also tend toward abstraction. This idea is actually another level of expansion. When we talk about market expansion, we are referring to markets built upon markets; examples of such expansion include stock futures and money/credit markets.

Black Nihilism

West characterizes this market expansion into the black community as a kind of *market saturation.* He first argues that the market saturation of the black population has stripped away community-based values and substituted market moralities. Above I mentioned that markets are amoral. But this is in a restricted sense only—markets can be used for anything. Markets do, however, convey some specific cultural ideas and sensibilities. In classical theory, for example, Max Weber was extremely interested in how markets and bureaucracies create rational rather than affective culture; and Georg Simmel saw markets as contributing to cultural signs becoming frivolous.

West argues that being a focus for market activity, commodification, and advertising has changed black culture in America. Prior to market saturation, blacks had a long history of community and tradition. They were equipped with a kind of cultural armor that came via black civic and religious institutions. This armor consisted of clear and strong structures of meaning and feeling that "embodied values of service and sacrifice, love and care, discipline and excellence" (West, 1993/2001, p. 24).

In some ways, this armor is specific to the black American community. However, it's important to note that this general shift from community-based traditional culture to less meaningful and more pliable culture was a concern of many social theorists of modernity. We find this idea of cultural shift repeatedly in classical theory. The basic idea is that culture has dramatically changed through processes accompanying urbanization and commodification. Rather than being meaningful, normative, and cohesive, culture is trivialized, anomic (without power to guide), and segmented.

West is making this same kind of argument, so he follows a strong theoretical tradition. West, however, is pointing out that while the dispersion of community and the emptying of culture may have affected much of modern society during the beginning and middle stages of modernity and capitalism, the black community in America wasn't strongly influenced by these changes until after the 1960s. Until then, blacks continued to rely on community-based relations and religiously influenced culture. As a result of the twin structural influences of civil rights and upward mobility, blacks moved out of black communities and churches. The structural bases for cultural armor were weakened as a result.

In place of cultural armor, blacks have since been inundated with the *market moralities* of conspicuous consumption and material calculus. The culture of consumption orients people to the present moment and to the intensification of pleasure. This culture of pleasure uses seduction to capitalize "on every opportunity to make money" (West, 1993/2001, p. 26). It overwhelms people in a moment where the past and future are swallowed up in a never-ending "repetition of hedonistically driven pleasure" (p. 26). Further, the material calculus argues that the greatest value comes from profit-driven calculations. Every other consideration, such as love and service to others, is hidden under the bushel of profit.

As I've mentioned, most of these cultural ramifications of markets and commodification have also been present in other groups. But in West's (1993/2001) opinion, two issues make these effects particularly destructive for blacks. First, black upward mobility and the presence of the black middle class concerns only a small sliver of the pie. Most of the black citizens of the United States still suffer under white oppression. The other issue that makes the black experience of market saturation distinct is the "accumulated effect of the black wounds and scars suffered in a white-dominated society" (p. 28). In other words, there is an historical and cultural heritage, no matter how much the immediacy of market saturation and pleasure tries to deny it—much of the history of the United States was built on the oppression of blacks, over the 188 years from 1776 to 1964.

Obviously, these two factors influence one another. Healing from past wounds can only take place in a present that is both nurturing and repentant, a place that does not replicate hurts from the past. According to West (1993/2001), this isn't happening for blacks in America: cultural beliefs and media images continue to attack "black intelligence, black ability, black beauty, and black character in subtle and not-so-subtle ways" (p. 27). And, as noted earlier, black upward mobility is still limited. For example, in 2002 over 30% of black children lived under the poverty line, compared to 12.3% of white children (Statistical Abstract of the United States, 2002).

In the abstract, West's argument so far looks like this: black upward mobility + increased civil rights → weakening of civic and religious community base → substitution of market moralities for cultural armor—all of which takes place within the framework of the black legacy in the United States and continued oppression. "Under these circumstances black existential *angst* derives from the lived experience of ontological wounds and emotional scars" (West, 1993/2001, p. 27, emphasis original). The "ontological wounds" that West is speaking of come from the ways in which black reality and existence have been denied throughout the history of the United States.

In general, *existential angst* refers to the deep and profound insecurity and dread that comes from living as a human being. Recall our discussion about existentialism: to be human is to suffer, to know we suffer, and to question the reason for suffering. West employs the idea of angst to describe the uniquely black experience of living under American capitalism and democracy. Black experience is deeply historical, yet the past and the future are buried under the market-driven pleasures of the moment; black experience is fundamentally communitarian, yet that civic and religious base is overwhelmed by market individualisms; black experience is painfully oppressive, yet it is countered only by increasing target marketing and consumerism. Thus, West argues that the result of market saturation and morality for blacks is a deeply spiritual condition of despair and insecurity.

Because blacks no longer have the necessary culture, community, or leadership, this angst cannot be used productively. It is instead turned inward as anger, and this anger is played out in violence against the weak. Righteous anger, turned against the oppressor in hopes of liberation, becomes increasingly difficult to express. **Black nihilism** denies the hope in which this anger is founded. With no viable path, this anger is turned inward and found in black-against-black violence, especially against black women and children.

Black Leadership Crisis

However, nihilism can be treated. West argues that it is a disease of the soul, one that cannot be cured completely as there is always the threat of relapse. This disease must be met with love and care, not arguments and analysis. What is required is a new kind of politics, a *politics of conversion,* which reaches into the subversive memory of black people to find modes of valuation and resistance. Politics of conversion is centered on a love ethic that is energized by concern for others and the recognition of one's own worth. This kind of politics requires prophetic black leaders who will bring "hope for the future and a meaning to struggle" (West, 1993/2001, p. 28). There is, however, a crisis in American black leadership.

For West, there is a relationship between community and leadership. Strong leaders come out of vibrant communities. With the breakdown of the black community, black leaders don't have a social base that is in touch with the real issues. There is thus no nurturing of critical consciousness in the heart of black America. Rather, much of the new black leadership in America comes out of the middle class, and black middle class life is "principally a matter of professional conscientiousness, personal accomplishment, and cautious adjustment" (West, 1993/2001, p. 57). West maintains that what is lacking in contemporary black leadership is anger and humility; what is present in overabundance is status anxiety and concerns for personal careers.

West divides contemporary black leaders into two general types—politicians and academics—with three kinds of leadership styles: race-effacing managerial leaders, race-identifying protest leaders, and race-transcending prophetic leaders. There are some differences between politicians and academics, but by and large they express the same leadership styles.

The managerial/elitist model is growing rapidly in the United States. This style of leadership is one that has been co-opted by bureaucratic norms. The leader navigates the political scene through political savvy and personal diplomacy, and race is downplayed in the hopes of gaining a white constituency. In academia, the elitist sees him- or herself as having a kind of monopoly over the sophisticated analysis of black America. But the analysis is flat and mediocre because of the intellectual's desire to fit into the university system. In both cases, whether under political savvy or academic abstraction, race is effaced.

The second type of leader, the protest leader, capitalizes on the race issue but in a very limited way, producing various "one-note racial analyses" (West, 1993/2001, p. 68). Here "black" becomes all-powerful. West characterizes these leaders as "confining themselves to the black turf, vowing to protect their leadership status over it, and serving as power brokers with powerful nonblack elites" (p. 60). In this context, racial reasoning reigns supreme.

Racial reasoning is a way of thinking that is more concerned with equality as a group right rather than a general social issue. For West, racial reasoning begins with an assumption of *the* black experience. The discourse of race then centers on black authenticity: the notion that some black experiences and people are really black while others aren't. Racial reasoning results in blacks closing ranks, but again it is around a one-note song rather than a symphony of color. Racial reasoning results in black nationalist sentiments that "promote and encourage black cultural conservatism, especially black patriarchal (and homophobic) power" (1993/2001, p. 37). Closing the ranks thus creates a hierarchy of acceptability within a black context: the black subordination of women, class divisions, and sexual orientation within black America.

Black Cultural Armor

These two kinds of black leaders have promoted political cynicism among black people, and have dampened "the fire of enraged local activists who have made a difference" (West, 1993/2001, p. 68). Part of black nihilism, or nothingness, is this sense of ineffectuality, of being lost in a storm too big to change. What is needed, according to West, are black leaders founded on moral reasoning rather than racial reasoning.

Moral reasoning is the stock and trade of race-transcending prophetic leaders. Prophetic leadership does not rest on any kind of racial supremacy, black or white. It uses a coalition strategy, which seeks out the antiracist traditions found in all peoples. It refuses to divide black people over other categories of distinction and rejects patriarchy and homophobia. This kind of approach promotes moral rather than racial reasoning.

This prophetic framework of moral reasoning is also based on a mature black identity of self-love and self-respect that refuses to put "any group of people on the pedestal or in the gutter" (West, 1993/2001, p. 43). Moral reasoning also uses subversive memory, "one of the most precious heritages [black people] have" (West, 1999, p. 221). It recalls the modes of struggling and resisting that affirmed community, faith, hope, and love, rather than the contemporary market morality of individualism, conspicuous consumption, and hedonistic indulgence.

Both the *coalition strategy* and mature black identity are built at the local level. West (1999) sees local communities as working "from below and sometimes beneath modernity" (p. 221), as if local communities can function below the radar of markets and commodification. It is within vibrant communities and through public discourse that local leaders are accountable and earn respect and love. Such leaders merit national attention from the black community and the general public, according to West.

In this framework, the liberal focus on economic issues is rejected as simplistic. Likewise, the conservative critique of black immorality is dismissed as ignoring public responsibility for the ethical state of the union. In their places, West proposes a democratic, pragmatically driven dialogue. As I mentioned earlier, West doesn't propose absolutes. His is a prophetic call to radical democracy and faith, to finally take seriously the declaration that all people are created equal.

Together, moral reasoning, coalition strategy, and mature black identity create the *black cultural armor*. West's use of "armor" is a biblical reference. Christians are told in Ephesians 6:13 (New International Version) to "put on the full armor of God, so that when the day of evil comes, you may be able to stand your ground, and after you have done everything, to stand." There the threat was the powers of darkness in heavenly places; here the threat is black nihilism in the heart of democracy. These two battles are at least parallel if not identical for West. The fight for true democracy is a spiritual battle for the souls of humankind that have been dulled by market saturation, especially the souls of black America. West (1993/2001) exhorts black America to put on its cultural armor—a return to community life and moral reasoning along with coalition strategy and mature black identity—so as to "beat back the demons of hopelessness, meaninglessness, and lovelessness" and create anew "cultural structures of meaning and feeling" (p. 23).

West Summary

- West draws on four traditions in organizing his thought: cultural Marxism, pragmatism, existentialism, and Christianity. Marx's theories of commodification and market expansion are key elements in West's theory as well. But West also draws upon the critical Marxist view of culture—culture is used to oppress in often hidden ways, but it can also be used to bring change. From pragmatism, West takes the idea of practical values emerging from collective dialogue. He sees pragmatism as a key philosophy in the American democratic experiment. From existentialism, West draws the ideas of subjective authenticity and passionate absurdity. Finally, from Christianity, he takes love for self and others and the prophetic outcry for mercy and justice.

- Since the 1960s, blacks in the United States have on the one hand enjoyed increased economic and political freedoms, but on the other have become the victims of market saturation, which has changed the primary orientations of blacks. Previously, blacks were strongly oriented to civic and religious institutions and the traditional ties of family and home. Market saturation has infested the black community with market moralities: fleeting hedonistic pleasure and monetary gain. The effects of markets are exaggerated for blacks because of the black heritage in America. The mix of past wounds, the continuing racial prejudice, and market moralities creates black nihilism—a sense of hopelessness and meaninglessness associated with living as a black person in the United States.

- West exposes a crisis in black leadership, arguing that most black leaders either fall under the managerial/elitist model or that of protest leaders. With protest leaders, racial reasoning is paramount, which promotes ethics based on skin color alone, rather than on moral or justice issues. West calls on prophetic leaders that will transcend race and return to moral reasoning. These leaders must begin in the community, at the grassroots level, where they can participate in pragmatic community dialogue, build up trust, and maintain accountability.

Patricia Hill Collins: Race and Matrices of Domination

Patricia Hill Collins will ask us to see two things. First, inequality in society is a complex matter. It can't simply be reduced to considerations of race or gender. Every person stands at a crossroads that distinguishes her or him from most others. For example, being black, female, middle class, and heterosexual is quite different than being black, female, working class, and lesbian. Collins wants us to see deeper into the workings of inequality than ever before.

The second thing that Collins will ask us to see is standpoint. Of course, Dorothy E. Smith asked us to do the same, but Collins wants us to see the value in the standpoint of black feminists. In my introduction to Smith, I said that gender crosscuts all other forms of inequality, and that's true. But in Collins's scheme, a single system isn't enough to explain inequality. Stratification works through matrices of domination, not single systems, and one of the most powerful intersectional standpoints is black women. It's at that point that race and gender meet. As such, it is probably the most powerful beginning point for intersectional analysis.

Photo: Courtesy of Patricia Hill Collins.

The Essential Collins

Biography

Patricia Hill Collins was born on May 1, 1948, in Philadelphia, Pennsylvania. Collins received her bachelor's degree and Ph.D. in sociology from Brandeis University and a master's degree in social science education from Harvard. Collins served as director of the African American Center at Tufts University before moving to the University of Cincinnati, where she was named the Charles Phelps Taft Distinguished Professor of Sociology in 1996. She is currently at the University of Maryland and holds the Wilson Elkins Professor of Sociology position. Her book, *Black Feminist Thought*, received the Association for Women in Psychology's Distinguished Publication Award, the Society for the Study of Social Problems' C. Wright Mills Award, and the Association of Black Women Historians' Letitia Woods Brown Memorial Book Prize.

Passionate Curiosity

Collins is a critical theorist who is motivated to understand and change the complex social relations surrounding inequality.

Keys to Knowing

Intersectionality; Eurocentric positivism; black feminist epistemology; common challenges/diverse responses; safe places; self-definition; rearticulation; black feminist intellectuals; matrix of domination; structural, disciplinary, hegemonic, and interpersonal domains of power

Seeing Further: Intersectionality and Black Feminist Epistemology

Patricia Hill Collins is centrally concerned with the relationships among empowerment, self-definition, and knowledge, and she is particularly concerned with black women—it is the oppression with which she is most intimately familiar. But Collins is also one of the few social thinkers who are able to rise above their own experiences and to challenge us with a significant view of oppression and identity politics that not only has the possibility of changing the world but also of opening up the prospect of continuous change.

For change to be continuous, it can't be exclusively focused on one social group. In other words, a social movement that is only concerned with racial inequality, for example, will end its influence once equality for that group is achieved. What Patricia Hill Collins gives us is a way of transcending group-specific politics that is based upon black feminist epistemology. However, it is vital to note that her intent is to place "U.S. Black women's experiences in the center of analysis without privileging those experiences" (P. H. Collins, 2000, p. 228). Collins is saying that there is something significant we can learn from black women's knowledge that can be applied to social issues generally.

Black women sit at a theoretically interesting point. Collins argues that black women are uniquely situated in that they stand at the focal point where two exceptionally powerful and prevalent systems of oppression come together: race and gender. Collins refers to this kind of social position as **intersectionality:** a place where different systems of domination crisscross. There are obviously other systems that Collins talks about, such as class, sexuality, ethnicity, nation, and age, but it is with black women where these different influences get played out most clearly. Seeing this intersectional position of black women, then, ought to compel us to see and look for other spaces where systems of inequality come together.

Just as important to this possibility of continuous change are the qualities of what Collins variously terms alternative or *black feminist epistemology.* This notion implies that one of the things that has hindered social reform is the emphasis on social, scientific knowledge. In this sense, Collins is a critical theorist (see Chapter 7 of this volume) who argues that all knowledge is political and can be used to serve specific group interests. Social science is particularly susceptible to this because it simultaneously objectifies its subjects and denies the validity of lived experience as a form of knowing.

Concepts and Theory: Black Feminist Epistemology

Epistemology is the study of knowledge, and we've been thinking a lot about knowledge in this chapter. Marx (1859/1978e) gave us eyes to see that epistemology is a sociological concern: "It is not the consciousness of men that determines their being, but, on the contrary, their social being that determines their consciousness" (p. 4). Patricia Hill Collins argues that the politics of race and gender influence knowledge. In Marxian terms, race and gender are part of our "social being." In order to talk about this issue, and specifically about black feminist knowledge, Collins juxtaposes it with Eurocentric, positivistic knowledge—the kind of knowledge in back of science. But before we get to that, I need to point out that there is more to knowledge than simply information. Knowledge—information and facts—can only exist within a context that is defined through specific ways of knowing and validation.

For example, the "fact" that God created the heavens and earth only exists within the context of a specific religious system. The same is true for any other "facts," scientific or otherwise. Thus, what we know is dependent upon how knowledge is produced and how it is validated as true. The question here becomes, what are the ways of knowing and methods of validation that are specific to Eurocentric, positivistic knowledge? Collins gives us four points. Note that sociology is generally defined as a social science, and insofar as it is a scientific inquiry into social life, it espouses these four points.

Eurocentric Positivism

First, according to the positivistic approach, true or correct knowledge only comes when the observer separates him- or herself from that which is being studied. You undoubtedly came across this idea in your methods class: the researcher must take an objective stand in order to safeguard against bias. Second, personal emotions must be set aside in the pursuit of pure knowledge. Third, no personal ethics or values must come into the research. Social science is to be value-free, not passing judgment or trying to impose values on others. And, fourth, knowledge progresses through cumulation and adversarial debate.

Recall our discussion of scientific theory in the introduction to Section II. Cumulation is that process by which theories are built up through testing and rejecting elements that don't correspond to the empirical world. The ideas that pass the test are carried on, and theory cumulates in abstract statements about the general properties of whatever is being investigated. The goal is to disassociate ideas from the people who spawned them and to end up with pure theory. Thus, scientific knowledge is validated because it is tested and argued against from every angle. The belief is that only that which is left standing is truth, and it is upon those remnants that objective, scientific knowledge will be built.

Four Tenets of Black Feminist Epistemology

Collins gives us four characteristics of alternative epistemologies, ways of knowing and validating knowledge that challenge the status quo. As we discuss these, notice how each point stands in opposition to the tenets of positivistic knowledge.

The first point is that alternative epistemologies are built upon lived experience, not upon an objectified position. Social science argues that, to truly understand society and group life, one must be removed from the particulars and concerns of the subjects being studied. In this way, subjects are turned into objects of study. Patricia Hill Collins's (2000) alternative epistemology claims that is it only those men and women who experience the consequences of living under an oppressed social position who can select "topics for investigation and methodologies used" (p. 258). Black feminist epistemology, then, begins with "connected knowers," those who know from personal experience.

The second dimension of Collins's alternative epistemology is the use of dialogue rather than adversarial debate. As we've seen, knowledge claims in social science are assessed through adversarial debate. Using dialogue to evaluate implies the presence of at least two subjects—thus, knowledge isn't seen as having an objective existence apart from lived experiences; knowledge ongoingly emerges through dialogue. In alternative epistemologies, then, we tend to see the use of personal pronouns such as "I" and "we" instead of the objectifying and distancing language of social science. Rather than disappearing, the author is central to and present in the text. In black feminist epistemology, the story is told and preserved in narrative form and not "torn apart in analysis" (P. H. Collins, 2000, p. 258).

Centering lived experiences and the use of dialogue imply that knowledge is built around ethics of caring, Collins's third characteristic of black feminist knowledge. Rather than believing that researchers can be value-free, Collins argues that all knowledge is intrinsically value-laden and should thus be tested by the presence of empathy and compassion. Collins sees this tenet as healing the binary break between the intellect and emotion that Eurocentric knowledge values. Alternative epistemology is thus holistic: it doesn't require the separation of the researcher from her or his own experiences nor does it require separation of our thoughts from our feelings, or even assume that it is possible to do so. In addition, Patricia Hill Collins (2000) argues that the presence of emotion validates the argument: "Emotion indicates that a speaker believes in the validity of an argument" (p. 263).

Fourth, black feminist epistemology requires personal accountability. Because knowledge is built upon lived experience, the assessment of knowledge is a simultaneous assessment of an individual's character, values, and ethics. This approach puts forth that all knowledge is based upon beliefs, things assumed to be true, and belief implies personal responsibility. Think about the implications of these two different approaches to knowing, information, and truth: On the one hand, information can be objective and truth exists apart from any observer, while on the other hand, all information finds its existence and "truth" within a preexisting knowledge system that must be believed in order to work. The first allows for, indeed demands, the separation of personal responsibility from knowledge—knowledge exists as an objective entity apart from the knower. The second places accountability directly on the knower. Collins would ask us, which form of knowing is more likely to lead to social justice, one that denies ethical and moral accountability or one that demands it?

Implications of Black Feminist Thought

By now we should see that, for Collins, ways of knowing and knowledge are not separable or sterile—they are not abstract entities that exist apart from the political values and beliefs of the individual. How we know and what we know have implications for who we see ourselves to be, how we live our lives, and how we treat others. Collins sees these connections as particularly important for black women in at least three ways.

First, there is a tension between common challenges and diverse experiences. Think for a moment about what it means to center the idea of lived experience. We've already touched upon several implications of this idea, but what problem might arise from this way of thinking? The notion of lived experience, if taken to an extreme, can privilege individual experience and knowledge to the exclusion of a collective standpoint. The possibility of this implication is particularly probable in a society like the United States that is built around the idea of individualism. However, this isn't what Collins has in mind. One doesn't overshadow the other in intersectionality. We'll explore this idea further later, but for now we want to see that each individual stands at a unique matrix of crosscutting interests. These interests and the diverse responses they motivate are defined through such social positions as race, class, gender, sexual identity, religion, nationality, and so on.

Thus, the lived experience of a middle class, pagan, single, gay black woman living in Los Angeles will undoubtedly be different from that of an impoverished, Catholic, married black woman living in a small town in Mississippi. As Patricia Hill Collins (2000) says, "it is important to stress that no homogeneous Black *woman's* standpoint exists" (p. 28, emphasis original). However, there are core themes or issues that come from living as a black woman such that "a Black *women's* collective standpoint does exist, one characterized by the tensions that accrue to different responses to common challenges" (p. 28, emphasis original). In other words, a black women's epistemology recognizes this tension between common challenges and diverse responses, which in turn is producing a growing sensibility that black women, because of their gendered racial identity, "may be victimized by racism, misogyny, and poverty" (P. H. Collins, 2000, p. 26). Thus, even though individual black women may respond differently based on different crosscutting interests, there are themes or core issues that all black women can acknowledge and integrate into their self-identity.

Another implication of black feminist epistemology is informed by this growing sensibility of diversity within commonality: understanding these issues leads to the creation of safe spaces. *Safe spaces* are "social spaces where Black women speak freely" (P. H. Collins, 2000, p. 100). These safe spaces are of course common occurrences for all oppressed groups. In order for an oppressed group to continue to exist as a viable social group, the members must have spaces where they can express themselves apart from the hegemonic or ruling ideology.

Collins identifies three primary safe spaces for black women. The first is black women's relationships with one another. These relationships can form and function within informal relationships such as family and friends, or they can occur within more formal and public spaces such as black churches and black women's organizations. In this context, Patricia Hill Collins (2000) also points to the importance of mentoring within black women's circles, mentoring that empowers black women "by passing on the everyday knowledge essential to survival as African-American women" (p. 102).

The other two safe spaces are cultural and are constituted by the black women's blues tradition and the voices of black women authors. Such cultural expressions have historically given voice to the voiceless. Those who were denied political or academic power could express their ideas and experiences through story and poetry. As long as the political majority could read these as "fictions"—that is, as long as they weren't faced with the facts of oppression—blacks were allowed these cultural outlets in "race markets." However, these books, stories, and poetry allowed oppressed people to communicate with one another and to produce a sense of shared identity.

There are several reasons why the musical form known as the blues is particularly important for constructing safe spaces and identities for black women. The blues originated out of the "call and response" of slaves working in the fields. It was born out of misery but simultaneously gave birth to hope. This hope wasn't simply expressed in words; it was also more powerfully felt in the rhythm and collectivity that made slave work less arduous. The blues thus expresses to even the illiterate the experience of black America, and it wraps individual suffering in a transcendent collective consciousness that enables the oppressed to persevere in hope without bitterness.

> The music of the classic blues singers of the 1920s—almost exclusively women—marks the early written record of this dimension of U.S. Black oral culture. The songs themselves were originally sung in small communities, where boundaries distinguishing singer from audience, call from response, and thought from action were fluid and permeable. (P. H. Collins, 2000, p. 106)

The importance of these safe spaces is that they provide opportunities for self-definition, and self-definition is the first step to empowerment—if a group is not defining itself, then it is being defined by and for the use of others. These safe spaces also allow black women to escape and resist "objectification as the Other" (P. H. Collins, 2000, p. 101), the images and ideas about black women found in the larger culture.

These safe spaces, then, are spaces of diversity, not homogeneity: "the resulting reality is much more complex than one of an all-powerful White majority objectifying Black women with a unified U.S. Black community staunchly challenging these external assaults" (P. H. Collins, 2000, p. 101). However, even though these spaces recognize diversity, they are nonetheless exclusionary (here we can clearly see the tension that Collins notes). If these spaces did not exclude, they would not be safe: "By definition, such spaces become less 'safe' if shared with those who were not Black and female" (p. 110). Although exclusionary, the intent of these spaces is to produce "a more inclusionary, just society" (p. 110).

This idea leads us to our third implication of black feminist thought: the struggles for self-identity take place within an ongoing dialogue between group knowledge or standpoint and experiences as a heterogeneous collective. Here Collins is reconceptualizing the tension noted above between common challenges and diverse responses. This is important to note because one of the central features of Collins's approach is complexity. Collins wants us to see that most social issues, factors, and processes have multiple faces. Understanding how the different facets of inequality work together is paramount for understanding any part of it. In this case, on the one hand we have a *tension* between common challenges and diverse responses, and on the other hand we have a *dialogue* between a common group standpoint and diverse experiences.

Collins is arguing that changes in thinking may alter behaviors, and altering behaviors may produce changes in thinking. Thus, for U.S. black women as a collective, "the struggle for a self-defined Black feminism occurs through an ongoing dialogue whereby action and thought inform one another" (P. H. Collins, 2000, p. 30). For example, because black Americans have been racially segregated, black feminist practice and thought have emerged within the context of black community development. Other ideas and practices, such as black nationalism, have also come about due to racial segregation. Thus, black feminism and nationalism inform one another in the context of the black community, yet they are both distinct. Moreover, the relationships are reciprocal in that black feminist and nationalist thought influences black community development.

Collins also sees this dialogue as a process of rearticulation rather than consciousness-raising. During the 1960s and 1970s, consciousness-raising was a principal method in the feminist movement. Consciousness-raising groups would generally meet weekly, consist of no more than 12 women, and would encourage women to share their *personal experiences as women*. The intent was a kind of Marxian class consciousness that would precede social change, except that it was oriented around gender rather than class.

Rearticulation, according to Collins, is a vehicle for re-expressing a consciousness that quite often already exists in the public sphere. In rearticulation, we can see the dialogic nature of Collins's perspective. Rather than a specific, limited method designed to motivate women toward social movement, Collins sees black feminism as part of an already existing national discourse. What black feminism can do is to take the core themes of black gendered oppression—such as racism, misogyny, and poverty—and infuse them with the lived experience of black women's taken-for-granted, everyday knowledge. This is brought back into the national discourse where practice and ideas are in a constant dialogue: "Rather than viewing consciousness as a fixed entity, a more useful approach sees it as continually evolving and negotiated. A dynamic consciousness is vital to both individual and group agency" (P. H. Collins, 2000, p. 285).

Black Intellectuals

Within this rearticulation, black feminist intellectuals have a specific place. To set ourselves up for this consideration, we can divide social intellectuals or academics into two broad groups: pure researchers and praxis researchers. Pure researchers hold to value-free sociology, the kind we noted above in considering Eurocentric thought. They are interested in simply discovering and explaining the social world. Praxis or critically oriented researchers are interested in ferreting out the processes of oppression and changing the social world. Black feminist intellectuals are of the latter kind, blending the lived experiences of black women with the highly specialized knowledge of intellectualism.

This dual intellectual citizenship gives black feminist scholars critical insights into the conditions of oppression. They both experience it as a lived reality and can think about it using the tools of critical analysis. Further, in studying oppression among black women, they are less likely to walk away "when the obstacles seem overwhelming or when the rewards for staying diminish" (P. H. Collins, 2000, p. 35). Black feminist intellectuals are also more motivated in this area because they are defining themselves while studying gendered racial inequality.

Finally, Patricia Hill Collins (2000) argues that black feminist intellectuals "alone can foster the group autonomy that fosters effective coalitions with other groups" (p. 36). In thinking about this, remember that Collins recognizes that intellectuals are found within all walks of life. Intellectual status isn't simply conferred as the result of academic credentials. Black feminist intellectuals think reflexively and publicly about their own lived experiences within the context of broader social issues and ideas.

Black feminist intellectuals, then, function like intermediary groups. On the one hand, they are very much in touch with their own and their peers' experiences as a disenfranchised group; on the other hand, they are also in touch with intellectual heritages, diverse groups, and broader social justice issues.

> By advocating, refining, and disseminating Black feminist thought, individuals from other groups who are engaged in similar social justice projects—Black men, African women, White men, Latinas, White women, and members of other U.S. racial/ethnic groups, for example—can identify points of connection that further social justice projects. (P. H. Collins, 2000, p. 37)

Collins notes, however, that coalition building with other groups and intellectuals can be costly. Privileged group members often have to become traitors to the "privileges that their race, class, gender, sexuality, or citizenship status provide them" (P. H. Collins, 2000, p. 37).

Concepts and Theory: Intersectionality and Matrices of Domination

Collins is best known for her ideas of intersectionality and the matrix of domination. Intersectionality is a particular way of understanding social location in terms of crisscrossing systems of oppression. Specifically, intersectionality is an "analysis claiming that systems of race, social class, gender, sexuality, ethnicity, nation, and age form mutually constructing features of social organization, which shape Black women's experiences and, in turn, are shaped by Black women" (P. H. Collins, 2000, p. 299).

This idea goes back to Max Weber and Georg Simmel. To refresh our memories, Weber's concern was to understand the complications that status and power brought to Marx's idea of class stratification. According to Weber, class consciousness and social change are more difficult to achieve than Marx first thought: status group affiliation and differences in power create concerns that may override class issues. And, as you'll remember, Simmel was interested in the way the motivations for and patterns of group memberships changed as a result of living in urban rather than rural settings. Simmel noted that people living in cities tend have greater freedom of choice and the opportunity to be members of more diverse groups than people in small towns. He was specifically concerned with the psychological and emotional effects that these different social network patterns have on people.

There is a way in which Collins blends these two approaches while at the same time going beyond them. Like Simmel, Collins is concerned with the influences of intersectionality on the individual. But the important issue for Collins is the way intersectionality creates different kinds of lived experiences and social realities. She is particularly concerned with how these interact with what passes as objective knowledge and how diverse voices of intersectionality are denied under scientism. Like Weber, she is concerned about how intersectionality creates different kinds of inequalities and how these crosscutting influences affect social change. But Collins brings Weber's notion of power into this analysis in a much more sophisticated way. Collins sees intersectionality working within a matrix of domination.

The **matrix of domination** refers to the overall organization of power in a society. There are two features to any matrix. First, any specific matrix has a particular arrangement of intersecting systems of oppression. Just what and how these systems come together is historically and socially specific. Second, intersecting systems of oppression are specifically organized through four interrelated domains of power: structural, disciplinary, hegemonic, and interpersonal.

The *structural domain* consists of such social structures as law, polity, religion, and the economy. This domain sets the structural parameters that organize power relations. For example, prior to February 3, 1870, blacks in the United States could not legally vote. Although constitutionally enabled to vote after that date, voting didn't become a reality for many African American people until almost a century later with the passage of the Voting Rights Act of 1965, which officially ended Jim Crow laws. Collins's point is that the structural domain sets the overall organization of power within a matrix of domination and that the structural domain is slow to change, often only yielding to large-scale social movements, such as the Civil War and the upheavals of the 1950s and 1960s in the United States.

The *disciplinary domain* manages oppression. Collins borrows this idea from both Weber and Michel Foucault (Chapter 14): the disciplinary domain consists of bureaucratic organizations whose task it is to control and organize human behavior through routinization, rationalization, and surveillance. Here the matrix of domination is expressed through organizational protocol that hides the effects of racism and sexism under the canopy of efficiency, rationality, and equal treatment.

If we think about the contours of black feminist thought that Collins gives us, we can see that the American university system and the methods of financing research are good examples. Sexism and racism never raise their ugly heads when certain kinds of knowledge are systematically excluded in the name of science and objectivity. This same kind of pattern is seen in the U.S. economy. According to the Bureau of Labor Statistics (2005), in the first quarter of 2005 the average weekly income for white men was $731.00, for white women $601.00, for black men $579.00, and for black women the average weekly wage was $506.00. In a country that has outlawed discrimination based on race and sex, black women still make on average about 31% less than white men.

In this domain, change can come through insider resistance. Collins uses the analogy of an egg. From a distance, the surface of the egg looks smooth and seamless. But upon closer inspection, the egg is revealed to be riddled with cracks. For those interested in social justice, working in a bureaucracy is like working the cracks, finding spaces and fissures to work in and expand. Again, change is slow and incremental.

The *hegemonic domain* legitimizes oppression. Max Weber was among the first to teach us that authority functions because people believe in it. This is the cultural sphere of influence where ideology and consciousness come together. The hegemonic domain links the structural, disciplinary, and interpersonal domains. It is made up of the language we use, the images we respond to, the values we hold, and the ideas we entertain. It is produced through school curricula and textbooks, religious teachings, mass media images and contexts, community cultures, and family histories. The black feminist priority of self-definition and critical, reflexive education are important steppingstones to deconstructing and dissuading the hegemonic domain. As Patricia Hill Collins (2000) puts it, "Racist and sexist ide-ologies, if they are disbelieved, lose their impact" (p. 284).

The *interpersonal domain* influences everyday life. It is made up of the personal relationships we maintain as well as the different interactions that make up our daily life. Collins points out that change in this domain begins with the *intra*per-sonal, that is, how an individual sees and understands her or his own self and expe-riences. In particular, people don't generally have a problem identifying ways in which they have been victimized. But the first step in changing the interpersonal domain of the matrix of domination is seeing how our own "thoughts and actions uphold *someone else's subordination*" (P. H. Collins, 2000, p. 287, emphasis added).

Part of this first step is seeing that people have a tendency to identify with an oppression, most likely the one they have experienced, and to consider all other oppressions as being of less importance. In the person's mind, his or her oppression has a tendency then to take on a master status. This leads to a kind of contradiction where the oppressed becomes the oppressor. For example, a black heterosexual woman may discriminate against lesbians without a second thought, or a black Southern Baptist woman may believe that every school classroom ought to display the Ten Commandments. "Oppression is filled with such contradictions because these approaches fail to recognize that a matrix of domination contains few pure victims or oppressors" (P. H. Collins, 2000, p. 287).

Black Feminist Thought, Intersectionality, and Activism

There are a number of implications for activism that Collins draws out from black feminist thought and the notions of intersectionality and the matrix of dom-ination. The first that I want to point out is the most immediate: Collins's approach to epistemology and intersectionality conceptualizes resistance as a complex inter-play of a variety of forces working at several levels—that is, resistance in the four interrelated domains of power that we've just discussed.

This point of Collins's isn't an incidental issue. Remember that part of what is meant by modernity is the search for social equality. In modernity, primary paths for these social changes correspond to Collins's first domain of power. For example, the United States Declaration of Independence, Constitution, and Bill of Rights together provide for principal mechanisms of structural change: the electoral process within a civil society guaranteed by the twin freedoms of press and speech and the upheaval or revolutionary process. Though we don't usually think of the latter as a legitimated means of social change, it is how this nation began and it is how much of the more dramatic changes that surround equality have come about (for example, the social movements behind women's suffrage and civil rights).

One of the ideas that comes out of postmodernism and considerations of late modernity is the notion that guided or rational social change is no longer possible (see the chapters on Niklas Luhmann, Anthony Giddens, and Jean Baudrillard). What Collins gives us is a different take on the issues of complexity and fragmentation. While recognizing the complexity of intersectionality and the different levels of the matrix of domination, Collins also sees the four domains of power as interrelated and thus influencing one another. By themselves, the structural and disciplinary domains are most resilient to change, but the hegemonic and interpersonal domains are open to individual agency and change. Bringing these domains together creates a more dynamic system, wherein the priorities of black feminist thought and understanding the contradictions of oppression can empower social justice causes.

Collins's approach also has other important implications. Her ideas of intersectionality and the matrix of domination challenge many of our political assumptions. Black feminist epistemology, for example, challenges our assumptions concerning the separation of the private and public spheres. What it means to be a mother in a traditional black community is very different from what it means in a white community: "Black women's experiences have never fit the logic of work in the public sphere juxtaposed to family obligations in the private sphere" (P. H. Collins, 2000, p. 228). Intersectionality also challenges the assumption that gender stratification affects all women in the same way; race and class matter, as does sexual identity.

In addition, Collins's approach untangles relationships among knowledge, empowerment, and power, and opens up conceptual space to identify new connections within the matrix of domination. The idea of the matrix emphasizes connections and interdependencies rather than single structures of inequality. The idea itself prompts us to wonder about how social categories are related and mutually constituted. For example, how do race and sexual preference work together? Asking such a question might lead us to discover that homosexuality is viewed and treated differently in different racial cultures—is the lived experience of a black gay male different from that of a white gay male? If so, we might take the next step and ask how does class influence those differences? Or, if these lived experiences are different, we might be provoked to ask another question: are there different masculinities in different racial or class cultures?

As you might be able to surmise from this example, Collins's approach discourages binary thinking and the labeling of one oppression or activism as more important or radical. From Collins's point of view, it would be much too simplistic to say that a white male living in poverty is enjoying white privilege. In the same way, it would be one-dimensional to say that any one group is more oppressed than another.

Collins's entire approach also shifts our understanding of social categories from bounded to fluid and highlights the processes of self-definition as constructed in conjunction with others. Intersectionality implies that social categories are not bounded or static. Your social nearness or distance to another changes as the matrix of domination shifts, depending on which scheme is salient at any given moment. You and the person next to you may both be women, but that social nearness may be severed as the indices change to include religion, race, ethnicity, sexual practices or identities, class, and so forth.

Groups are also constructed in connection to others. No group or identity stands alone. To state the obvious, the only way "white" as a social index can exist is if "black" exists. Intersectionality motivates us to look at just how our identities are constructed at the expense of others: "These examples suggest that moral positions as survivors of one expression of systemic violence become eroded in the absence of accepting responsibility of other expressions of systemic violence" (P. H. Collins, 2000, p. 247).

Here is one final implication of Collins's approach: Because groups' histories and inequalities are relational, understanding intersectionality and the matrix of domination means that some coalitions with some social groups are more difficult and less fruitful than others. Groups will more or less align on the issues of "victimization, access to positions of authority, unearned benefits, and traditions of resistance" (P. H. Collins, 2000, p. 248). The more closely aligned are these issues, the more likely and beneficial are the coalitions. Coalitions will also ebb and flow, "based on the perceived saliency of issues to group members" (p. 248). We end, then, with the insight that inequalities and dominations are complex and dynamic.

Collins Summary

- Collins argues that black women represent a powerful place to begin theorizing about social inequality. Studies and theories in inequality generally focus on one specific issue, such as race or gender. To understand the inequality of black women, however, forces us to be concerned with at least two systems of inequality (gender and race) and their intersections.

- According to Collins, the four qualities of Eurocentric positivism hamper our understanding of how systems of inequality work. These four characteristics are (1) the objective stand, (2) emotional divestment, (3) value-free theory and research, and (4) progress through cumulation and adversarial debate. The four characteristics of black feminist knowledge counter each of these points. Black feminist knowledge is (1) built upon lived experience, (2) emphasizes emotional investment and accountability, (3) honors ethically driven research and theory, and (4) understands intellectual progress through dialogue.

- There are three primary implications of black feminist thought: first, the recognition of the tension between common challenges and diverse responses; second, the creation of safe places that honor diversity; and third, self-identity formed within a continuing dialogue between common challenges and varied experiences. This last implication has importance for rearticulating the public discourse surrounding black women.

- Black feminist intellectuals have a specific place in the construction of black women's identities and the rearticulation of the public sphere. Specifically, black feminists hold a kind of dual intellectual citizenship: they are trained in positivistic methods yet they also have the lived experience of black women. This dual citizenship gives black feminists greater opportunities to forge coalitions with other social justice groups.

- Collins's approach to studying inequality is based on the concepts of intersectionality and a matrix of domination. Intersectionality captures the structured position of people living at the crossroads of two or more systems of inequality, such as race and gender. The different intersectionalities of a society influence the overall organization of power—what Collins refers to as the matrix of domination. These matrices are historically and socially specific, yet they are organized around four general domains of power: structural (the interrelationships of social structures), disciplinary (bureaucratic organization and protocol), hegemonic (cultural legitimations), and interpersonal (personal relationships).

- There are various implications of Collins's approach. First, activism is a complex enterprise that links together different practices in the four domains of power. In other words, activism can involve more than social movements aimed at changing the institutional arrangements of a society. Because there are four domains of power, there can be four fronts of activism. Second, her approach challenges many of the existing political assumptions and opens up new conceptual space for understanding how inequalities work. The ideas of intersectionality and a matrix of domination discourage simple, binary thinking in politics and research. Third, Collins's ideas sensitize us to the fluid nature of social categories, identities, and relations. And, fourth, political coalitions can ebb and flow as different groups align to varying degrees on the issues of power, victimization, and resistance.

Building Your Theory Toolbox

Knowing Identity Politics

After reading and understanding this chapter, you should be able to define the following terms theoretically and explain their theoretical importance to theories surrounding identity politics:

Standpoint; relations of ruling; new materialism; institutional ethnography; lived experience; text; fault line; bifurcated consciousness; pragmatism; existentialism; Christianity; culture; market saturation; black nihilism; market moralities; ontological wounds; black leadership crisis; politics of conversion; racial reasoning; moral reasoning; existential angst; black cultural armor; coalition strategy; market dynamics; market saturation; intersectionality; Eurocentric positivism; black feminist epistemology; common challenges/diverse responses; safe places; self-definition; rearticulation; black feminist intellectuals; matrix of domination; structural, disciplinary, hegemonic, and interpersonal domains of power

After reading and understanding this chapter, you should be able to

- Explain how standpoint is more method than theory. How does some feminist work actually defeat standpoint? How are standpoint and the general sociological approach different?

- Explicate how the relations of ruling are expressed through social science

- Define new materialism and discuss how it influences what people accept as true or factual

- Explain how the fault line perpetuates gender inequality

- Describe Smith's institutional ethnography. How is it dialectical? What do you think the benefits of institutional ethnography would be?

- Explain how West uses Marxist theory, pragmatism, existentialism, and Christianity to form his unique perspective

- Explain what market dynamics are and how and why market saturation has affected the black community in unique ways

- Discuss black nihilism, where it comes from, and how it is affecting blacks in the United States today

- Explain racial reasoning and how moral reasoning counters it

- Discuss the black leadership crisis and the politics of conversion

- Compare and contrast the characteristics of Eurocentric positivism and black feminist epistemology

- Explicate the implications of black feminist epistemology

- Explain how inequality can best be understood as intersectionality and matrices of domination. In your explanation, be certain to discuss the implications of such an approach.

- Discuss the kinds of activism that Patricia Hill Collins's approach includes. In your discussion, be certain to include the four domains of power and the unique place that black feminist intellectuals have in activism.

Learning More: Primary Sources

- Dorothy E. Smith

  Smith, D. E. (1987). *The everyday world as problematic: A feminist sociology.* Boston: Northeastern University Press.
  Smith, D. E. (1990). *The conceptual practices of power: A feminist sociology of knowledge.* Boston: Northeastern University Press.
  Smith, D. E. (2005). *Institutional ethnography: A sociology for people.* Walnut Creek, CA: Alta Mira Press.

- Cornel West

  West, C. (1993). *Race matters.* New York: Vintage.
  West, C. (2004). *Democracy matters: Winning the fight against imperialism.* New York: Penguin.

- Patricia Hill Collins

  Collins, P. H. (2000). *Black feminist thought* (2nd ed.). New York: Routledge.
  Collins, P. H. (2004). *Black sexual politics.* New York: Routledge.
  Collins, P. H. (2006). *From black power to hip hop: Racism, nationalism, and feminism.* Philadelphia: Temple University Press.

Theory You Can Use (Seeing Your World Differently)

- As a student, how do you see yourself being socialized to the relations of ruling? If you are a woman, do you see bifurcated consciousness in your life? As a sociologist, how will you avoid being trapped and controlled by the relations of ruling?

- Compare and contrast Smith's idea of the fault line and Chafetz's theory of male micro-resource power.

- Compare Wilson's and West's arguments about racism in the United States. Can they be brought together to form a more complete understanding?

- Become involved in campus efforts to end discrimination. Check and see if you have an office of multicultural affairs. Find out what other campus organizations are involved in ending discrimination.

Further Explorations: Web Links

- Explore feminism, beginning with a site such as this one: http://www.amazoncastle.com/feminism/ecocult.shtml

- Cornel West has put out a hip-hop CD. Find information at http://www.cornelwest.com

- Listen to an interview with Patricia Hill Collins on NPR: http://www.npr.org/templates/story/story.php?storyId=1802518

- Explore critical race theory in education: http://rpp.english.ucsb.edu/category/critical-race-theory-and-pedagogy/

CHAPTER 14

Post-Theories

Michel Foucault (1926–1984)

Jean Baudrillard (1929–)

> *But what if empirical knowledge . . . obeyed, at a given moment, the laws of a certain code of knowledge? . . . The fundamental codes of a culture . . . establish for every man, from the very first, the empirical orders with which he will be dealing and within which he will be at home.*
>
> (Foucault, 1966/1994b, pp. ix, xx)

> *The systems of reference for production, signification, the affect, substance and history, all this equivalence to a "real" content, loading the sign with the burden of "utility," with gravity—its form of representative equivalence—all this is over with.*
>
> (Baudrillard, 1976/1993b, p. 6)

These two quotes illustrate a shift that has occurred in the social disciplines. Sociology and the social disciplines in general have experienced what has been called a "linguistic turn." A good deal of this sea change toward language can be credited generally to the work of poststructuralism and postmodernism. Language itself wasn't part of classic theory, at least for most of sociology. However, classic sociologists did talk about culture. Harriet Martineau saw culture as a way to both measure a society's progress and encourage its progress through education. Herbert Spencer talked about categoric groups; Marx actually saw culture as a problem in the form of ideology; Durkheim's theory of the collective consciousness is utterly cultural; W. E. B. Du Bois gave significant place to cultural representation; Weber was concerned with cultural meaning; and a good deal of Georg Simmel's theory is based around the distinctions between subjective and objective culture. The place where language begins to come into sociology is through the work of George Herbert Mead. He exploited the reflexive properties of language in his theory of the mind and self.

But it wasn't until the 1960s that language became important for sociology as a whole. There are perhaps a number of reasons for this shift in the United States. For example, American sociology was just coming out of its functionalist phase and seeing the resurgence of Marxian theory. Structural functionalism's emphasis on culture and Marx's negative view of culture left a space in the intellectual field. American sociology was also opening up to more European intellectual influences.

At the time, French sociology was reeling under what seemed like an abrupt shift from linguistic structuralism to poststructuralism. Besides the conceptual issues, the shift itself meant that signs and language were part of an energetic debate, which splashed over into the States. American sociology in the sixties also saw the rise of phenomenological theory within its own borders, which brought increasing emphasis on language through ethnomethodology and the social constructivism of Peter Berger and Thomas Luckmann's *The Social Construction of Reality*.

In the theories of Michel Foucault and Jean Baudrillard, we see strong statements of this linguistic turn, in the form of poststructuralism with Foucault and postmodernism with Baudrillard. Both of these men approach the issues of signs and language differently, but what I want to point out here is that their work heralds a fundamental shift in the way social theorists think about culture. Rather than understanding culture and language within a social context, such as in symbolic interactions or symbolic markets, texts are understood on their own terms, as social factors in and of themselves. In this perspective, culture and cultural readings become fundamentally important. The radical postmodern thread in this linguistic turn is that readings of texts are themselves seen as texts, which means that since humans are defined through meaning, all we have are texts. Foucault's poststructuralism, on the other hand, sees texts as ways through which power is exercised over individual subjects.

Michel Foucault: Defining the Possible and Impossible

Foucault is a complex thinker and writer. As a result, trying to summarize Foucault's theory can be a frustrating experience. In writing this chapter, I had a continuing sense of incompletion. The more I wrote, the more I felt that I was leaving out. I mention this because I know that what I'm presenting in this book is a pared down version of Foucault. Yet I believe that in focusing on a select few of Foucault's major points, I can convey some sense of what he was trying to accomplish.

Stated succinctly, Foucault is interested in how power is exercised through knowledge or "truth," and how truth is formed through practice (note that with Foucault, we can use knowledge and truth interchangeably). His interest in truth isn't abstract or philosophical. Rather, Foucault is interested in analyzing what he calls *truth games*. His use of "games" isn't meant to imply that what passes as truth in any historical time is somehow false or simply a construction of language. Foucault feels that these kinds of questions can only be answered, let alone asked, after historically specific assumptions are made. In other words, something can only be "false" once a specific truth is assumed, and Foucault is involved in uncovering *how* truth is assumed. Specifically, Foucault's interest in truth concerns the game of truth: the rules, resources, and practices that go into making something true for humans.

The idea of *practice* is fairly broad and includes such things as institutional and organizational practices as well as those of academic disciplines—in these practices, truth is formed. The idea also refers to specific practices of the body and self—these are where power is exercised. Most of us use the word "practice" to talk about the behaviors we engage in to prepare for an event, such as band practice for a show. But practice has another meaning as well. This meaning is clear when we talk about a medical practice. When you go to your physician, you see someone who is "practicing" medicine. In this sense, practice refers to choreographed acts that interact with bodies—sets of behaviors that together define a way of doing something. This is the kind of practice in which Foucault is interested.

Photo: © Bettmann/Corbis.

The Essential Foucault

Biography

We should begin this brief biography by noting that Michel Foucault would balk at the idea that we need to know anything about the author in order to understand his or her work. Further, Foucault would say that any history of the author is something that we use in order to validate a particular reading or interpretation. Having said that, Foucault was born on October 15, 1926, in Poitiers, France. Foucault studied at the École Normale Supérieure and the Institut de Psychologie in Paris. In 1960, returning to France from teaching posts in Sweden, Warsaw, and Hamburg, Foucault published *Madness and Civilization*, for which he received France's highest academic degree, doctorat d'État. In 1966, Foucault published *The Order of Things*, which became a best-selling book in France. In 1970, Foucault received a permanent appointment at the Collège de France (France's most prestigious school) as chair of History of Systems of Thought. In 1975, Foucault published *Discipline and Punishment* and took his first trip to California, which came to hold an important place in Foucault's life, especially San Francisco. In 1976, Foucault published the first volume of his last major work, *The History of Sexuality*. The two other volumes of this history, *The Use of Pleasure* and *The Care of the Self,* were published shortly before Foucault's death in 1984.

Passionate Curiosity

In Foucault's (1984/1990b) own words,

> As for what motivated me. . . . It was curiosity—the only kind of curiosity, in any case, that is worth acting upon with a degree of obstinacy: not the curiosity that seeks to assimilate what it is proper for one to know, but that which enables one to get free of oneself. After all, what would be the value of the passion for knowledge if it resulted only in a certain amount of knowledgeableness and not, in one way or another and to the extent possible, in the knower's straying afield of himself? (p. 8)

In brief, Foucault is interested in how ideas and subjectivities come into existence and how they limit what is possible. But Foucault's search is not simply academic, though it is that. As the above quote tells us, Foucault (1984/1990b) seeks to understand his own practices "in relationship of self with self and the forming of oneself as a subject" (p. 6).

Keys to Knowing

Poststructuralism, practice, power, knowledge, order, truth games, discourse, counter-history, archaeology and genealogy, episteme, historical rupture, subject objectification, panopticon, human disciplines, clinical gaze, governmentality, microphysics of power, sexuality and subjectivity

Seeing Further: Truth Games and Counter-Histories

Foucault uncovers truth games by constructing what he calls counter-histories. When most of us think of history, we think of a factual telling of events from the past. We are aware, of course, that sometimes that telling can be politicized, which is one reason we have "Black History Month" here in the United States—we are trying to make up for having left people of color out of our telling of history. But most of us also think that the memory model is still intact; it's just getting a few tweaks. Foucault wants us to free history from the model of memory. He really doesn't say anything directly about whether any particular history is more or less true; that's not an issue for him. History in all its forms is both part of and generated by discourse. Thus, Foucault's concern is how the *idea* of true history is used. What Foucault wants to produce for us is a *counter-history*—a history told from a different point of view from the progressive, linear, memory model.

The important questions then become, why is one path taken rather than another? Why is the present filled with one kind of discourse rather than others? And what has been the cost of taking this path rather than all the other potentialities? Thus, a counter-history identifies

the accidents, the minute deviations—or conversely, the complete reversals— the errors, the false appraisals, and the faulty calculations that gave birth to those things that continue to exist and have value for us; it is to discover that truth or being does not lie at the root of what we know and what *we are,* but the exteriority of accidents. (Foucault, 1984a, p. 81)

Foucault uses two terms to talk about his counter-history: archaeology and genealogy. Though the distinctions are sometimes unclear, *archaeology* seems to be oriented toward uncovering the relationships among social institutions, practices, and knowledge that come to produce a particular kind of discourse or structure of thought. *Genealogy* may be better suited to describe Foucault's (1984a) work that is concerned with the actual inscription of discourse and power on the mind and body: "Genealogy, as an analysis of descent, is thus situated within the articulation of the body and history. Its task is to expose a body totally imprinted by history and the process of history's destruction of the body" (p. 83). We could say that archaeology is to text what genealogy is to the body. In both cases, there is an analogy to digging, searching, and uncovering the hidden history of order, thought, madness, sexuality, and so on. The hidden history isn't necessarily more accurate—it's simply a counter-story that is constructed more in an archaeological mode than an historical one.

Part of what Foucault wants to do with counter-histories is expose the contingencies of what we consider reality, but to what end? Many critical perspectives are based on assumptions of what would make a better society. In other words, there must be something to which the current situation is compared to demonstrate what it is lacking. But Foucault sees it otherwise. For him, the critical perspective in itself is sufficient because it opens up possibilities. In fact, Foucault would argue that a utopian scheme only attempts to replace one system of impoverishment with another. The point is to keep possibilities always open, to keep people critically examining their life and knowledge system so that they can perpetually be open to the possibility of something else.

According to Foucault's scheme, an important part of what creates knowledge, order, and discourse is the presence of "blank spaces." Foucault (1966/1994b) pictures knowledge as a kind of grid. The boxes in the grid are the actual linguistic categories, such as mammal, flora, mineral, human, black, white, male, and female. We are familiar with those sections; they form part of our everyday language. However, there is actually a more important part of the grid, the one that creates the order— the blank spaces between the categories. "It is only in the blank spaces of this grid that order manifests itself in depth as though already there, waiting in silence for the moment of its expression" (p. xx). The true power of a discourse or knowledge system is in the spaces between the categories. As Eviatar Zerubavel (1991) notes,

> separating one island of meaning from another entails the introduction of some mental void between them. . . . It is our perception of the void among these islands of meaning that makes them separate in our mind, and its magnitude reflects the degree of separateness we perceive among them. (pp. 21–22)

These spaces are revealed most clearly in transgression. As an illustration, let's think about a little boy of about 3 or 4 years of age. He is a playful boy, playing with the toys he's been given and emulating the role models he sees on TV and among the neighborhood children. But one day his father comes home and finds him playing with a doll. His father grabs the toy away and tells his son firmly that boys do not play with dolls. In this instance, the category of gender was almost invisible until the young boy unwittingly attempted to cross over the boundary or space between the categories. The meaning and power of gender waited "in silence for the moment of its expression."

This idea of space is provocative. A more Durkheimian way of thinking about categories would conceptualize the space between them as a boundary or wall. Using the idea of boundary to think about the division between categories is fruitful: walls separate and prevent passing. The young boy in our example certainly came up against a wall, and many of us have felt the walls of gender, race, or sexism. But the idea of walls makes the use of categories and knowledge seem objective, as if they somehow exist apart from us, and this is not what Foucault has in mind.

Notice that the boy in our example was unaware of the "wall" until his father showed it to him. From Foucault's position, the wall of gender was erected in the father's gendered practices. Foucault's idea of space helps us think about the practices of power. Space, in this sense, is empty until it is filled—seeing space between categories rather than a wall makes us wait to see what will go there and *how* it goes there. Space is undetermined. Something can be built in space, but the space itself calls our attention to potential. Foucault's research, his critical archaeology, fills in that potential—he tells us how that space became historically constructed in one way rather than any of the other potential ways.

Foucault's counter-history actually creates a space of its own. On one side, Foucault's archaeology of modernity uncovers the fundamental codes of thought that establish for all of us the order that we will use in our world. On the other side, Foucault sets the sciences and philosophical interpretations that explain why such an order exists. Between these two domains is a space of possibilities, a space wherein a critical culture can develop that sufficiently frees itself "to discover that these orders are perhaps not the only possible ones or the best ones" (Foucault, 1966/1994b, p. xx).

In other words, through the archaeology of knowledge, Foucault wants to not only expose the codes of knowledge that undergird everything we do, feel, and think, he also wants to set loose the idea that things might not be as they are. He wants to free the possibility of thinking something different. That possibility of thought exists in the critical space between—but in this case the space isn't specified, as it is in already existing orders. Foucault doesn't necessarily have a place he is taking us; he doesn't really have a utopian vision of what knowledge and practice ought to be. His critique is aimed at freeing knowledge and creating possibility; it's aimed at creating an empty space that is undetermined.

Defining Poststructuralism

Foucault is generally considered a poststructuralist, though some prefer to classify him as a postmodernist. As you'll see when we get to Baudrillard's postmodernism, the two perspectives are similar and at least complementary. Both are concerned with culture generally and language particularly, and both argue that culture and language function without any physical or objective reality in back of them. There are, I think, two main differences between them. First, they each locate the reasons for the lack of reality in different places. Poststructuralism generally considers the intrinsic characteristics of language itself, while social postmodernism usually looks to such factors as capitalism and mass media as the culprits.

Second, they each focus on different effects of the state of culture. For instance, Foucault argues that rather than referring to any physical reality, language contains political discourses that function to exercise power over the person. Baudrillard, on the other hand, sees culture as absolutely void of any significance at all, political or otherwise. Any meaning or power in culture has been stripped away by incessant commodification, advertising, and the trivializing effects of mass media. In a nutshell, Foucault says the individual subject is the product of historically specific discourses, and Baudrillard claims that the subject is dead.

Another quality that poststructuralism and postmodernism share is that they are hard to define. Poststructuralism doesn't refer to a clear body of knowledge that is universally accepted by everyone claiming to use the perspective. It is, rather, a loose amalgamation of ideas that in general define an approach. Overall, poststructuralism is defined in opposition to structuralism, an attempt to discover and explain the unobserved foundations for empirical phenomena. Implicit within such goals is the belief that a metalanguage (like science or mathematics) can be created that will adequately and objectively express the characteristics of such a structure. As the linguistic structuralist Claude Lévi-Strauss (1963) put it, "Structural linguistics will certainly play the same renovating role with respect to the social sciences that nuclear physics, for example, has played for the physical sciences" (p. 33). It is the criticism of this belief and hope that forms the core of poststructuralism.

Poststructuralism assumes that all languages are based on values, most of which are political in nature. Thus, a "pure science" of any social structure is impossible. Rather than language being able to represent an independent reality, language is inherently self-referential, creating a world of oppressive power relations. In addition, poststructuralism is characterized by several denials:

- Poststructuralism rejects the belief in essentializing ideas that conceptualize the social world or a portion of it as a universal totality—rather, the social world is fragmented and historically specific (general theories are thus impossible and oppressive).

- Poststructuralism denies the possibility of knowing an independent or objective reality—rather, the human world and knowledge are utterly textual.

- Poststructuralism discards the idea that texts or language have any true meaning—rather, texts are built around difference and carry a surplus of meaning (humanity is thus left with nothing but interpretation and interpretations of interpretation).

- Poststructuralism rejects the idea of universal human nature developed out of the Enlightenment—rather, the meaning of the human subject is historically specific and is an effect of discourse, with the discourses of an age producing the possible bodies and subjectivities of the person.

Concepts and Theory: The Practices of Power

According to Foucault, power isn't something that a person possesses; rather, it is something that is part of every relationship. Foucault tells us that there are three types of domains or practices within relationships: communicative, objective, and power. Communication is directed toward producing meaning; objective practices are directed toward controlling and transforming things—science and economy are two good examples; and practices of power, which Foucault (1982) defines as "a set of actions upon other actions" (p. 220), are directed toward controlling the actions and subjectivities of people. Notice where Foucault locates power—it's within the actions themselves, not within the powerful person or the social structure. Foucault uses the double meaning of "conduct" to get at this insight: conduct is a way of leading others (to conduct an orchestra, for example) and also a way of behaving (as in "Tommy conducted himself in a manner worthy of his position."). Thus, we conduct others through our conduct.

However, Foucault's intent is not to reduce power to the mundane, the simple organization of human behavior across time and place. Rather, Foucault's point is that power is exercised in a variety of ways, many of which we are unaware. Power, then, becomes insidious. Power acts in the normalcy of everyday life. It acts by imperceptible degrees, exerting gradual and hidden effects. In this way, the exercise of power entices us into a snare that feels of our own doing. But how is power exercised? Where does it exist and how are we enticed? Foucault argues that power is exercised through the epistemes (underlying order) and discourses found in what passes as knowledge. The potential and practice of power exists in these epistemes and discourses that set the limits of what is possible and impossible, which in turn are felt and expressed through a person's relationship with her- or himself, in subjectivities—the way we feel about and relate to our inner self—and the disposition of the body.

Epistemes and Order

Order is an interesting idea. We order our days and lives; we order our homes and offices; we order our files and our bank accounts; we order our yards and shopping centers; we order land and sea—in short, humans order everything. How do we order things? One of the ways is linguistically: "Indeed, things become meaningful only when placed in some category" (Zerubavel, 1991, p. 5). But a deeper and more fundamental question can be asked: how do we order the order of things? In other words, what scheme or system underlies and creates our categorical schemes? We may use categories to order the world around us, but where do the categories get their order?

To introduce us to this question, Foucault (1966/1994b) tells a delightful story of reading a book containing a Chinese categorical system that divides animals into those "(a) belonging to the Emperor, (b) embalmed, (c) tame, (d) sucking pigs, (e) sirens, (f) fabulous, (g) stray dogs, (h) included in the present classification,

(i) frenzied, (j) innumerable, (k) drawn with a very fine camelhair brush, (1) et cetera, (m) having just broken the water pitcher, (n) that from a long way off look like flies" (p. xv). The thing that struck Foucault about this system of categories was the limitation of his own thinking—"the stark impossibility of thinking *that*" (p. xv, emphasis original). In response, Foucault asks an important set of questions: What sets the boundaries of what is possible and impossible to think? Where do these boundaries originate? What is the price of these impossibilities—what is gained and what is lost?

Foucault argues that there is a fundamental code to culture, a code that orders language, perception, values, practices, and all that gives order to the world around us. He calls these fundamental codes epistemological fields, or the episteme of knowledge in any age. **Episteme** refers to the mode of thought's existence, or the way in which thought organizes itself in any historical moment. An episteme is the necessary precondition of thought. It is what exists before thought and that which makes thought possible. This foundation of thought is not held consciously. It is undoubtedly this preconscious character of the episteme that makes thought believable and ideas seem true.

Moreover, rather than seeing thought and knowledge as results of historical, linear processes, Foucault argues that discontinuity marks changes in knowledge. Most of us think that the knowledge we hold accumulated over time, that we have thrown out the false knowledge and replaced it with true knowledge as we have progressively learned how things work. This evolutionary view of knowledge actually comes from the culture of science. It is the way we *want* to see our knowledge, not necessarily the way it is. Foucault argues that knowledge doesn't progress linearly. Rather, what we know and how we know it is linked to historically specific patterns of behavior, institutional arrangements, and economic and social practices that set the rules and conditions of discourse and the limits of our possibilities. And that historical path is marked by rupture: discontinuities and sudden, radical changes.

Think about this: what is Foucault saying that hasn't been said before? Others have said that knowledge is socially constructed. But Foucault is saying that this idea of rupture implies that knowledge and truth are purely functions of institutional arrangements and practices and not the result of any real quest for truth. Thus, what counts as truth in any age—our own included—comes about through historically unique practices and institutional configurations. This implies not only that knowledge is socially constructed, but also, and more importantly, that *knowledge is nothing more and nothing less than the exercise of power*. This pure power is put into effect through discourse and the taken-for-granted ordering of human life.

Discourse

Discourse refers to languages and behaviors that are specific to a social issue, such as the discourse of race. In simple terms, a **discourse** is a way of talking about things. If you want to discuss music with a group of musicians, for instance, there is an acceptable discourse or language that you would use. It would include such words as key, modes, transposing, and so on. This discourse would be different from the one you would use to talk about baseball. You wouldn't normally tell your baseball team to hit the field and "tune up," nor would you tell a violinist to "bunt."

Foucault's interest in the idea of discourse is a bit more significant. First, he wants us to see beneath the surface of the word choice between "bunt" and "tune up." Foucault is interested in the rules and practices that underlie the words and ideas that we use. Discourse sets the possibilities of thought and existence. There is an obvious link between language and thought: we think in language. So, a discourse, with its underlying rules and practices, gives us a language with which to think and talk. That's a commonsensical statement, and we might be tempted just to accept it at face value. But using discourse as the basis of thought sets the boundaries of what is possible and impossible for us to think, so it is more profound than it might first appear.

The second thing that discourse does is to determine the position a person or object must occupy *in order to become the subject of a statement.* "I'm a man" is such a statement. For me to be a man, I must meet the conditions of existence that are set down in the discourse of gender. I not only have to meet those conditions for you; I must meet them for me as well, because the discourse sets out the conditions of subjectivity—how we think and feel about our self. Subjects, and the accompanying inner thoughts and feelings, are specific conditions within the discourse. As we locate ourselves within a discourse, we become subject to the discourse and thus subjectively answer ourselves through the discourse.

This work of positioning that discourse accomplishes is one of its most powerful acts. Think about it this way: it is extremely difficult to talk to someone about anything without positioning yourself within a discourse. There are discourses surrounding sports, family, gender, race, class, self-improvement, medicine, mass media, cars, trucks, and on and on. Once you begin to converse using a discourse, you automatically occupy a position within it that tells you how to think, feel, and act. For example, the modern discourse of gender tells me what I can and can't feel as a man. Here's a more provocative example: if you feel "sick," you have already positioned yourself within the modern medical discourse (compare this feeling to magical or religious discourses that define such things as spiritual possessions).

The third thing that Foucault wants us to see about discourse is that it is used instead of coercive force to impose order on a social group. Critically speaking, social order is always a problem for the elite in any society. One way to subjugate a population is through physical coercion. However, the use of force is costly and produces contrary effects. Discourse is used instead of force and is thus characterized by a will to truth and a will to power. In other words, there is political intention behind truth and power. What passes as truth and how truth is validated is dependent upon the discourse. And discourse intrinsically contains a will to power.

Let me give you a dramatic example. The attacks of September 11, 2001, were perpetrated by men who are considered either "terrorists" or "freedom fighters," depending on the discourse that is used. Within these discourses are legitimations and methods of reasoning that create these two different social meanings. Further, the discourses create the subjective experience of all the different peoples involved. The substance of one discourse is captured by the title of the report generated by the United States government: "The National Commission on Terrorist Attacks Upon the United States."

Clearly, the discourse in the United States defines the perpetrators as *terrorists* and the subjective experience of those in the United States as being innocently *attacked*. The substance of the other discourse is revealed in the title and opening lines of a document confiscated by the police in Manchester, England, during a search of an Al Qaeda member's home. The title of the document is "Declaration of Jihad," and the opening lines are addressed to "those champions who avowed the truth day and night" (*Al Qaeda Training Manual*, n.d.). One discourse creates the meaning of attack and terrorist; the other creates the meaning of holy war and champion.

There's something vitally important that I want you to notice here. Obviously, what we are talking about in this example is the creation of meaning. As we saw with symbolic interaction in Chapter 9, we experience events meaningfully. But notice that Foucault takes us further than symbolic interactionism's point about meaning and definition. Symbolic interaction, you will remember, says that meaning is negotiated through the back-and-forth interplays of conversation. Foucault's point is that this negotiation of meaning through interaction takes place within a framework that is set by the currently available episteme and discourses. These issues of why, what, and how we know things are exercises of social power, because they are all grounded in the epistemes and discourses of any particular society, which are themselves controlled by the elite. *How we know* something, or the basis upon which knowledge is created, is particularly insidious because it seems as if it is a simple function of the brain and our senses. Yet this is exactly what Marx had in mind with false consciousness and what Dorothy E. Smith (Chapter 13) intends when she talks about the new materialism, text, and the relations of ruling.

From Subject to Object

For Foucault, then, power is not so much a quality of social structures as it is the practices or techniques that become power as individuals are turned into subjects through discourse. Foucault intends us to see both meanings of the noun "subject": as someone to control, and as one's self-knowledge. Here Foucault's unique interest is quite clear—perhaps the most insidious form of power is that which is exercised by our self over how we think and feel; it is the power we exercise in the name of others over our self.

In an interesting analysis, Foucault uses the state to illustrate both meanings of subject. State rule is usually understood in terms of power over the masses. While this is a true characteristic of the state, Foucault argues that the modern state also exercises individualization techniques that exercise power over the subjectivity of the person. Foucault talks about this form of ruling as **governmentality**: "the government of the self by the self in its articulation with relations to others" (Foucault, 1989, as quoted in Davidson, 1994, p. 119). Governmentality was needed because of the shift from the power of the monarch to the power of the state.

Under a monarchy, the power of the queen or king was absolute and she or he required absolute obedience, but the scope of that control was fairly narrow. The nation-state "freed" people from the coercive control of the monarchy but at the same time broadened its scope of control. The nation-state is far more interested in controlling our behaviors today than monarchies were 300 years ago. In governmentality, the individual is enlisted by the state to exercise control over him- or herself. This is partly achieved through expert, professional knowledge that comes from medicine and the social and behavior sciences. The state supports such scientific research, and the findings are employed to extend control, particularly as the individual uses and consults medicine, psychology, and other sciences.

A fundamental part of Foucault's argument about the practices of power is the historical shift to *objectification*. Obviously, if power is intrinsic to human affairs of all kinds, then people have always exercised power. However, the practice of power became something different and more insidious due to historical changes that objectified the subject of power—the individual person. We can get a picture of this shift in how power is exercised over the person by comparing the roots, primary meanings, and transitive verb forms of *object* and *subject*.

I've listed the differences between them in Table 14.1, for easy comparison. All the references come from Merriam-Webster (2002). Notice the attitudes that the Latin roots imply: objects are things that can be thrown away, whereas subjects are things that are placed or thrown under. With this root meaning, subjects are controlled, but there is still a relationship between the subject and the one in charge—subjects are *under*, controlled but still present. Objects, on the other hand, are *thrown away*; there is no continuing relationship with whoever is doing the throwing. Notice also the first meaning of each word. Even though the definition of object is talking about something we perceive, like seeing a tree in the distance, *the object is regarded as in the way*. The first meaning of subject still carries with it the notion of connected but controlled.

Table 14.1   Object and Subject

	Object	*Subject*
Latin Root	*objectus* meaning "to throw away"	*subjectus* meaning "to bring under, throw under"
First Meaning	"something that is put or may be regarded as put in the way of some of the senses"	"one that is placed under the authority, dominion, control, or influence of someone or something"
Transitive Verb	"a: to cause to become or to assume the character of an object b: to render objective; *specifically*: to give the status of external or independent reality to"	"to identify with a subject or interpret in terms of subjective experience"

We are all probably familiar with the transitive verb "to objectify." It means to make something an object that isn't an object, and it also means to exist apart from any internal relationship. Interestingly, most of us are probably not familiar with the transitive verb "to subjectify." As a case in point, my word processor just highlighted "subjectify" as a misspelled word, yet it is a real word that appears in exhaustive dictionaries. We just rarely use it, nor do we think about things becoming subjectified—we assume that we subjectively relate to everything about ourselves on our own, without any outside influence. But, according to Foucault, that is not the case, especially in modernity. Today, we relate to our self, our body, and our sexuality as objects.

Foucault's work is found in a series of books that provide a counter-history to some of the objectifying power practices in Western societies. These books detail madness and rationality, abnormality and normality, medicine and the clinic, penal discipline and punishment, psychiatry and criminal justice, and the history of sexuality. In general, these works document how you and I exercise power over our bodies and subjectivities. While I don't have the luxury of introducing you to all of Foucault's archaeology and genealogy, it is important for us to talk about a few of his concepts so that you can get a sense of how his theoretical ideas get played out. We'll first be looking at how power is exercised over our body and then over our inner, subjective life.

Concepts and Theory: Power Over the Body

Foucault's intent in his book *Discipline and Punish* is to map a major shift in the way in which Western society handles crime and criminals. The shift is from punishment and torture to discipline. Foucault paints a graphic comparative picture in the first seven pages of this book. The first part of the picture is an account of the public torture and killing of a man named Damiens on March 2, 1757. Damiens had been convicted of murder and sentenced to having his flesh torn from his body with red-hot pincers, followed by various molten elements (such as lead, wax, and oil) poured into the open wounds. The hand that held the knife with which he had committed the murder was burnt with sulfur. Finally, he was drawn and quartered by four horses and his body burnt to ashes and the ashes scattered to the winds.

The second image in Foucault's picture is a set of 12 rules for the daily activities of prisoners in Paris. The rules covered the prisoners' entire day and included such things as prayer, bible reading, education, bathing, recreation, and work. These rules were in use a mere 80 years after the public torture of Damiens. The shortness of the time period indicates that the change isn't due to gradual adjustment and progress, but rather to abrupt shifts in knowledge, perception, and power.

Foucault uses this graphic comparison to point out a fundamental change that occurred in Europe and the United States. Most of us would look at these differences and attribute the change to a dawning of compassion and a desire to treat people more humanely. Foucault, on the other hand, looks deeper and more holistically at the shift. This change not only affected the penal system; it was a fundamental social change as well. During this period of time, from the eighteenth to nineteenth centuries (also known as the Enlightenment), science gained its foothold in society. Society as a whole began to embrace what we call *scientism*—the adaptation of the methods, mental attitudes, and modes of expression typical of scientists. Scientism values control, and control is achieved by objectifying the world and reducing it to its constituent parts. The gaze of the scientist is thus penetrating, particularizing, and objectifying. This kind of gaze results in universal technologies that allow humans to regularize and routinize their control of the world.

The shift, then, was not due to society becoming more compassionate and humane; the shift from punishment to discipline was a function of scientism and the desire to more uniformly control the social environment. As Foucault (1975/1995) says, the primary objective of this shift was

> to make of the punishment and repression of illegalities a regular function, coextensive with society; not to punish less, but to punish better; to punish with an attenuated severity perhaps, but in order to punish with more universality and necessity; to insert the *power to punish more deeply into the social body.* (p. 82, emphasis added)

This new way of discipline and control is best characterized by Jeremy Bentham's concept of a panopticon. The word **panopticon** is a combination of two Greek words. The first part, "pan," comes from the word *pantos* meaning "all." The second part comes from the word *optikos* meaning "to see." Together, panopticon literally means "all seeing." There is actually an optical instrument called the panopticon that combines features of both the microscope and telescope, allowing the viewer to see things both up close and far away, thus seeing all.

Jeremy Bentham developed a different kind of panopticon—a building design for prisons. Bentham's panopticon was a round building with an observation tower or core that optimized surveillance. The building was divided into individual prison cells that extended from the inner core to the outer wall. Each cell had inner and outer windows; thus, each prisoner was backlit by the outer window, allowing for easy viewing. "They are like so many cages, so many small theatres, in which each actor is alone, perfectly individualized and constantly visible" (Foucault, 1975/1995, p. 200). The tower itself was fitted with Venetian blinds, zigzag hallways, and partitioned intersections among the observation rooms in the tower. These made the tower guards invisible to the prisoners who were being observed. The purpose of the panopticon was to allow seeing without being seen. Here "inspection functions ceaselessly. The gaze is alert everywhere" (Foucault, 1975/1995, p. 195).

Foucault isn't really interested in the panopticon as such. Rather, he sees the idea of the panopticon as illustrative of a shift in the fundamental way people thought and the way in which power is practiced. In terms of crime and punishment, it involved a shift from the spectacle of torture (which fit well with monarchical power) to regulation in prison (which fits well with the nation-state); from seeing crime as an act against authority to viewing it as an act against society; from being focused on guilt (did he [or she] do it?) to looking at cause (what social or psychological factors influenced the person?); and, most importantly, from punishment to discipline—more specifically, to the self-discipline imposed by the ever-present but unseen surveillance of the panopticon.

Foucault (1975/1995) refers to this kind of control as the *microphysics of power* and sees this as the explicit link between knowledge and power: "There is no power relation without the correlative constitution of a field of knowledge, nor any knowledge that does not presuppose and constitute at the same time power relations" (p. 27). The microphysics of power is exercised or practiced as knowledge is produced, appropriated by groups for use, distributed to the population through education and mass media (such as books, magazines, and the Internet), and then retained internally by those that others want to control.

Obviously, all of society was not put into a physical panopticon, but society was placed within a symbolic or institutional system of surveillance. In another word-play, Foucault argues that the discipline associated with panopticon surveillance of the entire population comes from the "disciplines," in particular the human sciences. The modern episteme created the possibility of the human sciences, such as psychiatry, psychology, and sociology. The human has been the subject of thought and modes of control for quite some time, but in every case the human was seen holistically or as part of the universal scheme of things. In the modern episteme, however, mankind becomes the object of study—not as part of an aesthetic whole, but as a thing in its own right.

This discourse of science serves to objectify and control the individual. Psychiatry and psychology used the mechanical model of the universe to gaze inside the psyche of the person; sociology and political science looked at the external circumstances of humanity. Thus, the internal motivations and reasons behind action as well as the external factors became the objects of science in order to fulfill the chief goal of science, which is control. Statistics are used to quantify and categorize; psychotherapy and psychological testing are used to probe and catalog; all of the disciplines and their methodologies are brought into "discipline" in order to fulfill the primary goal of science: to control.

Foucault (1982) finds the human sciences particularly interesting because they are "modes of inquiry which try to give themselves the status of sciences" (p. 208). The human sciences are thus not true science; they only take on the guise of science. The human sciences did not grow out of scientific questions; they grew out of the modern episteme. Simply put, during the time that people began to talk about society and psychology, the kind of knowledge that was seen as real and valuable was science. So, in order to be accepted, the social and behavioral disciplines had to take on the guise of science.

More specifically, Foucault argues that there are three areas of knowledge in the modern episteme: mathematical and physical sciences, life and economic sciences, and philosophy. The human sciences grew out of the space created by these three knowledge systems. Asking scientific questions about things like biology and physics, which have some basis in the objective world, set the stage for those same questions to be asked about the questioner. Further, each of these sciences pursues knowledge in a distinctive manner, each with its own logic. The human sciences, on the other hand, must borrow from each of these because it has no unique domain or methodology. The human sciences stand in relation to all the other forms of knowledge . . . at one level or another, [they use] mathematical formalization; they proceed in accordance with models or concepts borrowed from biology, economics, and the sciences of language; and they address themselves to that mode of being of man which philosophy is attempting to conceive. (Foucault, 1966/1994b, p. 347)

Therefore, the precariousness or uncertainty of the human sciences isn't due to, "as is often stated, the extreme density of their object" (Foucault, 1966/1994b, p. 348); rather, their uncertainty of knowledge is due to the fact that they have no true method of their own—everything is borrowed. The validity of knowledge is in some way always related to methodology. *What* we know is an effect of *how* we know. Because the human sciences don't have their own methodology, the knowledge generated is without any basis—in the end, it is purely an expression of power that can be explicitly used by the state to control populations but is more generally part of the control people exercise over themselves in modernity.

As such, we generally see and understand ourselves in Western cultures from the human science model. We listen endlessly to public opinion polls and voting predictions, and they become constant topics of conversation for us. We understand the family in terms of such psychosocial models as the "functional family," and we raise our children according to the latest findings. Almost everything that we think, feel, and do is scrutinized by a human science, and we are provided with that knowledge so that we too can understand our own life and its circumstances.

But the human sciences are not alone in their objectification of humanity; they are aided by a culture produced by the medical gaze. The modern medical gaze is different from that of the eighteenth century. At that time, disease was organized into hierarchical categories such as families, genera, and species. The doctor's gaze was directed not so much at the patient as at the disease—the patient was in some ways superfluous. Diseases transferred to the body when their makeup combined with certain qualities of the patient, such as his or her temperament. Symptoms existed within the disease itself, not the patient. This way of seeing where symptoms live implies that the patient's body could actually get in the way of the doctor seeing the symptoms. For example, if the patient was old, then the symptoms associated with being elderly could obscure the doctor's view of the symptoms associated with the disease. The medical gaze, then, was directed at the disease, not the body.

However, by the nineteenth century, the modern medical gaze had come to locate disease within the patient. Disease was no longer seen to exist within its own world apart from the body; from this new clinical point of view, the patient can't get in the way of the symptoms because the symptoms and disease are the same and exist within the body. This shift in discourse created the *clinical gaze,* an objectifying way of seeing that looks within and dissects the patient. With the clinical gaze, "Western man could constitute himself in his own eyes as an object of science" (Foucault, 1963/1994a, p. 197).

Modern medicine is thus created through a gaze that makes the body an object, a thing to be dissected, either symbolically or actually, in order to find the disease within it. The culture of the clinical gaze helped to create a general disposition in

Western society to see the person as an object. This disposition, along with the human sciences, made the practices of power much more effective and treacherous—objects that can be thrown away are much easier to control than subjects that demand continuing emotional and psychic connections.

Concepts and Theory: Power Over the Subject

Thus, bodily regimens of exercise and diet, self-understanding, and regulation of feelings and behaviors all stem from medicine and the human sciences, which Foucault tells us make up the panopticon of modernity. But Foucault is interested in something deeper than the control of the body—he wants to document how we as individuals exercise social power over the way we relate to our own selves. Nowhere is this more clearly seen than in Foucault's counter-history of sexuality. In order to understand Foucault's intent, we will now briefly review Greek and modern ideas of sexuality.

Greek Sexuality

Ancient Greece was the birthplace of democracy and Western philosophy. There was, in fact, a connection between the two. In Athens, in response to an upheaval by the masses against their tyrannical leader, Isagoras, a politician named Cleisthenes introduced a completely new organization of political institutions called democracy (the rule of common people). Through democratic elections, the elite incrementally lost their advantage in the assemblies and the common people ruled. Unfortunately, the masses were susceptible to impassioned speech and ended up making several decisions that conflicted with one another or entailed high costs. In response, philosophers and the politically deposed elite began to search for absolute truth. To them, truth obviously couldn't be found simply through rhetoric; they believed there had to be some absolutes upon which decisions could be based.

Along with other factors, this impetus helped produce the Greek notion of the soul. For the Greek, the idea of the soul captured all that is meant by the inner person: the individual's mind, emotions, ethics, beliefs, and so on. But in reading Plato, it's also clear that the soul was seen to be hierarchically constructed. Within the soul, the mind is preeminent and alone is immortal. The emotions and appetites, though part of the soul, are of lesser import and are mortal. Thus, reason is godlike and education, especially philosophy, is essential for proper discipline.

It is important that we see the emphasis here. The mind, emotions, and bodily appetites are viewed hierarchically, but they are all seen as part of the soul. In order to get a sense of the relationships within the soul, let's take a look at a conversation that Plato (1993) sets up between Socrates and a group of students. These conversations are part of what are more generally referred to as the Socratic dialogues, a literary genre that emerged sometime around the turn of the fourth century BCE. Socrates speaks first:

"Do you think that it's a philosopher's business to concern himself with what people call pleasures—food and drink, for instance?"

"Certainly not, Socrates," said Simmias.

"What about those of sex?"

"Not in the least." . . .

"Then it is your opinion in general that a man of this kind is not preoccupied with the body, but keeps his attention directed as much as he can away from it and towards the soul?"

"Yes, it is." . . .

"Then when is it that the soul attains to truth? When it tries to investigate anything with the help of the body, it is obviously liable to be led astray."

"Quite so."

"Is it not in the course of reasoning, if at all, that the soul gets a clear view of reality?"

"Yes." (pp. 117–118)

Notice how Socrates views sex: it isn't something set aside and special. It is simply seen as a bodily appetite, on a par with eating and drinking. And these aren't a direct concern for the philosopher. If the bodily appetites get in the way of the search for reality or truth, only then are they of concern. The point is to keep the mind free. A person shouldn't be preoccupied with the body, because too much attention on the body and its appetites will take his or her attention away from the quest for truth. This bit of dialogue sets us up well for the way Foucault talks about sex in Greek society.

In Greek society, sexuality existed as *aphrodisia*. This Greek word is obviously where we get our term "aphrodisiac," but it had a much broader meaning for the Greeks. Foucault notes that neither the Greeks nor the Romans had an idea of "sexuality" or "the flesh" as distinct objects. When we think of sex, sexuality, or the flesh, we usually have in mind a single set of behaviors or desires. The Greeks, while they had words for different kinds of sexual acts and relations, didn't have a single word or concept under which they could all fit. The closest to that kind of umbrella term is *aphrodisia*, which might be translated as "sensual pleasures" or "pleasures of love," and more accurately the works and acts of Aphrodite, the goddess of love.

These works of Aphrodite, perhaps like the works of any god or goddess, cannot be fully categorized. To do so would limit the god. This lack of a catalog or objective specification of sexuality is exactly Foucault's point. In modern, Western society, particularly as expressed through Christianity, there is a definite way to index those things that are sexual, or the "works of the flesh." This identifiability is extremely important for the Western mind because sex is a moral issue; it, above all other things, defines immoral practices. So, what counts and doesn't count as sexual is imperative for us, but it wasn't for the Greeks.

The Greeks also employed the idea of *chrēsis aphrodisiōn* to sexuality: the phrase means "the use of pleasures." The Greeks' use of pleasure was guided by three strategies: need, timeliness, and status. The strategy of need once again highlights Socrates's approach to sexual practices. As we've seen, in Ancient Greece, the relationship to one's body was to be characterized by moderation, but every person's appetites and abilities to cope are different. Thus, the Greek strategy was for the individual to first know his need—to understand what the body wants, what its limits are, and how strong the mind is.

The second strategy is timeliness and simply refers to the idea that there are better and worse times to have sexual pleasures. There was a particularly good time in one's life, neither too young nor too old; a good time of the year; and good times during the day, usually connected with dietary habits. The issue of time "was one of the most important objectives, and one of the most delicate, in the art of making use of the pleasures" (Foucault, 1984/1990b, p. 57). The last strategy in the use of pleasures was status. The art of pleasure was adapted to the status of the person. The general rule was that the more an individual was in the public eye, the more he should "freely and deliberately" adapt rigorous standards regarding his use of pleasures.

Rather than seeing sexuality as moral, the Greeks saw it in terms of ascetics. *Ascetics* refers to one's attitude or relationship toward one's self, and for the Greek this was to be characterized through strength. The word comes from the Greek *askētikos,* which literally translated means exercise. The idea here is not simply something we do, as in exercising control; it also carries with it a picture of active training. Here we see the Greek link between masculinity and virility. The virile man in Greek society was someone who moderated his own appetites. He was the man who voluntarily wrestled with the needs of his body in order to discipline his mind. The picture we see is that of an athlete in training. For example, the athlete knows that eating chocolate or ice cream can be very pleasurable. But while in training, the athlete willingly forgoes those pleasures for what she or he sees as a higher good. The result of this training is *enkrateia,* the mastery of one's self. It's a position of internal strength rather than weakness.

Training is always associated with a goal; there is an end to be achieved or a contest to be won. In this case, the aim of the Greek attitude toward sexuality is a state of being, something that becomes true of the individual in the person's daily life. This is the *teleology* or ultimate goal of sexuality, the fourth structuring factor that defines a person's relationship to sex. The goal for the Greek was freedom. We can again see this idea in the conversation with Socrates. Truth and reality were things to be sought after. Too much emphasis on sex, just like eating and drinking, can get in the way of this search. As Socrates (Plato, 1993) said, "surely the soul can reason best when it is free of all distractions such as hearing or sight or pain or pleasure of any kind" (p. 118).

Modern Western Sexuality

The Western, modern view of sex is quite different from the Greek. It is, in fact, quite different from that which developed in the East. Where Eastern philosophy and religion developed a set of practices intended to guide sexual behavior to its highest and most spiritual expression and enjoyment (for instance, Kama Sutra), the West developed systems of external control and prohibitions. Of course, a great deal of the impetus toward this view of sex was provided by the Christian church.

Part of this movement came from Protestantism with its emphasis on individual righteousness and redemption. Rather than being worthy of God because of church membership and sacraments, Protestantism singled the individual out and made her or his moral conduct an expression of salvation and faith. But an important part was also played by the Counter-Reformation, a reform movement in the Catholic Church.

Confession and penance are sacraments in the Catholic Church. They are one of the ways through which salvation is imparted to Christians. The Counter-Reformation increased the frequency of confession and guided it to specific kinds of self-examination, designed to root out the sins of the flesh down to the minutest detail:

> sex . . . [in all] its aspects, its correlations, and its effects must be pursued down to their slenderest ramification: a shadow in a daydream, an image too slowly dispelled, a badly exorcised complicity between the body's mechanics and the mind's complacency: everything had to be told. (Foucault, 1976/1990a, p. 19)

This was the beginning of the Western idea that sex is a deeply embedded power, one that is intrinsic to the "flesh" (the vehicle of sin par excellence, as compared to the Greek idea of bodily appetites), and one that must be eradicated through inward searching using an external moral code and through outward confession.

While these Christian doctrines would have influenced the general culture, they would have remained connected to the fate of Christianity alone had it not been for other secular changes and institutions beginning in the eighteenth century, most particularly in politics, economics, and medicine. With the rise of the nation-state and science, population became an economic and political issue. Previous societies had always been aware of the people gathered together in society's name, but conceiving of the people as the *population* is a significant change. The idea of population transforms the people into an object that can be analyzed and controlled.

In this transformation, science provided the tools and the nation-state the motivation and control mechanisms (taxation, standing armies, and so on). The population could be numbered and analyzed statistically, and those statistics became important for governance and economic pursuit. The population represented the labor force, one that needed to be trained and, more fundamentally, born. At the center of these economic and political issues was sexuality:

> it was necessary to analyze the birthrate, the age of marriage, the legitimate and illegitimate births, the precocity and frequency of sexual relations, the ways of making them fertile or sterile, the effects of unmarried life or of the prohibitions, the impact of contraceptive practices [and so on]. (Foucault, 1976/1990a, pp. 25–26)

In the latter half of the nineteenth century, medicine and psychiatry took up the sex banner as well. Psychiatry, especially through the work of Freud, set out to discover the makeup of the human mind and emotion, and it began to catalog mental illnesses, especially those connected with sex. It conceptualized masturbation as a perversion at the core of many psychological and physical problems, homosexuality as a mental illness, and the maturation of a child in terms of successive sexual issues that the child must resolve on the way to healthy adulthood. In short, psychiatry "annexed the whole of the sexual perversions as its own province" (Foucault, 1976/1990a, p. 30). Law and criminal justice also bolstered the cause, as society sought to regulate individual and bedroom behaviors. Social controls popped up everywhere that "screened the sexuality of couples, parents and children, dangerous and endangered adolescents—undertaking to protect, separate, and forewarn, signaling perils everywhere, awakening people's attention, calling for diagnoses, piling up reports, organizing therapies. These sites radiated discourses aimed at sex" (pp. 30–31).

All of these factors worked to change the discourse of Western sexuality in the twentieth century. Sex went from the Greek model of a natural bodily appetite that could be satisfied in any number of ways, to the modern model of sex as the insidious power within. At the heart of this change is the confession, propagated by Catholicism and Protestantism and picked up by psychiatrists, medical doctors, educators, and other experts. Confessional rhetoric is found everywhere in a modern society that uses Victorian prudishness as its backdrop for incessant talk about sex in magazines, journals, books, movies, and reality television shows. Notice what Foucault is saying: repression is used as a source of discourse, and sex has become the topic of conversation—a central feature in Western discourse, and the defining feature of the human animal. Sex is suspected of "harboring a fundamental secret" concerning the truth of mankind (Foucault, 1976/1990a, p. 69).

In the modern discourse of sex, sexuality has become above all an object—a truth to discover and a thing to control. In this, sex has followed the use and development of science in general and the human sciences in particular: "the project of a science of the subject has gravitated, in ever narrowing circles, around the question of sex" (Foucault, 1976/1990a, p. 70). This form of objective control ("bio-control") over the intimacy of humanity came through science and is linked with the development of the nation-state and capitalism. While capitalism and the nation-state seem to be firmly established and the need for such control not as great, what we are left with is a way of constructing our self as the moral subject of our sexual behavior. We have inherited a certain kind of subjectivity from this discourse, a particular way of relating to our self and sexuality. This legacy of the modern discourse of sexuality sees sex as a central truth of the self, as an object that must be studied and understood. Further, the modern discourse of sexuality tells us that this part of us is intrinsically dangerous. It is at best an amoral creature and at worst a defiling beast that treads upon sacred and moral ground.

Foucault Summary

- Foucault takes the position that knowledge and power are wrapped up with one another; each produces and reinforces the other. Power as exercised and expressed through discourse creates the way in which we feel, act, think, and relate to our self. Likewise, the knowledge of any epoch defines what is mentally, emotionally, and physically possible. Foucault sees the practices involved in power and knowledge as games of truth—the use of specific rules and resources through which something is seen as truth in any given age. The games of truth that Foucault is particularly interested in are the ones that involve the practices through which we participate in the domination of our subjectivity.

- Much of Foucault's work is in the form of counter-history. The generally accepted model is that of history as memory: history is our collective memory of events. We also usually think of history as slowly progressing in a linear fashion. Foucault argues that history is far from a memory of linear events—it is power in use. It's a myth that is constructed according to specific values. Foucault proposes a counter-history, one that focuses on abrupt episodes of change and the way in which knowledge changes in response to various power regimes. Foucault uses an archaeological approach to uncover the practices that are associated with discourse and ways of thinking; and he uses a genealogical approach to uncover how discourse and power are inscribed on the body and mind.

- Foucault argues that the knowledge people hold is based upon historical epistemes, or underlying orders. Foucault uses the term "episteme" to refer to the way thought organizes itself in any historical period of time. Discourses are produced within historical epistemes. A discourse is a way of talking about something that is guided by specific rules and practices, that sets the conditions for our subjective awareness, and that subjugates through a will to truth and power.

- While power is found in all human practice, Foucault is particularly interested in the unique power of modernity. This expression of power is associated with changes in government, medicine, the institution of the human sciences, changes in the Western discourse of sexuality, and changes in the penal system. The change from rule by monarchy to rule by the nation-state demanded a new form of governmentality, one in which the individual watches over his or her own behaviors, and one that increases control while preserving the illusion of freedom. This governmentality was aided by the human sciences through the idea of population, an essentializing and mechanistic model of the person, and the value of expert knowledge. Governmentality was also produced through a new medical "gaze," which located symptoms and disease within the body; panoptical practices in controlling criminality; and changes in sexuality promoted by the Catholic confessional and Protestant individualism. Together, these created a discourse of governmentality that objectifies and controls the individual through her or his own practices.

Jean Baudrillard: The End of Everything

Photo: Copyright: Res Stolkiner. European Graduate School EGS, Saas-Fee, Switzerland, 2002.

The Essential Baudrillard

Biography

Jean Baudrillard was born in Reims, France, on July 29, 1929. Baudrillard studied German at the Sorbonne University, Paris, and was professor of German for eight years. During that time, he also worked as a translator and began his studies in sociology and philosophy. He completed his dissertation in sociology under Henri Lefebvre, a noted Marxist-humanist. Baudrillard began teaching sociology in 1966, eventually moving to the Université Paris-X Nanterre as professor of sociology. From 1986 to 1990, Baudrillard served as the director of science for the Institut de Recherche et d'Information Socio-Économique at the Université Paris-IX Dauphine. Since 2001, Baudrillard has been professor of the philosophy of culture and media criticism at the European Graduate School in Saas-Fee, Switzerland. He is the author of several international best-sellers, the most controversial of which is probably *The Gulf War Did Not Take Place* (1995).

Passionate Curiosity

Baudrillard is essentially concerned with the relationship between reality and appearance, an issue that has plagued philosophers and theorists for eons. Baudrillard's unique contributions concern the effects of mass media and advertising. He is deeply curious about the effects of mass media on culture and the problem of representation: Has capitalist-driven mass media pushed appearance to the front stage in such a way as to destroy reality? Is there a difference between image and reality in postmodernity?

Keys to Knowing

Postmodernism, symbolic exchange, excess, sign-value, fetish, four sign stages, consumer society, information entropy, sign-value and sign-vehicles, mass media, advertising, labor of consumption, simulacrum, hyperreality, free-floating signifiers, sign implosion, death of the subject, neo-tribes, play and spectacle

Seeing Further: Cultural Postmodernism

Baudrillard is both fun and frustrating, and usually at the same time. He, more than any other postmodern writer I know, exemplifies what he is writing about. "Baudrillard's writings are a kind of intellectual Disneyland, neither true nor false" (Danto, 1990, p. 48). In other words, Baudrillard is more than a postmodern theorist; he is part of the postmodern landscape. Just reading Baudrillard is an experience in postmodernity and cultural implosion. Yet at the same time, he tells us something about the society and people living in postmodernity, something very few say with as much insight or art. Baudrillard's writings are thus part play and part purpose.

Defining Postmodernism

"Postmodernism is . . ." and with that statement we have already run into trouble. One of the problems that postmodernism wants to point out is that of reference. The problem of reference has been around since people began to think about the nature and structure of language. We can phrase it like this: Does or can language represent the physical world? For example, does the word "tree" represent the physical object, or does it represent our *idea* of tree? In the first case, the word references the physical object, and in the other, it references our culture. One of the implications of this question concerns reality: Is human reality founded upon the physical world or is reality simply cultural?

For postmodernists, the problem of reference is much deeper. Previously, even if someone said that language only represents culture, culture was seen as a firm and cohesive base. For example, Georg Wilhelm Friedrich Hegel (the philosopher that Karl Marx drew on to theorize) basically argued that language references culture, but he added that true ideas come from God. For Hegel, then, language has an even firmer base than physicality. For postmodernists, however, while language references culture, culture in postmodernity is fragmented and free-floating and thus has no reference at all. This state of affairs generally leaves authors such as me in a pickle: how can we say anything about something that denies its ability to say anything?

That state of affairs is part of both the frustration and fun that one can have with the idea of postmodernism. If we accept that words don't have any specific cohesive references, then we either become dark in our writings and thoughts or we can be playful, intentionally using words and references that have multiple and perhaps contradictory meanings. A good illustration of the latter is Italian writer Umberto Eco. His novels contain words and ideas that are possibly contradictory and certainly require the reader to intellectually and emotionally engage the text; his books are thus intended to be open to interpretations rather than limited by clear, linear plots and text. However, for the most part, I'm going to proceed with our discussion in a fairly straightforward manner, which I admit up front is doomed to fail. Thus, we're going to say something definite and hopefully meaningful about Baudrillard and postmodernism in this chapter. We're going to focus on the purposeful part of his writings—but keep in mind the playful part too (think about it: I'm proposing to say something meaningful about meaninglessness!).

Postmodernity in Modernity

The word "postmodernism" was first used in Hispanic literary criticism in the 1930s, it gained currency in the visual arts and architecture by the 1960s and early 1970s, and it made its way into social theory with the French publication of Jean-François Lyotard's *The Postmodern Condition* in 1979 (see Anderson, 1998). As I said, postmodernism is a word that denies its ability to function as a word. There does, however, seem to be one organizing feature: no matter what field we're talking about, postmodernism is always understood in contrast to modernism.

Modernity and postmodernity are terms that are used in a number of disciplines. In each the meaning is somewhat different, but it is also generally the same: modernity is characterized by unity, and postmodernity is distinguished by disunity. In modern literature, for example, the unity of narrative is important. Novels move along according to their plot, and while there might be twists and turns, most readers are fairly confident in how time is moving and where the story is going. Postmodern literature, on the other hand, doesn't move in predictable patterns of plot and character. Stories will typically jump around and are filled with indirect and reflexive references to past styles or stories. The purpose of a modern novel is to convey a sense of continuity in story; the purpose of a postmodern work is to create a feeling of disorientation and ironic humor.

In the social world, modernity is associated with the Enlightenment and defined by progress; by grand narratives and beliefs; and through the structures of capitalism, science, technology, and the nation-state. To state the obvious, postmodernism is the opposite or critique of all that. According to most postmodern thought, things have changed. Society is no longer marked by a sense of hope in progress. People seem more discouraged than encouraged—more filled with a blasé attitude than optimism.

In a late capitalist society, rather than providing a basis of meaning, "culture has necessarily expanded to the point where it has become virtually coextensive with the economy itself . . . as every material object and immaterial service becomes [an] inseparably tractable sign and vendible commodity" (Anderson, 1998, p. 55). Thus, capitalism has colonized culture and turned meaning into goods that are bought and sold. At the same time, these processes make culture less real. As we saw in the quote from Baudrillard at the beginning of this chapter, signs in advanced capitalism don't represent anything; they have no utility or reality. They are commodified images that serve little more function than to seduce us into consumerism.

The same is true for self and identity: "as an older industrial order is churned up, traditional class formations have weakened, while segmented identities and localized groups, typically based on ethnic or sexual differences, multiply" (Anderson, 1998, p. 62). However, these segmented relationships "pull us in a myriad directions, inviting us to play such a variety of roles that the very concept of an 'authentic self' with knowable characteristics recedes from view. The fully saturated self becomes no self at all" (Gergen, 1991, p. 7).

Taken together, then, we can define social postmodernism as a critical form of theorizing that is concerned with the unique problems associated with culture and the subject in advanced capitalistic societies.

Concepts and Theory: Mediating the World

Pre-Capitalist Society and Symbolic Exchange

Baudrillard's (1981/1994) theory is based on a fundamental assumption: "Representation stems from the principle of the equivalence of the sign and of the real" (p. 6). What Baudrillard means is that it is possible for signs to represent reality, especially social reality. Think about it this way: what is the purpose of culture? In traditional social groups, culture was created and used in the same social context. For the sake of conversation, let's call this "grounded culture," the kind of culture that symbolic exchange is based on. In a society such as the United States, however, much of our culture is created or modified by capitalists, advertising agencies, and mass media. We'll call this "commodified culture."

Members in traditional social groups were surrounded by grounded culture; members in postmodern social groups are surrounded by commodified culture. There are vast differences in the reasons why grounded versus commodified culture is created. Grounded culture emerges out of face-to-face interaction and is intended to create meaning, moral boundaries, norms, values, beliefs, and so forth. Commodified culture is produced according to capitalist and mass media considerations and is intended to seduce the viewer to buy products. With grounded culture, people are moral actors; with commodified culture, people are consumers. Postmodernists argue that there are some pretty dramatic consequences, such as cultural fragmentation and unstable identities (for a concise statement, see Allan & Turner, 2000).

Beginning with this idea of grounded or representational culture, Baudrillard posits four phases of the sign. The first stage occurred in premodern societies. The important factor here is that language in premodern societies was not mediated. There were little or no written texts and all communication took place in real social situations in face-to-face encounters. In this first phase, the sign represented reality in a profound way. There was a strong correlation or relationship between the sign and the reality it signified, and the contexts wherein specific signs could be used were clear. In this stage, all communicative acts—including speech, gift giving, rituals, exchanges, and so on—were directly related to and expressive of social reality, in something that Baudrillard calls symbolic exchange.

Baudrillard understands **symbolic exchange** as the exchange of gifts, actions, signs, and so on for their symbolic rather than material value. In contrast to Marx, Baudrillard is making the same kind of argument about human nature that Durkheim did: humans are symbolic creatures oriented toward meaning rather than production. A good example of the value of symbolism in traditional societies is the prevalence of transition rituals, as in the transition from boyhood to manhood. In the ritual of attaining manhood, the actual behaviors themselves are immaterial, whether it is wrapping a sack of fire ants around the hands or mutilating the penis. Any object or set of behaviors can have symbolic value. What matters is what the ritualized actions symbolize for the group. Symbolic exchange formed part of daily life in pre-capitalist societies: the exchange of food, jewelry, titles, clothing, and so on were all involved in a symbolic "cycle of gifts and countergifts" (Baudrillard, 1973/1975, p. 83). Symbolic exchange thus established a community of symbolic meanings and reciprocal relations among a group of people.

Baudrillard also claims that human nature is wrapped up in excess. Like Marx, he sees humanity as capable of creating its own needs. That is, the needs of other animals are set, but the potential needs of humans are without limit. For example, today I "need" an iPod and an HDTV plasma screen, but a few years ago I didn't. And I can't even begin to imagine what I will need five years from now. However, unlike Marx, Baudrillard sees excess as an indicator of human boundlessness. Wrapped up in this excess is a sense of transcendence: the ability to reach above the mundane, which is in itself a symbolic move.

Thus, in place of Marx's species-being, Baudrillard proposes excess: rather than being bound up with survival, production, and materialism, human nature is found in excess and exuberance. Douglas Kellner (2003) summarizes Baudrillard's point of view nicely: "humans 'by nature' gain pleasure from such things as expenditure, waste, festivities, sacrifices and so on, in which they are sovereign and free to expend the excesses of their energy (and thus follow their 'real nature')" (p. 317). Baudrillard argues that these human characteristics were given license and support in pre-capitalist societies. However, the modernist demands of rationality and restraint in the beginning years of capitalism are the antithesis of symbolic exchange and excess.

The second phase of the sign marked a movement away from these direct kinds of symbolic relationships. This stage gained dominance, roughly speaking, during the time between the European Renaissance and the Industrial Revolution. While media such as written language began previous to the Renaissance, it was during this period that a specific way of understanding, relating to, and representing the world became organized. Direct representation was still present, but certain human ideals began to make inroads. Art, for example, was based on observation of the visible world and yet contained the values of mathematical balance and perspective. Nowhere is this desire for mathematical balance seen more clearly than in Leonardo Da Vinci's painting, *Proportions of the Human Figure.* Thinking of some of Da Vinci's other works, such as the *Mona Lisa* and the *The Last Supper,* we can also see that symbols were used to convey mystery and intrigue.

The Dawn of Capitalism and the Death of Meaning

The third phase of the sign began with the Industrial Revolution. This is the period of time generally thought of as modernity. The Industrial Age brought with it a proliferation of consumer goods never before seen in the history of humanity. It also increased leisure time and produced significant amounts of discretionary funds for more people than ever before. These kinds of changes dramatically altered the way produced goods were seen. Here is where we begin to see the widespread use of goods as symbols of status and power. Thorstein Veblen (1899) termed this phenomenon *conspicuous consumption.*

Baudrillard (1970/1998) characterizes this era as the beginning of the **consumer society.** The consumer society is distinctly different from the kind of capitalism that Marx saw himself critiquing. One of Marx's main criticisms was exploitation, and exploitation, you will remember, is based on use-value and exchange-value. *Use-value* refers to the actual function that a product contains, or its material makeup, while *exchange-value* refers to the rate of exchange one commodity bears when compared to other commodities. The interesting thing for Marx is that when reduced to monetary value, exchange-value is much higher than use-value. In other words, you get paid less to produce a product than it sells for, which is exploitation.

Baudrillard counters by arguing that Marx is ironically buying into the basic assumptions of capitalism. Use-value is completely bound up with the idea of products, oriented to a materialist world alone. It is filled with practical use that is used up in consumption and has no value or meaning other than material. Moreover, like species-being, the idea of use-value validates the basic tenets of capitalism—the truth of human life is rooted in economic production and consumption. Exchange-value is materialist as well, because exchange-value is based on human, economic production. The idea of exchange-value also legitimates and substantiates instrumental rationality, the utilitarian calculations of costs and benefits. Rather than critiquing capitalism, Baudrillard sees Marx as legitimating it.

> The Marxist seeks a *good use* of economy. Marxism is therefore only a limited petit bourgeois critique, one more step in the banalization of life toward the "good use" of the social! . . . Marxism is only the disenchanted horizon of capital—all that precedes or follows it is more radical than it is. (Baudrillard, 1987, p. 60, emphasis original)

In place of use- and exchange-values, Baudrillard proposes the idea of sign-value. Commodities are no longer purchased for their use-value, and exchange-value is no longer simply a reflection of human labor. Each of these capitalist, Marxist values has been trumped by signification. In postmodern societies, commodities are now purchased and used more for their sign-value than for anything else.

Baudrillard links sign-value with fetish, another idea from Marx. As you may recall, Marx was very critical of the process of commodification. A commodity is simply something that is sold in order to make a profit. Commodification as a process refers to the way more and more objects and experiences in the human world are turned into products for profit. Increasing commodification leads to *commodity fetish*. Marx used the term *fetish* in its pre-Freudian sense of idol worship. The idea here is that the worshipper's eyes are blinded to the falsity of the idol. Marx's provocative term has two implications that are related to one another. First, in commodity fetish people misrecognize what is truly present within a commodity. By this Marx meant that commodities and commodification are based on the exploitation of human labor, but most of us fail to see it.

Second, in commodity fetish there is a substitution. For Marx, the basic relationship between humans is that of production. But in commodity fetish, the market relations of commodity exchange are substituted for the productive or material relations of persons. The result is that, rather than being linked in a community of producers, human relationships are seen through commodities, either as buyers and sellers or as a group of like consumers. Commodification and its fetish are one of the primary bases of alienation, which, according to Marx, separates people from their own human nature as creative producers and from one another as social beings.

Again, Baudrillard argues that Marx's concern was misplaced and actually motivated by the capitalist economy. Marx's entire notion of the fetish is locked up with species-being and material production. With commodity fetish, we don't recognize the suppressive labor relations that underlie the product and its value, and we substitute an alienated commodity for what should be a product based in our own species-being. In focusing exclusively on materialism, "Marxism eliminates any real chance it has of analyzing the *actual process of ideological labor*" (Baudrillard, 1972/1981, p. 89, emphasis original). According to Baudrillard, ideology isn't based in or related to material relations of production, as Marx argued. Rather, ideology and fetishism are both based in a *"passion for the code"* (p. 92, emphasis original).

Human nature is symbolic and oriented toward meaning. In symbolic exchange, real meaning and social relationships are present. However, capitalism and changes in media have pushed aside symbolic exchange and in postmodernity have substituted sign-value. Moreover, sign-value is based on textual references to other signs, nothing else. The fetish, then, is the human infatuation with consuming sign-vehicles that are devoid of all meaning and reality. Thus, ideology "appears as a sort of cultural surf frothing on the beachhead of the economy" (Baudrillard, 1972/1981, p. 144). Signs keep proliferating without producing substance. This simulation of meaning is what constitutes ideology and fetish for Baudrillard. This implies that continuing to use the materialist Marxist ideas of fetish and ideology actually contributes to capitalist ideology, because it displaces analysis from the issues of signification.

Thus, in the consumer society, social relations are read through a system of commodified signs rather than symbolic exchange. Commodities become the *sign-vehicles* in modernity that carry identity and meaning (or its lack). For example, in modern society the automobile is a portable, personal status symbol. Driving an SUV means something different from driving a Volkswagen Beetle, which conveys something different from driving a hybrid. As this system becomes more important and elaborated, a new kind of labor eventually supplants physical labor, *the labor of consumption*. This doesn't mean the work involved in finding the best deal. The labor of consumption is the work a person does to place her- or himself within, or to "read" the signs of an identity that is established and understood in, a matrix of commodified signs.

The dynamics begun in the third stage of the sign are exacerbated in the fourth, which began shortly after WWII and continues through today. This fourth stage occurs in postindustrial societies. As such, there has been a shift away from manufacturing and toward information-based technologies. In addition, and perhaps more importantly for Baudrillard, these societies are marked by continual advances and an increasing presence of communication technologies and mass media. Mediated images and information, coupled with unbridled commodification and advertising, are the key influences in this postmodernity. The cultural logic has shifted from the logic of symbolic exchange in pre-capitalist societies, to the logic of production and consumption in capitalist societies, and finally now to the logic of simulation. For Baudrillard (1976/1993b), postmodernity marks the end of everything:

> The end of labor. The end of production. . . . The end of the signifier/signified dialectic which facilitates the accumulation of knowledge and of meaning. . . . And at the same time, the end simultaneously of the exchange value/use value dialectic which is the only thing that makes accumulation and social production possible. . . . The end of the classical era of the sign. (p. 8)

Baudrillard (1981/1994) posits that the postmodern sign has "no relation to any reality whatsoever: it is its own pure simulacrum" (p. 6). These kinds of signs are set free from any constraint of representation and become a "play of signifiers . . . in which the code no longer refers back to any subjective or objective 'reality,' but to its own logic" (Baudrillard, 1973/1975, p. 127). Thus, in postmodernity, a fundamental break has occurred between signs and reality. Signs reference nothing other than themselves; they are their own reality and the only reality to which humans refer. These seem like brash and bold claims, but let's look at Baudrillard's argument behind them.

Concepts and Theory: Losing the World

Entropy and Advertising

First, Baudrillard argues that there is something intrinsic in transferring information that breaks it down. This is an important point: Anytime we relate or convey information to another, there is a breakdown. So fundamental is this fact that Baudrillard makes it an equation: information = entropy. Baudrillard argues that information destroys its base. The reason for this is twofold. First, information is always *about* something; it isn't that thing or experience itself. Information by definition, then, is always something other than the thing itself. Second, anytime we convey information, we must use a medium, and it is impossible to put something through a medium without changing it in some way. Even talking to your friend about an event you experienced changes it. Some of the meaning will be lost because language can't convey your actual emotions, and some meaning will be added because of the way your friend individually understands the words you are using.

Mass media is the extreme case of both these processes: social information is removed innumerable times from the actual events, and capitalist mass media colors things more than any other form. One of the reasons behind this coloration is that mass media expends itself on the staging of information. Every medium has its own form of expression—for newspapers it's print and for television it's images. Every piece of information that is gleaned by the public from any media source has thus been selected and formed by the demands of the media. This is part of what is meant by the phrase "the media is the message."

Further, mass communication comes prepackaged in a meaning form. What I mean by that is that information is staged and the subject is told what constitutes his or her particular relationship to the information. The reason for this is that media in postmodernity exist to make a profit, not to convey information.

"Information" is presented more for entertainment purposes than for any intellectual ones. The concern in media is to appeal to and capture a specific market segment. That's why Fox News, CNN, and National Public Radio are so drastically different from one another—the information is secondary; the network's purpose is to draw an audience that will respond to appeals by capitalists to buy their goods. The presentation of information through the media is a system of self-referencing simulation or fantasy. Thus, "information devours its own content. It devours communication and the social" (Baudrillard, 1981/1994, p. 80).

There is yet another important factor in this decisive break—advertising: "Today what we are experiencing is the absorption of all virtual modes of expression into that of advertising" (Baudrillard, 1981/1994, p. 87). The act of advertising alone reduces objects from their use-value to their sign-value. For Marx, the movement from use-value to exchange-value entails an abstraction of the former; in other words, the exchange-value of a commodity is based on a representation of its possible uses. Baudrillard argues that advertising and mass media push this abstraction further. In advertising, the use-value of a commodity is overshadowed by a sign-value. Advertising does not seek to convey information about a product's use-value; rather, advertising places a product in a field of unrelated signs in order to enhance its cultural appearance. As a result of advertising, we tend to relate to the fragmented sign context rather than the use-value. Thus, in postmodern society, people purchase commodities more for the image than for the function they perform.

Let's think about the example of clothing. The actual use-value of clothing is to cover and keep warm. Yet right now I'm looking at an ad for clothing in *Rolling Stone* and it doesn't mention anything about protection from the elements or avoiding public nudity. The ad is rather interesting in that it doesn't even present itself as an advertisement at all. It looks more like a picture of a rock band on tour. We could say, then, that even advertising is advertising itself as something. On the sidebar of the "band" picture, it doesn't talk about the band. It says things such as "jacket, $599, by Avirex; T-shirt, $69, by Energie," and so on. Most of us won't pay $599 for a jacket to keep us warm, but we might pay that for the status image that we think the jacket projects. Baudrillard's point is that we aren't connected to the basic human reality of keeping warm and covered; we aren't even connected to the social reality of being in a rock band (which itself is a projected image of an idealized life). We are simply attracted to the images.

Simulacrum and Hyperreality

Baudrillard further argues that many of the things we do today in advanced capitalist societies are based largely on images from past lives. For example, most people in traditional and early industrial societies *worked with* their bodies. Today, increasing numbers of people in postindustrial societies don't work with their bodies; they *work out* their bodies. The body has thus become a cultural object rather than a means to an end. In postindustrial society, the body no longer serves the purpose of production; rather, it has become the subject of image creation: we work out in order to alter our body to meet some cultural representation.

Clothing too has changed from function to image. In previous eras, there was an explicit link between the real function of the body and the clothing worn—the clothing was serviceable with reference to the work performed or it was indicative of social status. For example, a farmer would wear sturdy clothing because of the labor performed, and his clothing indicated his work (if you saw him away from the field, you would still know what he did by his clothing). However, in the postmodern society, clothing itself has become the creator of image rather than something merely serviceable or directly linked to social status and function.

Let's review what we know so far: in times past, the body was used to produce and reproduce; clothing was serviceable and was a direct sign of work and social function. In postmodern society, however, bodies have been freed from the primary burden of labor and have instead become a conveyance of cultural image. The body's "condition" is itself an important symbol, and so is the decorative clothing placed upon the body. Now here is where it gets interesting: in the past, a fit body represented hard work and clothing signed the body's work, but what do our bodies and clothing symbolize today? Today, we take the body through a workout rather than actually working with the body. We work out so we can meet a cultural image—but what does this cultural image represent?

In times past, if you saw someone with a lean, hard body, it meant the person lived a mean, hard life. There was a real connection between the sign and what it referenced. But what do our spa-conditioned bodies reference? Baudrillard's point is that there is no real objective or social reference for what we are doing with our bodies today. The only reference to a real life is to that of the past—we used to have fit bodies because we worked. Thus, in terms of real social life, today's gym-produced bodies represent the past image of working bodies. Further, what does this imply about the clothes we wear? The clothes themselves are an image of an image that doesn't exist in any kind of reality. Baudrillard calls this **simulacrum**, an image of an image of a "reality" that never existed and never appears.

Thus, what we buy today aren't even commodities in the strictest sense. They are what Fredric Jameson calls **free-floating signifiers**—signs and symbols that have been cut loose from their social and linguistic contexts, and thus their meaning is at best problematic and generally nonexistent. In such a culture, tradition and family can be equated with paper plates (as in a recent television commercial) and infinite justice with military retribution (as in the U.S. president's initial characterization of the current Iraq conflict).

Rather than representing, as signs did in the first phase, and rather than creating meaningful and social relations, as symbolic exchange did, commodified signs do nothing and mean nothing. They have no referent and their sign-context—the only thing that could possibly impart meaning—is constantly shifting because of mass media and advertising. As they are, these signs cannot provoke an emotional response from us. Emotions, then, develop "a new kind of flatness or depthlessness, a new kind of superficiality in the most literal sense . . . [which is] perhaps the supreme formal feature of all the postmodernisms" (Jameson, 1984, p. 60).

These free-floating signs and images don't represent reality; they create hyperreality. Part of the hyperreality is composed of the commodities that we've been talking about. But a more significant part is provided by the extravagance of media entertainment, like Las Vegas and Disneyland. **Hyperreality** is a way of understanding and talking about the mass of disconnected culture. It comes to substitute for reality. In hyperreality, people are drawn to cultural images and signs for artificial stimulation. In other words, rather than being involved in social reality, people involve themselves with fake stimulations. Examples of such simulacrum include artificial Christmas trees, breast implants, airbrushed Playboy Bunnies, food and drink flavors that don't exist naturally, and so on.

A clear example of this kind of hyperreality is reality television. Though predated by *Candid Camera,* the first reality show in the contemporary sense was *An American Family,* shown on Public Broadcasting Service stations in the United States in 1973. It was a 12-installment show that documented an American family going through the turmoil of divorce. The show was heavily criticized in the press. Today, reality shows are prevalent. Though the numbers are difficult to document, between the year 2000 and 2005 there were some 170 new reality shows presented to the public in the United States and Great Britain, with the vast majority being shown in the United States. The year 2004 saw reality programming come of age, as there were nine reality shows nominated for a total of 23 Emmy awards.

The interesting thing about reality programming is that there is no reality. However, it presents itself as a *representation* of reality. For example, the show *Survivor* placed 16 "castaways" on a tropical island for 39 days and asked, "Deprived of basic comforts, exposed to the harsh natural elements, your fate at the mercy of strangers . . . who would you become?" (Survivor Show Concept, n.d.). But the "castaways" were never marooned nor were they in any danger (as real castaways would be). And the game rules and challenges read more like *Dungeons and Dragons* than a real survivor manual, with game "challenges" and changes in character attributes for winning (like being granted "immunity"). So, what do the images of reality programming represent? Perhaps they are representations of what a fantasy game would look like if human beings could really get in one. The interesting thing about reality programming, and fantasy games for that matter, is the level of involvement people generate around them. This kind of involvement in simulated images of non-reality is hyperreality.

The significance of the idea of hyperreality is that it lets us see that people in postmodernity seek stimulation and nothing more. Hyperreality itself is void of any significance, meaning, or emotion. But, within that hyperreality, people create unreal worlds of spectacle and seduction. Hyperreality is a postmodern condition, a virtual world that provides experiences more involving and spectacular than everyday life and reality.

Sign Implosion

Baudrillard characterizes the simulacrum and hyperreality of postmodernity as an implosion. Where in modernity there was an explosion of signs, commodities, and distinctions, postmodernity is an implosion of all that. In modernity, there were new sciences such as sociology and psychology; in postmodernity, the divisions between disciplines have collapsed and instead there is an increasing preference for and growth of multidisciplinary studies. In modernity, there were new distinctions of nationality, identity, race, and gender; in postmodernity, these distinctions have imploded and collapsed upon themselves. Postmodernity is fractal and fragmented, with everything seeming political, sexual, or valuable—and if everything is, then nothing is. Baudrillard claims that this implosion of signs, identities, institutions, and all firm boundaries of meaning has led to the end of the social.

What Baudrillard is saying is that the proliferation, appropriation, and circulation of signs by the media and advertising influence the condition of signs, signification, and meaning in general. In the first stage of the sign, signs had very clear and specific meanings. But as societies and economic systems changed, different kinds of media and ideas were added. In the postmodern age, the media used to communicate information becomes utterly disassociated from any kind of idea of representation. Everywhere a person turns today in a postmodern society, there are media. Cell phones, computers, the Internet, television, billboards, "billboard clothing" (clothing hocking brand names), and the like surround us. And every medium is commodified and inundated with advertising. In postmodernity, we are hard pressed to find any space or any object that isn't communicating or advertising something beyond itself.

All of this has a general, overall effect. Signs are no longer moored to any social or physical reality; all of them are fair game for the media's manipulation of desire. Any cultural idea, image, sign, or symbol is apt to be pulled out of its social context and used to advertise and to place the individual in the position of consumer. As these signs are lifted out of the social, they lose all possibility of stable reference. They may be used for anything, for any purpose. And the more media that are present, and the faster information is made available (like DSL versus dialup computer connections), the faster signs will circulate and the greater will be the appropriation of indigenous signs for capitalist gain, until there remains no sign that has not been set loose and colonized by capitalism run amok. All that remains is a yawning abyss of meaninglessness—a placeless surface that is incapable of holding personal identity, self, or society.

Let's take a single example—gender. Gender has been a category of distinction for a good part of modernity. Harriet Martineau, the first person to ever use the word "sociology" in print, saw some 80 years before women won the right to vote that the project of modernity necessarily entailed women's rights. So close is the relationship between the treatment of women and the project of modernity for Martineau (1838/2003), that to her it becomes one of the earmarks of civilization: "Each civilized society claims for itself the superiority in its treatment of women" (p. 183). When gender first came up as an issue of equality, specifically in terms of a woman's right to vote, there was little confusion about what gender and gender equality meant. It was common knowledge who women were and what that very distinct group wanted.

But as modernity went on, things changed. In the United States after the 1960s, the single category of gender broke down. Various claims to distinction began to emerge around gender: the experience of gender is different by race, by class, by sexuality, and so on. The category imploded, with all the implicit understandings that went along with it. Today, when confronted by a person who appears to be a woman, the observing individual may be unsure. Is the "woman" really a woman or a man trapped in a woman's body? Or, is the "woman" a transsexual, who has physically been altered or symbolically changed (as with someone like RuPaul)? In postmodernity, the given cues of any category or object or experience cannot be taken at face value as indicative of a firm reality. All of the signs are caught up in a whirlwind of hyperreality.

In postmodernity, very few if any things can be accepted at face value. Meaning and reality aren't necessarily what they appear, because signs have been tossed about by the media without constraint, driven by the need to squeeze every drop of profit out of a populace through the proliferation of new markets with ever-shifting directions, cues, signs, and meanings in order to present something "new."

I've pictured my take of Baudrillard's argument in Figure 14.1. Let me emphasize that my intention with this diagram is simply to give you a heuristic device; it's a way to order your thinking about Baudrillard's theory. One of the best things about postmodern theory is that it is provocative, partly because it isn't highly specified. In one sense, then, this kind of picture stifles the postmodern; there's also a way in which something like this is quite modernist: it seeks to reduce complexity by making generalizations. So, I present this figure with some trepidation, but with the intent of giving you a place to "hang your hat." (In other words, if you're feeling at all lost and uncomfortable, then you need a modernist moment.)

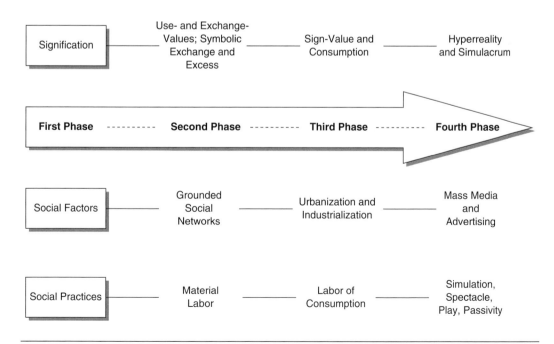

Figure 14.1   Baudrillard's Sign Stages

As you can see in the figure, I've divided Baudrillard's thought into three main groups: signification, social factors, and social practices. I see Baudrillard as weaving these themes throughout his opus. And I've organized these ideas around the notion of sign phases. Generally speaking, the first two phases of the sign were fairly well grounded in the social. People were clearly involved in face-to-face social networks, the principal form of labor was material, commodities basically contained use- and exchange-values, and humanity was deeply entrenched in symbolic exchange.

The third phase of the sign came about as the result of capitalism and the nation-state. The strong social factors in this era were urbanization and industrialization; people began to spend increasing amounts of time doing consumption labor; and the value of commodities shifted from use- to sign-value. As communication and transportation technologies increased, mass media and advertising became the important factors in social change. In a world of pure sign, signs no longer signify; it's all hyperreality and simulacrum, the fourth stage of the sign.

Concepts and Theory: The Postmodern Person

Fragmented Identities

Baudrillard envisions the "death of the subject." The subject he has in mind is that of modernity—the individual with strong and clear identities, able to carry on the work of democracy and capitalism. That subject, that person, is dead in postmodern culture. With the increase in mass media and advertising, there has been a corresponding decrease in the strength of all categories and meanings, including identities. What is left is a mediated person, rather than the subject of modernity. As Kenneth Gergen (1991) puts it, "one detects amid the hurly-burly of contemporary life a new constellation of feelings or sensibilities, a new pattern of self-consciousness. This syndrome may be termed *multiphrenia,* generally referring to the splitting of the individual into a multiplicity of self-investments" (pp. 73–74, emphasis original). This is an important idea, and one that appears in the work of a number of postmodernists. So let's take a moment to consider it fully.

One of the fundamental ways in which identity and difference are constructed is through exclusion. In psychology, this can by and large be taken for granted: I am by definition excluded from you because I am in my own body. For sociologists, exclusion is a cultural and social practice; it's something we *do,* not something we *are.* This fundamental point may sound elementary to the extreme, but it's important for us to understand it. In order for me to be me, I can't be you; in order for me to be male, I can't be female; in order for me to be white, I can't be black; in order to be a Christian, I can't be a Satanist; ad infinitum.

Cultural identity is defined in opposition to, or as it relates to, something else. Identity and self are based on exclusionary practices. The stronger the practices of exclusion, the stronger will be the identity; and the stronger my identities, the stronger will be my sense of self. Further, the greater the exclusionary practices, the more real will be my experience of identities and self. Hence, the early social movements for equality and democracy had clear practices of exclusion. This gave the people the strength of identity to make the sacrifices necessary to fight.

The twist that postmodernism gives to all this is found in the ideas of cultural fragmentation and de-centered selves and identities. In many ways, the ideas of gender and race are modernist: they collapse individualities into an all-encompassing identity. However, a person isn't simply female, for example—she also has many identities that crosscut that particular cultural interest and may shift her perceptions of self and other in one direction or another. For us to claim any of these identities—to claim to be female, black, male, or white—is really for us to put ourselves under the umbrella of a grand narrative, stories that deny individualities in favor of some broader social category. Grand narratives by their nature include very strong exclusionary tactics.

More to the point, the construction of centered identities is becoming increasingly difficult in postmodernity. Postmodernists argue that culture in postindustrial societies is fragmented. Since culture and identity are closely related, if the culture is fragmented, then so are identities. The idea of postmodernism, then, makes the issues of gender and race very complex. Clear racial and gender identities may thus be increasingly difficult to maintain. The culture has become more multifaceted, and so have identities. This means that we have greater freedom of choice, which we think we enjoy, but freedom of choice also implies that the distinctions between gender and racial identities aren't as clear or as real as they once were. Thus, social movements around race and gender become increasingly difficult to produce in postmodernity.

How, then, are postmodern identities constructed? Zygmunt Bauman (1992) argues that, as a result of de-institutionalization, people live in complex, chaotic systems. Complex systems differ from the mechanistic systems in that they are unpredictable and not controlled by statistically significant factors. In other words, the relationships among the parts are not predictable. For example, race, class, and gender in a complex system no longer produce strong or constant effects in the individual's life or self-concept.

Thus, being a woman, for instance, might be a disability in one social setting and not have any meaning at all in another; likewise, race, class, and gender might come together in a specific setting in unique and random ways. Within these complex systems, groups are formed through unguided self-formation. In other words, we join or leave groups simply because we want to. Moreover, the groups exist not because they reflect a central value system, as a modernist would argue; rather, they exist due to the whim and fancy of their members and the tide of market-driven public sentiment.

The absence of any central value system and firm, objective evaluative guides tends to create a demand for substitutes. These substitutes are symbolically, rather than actually or socially, created. The need for these symbolic group tokens results in what Bauman (1992, pp. 198–199) calls "tribal politics" and defines as self-constructing practices that are collectivized but not socially produced. These neo-tribes function solely as imagined communities and, unlike their premodern namesake, exist only in symbolic form through the commitment of individual "members" to the *idea* of an identity.

But this neo-tribal world functions without an actual group's powers of inclusion and exclusion. It is created through the repetitive and generally individual or imaginative performance of symbolic rituals and exists only so long as the rituals are performed. Neo-tribes are thus formed through concepts rather than actual social groups. They exist as imagined communities through a multitude of agent acts of self-identification and exist solely because people use them as vehicles of self-definition: "Neo-tribes are, in other words, the vehicles (and imaginary sediments) of individual self-definition" (Bauman, 1992, p. 137).

Play, Spectacle, and Passivity

All that we just covered is caught up in the social practices of postmodernity. Opposition is impossible because postmodern culture has no boundaries to push against; it is tantamount to pushing against smoke. Further, the acting subject is equally as amorphous. Thus, according to Baudrillard, there's little place in postmodernity to grab hold of and make a difference. Most things turn out to be innuendo, smoke, spam, mistakes . . . a smooth surface of meaninglessness and seduction. Baudrillard (1993a) leaves us with a few responses: play, spectacle, and passivity.

What can you do with objects that have no meaning? Well, you can play with them and not take them seriously. "So, all that are left are pieces. All that remains to be done is to play with the pieces. Playing with the pieces—that is postmodern" (Baudrillard 1993a, p. 95). How do we play in postmodernity? What would postmodern play look like? Here's an example: People in postmodern societies intentionally engage in fleeting contacts. Consider the case of "flash mobs." According to Wikipedia.com, "A flash mob is a group of people who assemble suddenly in a public place, do something unusual or notable, and then disperse. They are usually organized with the help of the Internet or other digital communications networks" (Flash Mob, n.d.). Sydmob, an Internet group facilitating flash mobs in Sydney, Australia, asks,

> Have you ever been walking down a busy city street and noticed the blank look on people's faces? How about on public transport? That look of total indifference is unmistakable; it's the face of [a] person feeling more like a worker bee than a human being. Have you ever felt like doing something out of the ordinary to see their reaction? (Sydmob, n.d.)

In this play, spectacle becomes important. Georg Simmel, a classical theorist ahead of his time, gives us the same insight. Simmel (1950) argues that

> life is composed more and more of these impersonal contents and offerings that tend to displace the genuine personal colorations and incomparabilities. This results in the individual's summoning the utmost in uniqueness and particularization, in order to preserve his most personal cores. He has to exaggerate this personal element in order to remain audible even to himself. (p. 422)

Echoing Simmel, Zygmunt Bauman (1992) notes that, "to catch the attention, displays must be ever more bizarre, condensed and (yes!) disturbing; perhaps ever more brutal, gory and threatening" (p. xx).

The remaining postmodern practice is a kind of resistance through passivity—refusing to play. There's an old American slogan from the Vietnam era that says, "Suppose they gave a war and nobody came?" This is similar to what Baudrillard has in mind with resistance through passivity. Rather than attempting to engage postmodern culture, or responding in frustration, or trying to change things, Baudrillard advocates refusal or passive resistance. And perhaps like the war that no one shows up for, postmodernity will simply cease.

Baudrillard Summary

- Baudrillard uses Marx's notions of use- and exchange-value to argue that commodities are principally understood in postmodernity in terms of their sign-value.

- Baudrillard proposes four stages of the sign. In the first two stages, the sign adequately represented reality, and social communities were held together through the reciprocity of symbolic exchanges. People were also able to practice "excess"—the boundless potential of humanity—through festivals, rituals, sacrifices, and so on.

- Modernity began in the third phase of the sign with the advent of capitalism and industrialization. Baudrillard characterizes modernity as the consumer society. Within such a society, labor shifts to techniques of consumption with an eye toward sign identification. Modernity also brought rationalization and constraint, the antithesis of symbolic exchange and excess.

- In postmodernity, the increasing presence and speed of mass media, along with ever-increasing levels of commodification and advertising, push all vestiges of meaning out of signs. Mass media tends to empty cultural signs because the natural entropy of information is multiplied and because signification is suppressed in favor of media concerns of production. Advertising pushes this process of emptying further: advertising sells by image rather than use, which implies that commodities are placed in unrelated sign-contexts in order to fit a media-produced image. These detached and redefined images are pure simulacrum. Postmodern society is inundated by media technology and thus an immense amount of this kind of signification and culture, most of which references and produces a hyperreality.

- Baudrillard's postmodern condition is found in simulation, spectacle, play, and passivity. Baudrillard claims that the central subject of modernity—the person as the nexus of national and economic rights and responsibilities—is dead. In the place of the subject stands a media terminal of fragmented images. What remains is play and spectacle. As signs move ever faster through the postmodern media, their ability to hold meaning continues to disintegrate. Thus, in order to make an impression, cultural displays must be more and more spectacular. In such a climate, the hyperreality of media becomes more enticing, with greater emotional satisfaction than real life—but these media images must continue to spin out to ever more radical displays. Playing with empty signs or intentionally disengaging are the only possible responses.

Building Your Theory Toolbox

Knowing Post-Theories

After reading and understanding this chapter, you should be able to define the following terms theoretically and explain their theoretical importance to Foucault's and Baudrillard's theories:

> Poststructuralism, practice, power, knowledge, order, truth games, discourse, counter-history, archaeology and genealogy, episteme, historical rupture, subject objectification, panopticon, human disciplines, clinical gaze, governmentality, microphysics of power, sexuality and subjectivity, postmodernism, symbolic exchange, excess, sign-value, fetish, four sign stages, consumer society, information entropy, sign-value and sign-vehicles, mass media, advertising, labor of consumption, simulacrum, hyperreality, free-floating signifiers, sign implosion, death of the subject, neo-tribes, play and spectacle

After reading and understanding this chapter, you should be able to

- Discuss truth games and Foucault's interest in them

- Explain Foucault's connection between power and knowledge. How does he conceptualize power? How does knowledge function as power?

- Describe what Foucault means by "the order of things" and explain how his "counter-histories" are used to expose this order

- Explain the differences between the Renaissance and the Enlightenment epistemes and how they order things

- Define discourse and explain how it provides a subjective position for the speaker

- Explain the place of the social sciences (human disciplines) in creating governmentality and the microphysics of power. In other words, what kinds of practices do we have in our daily lives that bring our selves under the dominion of this age?

- Describe how Baudrillard argues that Marx actually affirms and legitimates capitalism. How does Baudrillard invert Marx's argument?

- Explain how ideology and fetishism are based on a passion for the code

- Discuss the four stages of the sign. Be certain to explain their characteristics and the social factors that helped bring them about.

- Define the consumer society and explain the labor that is specific to the consumer society

- Explain how mass media and advertising empty the sign of all meaning and reference

- Describe how social identities have imploded in postmodernity and analyze the possible ramifications for political change

- Explain why play, spectacle, and passivity make sense in Baudrillard's postmodernity

Learning More: Primary Sources

- Michel Foucault

  Foucault, M. (1990). *The history of sexuality, volume I: An introduction.* New York: Vintage. (Original work published 1976)

  Foucault, M. (1994a). *The birth of the clinic: An archaeology of medical perception.* New York: Vintage. (Original work published 1963)

  Foucault, M. (1994b). *The order of things: An archaeology of the human sciences.* New York: Vintage. (Original work published 1966)

  Foucault, M. (1995). *Discipline and punish: The birth of the prison.* New York: Vintage. (Original work published 1975)

- Jean Baudrillard

  Baudrillard, J. (1975). *The mirror of production.* St. Louis, MO: Telos Press. (Original work published 1973)

  Baudrillard, J. (1981). *For a critique of the political economy of the sign.* St. Louis, MO: Telos Press. (Original work published 1972)

  Baudrillard, J. (1991). *Seduction.* New York: Palgrave Macmillan.

  Baudrillard, J. (1994). *Simulacra and simulation.* Ann Arbor: University of Michigan Press. (Original work published 1981)

Learning More: Secondary Sources

Deleuze, G. (1988). *Foucault.* Minneapolis: University of Minnesota Press.

Gutting, G. (Ed.). (1994). *The Cambridge companion to Foucault.* Cambridge, UK: Cambridge University Press.

Kellner, D. (1990). *Jean Baudrillard: From Marxism to postmodernism and beyond.* Stanford, CA: Stanford University Press.

Lane, R. (2000). *Jean Baudrillard* (Routledge Critical Thinkers). New York: Routledge.

Mills, S. (2003). *Michel Foucault.* New York: Routledge.

Theory You Can Use (Seeing Your World Differently)

- Using the index of this book, find the different ways power is defined and used theoretically. Evaluate each of these ways and create a theory of power that you think best explains it. Justify your answer.

- Compare and contrast Foucault's idea of governmentality and Anthony Giddens's notion of the reflexive project of the self. Explain why you think these two ideas are distinct. Together, what do they imply about how we relate to the "self" in modernity?

- Both Foucault and Baudrillard talk about what it means and what it's like to be you in this society. Go back through this chapter and make two lists, one for Foucault and the other for Baudrillard, that detail the ideas these theorists have about living during this time period. Now go back through the lists and think of examples from your life that illustrate each idea. What insights about your life did you glean? Were you able to think of examples for each idea? How do these different approaches compare and contrast with each other?

- Compare and contrast the idea of the death of the subject and postmodern identities (neo-tribes) with Giddens's theory of the reflexive project of the self. Which do you think more accurately describes the issues surrounding identities and selves in late and postmodernity? Why? (Remember to justify your answer theoretically.)

- Compare and contrast the political implications of Giddens's late modernity and Baudrillard's postmodernity. Which do you think more accurately reflects the current conditions? Why?

- Using Baudrillard's ideas of simulacrum and hyperreality, analyze the film *The Matrix* or the amusement park Disneyland.

Further Explorations: Web Links

- Foucault on the Web: http://www.theory.org.uk/ctr-fouc.htm
- Baudrillard on the Web: http://www.uta.edu/english/apt/collab/baudweb .html

Concluding Thoughts: Post-Thinking

Formerly, one could tell simply by looking at a person that he wanted to think . . . that he now wished to become wiser and prepared himself for a thought: he set his face as for prayer and stopped walking; yes, one even stood still for hours in the middle of the road when the thought arrived—on one leg or two legs. That seemed to be required by the dignity of the matter. (Nietzsche, 1974, p. 81)

Well, did you read the quote? If not, please do so. And let what Nietzsche says get inside of you. Have you ever had a thought? Of course, you think quite a bit. But have you ever had a thought in the way Nietzsche is describing it? This kind of thought is an event. It requires or perhaps captures the entire person. It's demanding and inspiring. Notice that Nietzsche says "formerly." He's referring to a time before the seduction of modern busy-ness. When life was slower, it was easier to have a thought, to be captured by an idea when you least expected it. Most of us today are too busy to be taken over by an idea—but we can still quite deliberately have a thought. And that's what this book is about. Throughout it, I have been inviting you to have a thought or two with me. It will require time and effort, but once you've had a thought, you'll never be the same again.

In this book, we have moved from classic to contemporary theory. Among the defining features of contemporary theory is modernity and all that it entails—not only the social and technical factors of nation-states, capitalism, urbanization, increasing communication and transportation technologies, mass education, and so on, but more specifically the idea of progress. Perhaps above everything else, modernity is the idea that human beings can, through reason, make their worlds better socially (through freedom and equality) and physically (through technology). Obviously, the idea of progress is intrinsically linked with every other feature of modernity. For example, the modern nation-state was specifically created to protect the inalienable rights of citizens. And capitalism is still seen by many as the great equalizer: the even playing field where neither birth nor social standing determines outcomes and life chances.

This link is part of what has produced the critical stands of contemporary theory. In fact, it is this critical perspective that is characteristic of contemporary theory, not modernity itself. Truly modernist theories are generally built upon the classics (such as Marx, Weber, Spencer, and Durkheim) and are part of the modern project itself. Contemporary theories, on the other hand, are generally critical of the modernist project in one way or another. Even contemporary theories of modernity modify the ideas of freedom and progress. Anthony Giddens's theory provides a clear example: in late modernity, there has been a shift from emancipatory politics to lifestyle politics and the reflexive project of the self.

The counterbalance to the critique of modernity is an emphasis on differing perspectives. Perspectives are made up of assumptions, values, sentiments, concepts, language, and norms—we see what we see because of the perspective we take. In contemporary theory, there is a growing sense of the power and politics of perspectives, rather than the unreflective acceptance of them. This understanding clearly comes out in the idea of discourse. Every discourse assumes a political world that situates all speakers and hearers with reference to one another.

Postcolonialism

Another place this idea of perspective is clearly seen is in *postcolonial theory*. Colonization has always had an interesting relationship to modernity. In a fundamental way, especially with the United States, we can see decolonization as part of what facilitated modernization. As we've consistently noted, the nation-state and the accompanying ideas of citizen and individual equality are central to modernity. In fact, it was the Revolutionary War (a "decolonization") that allowed the United States to become a nation-state. Yet, at the same time, many theories of modernization argue that colonization is a necessary precursor to modernity, helping to establish the needed economic and political infrastructures. These theories thus argue that the colonizing influence the United States has is necessary for the other nations to become "modernized."

Thus, both colonialism and decolonization have been prevalent in modernity. Up until about 1914, colonization continued to expand: the total geographic space occupied under colonial rule increased from about 55% in 1800 to 85% in 1914 (Said, 1994, p. 8). After WWII, decolonization took hold and continued to increase on through the 1960s. During that time, most of the colonies and protectorates in Asia and Africa won their independence. One of the most notable achievements during this period was the ultimately successful struggle for Indian independence led by Mahatma Gandhi.

Postcolonialism is an effort to understand the effects of colonization and decolonization on political systems, cultures, and individuals. Quite a bit of work in the field comes from literary criticism and is aimed at bringing the perspective of the colonized to the colonizer. In 1978, postcolonialism also became part of social theory with the publication of Edward Said's book *Orientalism*. Said argues that colonizing nations base much of their economic growth and stability on the markets and resources that the colonized provide. In this, Said echoes W. E. B. Du Bois' (1920/1996b) work some 60 years prior: "There is a chance for exploitation on an immense scale for inordinate profit, not simply to the very rich, but to the middle class and to the laborers. This chance lies in the exploitation of darker peoples" (pp. 504–505).

Said's insight is that the structural relationship between the colonized and colonizer is based upon specific kinds of cultures—in this case, "an undeterred and unrelenting Eurocentrism" (Said, 1994, p. 222). One of the things we've talked about in this book is that identities are formed through producing and maintaining cultural differences. The more pronounced the differences, the stronger the identity and the more dramatic the effects (the terrorist acts of the early twenty-first century against Western nations like the United States and England are examples).

One of the basic distinctions that has allowed Eurocentric identities to flourish is the East–West divide and the social construction of the "Orient." In many ways, this is similar to the distinction made in the United States between blacks and whites. That distinction and seeing differences based on "race" are what have allowed slavery, Jim Crow laws, and continued discrimination to exist. Fundamentally, "white," and all that identity entails, can only exist in reference to "black."

In the same way, the West became the "West" only in comparison to something intrinsically different and inferior. Only by seeing the Orient as inferior, underdeveloped, and deviant, and the Oriental people as "the Yellow Peril" or the "Mongol hordes," could the West see itself as superior, developed, humane, rational, and so forth. This superior view of Western civilization allowed and continues to allow European nations and the United States to force their brand of democracy, capitalism, and rationality upon other "uncivilized" nations. As Wallerstein points out, developing nations can only be "developing" in reference to some supposedly superior model, and this politically inspired model is invariably imposed from the outside. An important ramification here is that, while decolonized nations may no longer be directly dominated, they nonetheless continue to be colonized politically, economically, and culturally.

Frantz Fanon takes the idea of continued colonialism and applies it at the level of individual experience. Like Foucault's use of discourse, Fanon argues that the relationship between developed and developing nations creates a subjective position for the colonized: the colonized person is someone who experiences her or his being or existence through others. Demonstrating this idea of Fanon's, Robert J. C. Young (2003) asks, "Have you ever felt that the moment you said the word 'I,' that 'I' was someone else, not you? That in some obscure way, you were not the subject of your own sentence? . . . That you live in a world of others, a world that exists *for* others?" (p. 1). Yet colonialism is even more insidious. Fanon (1952/1967) argues that this cultural colonization becomes embodied:

> Below the corporeal schema I had sketched a historico-racial schema. The elements that I used had been provided for me not by "residual sensations and perceptions primarily of a tactile, vestibular, kinesthetic, and visual character," but by the other, the white man, who had woven me out of a thousand details, anecdotes, stories. (p. 111)

Thus, the struggle for decolonized peoples is not simply political and economic; it is also a struggle for psychological and cultural existence. Just as the white American idea of "manifest destiny" implied the genocide of Native Americans, so this colonization results in the loss of a people's cultural and psychological existence. Native American genocide didn't only mean physical death—though it did nearly wipe out the Native American population, from approximately 12 million prior to European contact to 250,000 by 1900—it also meant forced assimilation. If a Native American managed to physically survive the American holocaust, he or she was expected to die culturally and to become white. (See Brown, 2001; Maynard, 1996; Miller, 1996.)

In order to avoid cultural and psychological obliteration, Fanon argues, the colonized must place at risk their ontological existence. We've learned that human reality is cultural. This implies that the ontological source of human reality is culture—to exist as human is to exist culturally. Part of what Fanon (1952/1967) is arguing, then, is that human existence is dependent upon cultural recognition:

> Man is human only to the extent to which he tries to impose his existence on another man in order to be recognized by him. As long as he has not been effectively recognized by the other, that other will remain the theme of his actions. (pp. 216–217)

Thus, the task for the colonized is this: to keep, express, and demand recognition for indigenous cultures, identities, and subjectivities. To do that, the colonized must impose their existence upon a Westernizing world—not to dominate it, as is the intent of the colonizers, but to be recognized as fully and viably human. This move is risky. It risks ontological death, but the alternative is equally perilous: living out the themes and subjectivities determined by the colonizers. These themes originated in the Eurocentric East–West divide; were re-expressed in the vocabulary of first, second, and third worlds; and continue through the distinctions made between modern, developed nations and modernizing, developing nations.

What's my point in talking about postcolonial theory? First, as thinking democratic citizens and theoretical sociologists, postcolonialism and postcolonial theory are areas of knowledge that we should understand. I recommend you begin with Robert Young's *Postcolonialism: A Very Short Introduction.* Above all else, it will tell you why you should care. Then Said's *Orientalism* and Fanon's *Black Skins, White Masks* and *The Wretched of the Earth* should all be read. I then recommend that you follow up on Young's suggested further reading. It will remake your mind.

But I have another reason for mentioning postcolonialism. I said I was using it as an example of the idea of perspectives in contemporary theory, and that's partially true. My intent is a bit more significant, however. Earlier we entertained the idea that our ability to have a significant thought is an interesting theoretical question and not a personal issue. What we think and how we think are clearly influenced by the cultural milieu in which we live. So, for example, thinking in a postmodern society today is a different kind of enterprise from thinking in the traditional society of pre-Christian Jerusalem.

However, there is one thing that strikes me as I read even the most postmodern or poststructuralist of authors: every one of them assumes his or her reader to be capable of logical, critical, and reflexive thought. That our subjective existence is linked to the discourse of the age is beyond doubt, but within that existence, even within the mind-numbing world of postmodernism, we can have a thought. We can become aware and, because we are symbol-using creatures, we can take our thought a step further than its taken-for-granted base.

Thus, in bringing postcolonialism into our discussion, I want to emphasize a central point: *to have a thought is to be post.* I don't mean this statement as a catchphrase or truism. If we accept the implications of discourse, then having a thought as compared to being *handed* a thought always involves moving from a position of taken-for-grantedness. To even be aware of our discourse implies some movement out of it. How far we move depends on the distance between symbolic spaces. Though I admit there are limitations to analytic dichotomies, the basic distinction in symbolic space is between within and without. When we move outside our symbolic space in the West, it is usually in reference to such collective distinctions as gender, sexuality, race, ethnicity, class, religion, and so forth. Postcolonialism, however, asks us to take a larger step—a step from one world into another, from everything most Westerners believe about modernity, progress, technology, democracy, and capitalism to worlds that are defined and experienced in terms of diasporas, periphery, imperialism, dislocation, refugeeism, and terrorism.

What contemporary theory and this book are asking us to do is to *move*. In this sense, modernism is larger than an historical period or specific social factors—it is a perspective of the mind and an attitude of the heart. To be modern, then, is to uncritically believe in the hope of technology and government; it is to live within the confines of a materialized body, to accept a simplistic and political reduction of space and time, and to keep oneself within the confines of identities of certainty.

This idea of post-thinking that I'm advocating—thinking outside the box, moving away from comfortability and certainty—also implies that it is impossible to move backward or forward. To move backward implies trying to think in terms of cultures past, like trying to imagine what it was like to be a Hopi before the Europeans came. No matter how close we may think we get, whatever symbolic space we create is always and ever a "traditional-Hopi-as-imagined-by-a-twenty-first-century-person." The same is true about seemingly closer subjectivities, such as the idealized American family of the 1950s. Thus, we can't truly move backward in our thought and we can't move forward either. Moving "forward" exists only as the result of an ethical standard, and currently that standard is some form of the modern ideal of progress.

To have a thought, then, is to engage in post-thinking. It isn't a call to a bygone era, nor is it valued as a step forward; it is simply a step outside. Since today the most significant "inside" is the symbolic space wrapped around the idea of modernity, to have a thought is to be postmodern: to think outside the confines of modernist values and ideas. To have a thought is thus to be liminal—it is be outside but not yet arriving. In this way, Nietzsche misplaced his thinker. Rather than "in the middle of the road," perhaps we should see a person having a thought as being on a staircase—neither going up nor down but certainly situated between, in a conceptual space that is equally purposeful and unnerving.

Contemporary theory is vibrant and unruly precisely because it is post in this fashion. Theory today is also the most exciting it has been since the time of Marx, Durkheim, and Weber, where social theory was a main thread in public discourse and media. The book you hold in your hands invites you into this turbulent sea of ideas. Contemporary theory is an invitation to rethink society and the individual, power and possibilities, cultures and realities, time and space, human bodies and emotions, the earth and the risks we produce, and the ethics of it all. It is an invitation to move from our center and to see through different perspectives. How far are we willing to move, to be unsettled? What kinds of responsibilities are we willing to accept as a result of seeing differently? What will we make of our world? These are the questions of contemporary theory.

> My idea in *Orientalism* is to use humanistic critique to open up the fields of struggle, to introduce a longer sequence of thought and analysis to replace the short bursts of polemical, thought-stopping fury that so imprison us in labels and antagonistic debate whose goal is a belligerent collective identity rather than understanding and intellectual exchange. I have called what I try to do "humanism." . . . And lastly, most important, humanism is the only, and I would go so far as to say, the final resistance we have against the inhuman practices and injustices that disfigure human history. (Said, 2003, pp. xxii–xxix)

References

Al Qaeda Training Manual. Retrieved October 27, 2004, from http://www.usdoj.gov/ag/trainingmanual.htm

Alexander, J. C. (1985). Introduction. In J. C. Alexander (Ed.), *Neofunctionalism.* Beverly Hills, CA: Sage.

Allan, K. (1998). The meaning of culture: Pushing the postmodern critique forward. Westport, CT: Praeger.

Allan, K., & Turner, J. H. (2000). A formalization of postmodern theory. *Sociological Perspectives, 43*(3), 363–385.

Allen, B. (1996). *Rape warfare: The hidden genocide in Bosnia-Herzegovina and Croatia.* Minneapolis: University of Minnesota Press

Anderson, P. (1998). *The origins of postmodernity.* London: Verso.

Bandura, A. (1977). *Social learning theory.* New York: General Learning Press.

Barthes, R. (1967). *Elements of semiology* (A. Lavers & C. Smith, Trans.). London: Jonathan Cape. (Original work published 1964)

Baudrillard, J. (1972). *For a critique of the political economy of the sign.* St. Louis, MO: Telos Press. (Original work published 1981)

Baudrillard, J. (1975). *The mirror of production* (M. Poster, Trans.). St. Louis, MO: Telos Press. (Original work published 1973)

Baudrillard, J. (1987). When Bataille attacked the metaphysical principle of economy. *Canadian Journal of Political and Social Theory, 11,* 57–62.

Baudrillard, J. (1993a). *Baudrillard live: Selected interviews* (M. Gane, Ed.). New York: Routledge.

Baudrillard, J. (1993b). *Symbolic exchange and death* (I. H. Grant, Trans.). Newbury Park, CA: Sage. (Original work published 1976)

Baudrillard, J. (1994). *Simulacra and simulation* (S. F. Blaser, Trans.). Ann Arbor: University of Michigan Press. (Original work published 1981)

Baudrillard, J. (1995). *The Gulf War did not take place* (P. Patton, Trans.). Bloomington: Indiana University Press

Baudrillard, J. (1998). *The consumer society: Myths and structures* (C. Turner, Trans.). London: Sage. (Original work published 1970)

Bauman, Z. (1992). *Imitations of postmodernity.* New York: Routledge.

Becker, H. (1963). *Outsiders: Studies in the sociology of deviance.* New York: Free Press.

Bellah, R. N. (1970). *Beyond belief: Essays on religion in a post-traditional world.* New York: Harper & Row.

Bellah, R. N. (1975). *The broken covenant: American civil religion in time of trial.* New York: Seabury Press.

Bellah, R. N. (1980). *Varieties of civil religion.* San Francisco: Harper & Row.

Bellah, R. N., Madsen, R., Sullivan, W. M., Swidler, A., & Tipton, S. M. (1985). *Habits of the heart: Individualism and commitment in American life.* New York: Harper & Row.

Bellah, R. N., Madsen, R., Sullivan, W. M., Swidler, A., & Tipton, S. M. (1991). *The good society.* New York: Knopf.

Bentham, J. (1996). *An introduction to the principles of morals and legislation* (J. H. Burns & H. L. A. Hart, Eds.). New York: Oxford University Press. (Original work published 1789)

Berger, P. L. (1967). *The sacred canopy: Elements of a sociological theory of religion.* New York: Doubleday.

Blau, P. M. (1968). Social exchange. In David L. Sills (Ed.), *International encyclopedia of the social sciences.* New York: Macmillan.

Blau, P. M. (1995). A circuitous path to macrostructural theory. *Annual Review of Sociology, 21,* 1–19.

Blau, P. M. (2003). *Exchange and power in social life.* New Brunswick, NJ: Transaction.

Blau, P. M., & Meyer, M. W. (1987). *Bureaucracy in modern society* (3rd ed.). New York: McGraw-Hill.

Blumer, H. (1969). *Symbolic interactionism: Perspective and method.* Berkeley: University of California Press.

Bonacich, E. (1972). A theory of ethnic antagonism: The split labor market. *American Sociological Review, 37,* 547–559.

Bourdieu, P. (1984). *Distinction: A social critique of the judgment of taste* (R. Nice, Trans.). Cambridge, MA: Harvard University Press. (Original work published 1979)

Bourdieu, P. (1985). The genesis of the concepts of *habitus* and of *field. Sociocriticism, 2,* 11–29.

Bourdieu, P. (1989). Social space and symbolic power. *Sociological Theory, 7*(1), 14–25.

Bourdieu, P. (1990). *The logic of practice* (R. Nice, Trans.). Stanford, CA: Stanford University Press. (Original work published 1980)

Bourdieu, P. (1991). *Language and symbolic power* (J. B. Thompson, Ed.; G. Raymond & M. Adamson, Trans.). Cambridge, MA: Harvard University Press.

Bourdieu, P. (1993). *Outline of a theory of practice* (R. Nice, Trans.). Cambridge, UK: Cambridge University Press. (Original work published 1972)

Bourdieu, P., & Wacquant, L. J. D. (1992). *An invitation to reflexive sociology.* Chicago: University of Chicago Press.

Brown, D. (2001). *Bury my heart at Wounded Knee: An Indian history of the American West* (30th anniversary ed.). New York: Owl Books.

Business and Professional Women's Foundation. (2004). *101 facts on the status of women.* Retrieved May 9, 2005, from http://www.bpwusa.org/i4a/pages/index.cfm?pageid=3301

Calhoun, C. (2003). Pierre Bourdieu. In G. Ritzer (Ed.), *The Blackwell companion to major contemporary social theorists.* Malden, MA: Blackwell.

Cassirer, E. (1944). *An essay on man.* New Haven, CT: Yale University Press.

Chafetz, J. S. (1990). *Gender equity: An integrated theory of stability and change.* Newbury Park, CA: Sage.

Charon, J. M. (2001). *Symbolic interactionism: An introduction, an interpretation, an integration* (7th ed.). Upper Saddle River, NJ: Prentice Hall.

Chodorow, N. (1978). *The reproduction of mothering: Psychoanalysis and the sociology of gender.* Berkeley: University of California Press.

CNN (2003). *Commander in Chief lands on USS* Lincoln. Retrieved July 4, 2005, from http://www.cnn.com/2003/ALLPOLITICS/05/01/bush.carrier.landing/

Cohen, A., Baumohl, B., Buia, C., Roston, E., Ressner, J., & Thompson, M. (2001, January 8). This time it's different. *Time, 157*(1), 18–22.

Collins, P. H. (2000). *Black feminist thought: Knowledge, consciousness, and the politics of empowerment* (2nd ed.). New York: Routledge.

Collins, R. (1975). *Conflict sociology.* New York: Academic Press.

Collins, R. (1979). *The credential society: An historical sociology of education and stratification.* San Diego, CA: Academic Press.

Collins, R. (1986a). Is 1980s sociology in the doldrums? *American Journal of Sociology, 91,* 1336–1355.

Collins, R. (1986b). *Max Weber: A skeleton key.* Newbury Park, CA: Sage.

Collins, R. (1986c). *Weberian sociological theory.* Cambridge: Cambridge University Press.

Collins, R. (1987). Interaction ritual chains, power and property: The micro–macro connection as an empirically based theoretical problem. In J. C. Alexander, B. Giesen, R. Münch, & N. J. Smelser (Eds.), *The micro–macro link.* Berkeley: University of California Press.

Collins, R. (1988). *Theoretical sociology.* San Diego, CA: Harcourt Brace Jovanovich.

Collins, R. (1989). Sociology: Proscience or antiscience? *American Sociological Review, 54,* 124–139.

Collins, R. (1993a). What does conflict theory predict about America's future? *Sociological Perspectives, 36*(4), 289–313.

Collins, R. (1993b). Emotional energy as the common denominator of rational action. *Rationality and society, 5,* 203–230.

Collins, R. (2004). *Interaction ritual chains.* Princeton, NJ: Princeton University Press.

Comte, A. (1896). *The positive philosophy of Auguste Comte* (H. Martineau, Trans.). London: G. Bell.

Condon, W. S., & Ogston, W. D. (1971). Speech and body motion synchrony of the speaker-hearer. In D. D. Horton & J. J. Jenkins (Eds.), *Perception of language.* Columbus, OH: Merrill.

Cook, K. S., & Emerson, R. M. (1978). Power, equity and commitment in exchange networks. *American Sociological Review, 43,* 721–739.

Cook, K. S., Emerson, R. M., Gillmore, M. R., & T. Yamagishi. (1983). The distribution of power in exchange networks: Theory and experimental results. *American Journal of Sociology, 89,* 275–305.

Cook, K. S., & Gillmore. M. R. (1984). Power, dependence, and coalitions. In E. J. Lawler (Ed.), *Advances in group processes.* Greenwich, CT: JAI Press.

Cook, K. S., & Rice, E. R. W. (2001). Exchange and power: Issues of structure and agency. In Jonathan H. Turner (Ed.), *Handbook of sociological theory.* New York: Kluwer.

Cooley, C. H. (1998). *On self and social organization* (H. Schubert, Ed.). Chicago: University of Chicago Press.

Coser, L. A. (1956). *The functions of social conflict.* Glencoe, IL: Free Press.

Coser, L. A. (2003). *Masters of sociological thought: Ideas in historical and social context* (2nd ed.). Prospect Heights: IL: Waveland Press.

Dahrendorf, R. (1959). *Class and class conflict in industrial society.* Stanford, CA: Stanford University Press. (Original work published 1957)

Dahrendorf, R. (1968). *Essays in the theory of society.* Stanford, CA: Stanford University Press.

Dahrendorf, R. (1989). *Straddling theory and practice: Conversation with Sir Ralf Dahrendorf.* Retrieved May 5, 2006, from http://globetrotter.berkeley.edu/Elberg/Dahrendorf/dahrendorf2.html

Danto, A. C. (1990). The hyper-intellectual. *New Republic, 203*(11–12), 44–48.

Davidson, A. I. (1994). Ethics as ascetics: Foucault, the history of ethics, and ancient thought. In G. Gutting (Ed.), *The Cambridge companion to Foucault.* Cambridge, UK: Cambridge University Press.

Denzin, N. (1992). *Symbolic interactionism and cultural studies: The politics of interpretation.* Oxford, UK: Blackwell.

Denzin, N. (1993). *The alcoholic society: Addiction and recovery of the self.* New York: Transaction.

Drake, J. (1997). Review essay: Third Wave feminisms. *Feminist Studies, 23*(1), 97–108.

Du Bois, W. E. B. (1996a). The souls of black folk. In E. J. Sundquist (Ed.), *The Oxford W. E. B. Du Bois reader.* New York: Oxford. (Original work published 1903)

Du Bois, W. E. B. (1996b). Darkwater. In E. J. Sundquist (Ed.), *The Oxford W. E. B. Du Bois reader.* New York: Oxford. (Original work published 1920)

Du Bois, W. E. B. (1996c). The propaganda of history. In E. J. Sundquist (Ed.), *The Oxford W. E. B. Du Bois reader.* New York: Oxford. (Original work published 1935)

Du Bois, W. E. B. (1996d). In black. In E. J. Sundquist (Ed.), *The Oxford W. E. B. Du Bois reader.* New York: Oxford. (Original work published 1920)

Dunn, S. (1996). *Loosestrife: Poems.* New York: W.W. Norton.

Durkheim, É. (1938). *The rules of sociological method* (G. E. G. Catlin, Ed.; S. A. Solovay & J. H. Mueller, Trans.). Glencoe, IL: Free Press. (Original work published 1895)

Durkheim, É. (1957). *Professional ethics and civic morals* (C. Brookfield, Trans.). London: Routledge.

Durkheim, É. (1961). *Moral education: A study in the theory and application of the sociology of education* (E. K. Wilson, Trans.). New York: Free Press. (Original work published 1903)

Durkheim, É. (1984). *The division of labor in society* (W. D. Halls, Trans.). New York: Free Press. (Original work published 1893)

Durkheim, É. (1993). *Ethics and the sociology of morals* (R. T. Hall, Trans.). Buffalo, NY: Prometheus. (Original work published 1887)

Durkheim, É. (1995). *The elementary forms of the religious life* (K. E. Fields, Trans.). New York: Free Press. (Original work published 1912)

Engels, F. (1978). The origin of the family, private property, and the state. In R. C. Tucker (Ed.), *The Marx–Engels reader.* New York: W.W. Norton. (Original work published 1884)

Fanon, F. (1967). *Black skins, white masks* (C. L. Markmann, Trans.). New York: Grove Press. (Original work published 1952)

Fine, G. A. (1987). *With the boys: Little league baseball and preadolescent culture.* Chicago: University of Chicago Press.

Fisher, S. (1973). *Body consciousness.* London: Calder & Boyars.

Flash mob. (n.d.). *Wikipedia.* Retrieved June 8, 2005, from http://en.wikipedia.org/wiki/Flash_mob

Flyvbjerg, B. (2001). *Making social science matter: Why social inquiry fails and how it can succeed again.* Cambridge, UK: Cambridge University Press.

Foucault, M. (1982). The subject and power. In H. L. Dreyfus & P. Rabinow (Eds.), *Michel Foucault: Beyond structuralism and hermeneutics.* Brighton, UK: Harvester Press.

Foucault, M. (1984a). Nietzsche, genealogy, history. In P. Rabinow (Ed.), *The Foucault reader.* New York: Pantheon.

Foucault, M. (1984b). Space, knowledge, and power. In P. Rabinow (Ed.), *The Foucault reader.* New York: Pantheon.

Foucault, M. (1984c). On the genealogy of ethics: An overview of work in progress. In P. Rabinow (Ed.), *The Foucault reader.* New York: Pantheon.

Foucault, M. (1990a). *The history of sexuality, volume I: An introduction* (R. Hurley, Trans.). New York: Vintage. (Original work published 1976)

Foucault, M. (1990b). *The history of sexuality, volume 2: The use of pleasure* (R. Hurley, Trans.). New York: Vintage. (Original work published 1984)

Foucault, M. (1994a). *The birth of the clinic: An archaeology of medical perception* (A. M. Sheridan Smith, Trans.). New York: Vintage. (Original work published 1963)

Foucault, M. (1994b). *The order of things: An archaeology of the human sciences.* New York: Vintage. (Original work published 1966)

Foucault, M. (1995). *Discipline and punish: The birth of the prison* (A. Sheridan, Trans.). New York: Vintage. (Original work published 1975)

Fromm, E. (1955). *The sane society.* New York: Henry Holt

Fromm, E. (1961). *Marx's concept of man.* New York: Continuum.

Garfinkel, H. (1967). *Studies in ethnomethodology.* Cambridge, UK: Polity Press.

Garfinkel, H. (1974). On the origins of the term "ethnomethodology." In R. Turner (Ed.), *Ethnomethodology: Selected readings.* Harmondsworth, UK: Penguin Education.

Garfinkel, H. (1996). Ethnomethodology's program. *Social Psychology Quarterly, 59,* 5–21.

Gergen, K. J. (1991). *The saturated self: Dilemmas of identity in contemporary life.* New York: Basic Books.

Giddens, A. (1986). *The constitution of society.* Berkeley: University of California Press.

Giddens, A. (1990). *The consequences of modernity.* Stanford, CA: Stanford University Press.

Giddens, A. (1991). *Modernity and self-identity: Self and society in the late modern age.* Stanford, CA: Stanford University Press.

Giddens, A. (1992). *The transformation of intimacy: Sexuality, love, and eroticism in modern societies.* Stanford, CA: Stanford University Press.

Gilman, C. P. (1975). *Women and economics: A study of the economic relation between men and women as a factor in social evolution.* New York: Gordon Press. (Original work published 1899)

Gilman, C. P. (2001). *The man-made world.* New York: Humanity Books. (Original work published 1911)

Global Exchange. (1998, September). *Wages and living expenses for Nike workers in Indonesia, September 1998.* Retrieved December 22, 2004, from http://www.globalexchange.org/campaigns/sweatshops/nike/

Goffman, E. (1959). *The presentation of self in everyday life.* Garden City, NY: Anchor Books.

Goffman, E. (1961). *Asylums: Essays on the social situation of mental patients and other inmates.* Garden City, NY: Anchor Books.

Goffman, E. (1963a). *Stigma.* New York: Touchstone.

Goffman, E. (1963b). *Behavior in public places.* New York: Free Press.

Goffman, E. (1967). *Interaction ritual: Essays on face-to-face behavior.* New York: Pantheon.

Goffman, E. (1977). The arrangement between the sexes. *Theory and Society, 4,* 301–331.

Gramsci, A. (1971). *Selections from the prison notebooks.* London: Lawrence & Heinemann. (Original work published 1928)

Habermas, J. (1984). *The theory of communicative action, vol. 1: Reason and the rationalization of society* (T. McCarthy, Trans.). Boston: Beacon. (Original work published 1981)

Habermas, J. (1987). *The theory of communicative action, vol. 2: Lifeworld and system: A critique of functionalist reason* (T. McCarthy, Trans.). Boston: Beacon. (Original work published 1981)

Harrison, F. (1913). Introduction. In H. Martineau (Trans.), *The positive philosophy of Auguste Comte.* London: G. Bell.

Hart, J. A. (1993). *Rival capitalists: International competitiveness in the United States, Japan, and Western Europe.* Ithaca, NY: Cornell University Press.

Havens, J. J., & Schervish, P. G. (2003). *Why the $41 trillion wealth transfer estimate is still valid: A review of challenges and questions.* Retrieved August 31, 2004, from http://www.bc.edu/research/swri/meta-elements/pdf/41trillionreview.pdf

Heritage, J. (1984). *Garfinkel and ethnomethodology.* Cambridge, UK: Polity Press.

Hewitt, J. P. (1998). *The myth of self-esteem: Finding happiness and solving problems in America.* New York: St. Martin's Press.

Hochschild, A. R. (1983). *The managed heart: Commercialization of human feeling.* Berkeley: University of California Press.

Hofstadter, D. R. (1985). *Metamagical themas: Questing for the essence of mind and pattern.* New York: Basic Books.

hooks, b. (1989). *Talking back: Thinking feminist, thinking black.* Boston: South End Press.

Horkheimer, M. (1993). *Between philosophy and social science: Selective early writings* (G. F. Hunter, M. S. Kramer, and J. Torpey, Trans.). Cambridge, MA: MIT Press.

Jameson, F. (1984). *The postmodern condition.* Minneapolis: University of Minnesota Press.

Johnson, A. G. (2000). *The Blackwell dictionary of sociology: A user's guide to sociological language* (2nd ed.). Malden, MA: Blackwell.

Kanter, R. M. (1977). *Men and women of the corporation.* New York: Basic Books.

Kellner, D. (2003). Jean Baudrillard. In G. Ritzer (Ed.), *The Blackwell companion to major contemporary social theorists.* Malden, MA: Blackwell.

Kennickell, A. B. (2003). *A rolling tide: Changes in the distribution of wealth in the U.S., 1989–2001.* Retrieved August 27, 2004, from http://www.federalreserve.gov/pubs/feds/2003/200324/200324pap.pdf

Kozol, J. (1991). *Savage inequalities: Children in America's schools.* New York: Crown.

Lemert, C. (2000). W. E. B Du Bois. In G. Ritzer (Ed.), *The Blackwell companion to major social theorists.* Malden, MA: Blackwell.

Lemert, E. M. (1951). *Social pathology: A systematic approach to the theory of sociopathic behavior.* New York: McGraw-Hill.

Lemert, E. M. (1967). *Human deviance, social problems, and social control.* Englewood Cliffs, NJ: Prentice Hall.

Lengermann, P. M., & Niebrugge-Brantley, J. (2000). Early women sociologists and classical sociological theory: 1830–1930. In G. Ritzer (Ed.), *Classical sociological theory* (3rd ed.). Boston: McGraw-Hill.

Lévi-Strauss, C. (1963). *Structural anthropology.* New York: Basic Books.

Lidz, V. (2000). Talcott Parsons. In G. Ritzer (Ed.), *The Blackwell companion to major social theorists.* Malden, MA: Blackwell.

Lilley, S. J., & Platt, G. M. (1994). Correspondents' images of Martin Luther King, Jr: An interpretive theory of movement leadership. In T. R. Sarbin and J. I. Kitsuse (Eds.), *Constructing the social.* Newbury Park, CA: Sage.

Luckmann, T. (1973). Philosophy, science, and everyday life. In M. Natanson (Ed.), *Phenomenology and the social sciences.* Evanston, IL: Northwestern University Press.

Luhmann, N. (1982). *The differentiation of society* (S. Holmes and C. Larmore, Trans.). New York: Columbia University Press.

Luhmann, N. (1989). *Ecological communication* (H. Bednarz Jr., Trans.). Chicago: Chicago University Press. (Original work published 1986)

Luhmann, N. (1995). *Social systems* (J. Bednarz Jr. & D. Baecker, Trans.). Stanford, CA: Stanford University Press. (Original work published 1984)

Lukács, G. (1971). *History and class consciousness: Studies in Marxist dialectics* (R. Livingstone, Trans.). London: Merlin. (Original work published 1923)

Macionis, J. J. (2005). *Sociology* (10th ed.). Upper Saddle River, NJ: Prentice Hall.

Mahar, C., Harker, R., & Wilkes, C. (1990). The basic theoretical position. In R. Harker, C. Mahar, & C. Wilkes (Eds.), *An introduction to the work of Pierre Bourdieu.* London: Macmillan.

Marcus, M. (2005). Indiana's manufacturing advantage. *In context,* 6(4). Retrieved August 11, 2006, from http://www.incontext.indiana.edu/2005/july/1.html

Marshall, G. (1998). *A dictionary of sociology* (2nd ed.). Oxford, UK: Oxford University Press.

Martineau, H. (2003). *How to observe morals and manners.* New Brunswick, NJ: Transaction Press. (Original work published 1838)

Marx, K. (1977). *Capital: A critique of political economy,* Vol. 1 (E. Mandel, Trans.). New York: Vintage. (Original work published 1867)

Marx, K. (1978a). Economic and philosophic manuscripts of 1844. In R. C. Tucker (Ed.), *The Marx-Engels reader.* New York: Norton. (Original work published 1932)

Marx, K. (1978b). The German ideology. In R. C. Tucker (Ed.), *The Marx-Engels reader.* New York: Norton. (Original work published 1932)

Marx, K. (1978c). Contribution to the critique: Introduction. In R. C. Tucker (Ed.), *The Marx-Engels reader.* New York: Norton. (Original work published 1844)

Marx, K. (1978d). Theses on Feuerbach. In R. C. Tucker (Ed.), *The Marx-Engels reader.* New York: Norton. (Original work published 1888)

Marx, K. (1978e). A contribution to the critique of political economy. In R. C. Tucker (Ed.), *The Marx-Engels reader.* New York: Norton. (Original work published 1859)

Marx, K. (1995). Economic and philosophic manuscripts of 1844. In E. Fromm (Trans.), *Marx's concept of man.* New York: Continuum. (Original work published 1932)

Marx, K., & Engels, F. (1978). Manifesto of the Communist Party. In R. C. Tucker (Ed.), *The Marx-Engels reader.* New York: Norton. (Original work published 1848)

Maturana, H. R., & Varela, F. J. (1991). *Autopoiesis and cognition: The realization of the living.* New York: Springer.

Maynard, J. (Ed.). (1996). *Through Indian eyes.* Washington, DC: Reader's Digest.

McCready, S. (Ed.). (2001). *The discovery of time.* Naperville, IL: Sourcebooks.

Mead, G. H. (1925). The genesis of the self and social control. *International Journal of Ethics, 55,* 255–277.

Mead, G. H. (1934). *Mind, self, and society: From the standpoint of a social behaviorist* (C. W. Morris, Ed.). Chicago: University of Chicago Press.

Mead, G. H. (1938). *The philosophy of the act.* Chicago: University of Chicago Press.

Mehan, H., & Wood, H. (1975). *The reality of ethnomethodology.* New York: John Wiley.

Menand, L. (2001). *The metaphysical club: A story of ideas in America.* New York: Farrar, Straus and Giroux.

Merriam-Webster. (2002). *Webster's third new international dictionary, unabridged* [online version]. Retrieved March 31, 2006, from http://unabridged.merriam-webster.com

Merton, R. K. (1967). *On theoretical sociology: Five essays, old and new.* New York: Free Press.

Merton, R. K. (1976). *Sociological ambivalence and other essays.* New York: Free Press.

Miller, L. (1996). *From the heart: Voices of the American Indian.* New York: Vintage.

Mills, C. W. (1956). *The power elite.* London: Oxford University Press.

Morris, C. W. (1962). Introduction: George H. Mead as social psychologist and social philosopher. In C. W. Morris (Ed.), *Mind, self, and society from the standpoint of a social behaviorist.* Chicago: University of Chicago Press.

National Academy of Sciences. (1995). *National science education standards.* Retrieved August 20, 2004, from http://www.nap.edu/readingroom/books/nses/html/action.html

Nietzsche, F. (1968). On the genealogy of morals. In W. Kaufmann (Ed. & Trans.), *Basic writings of Nietzsche.* New York: The Modern Library. (Original work published 1887)

Nietzsche, F. (1974). *The gay science.* New York: Vintage.

Nöth, W. (1995). *Handbook of semiotics.* Bloomington: Indiana University Press. (Original work published 1985)

Oakes, G. (1984). Introduction. In *Georg Simmel on Women, Sexuality, and Love.* New Haven: Yale University Press.

Orwell, G. (1946). *Shooting an elephant and other essays.* New York: Harcourt.

Outhwaite, W. (2003). Jürgen Habermas. In G. Ritzer (Ed.), *The Blackwell companion to major contemporary social theorists.* Malden, MA: Blackwell.

Parsons, T. (1949). *The structure of social action* (2nd ed.). New York and London: Free Press.

Parsons, T. (1951). *The social system.* London: Free Press.

Parsons, T. (1961). Culture and the social system. In Talcott Parsons (Ed.), *Theories of society: Foundations of modern sociological theory* (pp. 963–993). New York: Free Press.

Parsons, T. (1966). *Societies: Evolutionary and comparative perspectives.* Englewood Cliffs, NJ: Prentice Hall.

Parsons, T. (1990). Prolegomena to a theory of social institutions. *American Sociological Review, 55*(3), 319–339.

Parsons, T., & Shils, E. (Eds.). (1951). *Toward a general theory of action.* Cambridge, MA: Harvard University Press.

Pianin, E. (2001, July 10). Superfund cleanup effort shows results, study reports. *The Washington Post,* p. A19.

Plato. (1993). *The last days of Socrates* (H. Tredennick & H. Tarrant, Trans.). London: Penguin.

Rawls, A. (2003). Harold Garfinkel. In G. Ritzer (Ed.), *The Blackwell companion to major contemporary social theorists.* Malden, MA: Blackwell.

Ray, L. (Ed.). (1991). *Formal sociology: The sociology of Georg Simmel.* Aldershot, Hampshire, UK: Elgar.

Ritzer, G. (1998). *The McDonaldization thesis.* London: Sage.

Ritzer, G. (2004). *The McDonaldization of society* (Rev. ed.). Thousand Oaks, CA: Pine Forge Press.

Ritzer, G., & Goodman, D. (2000). Introduction: Toward a more open canon. In G. Ritzer (Ed.), *The Blackwell Companion to major social theorists.* Malden, MA: Blackwell.

Roy, W. G. (2001). *Making societies.* Thousand Oaks, CA: Pine Forge.

Rule, J. B. (2003). Lewis Coser: 1913–2003. *Dissent.* Retrieved May 5, 2006, from http://www.dissentmagazine.org/article/?article=470

Said, E. W. (1994). *Culture and imperialism.* New York: Vintage.

Said, E. W. (2003). *Orientalism: Western concepts of the Orient* (Preface to the twenty-fifth anniversary ed.). New York: Vintage.

Sanderson, S. K. (2005). Reforming theoretical work in sociology: A modest proposal. *Perspectives 28*(2), 1–4.

Schutz, A. (1967). *The phenomenology of the social world* (G. Walsh & F. Lehnert, Trans.). Evanston, IL: Northwestern University Press.

Simmel, G. (1950). *The sociology of Georg Simmel* (K. H. Wolff, Ed. & Trans.). Glencoe, IL: Free Press.

Simmel, G. (1959). *Essays on sociology, philosophy, and aesthetics [by] Georg Simmel [and others]: Georg Simmel, 1858–1918* (K. H. Wolfe, Ed.). New York, Harper & Row.

Simmel, G. (1971). *Georg Simmel: On individuality and social forms* (D. N. Levine, Ed.). Chicago: University of Chicago Press.

Simmel, G. (1978). *The philosophy of money* (T. Bottomore, & D. Frisby, Trans.). London: Routledge and Kegan Paul.

Simmel, G. (1984). *Georg Simmel on women, sexuality, and love* (G. Oakes, Trans.). New Haven, CT: Yale University Press. (Original work published 1858–1918)

Simmel, G. (1997). *Essays on religion* (H. J. Helle, Trans. & Ed.). New Haven, CT: Yale University Press.

Smith, A. (1937). *An inquiry into the nature and causes of the wealth of nations.* New York: The Modern Library. (Original work published 1776)

Smith, D. E. (1987). *The everyday world as problematic: A feminist sociology.* Boston: Northwestern University Press.

Smith, D. F. (1990). *The conceptual practices of power: A feminist sociology of knowledge.* Boston: Northeastern University Press.

Smith, D. E. (1992). Sociology from women's experience: A reaffirmation. *Sociological Theory, 11,* 1.

Smith, D. E. (2005). *Institutional ethnography: A sociology for people.* Lanham, MD: AltaMira Press.

Statistical abstract of the United States (2002). *The national data book.* Washington, DC: U.S. Department of Commerce.

Stiglitz, J. E. (2003). *Globalization and its discontents.* New York: Norton.

Stiglmayer, A. (1994). *Mass rape: The war against women in Bosnia-Herzegovina.* Lincoln: University of Nebraska Press.

Stryker, S. (1980). *Symbolic interactionism: A social structural version.* Menlo Park, CA: Benjamin/Cummings.

Survivor show concept. (n.d.). Retrieved July 27, 2005, from http://www.cbs.com/prime-time/survivor/show/concept.shtml

Sydmob. (n.d.). Retrieved June 8, 2005, from http://www.sydmob.com

Timasheff, N. S. (1967). *Sociological theory: Its nature and growth* (3rd ed.). New York: Random House.

Truman, H. S. (1945a). *Statement by the president announcing the use of the A-Bomb at Hiroshima.* Retrieved August 1, 2004, from http://www.trumanlibrary.org/publicpapers/

Truman, H. S. (1945b). *Radio report to the American people on the Potsdam Conference.* Retrieved August 1, 2004, from http://www.trumanlibrary.org/publicpapers/

Tucker, R. C. (1978). *The Marx-Engels reader.* New York: Norton.

Turner, R. (2002). Role theory. In J. Turner (Ed.), *Handbook of sociological theory.* New York: Plenum.

Turner, J. H. (1991). *The structure of sociological theory* (5th ed.). Belmont, CA: Wadsworth.

Turner, J. H. (1993). *Classical sociological theory: A positivist's perspective.* Chicago: Nelson-Hall.

Turner, J. H. (1998). *The structure of sociological theory* (6th ed.). Belmont, CA: Wadsworth.

Turner, J. H. (1988). *A theory of social interaction.* Stanford, CA: Stanford University Press.

Turner, V. W. (1969). *The ritual process: Structure and anti-structure.* Chicago: Aldine.

Veblen, T. (1899). *The theory of the leisure class: An economic study of institutions.* New York: Macmillan.

Vitousek, P., Ehrlich, P. R., Ehrlich, A. H., & Matson, P. (1986). Human appropriation of the products of photosynthesis. *BioScience, 34*(6).

Wacquant, L. J. D. (1992). Preface. In P. Bourdieu & L. J. D. Wacquant, *An invitation to reflexive sociology.* Chicago: University of Chicago Press.

Wallerstein, I. (1974). *The modern world-system I: Capitalist agriculture and the origins of the European world-economy in the sixteenth century.* New York: Academic Press.

Wallerstein, I. (1980). *The modern world-system II: Mercantilism and the consolidation of the European world-economy, 1600–1750.* New York: Academic Press.

Wallerstein, I. (1989). *The modern world-system III: The second era of great expansion of the capitalist world-economy, 1730–1840.* New York: Academic Press.

Wallerstein, I. (1995). The end of what modernity? *Theory and Society, 24,* 471–488.

Wallerstein, I. (1999). *The end of the world as we know it: Social science for the twenty-first century.* Minneapolis: University of Minnesota Press.

Wallerstein, I. (2000). *The essential Wallerstein.* New York: The New Press.

Wallerstein, I. (2004). *World-systems analysis: An introduction.* Durham, NC: Duke University Press.

Weber, M. (1948). *From Max Weber: Essays in sociology* (H. H. Gerth & C. Wright Mills, Eds. & Trans.). London: Routledge and Kegan Paul.

Weber, M. (1949). *The methodology of the social sciences* (E. A. Shils & H. A. Finch, Eds. & Trans.). New York: Free Press. (Originally published 1904, 1906, 1917–1919)

Weber, M. (1961). *General economic history* (F. H. Knight, Trans.). New York: Bedminster. (Originally published 1923)

Weber, M. (1963). *The sociology of religion* (T. Parsons, Trans.). Boston: Beacon. (Original work published 1922)

Weber, M. (1968). *Economy and society* (G. Roth & C. Wittich, Eds.). Berkeley: University of California Press. (Originally published 1922)

Weber, M. (1993). *The sociology of religion* (E. Fischoff, Trans.). Boston: Beacon. (Originally published 1922)

Weber, M. (2002). *The Protestant ethic and the spirit of capitalism* (S. Kalberg, Trans.). Los Angeles: Roxbury. (Original work published 1904–1905)

Webster's new universal unabridged dictionary (2nd ed.). (1983). New York: Simon & Schuster.

Weinstein, D., & Weinstein, M. A. (1993). *Postmodern(ized) Simmel.* London: Routledge.

West, C. (1999). *The Cornel West reader.* New York: Basic Civitas.

West, C. (2001). *Race matters.* New York: Vintage. (Original work published 1993)

West, C., & Zimmerman, D. H. (1987). Doing gender. *Gender and Society, 1,* 125–151.

Wiggershaus, R. (1995). *The Frankfurt School: Its history, theories, and political significance* (M. Robertson, Trans.). Cambridge, MA: MIT Press. (Original work published 1986)

Williams, R. (1985). *Keywords: A vocabulary of culture and society* (Rev. ed.). New York: Oxford University Press.

Wilson, W. J. (1980). *The declining significance of race: Blacks and changing American institutions* (2nd ed.). Chicago: University of Chicago Press.

Wilson, W. J. (1996–1997). When work disappears. *Political science quarterly, 111*(4): 567–595.

Wilson, W. J. (1997). *When work disappears: The world of the new urban poor.* New York: Vintage.

Young, R. J. C. (2003). *Postcolonialism: A very short introduction.* Oxford, UK: Oxford University Press.

Zerubavel, E. (1991). *The fine line: Making distinctions in everyday life.* Chicago: University of Chicago Press.

Glossary

Account-able: A theoretical idea in ethnomethodology, the term implies that the basic requirement of all social settings is that they be recognizable or accountable as whatever social setting they are supposed to be. Members visibly and knowingly work at making their scenes accountable; this work, in turn, organizes the situation and renders it meaningful and real.

Action theory: In general, any theory that begins and is concerned with individual, social action rather than structure. For Max Weber, action is social insofar as the individual takes into account other people. Weber is specifically concerned with rational action (in his typology—traditional, affective, value-rational, instrumental rational—only value and instrumental are truly rational). Parsons is interested in the conditions under which action takes place (the choice of means and ends constrained by the physical and social environments). Parsons's theory of the social system is built upon his notion of voluntaristic action.

Adaptation: One of four sub-systems in Parsons's AGIL conception of requisite needs; the subsystem that converts raw materials from the environment into usable stuffs (in the body, the digestive system; in society, the economy).

Alienation: A theoretical concept in Marxian theory. The word itself means "to be separated from"; it also implies that there is something that faces humans as an unknown or alien object. For Marx, there are four different kinds of alienation: alienation from one's own species-being, alienation from the work process, alienation from the product, and alienation from other social beings.

Anomie: Used by both Durkheim and Simmel; literally, to be without norms or laws. Because human beings are not instinctually driven, they require behavioral regulation. Without norms guiding behavior, life becomes meaningless. High levels of anomie can lead to anomic suicide.

Black nihilism: Black nihilism is a theoretical concept from the work of Cornel West. Nihilism in general refers to the idea that human life and existence are meaningless and useless. West uses the idea to describe the African American experience as the result of market saturation. Up until the 1950s, blacks were not considered a viable or large market for capitalism. Since the era of civil rights, however, African

Americans have been increasingly targeted: blacks have been inundated with new commodities and enticed by values of consumerism. Consumerism along with upward mobility for some blacks has weakened the civic and religious community base that has traditionally been the source of strength and validation for blacks living under American apartheid. For West, the main problem is that true equality and justice still do not exist for African Americans: the black life is still one of oppression, yet it is experienced under the shroud of consumerism and the belief that owning more and more commodities is the source of happiness. The loss of community coupled with consumerism and continued oppression has contributed to a growing sense of meaninglessness and loss of purpose for African Americans.

Blasé attitude: The blasé attitude is a concept from Simmel's theory of urbanization. The word "blasé" means "to be uninterested in pleasure or life." According to Simmel, the blasé attitude is the typical emotional state of the modern city dweller. It is the direct result of the increased emotional work and anomie associated with diverse group memberships and increased cognitive stimulation due to the rapidness of change and flow of information.

Bureaucratic personality: The bureaucratic personality is an extension of Weber's theory of the rationalization processes intrinsic within bureaucracies, and is the result of extensive use of bureaucratic methods for organization within a society. The bureaucratic personality is characterized by rational living, identification with organizational goals, reliance on expert systems of knowledge, and sequestered experiences (experiences that are removed from social or family life and placed in institutionalized settings).

Civil society: A concept from Jürgen Habermas's critical theory, civil society is a social network of voluntary associations, organizations (especially mass media), and social movements; its purpose is to inform and actualize the public sphere. To function properly, the civil society must be free from control by the state, economy, and religion, and it must exist within a liberal culture that values equality and freedom.

Class: Class is a theoretical concept, broadly used to talk about a social structure that is built around issues of economic production. Class is specifically an issue in modernity and capitalism in that other social positions tended to be more significant in previous eras, such as in feudalism. Karl Marx argued that, under capitalism, class became explicit and the most powerful issue in the stratification of unequal resources. The issue in classical theory is central for Marx and Weber. For Marx, class is the only structure that matters, and it is specifically defined by the ownership of the means of production. In capitalist societies, class bifurcates into owners (bourgeoisie) and workers (proletariat). For Weber, class is one of three systems of stratification, status and power being the other two. Weber also defines class more complexly than does Marx. According to Weber, one's class is defined by the probability of acquiring the goods and positions that are seen to bring inner satisfaction. Class position is determined by the control of property or market position, both of which may be positively, negatively, or medially possessed. In contemporary theory, class is a central issue in Bourdieu's theory of habitus and symbolic violence.

Class consciousness: A concept from Marx's theory of capitalism, it is the subjective awareness that class determines life chances and the group identity that results. With class consciousness, a class moves from a class in itself to a class *for* itself, a shift from a structured position to a political one. Class consciousness is one of the prerequisites to social change in Marxian theory and varies directly by the levels of worker communication, exploitation, and alienation.

Collective consciousness: The collective consciousness is a central theoretical issue in Durkheim's theory of social solidarity. It refers to the collective representations (cognitive elements) and sentiments (emotional elements) that guide and bind together any social group. The collective consciousness varies by three elements: the degree to which culture is shared; the amount of power the culture has to guide an individual's thoughts, feelings, and actions; its degree of clarity; and its relative levels of religious versus secular content. Each of these is related to the amount of ritual performance in a collective. According to Durkheim, the collective consciousness takes on a life and reality of its own and independently influences human thought, emotion, and behavior, particularly in response to high levels of ritual.

Colonization of the life-world: A theoretical concept used by Jürgen Habermas to describe the process through which the everyday life of people is taken over by economic and political structures. By definition, the life-world is the primary place of intimate communication and social connections. These functions are overshadowed by the values of money and power that come from the economic and political realms, which in turn reduce true communication and sociability. Life-world colonization tends to increase as social structures become more complex and bureaucratized, and when the state creates a climate of entitlement.

Commodification: Commodification is a theoretical idea in Marxian theory that captures the process through which material and nonmaterial goods are turned into products for sale. From this point of view, nothing is by its nature a product for sale; goods must be placed in markets in order to be commodities. In Marxian theory, commodification is seen as an ever-increasing force in capitalism.

Commodity fetish: A central concept in Marxian theory. In commodity fetish, workers fail to recognize the human factor in products. Creative production is the distinctive trait of humanity. Therefore, all products have an intrinsic relationship to the people who make them. However, in capitalism, the product is owned and controlled by another, and it thus faces the worker as something alien that must be bought and appropriated. In misrecognizing their own nature in the product, workers also fail to see that there are sets of oppressive social relations in back of both the perceived need and the simple exchange of money for a commodity. Moreover, commodities come to substitute for real social relations—under capitalism, people see their lives being defined through commodity acquisition rather than community relations.

Consumer society: A theoretical idea from the work of Jean Baudrillard. Goods and services have always been consumed; it's a basic fact of economic existence.

However, in postmodernity, consumption itself drives the economy rather than basic needs and production. Pure consumption has no natural limits—that is, consumption for consumption's sake cannot be satisfied and produces a never-ending demand for new and different commodities. Further, consumption becomes a primary, if not *the* primary, way people define themselves and experience life.

Cultural capital: Cultural capital is a theoretical concept used in Randall Collins's theory of interaction ritual chains and Pierre Bourdieu's theory of class replication. For Collins, cultural capital is defined as the amount of cultural goods—such as knowledge and symbols—that a person has at his or her disposal to engage others in interaction rituals. The more cultural capital is held in common, the more likely people are to interact with one another. But because it is a form of capital, people are also looking for a return on their investment—they hope to take more culture away than what they brought into the interaction. Collins conceptualizes three types of cultural capital: generalized (group specific), particularized (specific to relationships between individuals), and reputational (what is known about the individual). For Bourdieu, there are three kinds of cultural capital: objective (material goods that vary with class), institutionalized (official recognition of knowledge and skills), and embodied (the result of class-based socialization; habitus). Generally, the concept refers to the social skills, linguistic styles, and tastes that one accrues through education and distance from necessity (a measure of how far removed someone is from basic sustenance living). As the levels of education and distance from necessity increase, language, taste, and social skills tend to become more complex and abstract.

Cybernetic hierarchy of control: In Parsonian theory, the idea that all systems are controlled by information; thus, it is the subsystem that controls and provides needed information for the rest of the system (in the social system: culture).

Definition of the situation: Part of symbolic interactionist theory, it refers to the primary meaning given to a social interaction. The definition of the situation is important because it implicitly contains identities and scripts for behavior. Because the definition of the situation is a meaning attribution, it is flexible and negotiable—in other words, people can change it at a moment's notice and with it the available roles and selves.

Dialectic: Dialectic is a theoretical concept that describes the intrinsic dynamic relations within a phenomenon, such as the economy. The term is generally, though not exclusively, used in conflict or critical perspectives. The idea of dialecticism came to sociology through Karl Marx, and Marx, for his part, got the idea from the philosopher Georg Wilhelm Hegel. A dialectic contains different elements that are naturally antagonistic or in tension with one another—this antagonism is what energizes and brings change. Dialectics are cyclical in nature, with each new cycle bringing a different and generally unpredictable resolve. The resolve, or new set of social relations, contains its own antagonistic elements, and the cycle continues.

Discourse: Discourse is a theoretical concept that is widely used but is most specifically associated with the work of Michel Foucault. A discourse is an institutionalized way of thinking and speaking. It sets the limits of what can be spoken and, more importantly, *how* something may be spoken of. In setting these limits, discourses delineate the actors of a field, their relationships to one another, and their subjectivities. Discourses are thus an exercise of power.

Discursive consciousness: A theoretical concept used to understand the hierarchy of the agent (different levels of awareness and thus action) in Anthony Giddens's theory. Discursive consciousness refers specifically to the ability to give verbal accounts or rationalizations of action. It's what we are able to say about the social situation. Discursive consciousness is the awareness of social situations in verbal form.

Disembedding mechanisms: The idea of disembedding mechanisms is a theoretical concept from Anthony Giddens's theory of modernity. Disembedding speaks of processes that lift social relations and interactions out from local contexts, which has implications for time–space distanciation and ontological security. There are two types of mechanisms: symbolic tokens and expert systems.

Division of labor: The way in which work is divided in any economy. It can vary from everyone doing similar tasks to each person having a specialized job. This concept is an important one for most theories of modernity. In previous epochs, labor was more holistic in the sense that the worker was invested in a product from beginning to end. Thus, a shoemaker made the entire shoe. One of the distinctive traits of modernity is the use of scientific management, or *Fordism*, to divide work up into the smallest manageable parts. The division of labor is an important variable for Durkheim, Simmel, Marx, and others. For Durkheim, the division of labor creates specialized cultures for each group that may in turn threaten the cohesiveness of the general culture; for Simmel, the division of labor increases the level of objective culture in any society and it trivializes the meaning surrounding products; for Marx, the division of labor is understood as potentially separating people from species-being (natural or forced; material versus mental).

Documentary method: A theoretical term from ethnomethodology. The documentary method is the activity through which a link is created between an event or object and an assumed meaning structure. It is more than interpretation. Documentary method refers to the actual work that people perform in a social situation that links an event to its interpretation in such a way as to authenticate the correspondence.

Double consciousness: Du Bois' argument that members of a disenfranchised group have two ways to understand and be aware of themselves: as disenfranchised and as full members of society. These two awarenesses are at war with and negate one another so that the disenfranchised are left with no true consciousness.

Emergent/emergence: Emergence is a theoretical idea used to understand meaning and interaction in symbolic interactionism. To emerge means to rise from or come out into view, like steam from boiling water. Emergent meaning (or self), then, implies that meaning is not intrinsic to any sign or object. Meaning, rather, is a function of social interaction: the meaning of a sign, symbol, nonverbal cue, social object, and so on comes out of (emerges from) social interaction. In this perspective, meaning is thus a very supple thing and is controlled by people in face-to-face interactions, not social structure.

Emotional energy: A theoretical idea that originated with Émile Durkheim and is used by Randall Collins in his theory of interaction rituals. It is defined as the level of motivational energy an individual feels after leaving an interaction. Emotional energy is specifically linked to the level of collective emotion formed in a ritual, and it predicts the likelihood of further ritual performance and the individual's initial involvement within the ritual.

Episteme: Episteme is a theoretical concept from Michel Foucault's theory of knowledge and power. It refers to the fundamental notions of truth and validity that underlie knowledge—it's the hidden order of knowledge. Epistemes organize and are a necessary precondition for thought; they set the boundaries of what is possible and knowable. One important implication is that an episteme will maintain the order produced by a knowledge system even in the face of contradictory events or findings. Epistemes are historically specific and change through rupture rather than linear progress.

Equilibrium: A concern of functionalist theory, equilibrium refers to a state of balance between integrative and disintegrative forces; it implies a social system that maintains certain constancies of patterns relative to its environment. The idea is that a social system must maintain a balance, whether internally or between itself and its environment.

Exploitation: Exploitation is a central concept in Marxian economic theory. In its sparsest terms, exploitation refers to the measurable difference between what a worker gets paid and the amount of product she or he produces. Capitalist profit is based on exploitation (paying the worker less than he or she actually earns). Capitalists are thus dependent upon workers, which, in turn, gives workers a bargaining tool to push for higher wages and benefits. Because of the rising cost of labor in advanced capitalist countries, capitalists are forced to export exploitation to "less developed" countries, where workers will labor for less. These workers, however, will push for better wages and benefits, and the capitalists eventually will have to find other labor markets with cheap labor. Ultimately, according to Marxian theory, this dynamic of exploitation will lead to the demise of capitalism.

Facticity: A theoretical concept found in a number of different social constructivist approaches, facticity refers to the quality of being a fact. Implicit in this concept is the idea that facts are produced; they don't intrinsically exist as facts. Events, objects, and phenomena in general become facts under certain knowledge systems.

False consciousness: According to Marx, false consciousness is consciousness that is false in its very foundations. True consciousness is founded on species-being: recognizing the intrinsic human nature of creative production. Any idea of humanity apart from species-being is laid on a false foundation.

Fault line: A theoretical concept specific to Dorothy E. Smith's feminist theory. Smith argues that a gap exists between official knowledge—especially knowledge generated through the social sciences—and the experience of women. This fault line is particularly important for understanding how men are unable to see the differences between objective, public knowledge and the reality of day-to-day existence: women traditionally negotiate or obscure the disjunction for men through their caring for the daily administration of the household, including meeting the sustenance and comfort needs of both men and children.

Field: The field is a theoretical concept from Pierre Bourdieu's constructivist structuralism approach. The concept functions to orient the researcher to an arena of study. Field denotes a set of objective positions and relations that are tied together by the rules of the game and by the distribution of four fundamental powers or capitals: economic, cultural (informal social skills, habits, linguistic styles, and tastes), social (networks), and symbolic (the use of symbols to recognize and thus legitimate the other powers). People and positions are hierarchically distributed in the field through the overall volume of capital they possess. Symbolic and cultural capitals are specifically important: cultural capital helps to form habitus and thus contributes to the replication of the field, and symbolic capital gives legitimacy and meaning to the empirical field.

Free-floating signifiers: The idea of free-floating signifiers comes from the postmodern writings of Fredric Jameson. In traditional and modern societies, signifiers, or signs, are set and understood within a social or linguistic structure, what is sometimes referred to as a signification chain. In postmodernity, however, the chain or context of signs has been disrupted by machines of reproduction, or mass media. This break in the signification chain indicates that each sign stands alone, or in relatively loose association with fragmented groups of other signs. Signs, then, become free-floating signifiers.

Front: The theoretical idea of a front comes from Erving Goffman's dramaturgy. Front refers to the totality of identity cues offered by an individual in a social encounter. These cues come from setting, manner, and appearance.

Function: The idea of social function comes through analogy from biology. Classic thinkers like Spencer and Durkheim used the analogy of how an organism works to think about society (the *organismic* analogy). A biological creature is made up of different organs, each of which fulfills a need that the animal has, with each and every organ purposefully related to other organs within the body. The idea of social function, then, denotes the necessary contribution a social structure makes, as well as the way the structure is related to other elements in society.

Generalized media of exchange: A concern for functional theory, specifically Durkheim and Parsons, generalized media of exchange refers to symbolic goods that are used to facilitate interactions across institutional domains. Thus, they are values and symbols that allow different structures to relate one to another; the goods and services exchanged by different institutions.

Generalized other: A theoretical concept used in symbolic interactionism that refers to sets of perspectives and attitudes indicative of a specific group or social type with which the individual can role-take. In the formation of the self, the generalized other represents the last stage in which the child can place him- or herself in a collective role from which to view his or her own behaviors. It is the time when the self is fully formed as the person takes all of society inside.

Goal attainment: One of four subsystems in Parsons's AGIL conception of requisite needs; the subsystem that motivates and guides the system as a whole (in the body, the mind; in society, government).

Governmentality: Governmentality is a theoretical term from Michel Foucault's theory of knowledge and power. The term refers to a specific kind of institutionalized power. In Foucault's scheme, modernity is unique because of the manner through which states control populations. Generally speaking, in previous ages power and control were exercised upon the individual from without. Governments would actively and directly control the person, when domination was necessary or desired. In contrast, modern states exercise power from within the individual. Governmentality, then, captures the process through which the person participates in her or his own domination—control is exercised within the person by the person.

Gynaecocentric theory: Originated with Lester F. Ward and used by Gilman, gynaecocentric theory argues that women—not men, as was and is commonly assumed—are the general species type for humans: it is through women that the race is born and the first social connections created. In gynaecocentric theory, men and women have essential, sex-specific energy, and the dominance of the masculine energy in society makes its evolutionary path unbalanced.

Habitus: A central theoretical term from Pierre Bourdieu's theory of class replication. Habitus refers to the cultural capital an individual possesses as a result of his or her class position. Habitus is embodied; that is, it works through the body at a non-conscious level. Cultural capital and thus class position are expressed unthinkingly. On the one hand, this embodied expression structures class; on the other hand, habitus works creatively as it is possible to make intuitive leaps. Habitus varies by cultural capital and is expressed in linguistic markets.

Hyperreality: Hyperreality is a theoretical concept from the work of Jean Baudrillard and refers to the preference in postmodern societies for simulated experiences rather than real ones. In Baudrillard's scheme, the real has been stripped of all meaning, authenticity, and subjective experience through the proliferation of simulacrum. People thus seek emotional experiences in images of reality that have been exaggerated to the grotesque and fantastic (breast implants and video games are good examples).

Ideal speech communities: The concept of ideal speech communities comes from the work of Jürgen Habermas. Ideal speech refers to communicative acts or interactions where every member is granted full participation and is free from any type of coercion. Each member is responsible for expressing her or his opinion, keeping her or his speech free from ideology and objective standings (such as educational credentials). The goal of ideal speech is consensus. Habermas argues that reasoned consensus—and thus emancipation—is possible because of the inherent assumptions of communication. The act of communication assumes that communication, intersubjectivity, and validity are all possible. The latter implies the possibility of universal norms or morals, and the fact that validity claims can be criticized implies that they are in some sense active and accountable to reason.

Ideology: In Marxian theory, ideology is any system of ideas that blinds us to the truth of economic relations. In a more general sense, ideology is any set of norms, values, and beliefs that undergird and legitimate a culture or specific knowledge system.

Imperatively coordinated associations (ICA): This concept is central to Ralf Dahrendorf's dialectical theory of class power relations. Dahrendorf argues that society is organized around social relations with differential power. In other words, associations between and among groups of people are coordinated through the use of legitimated power (authority). The issue for Dahrendorf is that society is built around the unequal distribution of power, and thus conflict is always a real potential.

Incorrigible assumptions: The theoretical idea of incorrigible assumptions comes from ethnomethodology and refers to the assumptions upon which every version of human reality is based. These assumptions cannot be questioned because to do so would erode the foundation of reality. Thus, the assumptions are protected through secondary elaborations of belief.

Indexical expressions: The idea of indexical expressions comes from ethno-methodology and conceptualizes the notion that the meaning of all conversation is dependent upon a context that is assumed to be shared by all interactants. Talk, then, indexes contexts in order to be sensible and taken-for-granted.

Industrialization: In general, industrialization refers to the process through which machines replace direct human manipulation of objects in work. Industrialization is also an intrinsic part of what we mean by modernity. As such, it carries with it other factors such as high division of labor, rationalization, scientific management, urbanization, markets, and so on. Industrialization is specifically a concern for Marxian theory.

Institutional ethnography: A methodological approach developed by Dorothy E. Smith. Its focus is to understand how everyday experience is socially organized through authoritative texts. Texts, such as written documentation or research articles, are generated through professional practices and governmental policy making. These texts then come to organize local activities. The texts further provide a scheme for people within the activities to assess and experience them-selves as members of the social situation. Institutional ethnography thus provides a way of mapping the social relations of power that govern daily life, particularly that of women or other minorities.

Integration: In general, the degree to which subsystems or units within a larger sys-tem work collectively for the common good; in social systems, integration is often related to cultural consensus, particularly in the work of Durkheim. In Parsons, integration refers to one of four subsystems in his AGIL conception of requisite needs, the subsystem that regulates the activities of the system's diverse members (in the body, the central nervous system; in society, the law).

Interaction: Interaction is a central theoretical idea in symbolic interactionism. Interaction is the intertwining of individual human actions. In symbolic interactionism, the interaction is the true acting unit in society. According to this view, interaction is not simply the means of expressing social structure or the individual's personality traits; rather, the interaction is the premier social domain and is thus the chief factor through which social structures and individual personalities are created and sustained. Interaction occurs via a three-part process through which meaning, society, and self emerge: the presentation of a cue, the initial response to the cue, and the response to the response. However, the end point of the process generally becomes itself an initial cue for further interaction, starting the process over again.

Intersectionality: Intersectionality is a theoretical concept in Patricia Hill Collins's theory of structured inequality. The basic idea is that systems of race, gender, class, nationality, sexuality, and age crisscross one another at specific social locations that form a matrix of domination. The analytical issue, then, is to empirically discover how these various systems come together for any specific person or group. Collins is specifically concerned with how these systems shape black women's experiences and are, in turn, influenced by black women.

Iron cage of bureaucracy: An idea from Max Weber's theory of bureaucracy. The iron cage is the result of extensive use of bureaucratic methods for organization. Weber argues that, once in place, bureaucracies are difficult if not impossible to replace because rulers cannot easily get rid of the need for expert knowledge or a professionally trained staff. In addition, professionals tend to perpetuate their job status and make knowledge secret in order to control the organization and avoid public scrutiny.

Joint action: Joint action is a theoretical term from Blumer's symbolic interactionism that describes the process through which various individual and discrete actions and interactions are brought together to form a meaningful whole. This joining is accomplished symbolically by both individuals and groups and constitutes a significant portion of what is meant by "society." The insight of this concept is that at every point of interaction or joint action there is uncertainty. The implication is that society is emergent.

Latent pattern maintenance: One of four subsystems in Parsons's AGIL conception of requisite needs; the subsystem that indirectly preserves patterns of behavior that are needed for survival (in the body, the autonomic nervous system; in society, education, religion, and family).

Legitimation: Legitimation is a theoretical concept that describes the effects that specific stories, histories, and myths have in granting ethical, moral, or legal status or authorization to social power and relations. Legitimation is a central concern of Max Weber's and is particularly important in contemporary theories of the social construction of reality. In social constructivist theory, there are three levels of legitimation (self-evident, theoretical, and symbolic universes [i.e., religious systems]), each more powerful than the previous.

Life politics: A theoretical idea from Anthony Giddens's theory of modernity, life politics is an outgrowth of emancipatory politics. Emancipatory politics is concerned with liberating individuals and groups from the constraints that adversely affect their lives—people are thus liberated to make choices. Life politics is the politics of choice and lifestyle. It is concerned with issues that flow from the practices of self-actualization within the dialectic of the local/global, where self-realization affects global strategies. Life politics is dependent upon the individual creating and maintaining inner authenticity.

Life-world: The life-world is a theoretical concept that came into sociology through the work of Alfred Schutz and is prominent in Jürgen Habermas's work (colonization of the life-world). The life-world refers to the world as it is experienced immediately by each person. It is a cultural world filled with meaning and is made up of the sets of assumptions, beliefs, and meanings against which an individual judges and interprets everyday experiences.

Linguistic market: A theoretical concept from Pierre Bourdieu's theory of class replication, the idea emphasizes the importance of language and social skills in reproducing class. Every time a person speaks with another, there is a linguistic market within which each person has different skill levels and knowledge. These differing levels lead to distinct rewards in the market that in turn announce one's class position. The power of the market is that people tend to sanction themselves because they intuitively understand how their culture and language skills will play out in any given market.

Manifest and latent functions: Elements of functional theorizing specifically introduced by Robert K. Merton. Manifest functions are those known effects of a social structure that lead to integration and equilibrium; latent functions are the hidden or unacknowledged contributions of social structure that lead to integration and equilibrium.

Matrix of domination: The idea of matrices of domination comes from Patricia Hill Collins's theory of black feminist epistemology. In general, the notion of a matrix refers to a mass within which something is enclosed. This enclosure is a point of origin or growth (as in the cradle or matrix of civilization). For Collins, the issue is that current society is built of matrices of domination that form around the intersecting issues of race, gender, age, sexuality, nationality, and so on. Any matrix of domination works through four different domains: structural, disciplinary, hegemonic, and intrapersonal.

Means of production: A central concept in Marx's theory of capitalism. Simply, the means of production refers to the way in which commodities are produced. However, in Marx's hands the concept comes to connote quite a bit more. Because the basic social fabric is economic in Marxian theory, relations of production are inherent within the means of production. Thus, the means of production—such as capitalism, feudalism, and socialism—determines how people relate to self and others (the relations of production).

Mechanical solidarity: Mechanical solidarity is part of Durkheim's typology of society and refers to the kind of social solidarity experienced in traditional societies. Durkheim uses the term "social solidarity" to refer to the level of cultural integration in a society, as measured by the subjective sense of being part of a group (cohesion), the constraint of individual behaviors for the group good (normative regulation), and the level of coordination and control among various social units. In mechanical solidarity, all three of these variables are high. Specifically, people are united by sharing a clear and limited set of common beliefs and sentiments, and through a sense that society (or the group) is more important and viable than the individual person. In addition, people's behaviors are regulated through strong feelings of morality and through repressive law. Mechanical solidarity is dependent upon a limited social network with frequent face-to-face interactions.

Middle-range theories: Middle-range theories are a central component in Robert K. Merton's empirical functionalism. These are theories that are empirically based yet generalized to an intermediate level; that is, they are not completely abstract or empirical. Merton proposed the use of middle-range theories as a method of building sociologically relevant theory, in comparison to the more abstract schemes of such theorists as Talcott Parsons and Karl Marx. Merton's idea was that middle-range theories could be linked together to form more general yet empirically relevant theories.

Misrecognize: A central concept in Marxian theory. Marxists argue that people fail to see or recognize the relations of production within a commodity or the means of production. Misrecognition is thus a function of ideology and implies that the dominated are blind to their own oppression.

Modalities of structuration: A concept used in Anthony Giddens's structuration, modalities of structuration are the paths through which structure is expressed in social encounters. There are three modalities (interpretive schemes, facilities, and norms) that are linked on the one hand to structures (signification, domination, and legitimation) and on the other to social practices (communication, power, and sanctions). These modalities are socially and culturally specific.

Modes of orientation: The beginning element in Parsons's theory of social action and institutionalization, modes of orientation are the different values (cognitive, appreciative, and moral) and motivations (cognitive, cathectic, and evaluative) that people bring into a situation. They result in identifiable types of action (strategic, expressive, or moral) that in the long run pattern interactions across time and space and result in the taken-for-granted norms, roles, and status positions that constitute social structure and society.

Morbid excess in sex distinction: A theoretical concept in Charlotte Perkins Gilman's evolutionary theory of gender. Sex distinctions are those physical and visual cues that make sexes different from one another. Gilman argues that the sex distinctions in humans have been carried to a harmful extreme due to women no longer living in the natural environment of economic pursuit, but, rather, living in the artificial environment of the home that has been established by men. Women have thus been physically and sexually changed by evolution so that they can better survive under patriarchy.

Nation-states: A defining feature of modernity; a nation-state is a socially diverse collective that occupies a specific territory, creates a common history and identity, and whose members sees themselves as sharing a common fate. Nation-states are bureaucratically organized and emphasize mass democracy.

Normative specificity: Norms are behaviors that have some sanction attached to them. The sanctions can be positive (reward) or negative (punishment), and they can be formal (like laws) or informal (generally understood). Normative specificity, then, refers to the degree to which behaviors are specifically regulated (explicit norms for specific behaviors). Normative specificity is high when a group is regulating a large proportion of a person's behaviors. We find this issue in Durkheim, Simmel, and Parsons's theories.

Ontological security/insecurity: Ontological security is a theoretical concept from Anthony Giddens's structuration theory. Ontology refers to the way things exist and specifically implies that human existence is different from all other forms of existence because of meaning. Meaning is not a necessary or intrinsic feature of any event or object, which implies that human reality is unstable. This intrinsic instability creates an unconscious need for ontological security—a sense of trust in the taken-for-grantedness of the human world. This need motivates humans to routinize and regionalize their practices.

Ontology: A branch of philosophy that is concerned with how things exist. An obvious example is that rocks and people exist differently: one is biological and the other isn't. But ontology is concerned with less obvious, more fundamental questions—in particular, how categories exist. In our case, we are concerned with how society exists—what is its ontological source? Does society exist as an object made up of social facts and structures? Or, does society exist symbolically through the way we create meaning around the idea of society?

Organic solidarity: Organic solidarity is part of Durkheim's typology of social integration, which is measured by the subjective sense of being part of a group (cohesion), the constraint of individual behaviors for the group good (normative regulation), and the level of coordination and control among various social units. Organic solidarity is characteristic of modern societies with a high division of labor, and is created through high levels of mutual dependency (structural, group, and individual), generalized culture, and the presence of intermediate (between the level of society and the level of face-to-face interaction) groups and organizations. Relationships are regulated more through rational-legal means (restitutive law) than moral codes, and people generally have weak attachments to family and tradition. In Durkheim's mind, organic solidarity is more precarious than mechanical and susceptible to various "pathologies" (such as anomie).

Organismic analogy: A way of understanding how society functions by comparing it to organisms. The organismic analogy is specific to functionalist theorizing.

Panopticon: Panopticon is a theoretical concept from Michel Foucault's theory of knowledge and power. The panopticon was an architectural design for a prison that allowed for the unobserved observation of prisoners. The idea behind the panopticon is that if prisoners thought they were being watched, even if they weren't, the prisoners themselves would exercise control over their own behaviors. Foucault sees this physical prison as a metaphor for the way control and power are exercised in modernity. He specifically has in mind the self-administered control that comes through the knowledge produced by the social and behavioral sciences as well as medical science.

Phenomenology: Phenomenology is a school of philosophy developed by Edmund Husserl that argues that consciousness is the only experience or phenomenon of which humans can be certain. It seeks to discover the natural and primary processes of consciousness apart from the influence of culture or society. Husserl hoped to create "a descriptive account of the essential structures of the directly given"—that is, the immediate experience of the world apart from preexisting values or beliefs. Social phenomenology, on the other hand, argues that nothing is directly given to humans; we experience everything through stocks of language, typifications, and so on. Social phenomenology, then, seeks to explain the phenomena of the social world as they are presented to us. This approach is distinct from most sociology in that phenomenologists reduce phenomena to the simplest terms possible. For example, a structural sociologist will study racial inequality and its effects. Race is simply a given in this kind of approach. A phenomenological approach will take race itself as the most basic problem to explain: how is it that race can be experienced as a given, as something that can be taken-for-granted, as an intersubjective reality? How does "race" present itself to us in just such a way as to appear real, taken-for-granted, and intersubjective?

Postindustrial: A society wherein the economy has moved from having its base in industry and manufacturing to service and knowledge. In a postindustrial society, knowledge is more important than property, and professionals and technicians are the most important social type. The idea of postindustrial society was popularized by Daniel Bell in his book, *The Coming of Post-Industrial Society*.

Power: Power is a theoretical concept found in many social and sociological theories. The short definition of power is the ability to get others to do what you want. Yet, in terms of how it works and where it resides, power is one of our most difficult and controversial terms. For Anthony Giddens, power is part of every interaction and is defined in terms of autonomy and dependence. The greater our level of autonomy and the greater others' level of dependence, the greater will be our power in an encounter. Thinking of power in this way makes it an outcome rather than a resource that is possessed. It also makes power part of face-to-face encounters rather than part of an obdurate structure or linked to group politics. In structuration theory, power is one result that comes from the use of allocative and authoritative resources. For exchange theorists, power is the result of unequal exchange relations. Power refers to the other individual's or group's ability to recurrently impose its will on a person. There are four conditions of power in Peter Blau's exchange scheme: the power of actor A over actor B is contingent upon (1) B having limited needs; (2) B having few or no alternatives; (3) B being unable or unwilling to use force; and (4) B continuing to value the good or service that A controls. Michel Foucault sees power in two ways; both are hidden rather than overt. First, and most importantly, power is exercised through knowledge. The knowledge that any person holds at any given time is the result of historically specific institutional arrangements and practices. For Foucault, knowledge isn't simply held; it is applied to every aspect of the person's life by the individual. Thus, knowledge exercises control over people's bodies, minds, and subjectivities. The second way Foucault uses the notion of power is in daily encounters with others. In every social encounter, people's actions influence other actions; these practices enact the social discourse or knowledge and serve to guide and reinforce one another.

Practical consciousness: Practical consciousness is a concept from Anthony Giddens's theory and refers to what people know or believe about social situations and practices but can't verbally explain. It is the basis for the routinization of daily life, which, in turn, provides ontological security. One important ramification of practical consciousness is that it implies that behavior is often directed by nonconscious intuition and that the reasons given for action (discursive consciousness) can have a separate interactional function.

Pragmatism: Pragmatism is a school of philosophy that argues that the only values, meanings, and truths humans hold onto are the ones that have practical benefits. These values, meanings, and truths shift and change in response to different concrete experiences. Pragmatism forms the base for many American social theories, most specifically symbolic interactionism.

Primary groups: A theoretical concept from Charles Horton Cooley and symbolic interactionism. Primary and secondary groups vary by length of time of association, purpose, degree of involvement, and intimacy. Primary groups stay together for long periods of time and tend to be open, honest, and emotionally based. Primary groups will have stronger influence on individual thoughts, feelings, and behaviors.

Problem of routinization: From Weber's theory of social change and legitimation, the problem of routinization occurs after the charismatic leader dies. Because charismatic authority is rooted in an individual, when that individual is gone the authority must be made routine, either through traditional or rational-legal authority.

Public sphere: The public sphere is a theoretical construct from the critical work of Jürgen Habermas. It is an imaginary community or virtual space where a democratic public "gathers" for dialogue. With the idea of the public sphere, Habermas is arguing that a true democratic process demands an active, public dialogue that takes place outside the influence of government or the economy. To function properly, the public sphere demands unrestricted access to information and equal participation of all members.

Race-preservation: The idea of race-preservation is part of Gilman's evolutionary theory of gender relations. Race-preservation stimulates the natural selection of skills that promote the general welfare of the collective; however, because women have been removed from the natural environment by male-dominated society, race-preservation skills are deemphasized and the evolutionary system is out of balance—motivations for self-preservation dominate.

Rationalization: According to Weber, rationalization is the main defining dynamic of modernity. It is the process through which spirituality, tradition, moral values, and affective social ties are replaced by rational calculation, efficiency, and control.

Reification: Reification captures the idea that concepts and ideas may be treated as objectively real things. In social science generally, reification can be seen as a methodological problem because of the tendency to ascribe causation to ideas, as in gender causing inequality. In critical theory, reification is taken further and refers to the process through which human beings become dominated by things and become more thing-like themselves. Marx specifically argued that ideas that do not naturally spring out of species-being can only appear real through reification (making something appear real that isn't); all ideology is reified and the furthest reach of reification is the idea of God.

Role conflict: According to Simmel, role conflict occurs when behavioral expectations of two or more roles that an individual holds clash with one another.

Secondary elaborations of belief: A theoretical concept from ethnomethodology's understanding of the reflexive nature of human organization and reality. Secondary elaborations of belief are prescribed, legitimating accounts that explain away any piece of empirical data that contradicts how reality is assumed to function, as when the Azande's oracle fails, or the Christian doesn't received an answer to prayer, or the scientist's experiment doesn't yield expected results.

Secondary groups: From Charles Horton Cooley and symbolic interaction. Primary and secondary groups differ by length of time of association, purpose, degree of involvement, and intimacy. Secondary groups are those that do not last long, that tend to be goal rather than emotion based, and people in the groups tend not to reveal personal matters.

Self: The self is a theoretical idea that describes various features of the individual. According to symbolic interactionist theory, the self is a social object, a perspective, a conversation, and a story. The self is seen as arising from diverse role-taking experiences. It is thus a social object in that it is formed through definitions given by others, especially the generalized other, and is a central meaningful feature in interactions. The self is a perspective in the sense that it is the place from which we view our own behaviors, thoughts, and feelings. It is an internal conversation through which we arrive at the meaning and evaluation we will give to our own behaviors, thoughts, and feelings. This conversation is ongoing and produces a story we tell ourselves and others about who we are (the meaning of this particular social object). The self is initially created through successive stages of role-taking and the internalization of language. The self continually emerges, is given meaning, and is furnished with stability or flexibility through patterns of interactions with distinct groups and generalized others. The idea of the self is also prominent in Goffman's dramaturgy (the presentation of self) and Giddens's understanding of modernity (the reflexive project of the self).

Self-preservation: Part of Gilman's evolutionary scheme, self-preservation motivates the natural selection of skills that protect the individual. Self-preservation and race-preservation skills ought to balance one another; however, because women have been removed from the natural environment by male-dominated society, the self-preservation skills of women are overemphasized and the evolutionary system is disequilibrated.

Setting: Setting is a theoretical concept from Goffman's dramaturgical perspective. Settings are composed of physical sign equipment that is semi-permanently attached to physical locations. The physical props of the setting cue people to a limited number of possible definitions and self-identities. Settings thus stabilize encounters. They are part of the front that people manage.

Sexuo-economic relations: Gilman's characterization of the human system when women are forbidden to labor in the economy; in such a society, the structures of economy and family are confounded, the sex distinctions between men and women are accentuated and unbalanced, sex itself becomes pathologically important to people, women become consumers par excellence, and men are alienated from their work.

Simulacrum: A theoretical concept most closely associated with the postmodern work of Jean Baudrillard. The word itself simply refers to a representation or simulation. For example, a statue of Karl Marx has a similar appearance to, and thus represents or simulates, Marx. Baudrillard argues that, in postmodernity, most cultural images and signs do not have an actual reference and thus don't represent or simulate anything. Further, simulacra in postmodernity are simply images of images that never existed in the first place. The Disneyland ride "Pirates of the Caribbean" is a good example. Pirates certainly existed, but never in the form presented at the Disney ride: the ride is an idealized version of the fantasy of pirates given life in such novels as Peter Pan. So, the ride is an image of an image that doesn't truly represent. But the slide of simulacrum doesn't end there: Disney produced a movie based on the ride. The movie is thus an image of an image of an image, which never referred to anything real in the first place.

Social action typology: Social action is defined by Max Weber as any individual action that takes into account other people or has some value attached to it; value in this case is socially defined. In order to understand action, Weber constructed an ideal typology with four categories: instrumental-rational action is behavior that is guided by means–ends considerations; value-rational action is behavior that is motivated by and makes sense from the perspective of some system of values (like religion); traditional action is behavior that is motivated by custom and long-practiced routines; and, finally, affectual action is motivated and guided by situationally provoked emotions. This general concern of Weber's forms the basis of action theory and Talcott Parsons's explication of the unit act.

Social fact: Durkheim's term to describe the effects that the collective consciousness or moral culture has. Durkheim uses this term to underscore the argument that culture has the same attributes as any empirical object: it exists external to and coercive of the individual. The existence of social facts is used to argue in favor of a scientific approach to understanding society.

Social form: From Simmel: a patterned mode of interaction through which people obtain goals (examples: dyad/triad, conflict, exchange, sociability); forms exist prior to the interaction and provide rules and values that guide the interaction and contribute to the subjective experience of the individual; forms also imply social types—types of people that occupy positions within a social form (examples: stranger, adventurer, competitor, miser).

Social objects: The idea of social objects is a theoretical concept in symbolic inter-actionist theory. Social objects are anything in an interaction that we call attention to, attach legitimate lines of behavior to, and name. In this sense, the self and one's own feelings and thoughts can become social objects, as well as the more obvious "objects" in the environment.

Social system: Seeing society as a set of interrelated parts that function together to create integration and equilibrium. Systems can be smart or dumb (ability to take in and adjust to information) and open or closed (exposure to external forces); systems may also contain feedback and feed-forward effects as well as mechanisms that produce equilibrium; according to Parsons, social systems may be large (entire societies) or small (face-to-face interactions), are distinguished by boundaries between the system and its environment, and are controlled cybernetically.

Sociological ambivalence: A theoretical concept from Robert K. Merton's understanding of functionalism, sociological ambivalence is an effect of structural relations. In sociological ambivalence, the structured, normative role expectations of a social position are contradictory.

Species-being: One of Karl Marx's basic assumptions about human nature. The idea literally refers to species existence and awareness. In particular, species-being refers to how humans are conscious of being human. According to Marx, every species is unique because of the way it as a biological organism physically exists. Humans exist and survive through creative production. Thus, we become conscious of our own existence and the world around us through material production and the economy.

Spirit of capitalism: Weber's term to capture the cultural values and beliefs that undergird rational capitalism. There are at least three values and beliefs in the spirit of capitalism: life should be rationally organized to maximize profit; economic work is the most important thing we can do with our time and we must be diligent in our work; things of true value are quantifiably measured.

Structural differentiation: Structural differentiation is a theoretical concept from functionalism that captures the process through which the behaviors associated with social networks of roles, norms, and status positions (social structure) are acted out in different places and different times. In societies with high levels of structural differentiation, the requisite functions are carried out in distinct and separate institutions.

Structuration theory: A theoretical perspective developed by Anthony Giddens. The perspective is more an analytical framework—or ontological scheme—than a complete theory. Structuration tells the theorist-researcher what kinds of things exist socially and what to pay attention to. Structuration denies the existence of structure and free agency (seen as a false dualism) and argues that these generally reified concepts form an active duality: two parts of the same thing. In order to act, social actors must use known rules and resources. Giddens conceptualizes rules (normative rules and codes of signification) and resources (authoritative resources and allocative resources) as structural elements, which means that actors constitute or enact structure through their agency. Giddens applies this concept to the issue of time–space distanciation. Thus, in structuration, local interactions are linked with distant others through the use of known rules and resources. This way of seeing things avoids the reification of agency or structure, places the structuring (patterning) elements within the observable interaction, and encourages an historical sociology. The latter is important for Giddens as he explains the ways through which modern society stretches out time and space, or links local interactions with distant ones, and allows him to explicate some of the unique features of modernity.

Surplus labor: A theoretical concept from Marx's understanding of capitalism. According to Marx, surplus labor is the amount of labor a worker performs for which she or he does not get paid. It's the difference between the worker's pay and the value of the products he or she produces. Surplus labor is equivalent to the level of exploitation and is the source of capitalist profit. There are two main types of surplus labor: absolute surplus labor, a method of increasing profit by increasing the number of hours a worker works; and relative surplus labor, increased exploitation through industrialization.

Symbolic capital: Symbolic capital is a theoretical concept from Pierre Bourdieu's theory of class replication and refers to socially legitimated symbolic power of definition. Bourdieu argues that social groups and status positions exist empirically and symbolically; but it is the symbolic that gives groups and positions meaning and legitimacy. This power of definition is thus an important factor in creating the social world. Symbolic capital varies by social credentials, which generally come through education and political office.

Symbolic exchange: Symbolic exchange is an idea from Jean Baudrillard's theory of postmodernism that he uses as a comparison point for the pure sign-value of postmodern culture. In more traditional societies, when communication was closely tied to human interaction in social groups, symbols carried meaning. The symbols that people used arose from and represented real-life concerns and experiences. In contrast, much of postmodern culture has been created for or trivialized by mass media and advertising. These signs, rather than symbols, do not and cannot represent any kind of real social existence. Thus, cultural signs in postmodernity carry no meaning, are free-floating (released from all social context and linguistic structure), and are simply used by mass media and people in endless fields of play.

Symbolic violence: Symbolic violence is a theoretical concept from Pierre Bourdieu's theory of class replication that refers to the self-sanctioning that occurs in linguistic markets.

Time–space distanciation: Time–space distanciation is a theoretical concept from Anthony Giddens's structuration approach. This concept reformulates the problem of social order by focusing on the stretching out of time and space rather than the patterning of behaviors. Generally speaking, concentrating on how behaviors are patterned has resulted in an emphasis on either structure or agency. Giddens sees this distinction as a false dualism, and focusing on time–space distanciation avoids this issue. Specifically, time–space distanciation refers to the process through which local social interactions are linked to distant ones either through time (as in future with past interactions) or geographic space (as in an interaction in New York City with one in Los Angeles), thus ordering society.

Unanticipated consequences: The idea of unanticipated consequences is a theoretical concept from Merton's functionalism. According to Merton, purposeful social action can have unexpected consequences. These kinds of effects of social action can accumulate and lead to structural change. Ideas like this allow Merton to move functionalism into a more dynamic realm.

Urbanization: The process through which more and more of a given population moves from rural settings to the city. A distinct process of modernity that is associated with industrialization and capitalism, it results in such things as higher levels in the division of labor, increases in the use of money and the size and velocity of exchanges and markets, rational versus organic group memberships, overstimulation, increasing social diversity, and so on. Urbanization is specifically a concern of Simmel's.

Use-value: According to Marx, use-value is determined by the use or utility any commodity has. It is expended or consumed through use. The term is understood in relation to exchange-value, and later in relation to Baudrillard's notion of sign-value.

Web of group affiliations: The web of group affiliations is a theoretical concept developed by Georg Simmel to describe social networks: the number, frequency, and intensiveness of relationships among people. Simmel is particularly concerned with why people join groups—variation by motivation. Simmel argues that the motivation behind group membership changes as a society moves from traditional to modern. Organic motivation is prevalent in traditional society and indicates that group membership that is based on either family ties or previously established social relations. In modern society, however, motivation for group membership is based on free choice rather than family ties or previously established social relations.

About the Author

Kenneth Allan received his Ph.D. in sociology from the University of California, Riverside (1995), and is currently Associate Professor of Sociology at the University of North Carolina at Greensboro (UNCG). Before moving to UNCG, he directed the Teaching Assistant Development Program at the University of California, Riverside, and coedited *Training Teaching Assistants*, 2nd edition, published by the American Sociological Association. He has also published several works in the area of theory, including the 1998 monograph *The Meaning of Culture* and the more recent text *Explorations in Classical Sociological Theory: Seeing the Social World*.